Seaweeds of the British Isles

A collaborative project of the British Phycological Society and the Natural History Museum with financial support from the Joint Nature Conservation Committee

Volume 1 Rhodophyta

Part 3A Ceramiales

Christine A Maggs & Max H Hommersand

Natural History Museum, London

First published by the Natural History Museum,
Cromwell Road, London SW7 5BD

This edition printed and published by Pelagic Publishing, 2013,
in association with the Natural History Museum, London

ISBN 978-1-907807-71-8

This book is a reprint edition of 1-898298-81-5.

A catalogue record for this book is available from the British Library.

Contents

Foreword

There was a time when I managed to spend some weeks away from the desk each year undertaking surveys of Britain's marine benthic communities for the former Nature Conservancy Council (NCC). On some of these surveys we were fortunate to have available the expertise of Dr Christine Maggs, one of the authors of this volume. Although struck with the beauty of the Ceramiales, I did have problems with their identification. Christine consoled me by pointing out that their taxonomy was very complex although she seemed to be able to put names to most! I was therefore delighted when Christine decided to take on this most difficult task of sorting out the Ceramiales! However, the users of this volume are doubly fortunate in that Dr Max Hommersand, the other author, is undoubtably one of the most knowledgeable people concerned with red algal taxonomy and phylogeny.

It was such uncertainties in the identification of the apparently common and ubiquitous algae, such as the Ceramiales, that indicated the good sense of The Natural History Museum and the British Phycological Society in starting the 'Seaweeds of the British Isles' series. This also met the urgent requirement for the provision of a working text for scientists engaged in floristic surveys, experimental studies, teaching, monitoring and surveillance. Such information is particularly vital in marine nature conservation where knowledge of the distribution of species is essential in the description and classification of marine ecosystems. This provides the framework for the process of identifying important sites and species, and in detecting and assessing natural and man- made changes.

Because of the immediate needs of the former NCC for up-to-date information on the marine algae, it was agreed that they would commission the British Phycological Society to research and draft this volume. Research for the volume coincided with the wider recognition of the true worth of taxonomy, sometimes called the 'Cinderella' of the biological sciences. In particular, the Natural Environment Research Council's report on 'The New Taxonomy' (1992) emphasised the importance of this science, the increased impetus from new techniques and applications and an increasing synergism with other disciplines, especially in explaining contemporary changes in biodiversity. Further development of taxonomic-geographic databases for the design of conservation strategies in the marine environment is identified as a high priority. I trust that this is what is being achieved in this association between the British Phycological Society, The Natural History Museum and the statutory nature conservation bodies; that was certainly the objective.

DR ROGER MITCHELL (Joint Nature Conservation Committee) presently at English Nature.

Acknowledgements

We would first like to thank several people without whom this book could not have been written within the proposed time period. Mrs L. M. Irvine has generously helped us with several difficult nomenclatural problems and with the loan of a set of relevant publications. Dr M.-T. L'Hardy-Halos kindly shared her invaluable knowledge of the Ceramiaceae of France with us. For helpful discussions on various aspects of the flora, we are grateful to: Dr F. Ardré, Dr A. Athanasiadis, Dr J. Cremades, Dr M. D. Guiry, Dr D. F. Kapraun, Prof. F. Magne, Ms H. M. Parkes, Mr J. Price, Dr W.F. Prud'homme van Reine, Dr J. Rueness, Dr R. Schmid, Dr P. C. Silva, Dr A.R.A. Taylor, Mr I. Tittley, Ms B. A. Ward, Dr A. Whittick, and Dr M. Wynne. Dr. G. W. Lawson provided many useful suggestions on preparation of the manuscript and Mrs Angela Shipman kindly prepared the Latin diagnoses.

We are very grateful to the following for loans or gifts of specimens: Mr F. Bunker, Ms T. Bennett, Dr J. Cabioch, Dr J. Cloakey, Dr L. M. Davies, Dr M. J. Dring, Dr W. F. Farnham, Dr R. L. Fletcher, Dr R. Govier, Dr R. Forster, Dr K. Hiscock, Ms C. M. Howson, Dr R. Huxley, Dr J.M. Kain, Mrs A. E. Little, Dr J. Nunn, Mr B. E. Picton, Ms J. Picton, Mrs S. Scott (formerly Hiscock), Dr W. Woelkerling and Dr H. B. S. Womersley. We would also like to thank the Joint Nature Conservation Council (formerly Nature Conservancy Council), Peterborough, for permission to use records from field surveys, including those carried out by the Field Studies Council and the University Marine Biological Station Millport.

We would like to express our appreciation of the financial support provided by the JNCC, British Petroleum plc, the British Phycological Society, the Royal Irish Academy and the British Council. We would also like to acknowledge the contribution made by the Natural Environment Research Council (Grant GR3/7945) to taxonomic research on the genus *Ceramium*, some results of which are incorporated in this volume.

We are indebted to the Directors and Curators of the following institutions for working facilities, for permission to examine specimens or borrow material:

Ulster Museum, Belfast **BEL**

The Natural History Museum, London **BM**

Royal Botanic Gardens, Kew. The algae previously at **K** are now on permanent loan to **BM** as **BM-K**.
Botanical Museum and Herbarium, Copenhagen (Denmark) **C**
Royal Botanic Garden, Edinburgh **E**
Department of Botany, University College, Galway **GALW**
Rijksherbarium, Leiden (Netherlands) **L**
Botanical Museum, Lund (Sweden) **LD**
Linnean Society, London **LINN**
Marine Biological Association, Plymouth **MBA**
Laboratoire de Cryptogamie, Museum Nationale d'Histoire Naturelle, Paris (France) **PC**
Department of Botany, Trinity College, Dublin **TCD**

The standard herbarium abbreviations given above are used throughout the text to indicate the location of type specimens. For a complete list of abbreviations, see Holmgren, Keuken & Schofield (1981).

Much of the final photographic preparation was carried out by Ms C. McLaughlin (who also drew the illustration on the cover) and Ms D. Stengel, and we sincerely appreciate their dedication to the project. CAM would also like to thank Dr M. J. Dring for his support over several years, including the period during which this flora was prepared, and she thanks her husband Colm Campbell for assistance during field collections. MHH is very grateful for all the support and encouragement provided by his wife Fran, and thanks the University of North Carolina for a leave of absence in autumn 1989.

Finally, we much appreciate the involvement of other members of the Flora Committee, particularly Dr R. L. Fletcher and Dr D. M. John, in the preparation of this flora volume.

Arrangement of the Work

The structure of this book has been based on the previous volumes of the Rhodophyta. There are differences, partly arising from the involvement of two authors; our approach to the entries for higher taxa, genera and species is described below.

Dr Hommersand was responsible for all considerations regarding taxa above the level of species. Our classification and descriptions of the families, subfamilies, tribes and Delesseriacean groups of the Ceramiales, and our circumscriptions of the genera, are based partly on published studies but incorporate a great deal of Dr Hommersand's unpublished research. For each of the higher taxa, key features of systematic significance are described, with particular emphasis on female reproductive development. References to any published work used in the preparation of each description are listed at the end of the diagnostic information. We have not attempted to evaluate or list generic synonyms.

Dr Maggs had responsibility for evaluation of the species present in the British Isles, and prepared the species entries and illustrations in consultation with Dr Hommersand; she also wrote the key to genera. For each genus, a brief discussion on useful features for identification of species is provided, with an explanation of changes in species names and cirumscriptions since the most recent published checklists for the red seaweeds of the British Isles. The species entries use the earliest valid name that we have been able to find, and list the location of type material if known, with explanatory footnotes where necessary. The basionym is then provided, and synonyms are given only if they have been used in a British Isles context. The descriptions are based entirely on examination of material collected in the British Isles, unless otherwise stated, and concentrate on features that are relatively easy to observe, thus there are few details of female reproductive morphology.

Habitat and distribution in the British Isles and elsewhere follow each species description. Habitat data is based mainly on personal observations, with reference to published information where it differs significantly. British Isles distributions include authenticated records, especially the unpublished data of Mrs M. A.Wilson (née Westbrook), and those from herbarium specimens. Distributions for the rest of the world are based on published information (cited) with occasional references to specimens we have examined. Where geographical records were reported by synonyms we have cited, no reference is made to the name used in the original publication, but if a misidentification or other synonym is involved, the name is included. A section on seasonality and life histories, with chromosome numbers where known, is based on our observations except where otherwise acknowledged. The section on morphological variation is derived from observations and published literature. Finally, possible identification problems are discussed, with notes on key features.

The illustrations, except one plate of reproduced drawings, are all photographs of material collected in the British Isles. The legends generally include the county and month in which the collection was made. A set of specimens used in the preparation of this flora volume are in the herbarium of The Natural History Museum, London.

Key to Genera

The following is a key to the genera of the Ceramiales in the British Isles. It is designed only for use within our geographical area, and it frequently uses trivial characters because important generic characters may be difficult to see. The number of the previous dichotomy is given in parentheses after the key number in places where the sequence of dichotomies is not direct.

For the non-specialist reader, we recommend the use of Hiscock's (1986) key to all the red algae of the British Isles. This should provide sufficient information to determine which of the red algae encountered are members of the Ceramiales. Two seaweed guides for areas with related floras, illustrated with photographs, have appeared recently. Bird & McLachlan (1992) include all the common red algae of the Canadian Maritime Provinces, many of which are also found in the British Isles. A guide to some of the algae of the French Atlantic and Mediterranean coasts by Cabioch & Boudouresque (1992) has colour photographs of some of our Ceramiales. Further information on all aspects of the red algae, including the Ceramiales, can be obtained from Cole & Sheath (1990).

Key

1 Parasitic on members of the Delesseriaceae; thalli <5 mm in diameter 2
  Not parasitic on members of the Delesseriaceae 4
2 Parasitic on *Apoglossum* .... *Apoglossocolax*
  Parasitic on another species of the Delesseriaceae 3
3 Parasitic on *Erythroglossum* .... *Asterocolax*
  Parasitic on *Cryptopleura* or *Acrosorium* .... *Gonimophyllum*
4 Thalli leafy, membranous and translucent; blades one to a few cells thick, sometimes with thicker midribs and/or veins 5
  Thalli not membranous; either filamentous or solid and cartilaginous 17
5 Midribs conspicuous, running from base to apex of blades 6
  Midribs absent, but veins or thickenings may be present 10
6 Lateral veins conspicuous, extending from midrib towards margins 7

Lateral veins absent or microscopic only 8

7 Blade margins entire or torn but not lobed or dentate; reproductive structures borne on old midribs in winter .... *Delesseria*

Blade margins lobed or dentate; reproductive structures borne on mature blades and marginal outgrowths throughout the year .... *Phycodrys*

8 Each blade branching alternately from margins; reproductive structures borne at blade tips and on axillary bladelets .... *Membranoptera*

Individual blades unbranched, with entire margins; new blades formed by midribs only 9

9 Microscopic lateral veins present; cells of blade < 20 µm long .... *Apoglossum*

Microscopic lateral veins absent; cells of blade > 60 µm long .... *Hypoglossum*

10 (5) Veins absent but base of blade may be thickened, sometimes in rays 11

Veins present throughout blade, macroscopic or microscopic 13

11 Plastids large and convoluted; tetrasporangial sori either minute and inconspicuous or borne only in winter on special bladelets .... *Drachiella*

Plastids numerous, small, discoid or bacilloid; reproductive structures conspicuous, borne on mature blades 12

12 Cystocarps conical with conspicuous projecting ostiole; tetrasporangia > 100 µm in diameter .... *Nitophyllum*

Cystocarps subspherical, lacking projecting ostiole; tetrasporangia < 70 µm in diameter .... *Haraldiophyllum*

13 Macroscopic veins conspicuous, at least near base of blade 14

Macroscopic veins absent; microscopic veins forming a network 16

14 Sori (male and tetrasporangial) scattered between veins on mature blades; cystocarps with spiny ornamentation .... *Polyneura*

Sori (male and tetrasporangial) borne along margins of mature blades or in young outgrowths from blade remnants; cystocarps lacking spines 15

15 Margins of young blades smooth or lobed, lacking triangular projections; prostrate blades with rounded apices, attached by peg-like holdfasts from lower surfaces .... *Cryptopleura*

Margins of young blades with teeth or triangular projections; prostrate blades narrow, with pointed apices, attached by marginal rhizoidal holdfasts .... *Erythroglossum*

16 Thalli attached by small inrolled marginal projections that develop at regular intervals; plastids ribbon-like or beaded in mature blades .... *Radicilingua*

Thalli attached by peg-like holdfasts from lower surfaces of prostrate blades; plastids discoid to bacilloid .... *Acrosorium*

17 (4) Thalli mucilaginous; axes terete, 150-400 µm in diameter, composed of axial filaments bearing whorls of 3 branches on each cell .... *Crouania*

Thalli not mucilaginous 18

18 Thalli monosiphonous and ecorticate throughout; mature axes composed of branched or unbranched axial filaments, with or without shorter branches in various arrangements 19

Thalli **not** monosiphonous and ecorticate throughout, either monosiphonous and partly corticated, or polysiphonous, or with construction not obvious externally 36

19 Paired or whorled branches present 20

Paired or whorled branches absent; branching alternate or pseudodichotomous 26

20 Axial cells > 500 μm in diameter .... *Sphondylothamnion*

Axial cells < 150 μm in diameter 21

21 Gland cells present (usually conspicuous, rarely small and vestigial); subapical cells less than half length of apical cells 22

Gland cells absent; subapical and apical cells more-or-less equal in length 25

22 Whorl-branches simple or very sparsely branched .... *Antithamnionella*

Whorl-branches bearing numerous branchlets 23

23 Main axes branched pseudodichotomously, with paired, incurved tips; gland cells cylindrical or conical, with their long axis at right-angles to branchlet .... *Pterothamnion*

Main axes branched alternately, with straight or sinusoidal tips; gland cells with long axis parallel to branchlet 24

24 Gland cells lying alongside 2-3 cells of special short branchlets (rarely small and vestigial on short branchlets) .... *Antithamnion*

Gland cells lying alongside single cell of ordinary branchlet .... *Scagelia*

25 (21) Cells of prostrate axes bearing rhizoids and erect axes in a median position, often opposite each other; tetrasporangia formed singly, terminally or laterally on branchlets .... *Ptilothamnion*

Cells of prostrate axes bearing rhizoids near posterior end, and erect axes near anterior end, never opposite each other; tetrasporangia clustered and pedicellate .... *Spermothamnion*

26 (19) Propagules present, consisting of dense ovoid bodies (resembling monosporangia) borne on non-pigmented pedicels; other reproductive structures absent ... . *Monosporus*

Propagules absent; other reproductive structures often present 27

27 Main axes distinct, obviously wider than branches, bearing laterals in various arrangements 28

All axes more-or-less equal in width 31

28 Main axes bearing laterals in an alternate arrangement in one plane 29

Main axes bearing laterals in a spiral or irregular arrangement, not all in one plane ... . *Aglaothamnion*

29 Cells uninucleate; rhizoidal filaments not forming secondary connections to other cells; cystocarps borne laterally on main axes; spermatangial branchlets forming small tufts .... *Aglaothamnion*

Cells multinucleate; rhizoidal filaments forming conspicuous secondary connections to other cells; cystocarps borne terminally or laterally on lateral branches; spermatangia formed by large cylindrical heads 30

30 First-order laterals unbranched below, for about half their length; polysporangia formed instead of tetrasporangia; spermatangial heads lateral, adaxial .... *Pleonosporium*
First-order laterals branched from every cell or from all but a few basal cells; tetrasporangia and octosporangia present but not polysporangia; spermatangial heads terminal .... *Compsothamnion*
31 (27) Thalli turf-like, consisting of extensive prostrate axes bearing numerous erect axes 32
Thalli erect 33
32 Cells of prostrate axes bearing rhizoids and erect axes in a median position, often opposite each other; tetrasporangia formed singly, terminally or laterally on branchlets .... *Ptilothamnion*
Cells of prostrate axes bearing rhizoids near posterior end, and erect axes near anterior end, never opposite each other; tetrasporangia clustered and pedicellate ... . *Spermothamnion*
33 Apical cells conical; reproductive structures borne on whorls of pigmented branchlets in short lateral outgrowths .... *Halurus*
Apical cells cylindrical or globose; reproductive structures formed either at nodes of main axes or singly on pedicels or on non-pigmented trichoblasts, but not in whorls of pigmented branchlets 34
34 Reproductive structures usually present; tetrasporangia and spermatangia formed in rings around nodes .... *Griffithsia*
Reproductive structures rarely present, if so then not formed in rings around nodes but either singly on pedicels or on non-pigmented trichoblasts 35
35 Axial cells > 400 µm in diameter; thalli rigid and robust .... *Bornetia*
Axial cells < 200 µm in diameter; thalli extremely flaccid and delicate ..... *Anotrichium*
36 (18) Last-order branches obviously monosiphonous although rings of small cells may be present at nodes; mature main axes partly or entirely corticate 37
Last-order branches **not** obviously monosiphonous, either fully corticated, or polysiphonous, or construction not obvious externally 43
37 Branches whorled, 5-8 per axial cell. .... *Halurus*
Branches not whorled 38
38 Last-order branchlets consisting only of single filaments, lacking any cortication or rings of periaxial cells 39
Last-order branchlets consisting of filaments with rings of smaller cells around nodes 42
39 Last-order branchlets paired .... *Plumaria*
Last-order branchlets borne singly 40
40 Cells multinucleate (visible in live plants and stained material); thalli usually iridescent .... *Callithamnion*
Cells uninucleate (visible only in stained material); thalli not iridescent 41
41 Chains of ovoid seirosporangia formed terminally; spermatangia with apical nucleus; cystocarps consisting of separate chains of carposporangia .... *Seirospora*

Lacking terminal chains of aggregated ovoid sporangia; spermatangia with median nuclei; cystocarps consisting of coalesced mass of carposporangia .... *Aglaothamnion*

42 (38) Thalli composed of two distinctly different types of axes: major axes corticated by alternate bands of long and short cells; branchlets mostly ecorticate, with rings of small cells at nodes .... *Spyridia*

Thalli composed of only one type of axis, variably corticated but not with regularly alternating bands of long and short cells .... *Ceramium*

43 (36) Thalli feathery, branching all in one plane, with paired long and short branches and wing-like last-order branchlets .... *Ptilota*

Thalli not feathery, branches, if paired, more-or-less equal in length 44

44 Axes with partial or complete cortication 45

Axes ecorticate, obviously polysiphonous, with each segment composed of an axial cell and a ring of periaxial cells equal to it in length 48

45 Axial cells visible through cortex, large, round or oval in surface view, constituting the majority of the axis 46

Axial cells either indistinguishable externally or comprising <1/3 of the axis diameter 47

46 Axes terete .... *Ceramium*

Axes compressed .... *Microcladia*

47 Major axes of simple polysiphonous construction, consisting in section of axial and periaxial cells surrounded by small-celled cortex but lacking large-celled medulla 48

Major axes consisting in section of large-celled medulla, sometimes with distinct central axial and periaxial cells, surrounded by small-celled cortex.................... 58

48 (44, 47) Axes bearing pigmented monosiphonous branchlets spirally or alternately..... 49

Axes lacking pigmented monosiphonous branchlets 51

49 Main axes compressed when young, bearing branches alternately in one plane; periaxial cells 9-10 .... *Heterosiphonia*

Main axes terete throughout, bearing branches in a spiral arrangement; periaxial cells 5-7 50

50 Mature axes corticated; tetrasporangia borne in specialized pod-like stichidia ... *Dasya*

All axes ecorticate; tetrasporangia borne in ordinary branches .... *Brongniartella*

51 (48) Growing tips tightly inrolled .... *Bostrychia*

Growing tips straight or slightly curved 52

52 Thalli complanate; main axes bearing regularly alternate arrangement of shorter laterals in one plane 53

Thalli branched 3-dimensionally 56

53 Axes ecorticate throughout 54

Axes corticate throughout 55

54 Major axes bearing branches regularly on every alternate segment .... *Pterosiphonia*

Major axes bearing branches at intervals of 4 or more segments .... *Polysiphonia*

55 Periaxial cells 8 or more .... *Boergeseniella*

Periaxial cells 5 .... *Pterosiphonia*
56 (52) Thalli erect, more than 5 cm high .... *Polysiphonia*
Thalli forming turfs less than 2 cm high 57
57 Periaxial cells 10 or more .... *Lophosiphonia*
Periaxial cells 4 .... *Polysiphonia*
58 (47) Axes compressed or blade-like 59
Axes terete 60
59 Axes compressed, less than 5 mm wide .... *Laurencia*
Axes flat, blade-like, with alternate, wing-like last-order branches .... *Odonthalia*
60 Growing tips of main axes tightly inrolled 61
Growing tips of main axes straight or slightly incurved 62
61 Thalli turf-forming, less than 6 cm high; last-order branches alternate .... *Bostrychia*
Thalli solitary, erect, at least 10 cm high; last-order branches paired .... *Halopitys*
62 Axial and periaxial cells not distinguishable in section from other medullary cells .... *Laurencia*
Axial cell and ring of 5-6 periaxial cells clearly distinguishable in transverse sections 63
63 Young branches markedly contricted at bases; spermatangia borne on specialized plates; tetrasporangia formed spirally .... *Chondria*
Young branches not constricted at bases; spermatangial sori formed on ordinary branches; tetrasporangia formed in pairs .... *Rhodomela*

CERAMIALES Oltmanns

CERAMIALES Oltmanns (1904), p. 683.

Having a triphasic *Polysiphonia*-type life history, consisting of a haploid sexual phase, the gametophyte, a diploid phase that develops directly on the gametophyte, the carposporophyte, and a free-living diploid phase, the tetrasporophyte, with gametophytes and tetrasporophytes morphologically similar; gametophytes typically dioecious, but monoecious thalli and thalli with mixed gametophytic and sporophytic phases reported in many species. Typical monosporangia absent. Carpospores and tetraspores spherical, densely filled with cytoplasm; germination mostly bipolar (*Ceramium*-type) with the formation of a narrow primary rhizoid initial and a broader upper cell that divides transversely forming the apical initial of the primary axis, rarely unipolar with the primary rhizoid absent; primary axis usually erect, occasionally prostrate, attached by rhizoids or a rhizoidal disc; secondary axes erect or prostrate. Growth uniaxial, maintained by transverse division of a uninucleate or, less often, a multinucleate apical cell; body plan consisting fundamentally of ecorticate axes or axes corticated by rhizoidal filaments, and bearing four lateral determinate assimilatory filaments (whorl-branches) in opposite pairs from the upper ends of axial cells; basic body plan modified by: (1) reduction in the number of lateral determinate filaments to 3, 2 or 1; (2) condensation of the lateral filaments to form a well-defined cortex around the axis; (3) condensation of the lateral filaments in the plane of bilateral symmetry and linkage between cells by secondary pit connections to form a membrane; (4) loss of all but the basal cells of the lateral filaments (periaxial cells), usually accompanied by an increase in periaxial cell number and a shift to an alternate arrangement.

Spermatangial mother cells terminal, solitary, clustered, organized in sori, or borne on special branches (spermatangial axes or heads); spermatangia formed singly or in clusters of 2-3(-4), either free and separate or confluent within a common matrix and sometimes covered by a firm outer layer; spermatia released individually or together in a mucilaginous matrix; additional spermatangia commonly proliferating in succession from the spermatangial mother cell, sometimes surrounded by previous spermatangial walls.

Female reproductive apparatus procarpic, the auxiliary cell cut off after fertilization in close proximity to the carpogonium; procarp typically consisting of a fertile periaxial cell (supporting cell), a sterile vegetative filament (sterile group-1), a (3-)4 celled carpogonial branch terminated by a carpogonium bearing a trichogyne; a second sterile vegetative filament (sterile group-2) present or absent, or replaced by a second carpogonial branch. Sperm and egg nucleus uniting inside the carpogonium after fertilization, the resulting diploid nucleus dividing twice to produce a carpogonial nucleus and: (1) a terminal capping cell and a connecting cell, (2) two connecting cells, (3) three connecting cells, or (4) 3(-4) nuclei that remain inside the carpogonium; nuclei within the connecting cells minute, with condensed DNA, and surrounded by a hyaline region and external membrane or thin wall; supporting cell and sometimes also 1-2 additional periaxial cells in the fertile whorl cutting off an auxiliary cell after fertilization; each functional auxiliary cell expanding or forming a process that contacts a connecting cell or an extension from the carpogonium, effecting transfer of the diploid nucleus; haploid nucleus inside the auxiliary cell remaining undivided or dividing one or more times, and either cut off inside a basal cell (foot cell) or lateral cell (disposal cell), or quiescent and ultimately degenerating inside the auxiliary cell; diploid nucleus enlarging inside the auxiliary cell and dividing, with one derivative remaining inside the residual auxiliary cell, foot cell, or disposal cell, and the other cut off along with the gonimoblast initial; gonimoblasts at first monopodially branched, and either continuing to grow monopodially or shifting to sympodial growth after carposporangial initiation; carposporangia clustered in globose masses, formed in simple or branched chains, or terminal and solitary; fusion cell absent or present and small, incorporating few cells, or large, incorporating many gametophytic and carposporophytic cells; gonimoblasts naked or surrounded by sterile involucral filaments, covered by a cortical pericarp, or enclosed within an external pericarp consisting of modified axes united laterally by secondary pit connections.

Tetrasporangia terminal, sessile or stipitate, solitary or in clusters on determinate lateral filaments (whorl-branches) or borne on periaxial or central cells, and then either solitary or organized in sori or in whorls on specially modified axes (stichidia). Tetrasporangia spherical to subspherical, divided by successive or simultaneous cleavages into four tetrahedrally arranged tetraspores, or the tetrasporangia ellipsoid, cleaved first along the short axis and then twice along the long axis into four cruciately arranged tetraspores.

Some of the variations included in this description reflect putatively primitive characters encountered mainly in the Ceramiaceae. The rest involve specialized features found mainly in the more advanced families Delesseriaceae, Dasyaceae, and Rhodomelaceae.

References: Kylin (1956), Dixon (1973), Coomans & Hommersand (1990), Hommersand & Fredericq (1990), Guiry (1990).

CERAMIACEAE Dumortier

CERAMIACEAE Dumortier (1822), pp. 73, 100 [as Ceramineae].

Germination bipolar, rarely unipolar; thallus erect, decumbent or prostrate, attached by a primary basal rhizoid and often also by secondary rhizoids originating from axial cells or cells of determinate lateral filaments (whorl-branches); erect axes ecorticate, or corticated by descending rhizoidal filaments; determinate lateral filaments 4, 3, 2, or 1 per axial cell, formed opposite one another from upper ends of axial cells, sometimes scarcely distinguishable from the indeterminate axes; periaxial cells sometimes more than 4 per axial cell, and then usually arranged in an alternating (rhodomelacean) sequence; axial cells often elongating at a distance below the thallus apex, separating the determinate lateral filaments (whorl-branches), or cortical bands; indeterminate branches originating near the apex, replacing determinate branches, or borne on basal cells of determinate branches; adventitious branches rare to abundant, originating from axial cells, periaxial cells, cortical cells, or rhizoidal cells.

Gametophytes dioecious or monoecious, often with mixed gametophytic and sporophytic phases occurring on the same plant. Spermatangial mother cells solitary on unspecialized determinate filaments, superficial on cortical filaments, densely clustered on modified, short-celled filaments, or condensed in heads; spermatangia usually 2-3 per spermatangial mother cell, sometimes more numerous; spermatia superficial, released individually, or embedded in a common matrix, sometimes covered by a firm outer layer, and shed *en masse*. Procarps borne on periaxial cells along the length of ordinary or adventitious indeterminate axes, or restricted to periaxial cells localized at the tips of modified indeterminate axes; consisting of an unmodified or modified vegetative sterile filament (sterile group-1) and a 4-celled lateral carpogonial branch, or the carpogonial branch replacing the terminal filament and a vegetative sterile filament absent; a second, lateral carpogonial branch present in a few species, however a second sterile group absent. Supporting cell and sometimes 1-2 periaxial cells in the fertile whorl cutting off auxiliary cells after fertilization or, less often, the supporting cell remaining undivided and functioning directly as an auxiliary cell; fertilized carpogonium cutting off: (1) a terminal capping cell and a posteriolateral connecting cell, (2) two connecting cells, (3) three connecting cells, (4) 3(-4) nuclei which remain inside the carpogonium; diploidization of the auxiliary cell effected either by contact with a connecting cell or by direct fusion with the carpogonium; diploidized auxiliary cell (1) dividing into a foot cell containing a haploid and a diploid nucleus and a gonimoblast initial, (2) a foot cell containing only a haploid nucleus and a diploid gonimoblast initial, or (3) a lateral cell (disposal cell) containing a haploid nucleus and an auxiliary cell containing a diploid nucleus which either cuts off a gonimoblast initial or functions directly as the primary gonimoblast cell; gonimoblasts cutting off 2-6(-7) gonimolobe initials that branch monopodially producing compact clusters of gonimoblast cells in which most of the cells are transformed into carposporangia, or the gonimoblasts loosely branched, bearing terminal carposporangia or carposporangia in terminal chains; gonimoblasts naked or surrounded loosely by involucral filaments or lateral involucral axes, or by a loosely organized external pericarp composed of anastomosing determinate lateral filaments.

Tetrasporangia sessile or pedicellate, solitary or clustered on ordinary or specialized determinate filaments or adventitious branches, or cut off laterally from periaxial cells or cortical cells, tetrahedrally or cruciately divided; polysporangia with multiples of four spores infrequent and quadrinucleate bisporangia rare; asexual reproduction by uninucleate or multinucleate monosporangia-like bodies (seirosporangia, parasporangia, propagules), or by fragmentation.

References: Kylin (1956), Hommersand (1963), Wollaston (1968), Gordon (1972), Dixon (1973), Itono (1977), Stegenga (1986), Millar (1990), Huisman & Kraft (1992).

Tribe ANTITHAMNIEAE Hommersand (1963), p. 330.

ANTITHAMNION Nägeli

ANTITHAMNION Nägeli (1847), p. 202.

Type species: *A. cruciatum* (C. Agardh) Nägeli (1847), p. 202.

Thalli uniseriate, ecorticate, consisting of branched prostrate axes and unattached erect tips up to 2 cm high, or differentiated into prostrate and erect axes, or wholly erect to 5 cm high; prostrate axes and basal parts attached by rhizoids that arise from basal and sometimes suprabasal cells of determinate branches and produce digitate holdfasts upon contact with the substratum; determinate lateral branches (= whorl-branches) alternate or opposite in prostrate axes and arranged in distichous or decussate opposite pairs in erect axes, consisting of 1-3 orders of successively shorter filaments that may be alternate, opposite, or secund; basal cells of determinate branches short, usually as broad as long, naked, lacking determinate laterals; gland cells absent or present, 1-few, on modified filaments up to 5 (-6) cells long; indeterminate branches either replacing determinate branches or arising from their basal cells.

Growth of indeterminate axes by transverse division of apical cells; lateral initials of determinate branches formed close to apex, often in a prescribed pattern with branching of the filaments characteristic of the species; gland cells initiated by oblique division of apical cell, displaced secondarily to a lateral position by overgrowth of filaments from subapical cells; vegetative cells uninucleate with the nuclei enlarged in elongated axial cells; plastids parietal, dissected into a network of bead-like to ribbon-shaped bands.

Gametophytes dioecious, rarely with mixed phases. Spermatangial mother cells formed in whorls on short-celled axes 4-10 cells long borne on determinate branches; spermatangia 2-5 (-6) per mother cell. Procarps formed singly in series of 2-20 near tips of immature indeterminate axes which convert facultatively to determinate branches, consisting of a basal supporting cell, an unmodified or reduced terminal vegetative branch, and a 4-celled, lateral carpogonial branch; auxiliary cell cut off distally from the supporting cell after fertilization and containing a conspicuous protein body; the fertilized carpogonium cutting off a terminal cell and forming a lateral extension that fuses directly with the auxiliary cell depositing a derivative of the fertilization nucleus, or diploidization of the auxiliary cell possibly mediated by a connecting cell; auxiliary cell dividing into a foot cell containing the haploid nucleus and the gonimoblast initial containing the diploid nucleus and protein body; gonimoblast initial producing a distal primary gonimolobe initial and 1-3 (-4) lateral gonimolobe initials that develop successively forming globose clusters of carposporangia; fusion cell present, formed primarily by fusions around pit connections and incorporating the central cell, supporting cell, foot cell and primary gonimoblast cell; gonimoblasts naked or surrounded by a loose involucre of determinate branches derived from axial cells below the fertile segment; only one cystocarp maturing per fertile axis. Tetrasporangia sessile or pedicellate, formed singly or in clusters from upper ends of cells of determinate branchlets, ellipsoid to ovoid, cruciately divided.

References: Wollaston (1968, 1972a), L'Hardy-Halos (1968b), Athanasiadis (1986, 1988).

The most useful taxonomic characters discriminating between British Isles representatives of this genus are the branching pattern of whorl-branches and the positions of gland cells.

Only one species of *Antithamnion*, *A. cruciatum*, is listed in the most recent algal checklist for the British Isles (South & Tittley, 1986). Several varieties of *A. cruciatum* have been described; one of these, var. *villosum* (= var. *scandinavicum*), is recognized here as a separate species, *A. villosum*. In addition, a third species, *A. densum* (= *A. defectum*), has recently been discovered in the British Isles (Guiry & Maggs, 1991).

KEY TO SPECIES

1 All whorl-branches in one plane, distichous, bearing secund (= comb-like) series of adaxial branchlets .... *A. densum*
Each pair of whorl-branches formed at 90° to the next pair (= decussate), bearing opposite, irregularly alternate or secund series of branchlets 2
2 Gland cells abundant, lying alongside 3 cells of reduced branchlet (see Fig. 1D); first few branchlets on most whorl-branches borne in opposite pairs .... *A. cruciatum*

Gland cells absent or sparse, lying alongside 2 cells of reduced branchlet; whorl-branches bearing branchlets in irregularly alternate or secund arrangement .... *A. villosum*

Antithamnion cruciatum (C. Agardh) Nägeli (1847), p. 200.

Lectotype: LD 18774* (Athanasiadis, 1986, fig. 1). Syntypes: LD, BM. Italy (Trieste).

Callithamnion cruciatum C. Agardh (1827), p. 637.
Callithamnion pumilum Harvey in W. J. Hooker (1833), p. 339.
Callithamnion cruciatum var. *pumilum* (Harvey in W. J. Hooker) Harvey (1841), p. 104.
Antithamnion cruciatum var. *pumilum* (Harvey in W. J. Hooker) Reinke (1891), p. 273.

Thalli spreading over the substratum by prostrate axes that give rise to erect axes 0.5-5 cm high; prostrate and erect axes branched, 0.8-3 mm wide including whorl-branches, with densely tufted apices; dull brown to brownish-red in colour, delicate and flaccid in texture.

Main axes enlarging from apical cells 10-12 µm wide to axial cells 100-120 µm diameter and 2.5-4 diameters in length; *whorl-branches* borne in an opposite pair on each axial cell, sequential pairs arranged at 90° to each other, 800-1800 µm long and 30 µm wide, the basal cell isodiametric and unbranched, each of the next few cells elongate, 2-3 diameters long, and branched; these *branchlets* usually paired below, in a distichous arrangement, the more distal ones formed singly in an irregularly alternate or irregularly secund pattern on successive cells, mostly 7-10 cells long, occasionally bearing a few short branchlets; terminal cells of whorl-branches and branchlets conical; *gland cells* borne on reduced branchlets 3-4(-5) cells in length, covering the adaxial face of 3 cells, ovoid to spherical, 25-40 x 12-40 µm; *indeterminate laterals* formed on alternate sides of main axes at intervals of 7-8 cells, replacing whorl-branches, completely suppressing the opposite whorl-branch; further indeterminate axes developing from basal cells of whorl-branches; *rhizoids* formed by basal cells of whorl-branches, 1-3 per cell, 16-30 µm in diameter, multicellular, terminating in multicellular discoid attachment pads; *plastids* initially discoid, becoming filiform in axial cells.

Gametophytes unknown in British Isles. *Tetrasporangia* pedicellate, formed on

Fig. 1. *Antithamnion cruciatum*

(A) Erect axes with decussate pairs of whorl-branches and tufted apices (A-C Donegal, Dec.). (B) Prostrate axis with paired whorl-branches bearing paired branchlets. (C) Tufted apex, with reduced branchlets bearing gland cells (arrow). (D) Gland cells (arrow) on reduced branchlets (Devon, Oct.). (E) Pedicellate cruciately divided tetrasporangia (as D).

* Athanasiadis (1986) selected specimens 18773-18778 as lectotype. We have chosen 18774 as lectotype, the others being syntypes.

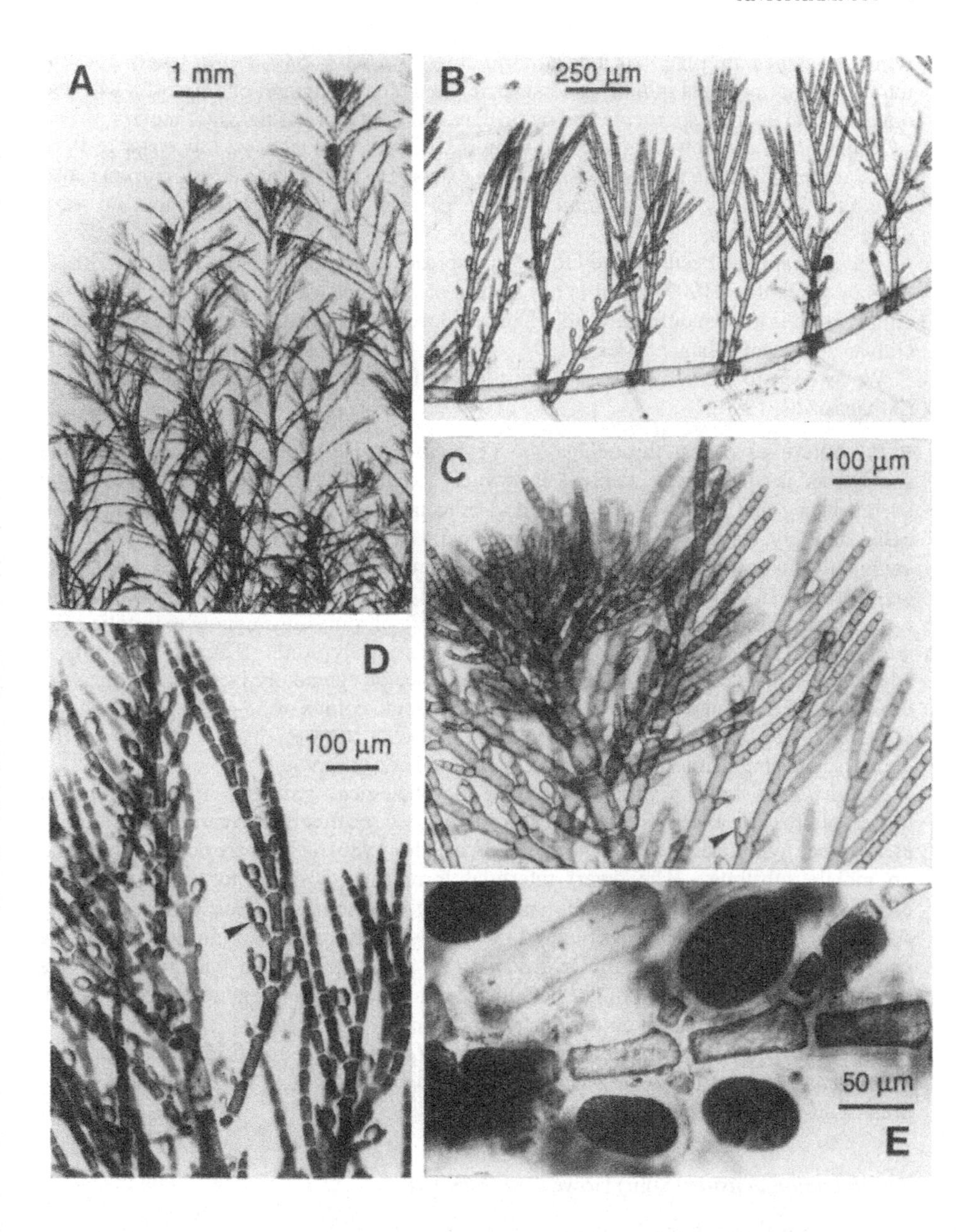
A
1 mm
B
250 µm
C
100 µm
D
100 µm
50 µm
E

whorl-branches in place of lateral branchlet, 80-92 x 64-68 μm when mature; tetrasporangial pedicels initially isodiametric, then elongating and cutting off a cell that develops into another sporocyte and pedicel, resulting in pairs of tetrasporangia.

Epiphytic on a variety of algae, including maerl, from near extreme low water to 10 m depth at sites with little wave-exposure, often exposed to moderate to strong currents; also epiphytic, especially on *Corallina* species, in mid- to lower-shore pools on very wave-exposed coasts.

A. cruciatum has been reported to be widespread, although uncommon, in the British Isles (e. g., Batters, 1902) but it is not always clear which variety of *A. cruciatum* was collected. It is known definitely from Dorset, Devon, S. Cornwall, Orkney; Cork, Clare, Galway, Donegal; Channel Isles.

W. Sweden to N. W. Africa; E. Canada and USA; Mediterranean and Black Seas (Athanasiadis, 1990).

Prostrate axes persist throughout the year, giving rise to erect axes whenever the substratum is sufficiently stable. In winter, axial cells of prostrate thalli become barrel-shaped and starch-filled, and whorl-branches are simple, composed of globular cells; new erect axes arise from the basal cells of whorl-branches. Gametangia are apparently unknown in the British Isles; tetrasporangia have been recorded for April, May and October. In Newfoundland, only tetrasporophytes occur in the field, and isolates did not reproduce in culture (Whittick & Hooper, 1977); chromosome counts in mitotic cells were of 85-110, indicating that the plants were polyploid. Tetraspores from a Mediterranean isolate (from Greece) grew into dioecious gametophytes, and fertilization took place (Athanasiadis, 1986). Mitotic chromosome counts of 55-65 in gametophytes appeared to be diploid by comparison with chromosome numbers of other members of the genus, suggesting that tetrasporophytes may have been tetraploid.

A. cruciatum shows a wide spectrum of morphological variation. Plants composed largely of prostrate axes were previously known as var. *radicans* (J. Agardh) F. Collins & Hervey (see Whittick & Hooper, 1977; Athanasiadis, 1986). These are commonly found on mobile substrata. The dwarf intertidal form corresponding to var. *pumilum* is characterized by shorter axial cells and a stocky appearance (L'Hardy-Halos, 1968b; Athanasiadis, 1988). It was considered by Harvey (1846, pl. 54) to intergrade with the usual form. We have made similar observations, and conclude that 'var. *pumilum*' is a form of *A. cruciatum* which might be either environmentally or genetically determined, but does not appear to merit taxonomic recognition.

A. cruciatum could be misidentified as *Scagelia pusilla* (q. v. for differences). Species of *Antithamnion* are occasionally confused with *Spermothamnion* spp., but gland cells are present only in *Antithamnion*.

Antithamnion densum (Suhr) Howe (1914), p. 151.

Holotype: MEL 692189 (see Athanasiadis, 1990; Athanasiadis, pers. comm.). Peru.

Callithamnion densum Suhr (1840), p. 281.
Antithamnion defectum Kylin (1925), p. 46.

Thalli to 2 cm in extent, consisting of prostrate axes spreading over the substratum, attached by rhizoids, with their apices growing erect to a height of 3-10 mm; prostrate and erect axes branched, complanate, 550-850 μm wide including whorl-branches, bright red in colour, of a delicate and flaccid texture.

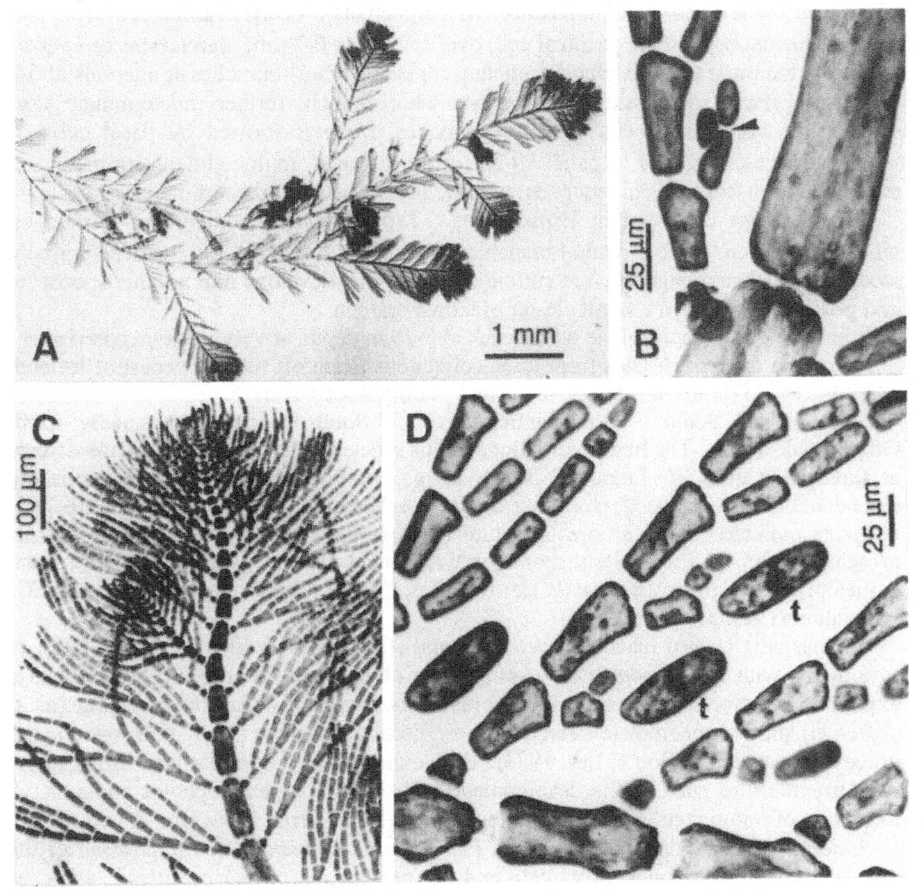

Fig. 2. *Antithamnion densum*

(A) Habit of prostrate thallus with complanate arrangement of paired whorl-branches (A-D Mayo, June). (B) Gland cell (arrow) on reduced branchlet. (C) Apex with paired whorl-branches bearing comb-like branchlets. (D) Developing pedicellate tetrasporangia (t).

Main axes enlarging from apical cells 6-7 μm wide to axial cells 40-68 μm diameter and 2.5-4 diameters in length; *whorl-branches* borne in an opposite pair on each axial cell, all in one plane, curved adaxially, 8-12 cells long and 20 μm wide, the basal cell isodiametric and unbranched, the other cells elongate, with conical or blunt apical cells terminating when young in hairs 14-16 μm long; *branchlets* borne adaxially on the suprabasal and next few cells of whorl-branch in a secund pattern, mostly 4-8 cells long; *gland cells* borne occasionally on reduced first branchlets 1-4 cells in length, covering part of two adjacent cells or the terminal cell, ovoid, 12-14 x 6-7 μm; *indeterminate laterals* formed in the same plane as whorl-branches, replacing whorl-branches at intervals of 6-7 cells, completely suppressing the opposite whorl-branch; further indeterminate axes developing from basal cells of whorl-branches; *rhizoids* formed by basal cells of whorl-branches, singly or in pairs, 10-30 μm in diameter, multicellular, terminating in multicellular discoid attachment pads; *plastids* initially discoid, becoming elongate.

Gametophytes unknown in British Isles. *Tetrasporangia* pedicellate, formed on whorl-branches in place of lateral branchlet, ovoid, 60-68 x 32-34 μm; pedicels initially isodiametric, then elongating and cutting off a cell that develops into another sporocyte and pedicel, resulting in a small cluster of tetrasporangia.

Epiphytic on perennial algae on bedrock at 5-20 m depth, at very wave-exposed sites.

Known in the British Isles from three collections made off the west coast of Ireland: Clare Is., Mayo (Guiry & Maggs, 1991) and Skellig Rocks, Kerry.

N. France, N. Spain; South Atlantic, North and South Pacific (Athanasiadis, 1990; Granja et al., 1992). The first observations of this species in the North Atlantic were made at three sites in N.W. France (L'Hardy-Halos, 1968b). European populations are considered to be introduced, possibly from the southern Atlantic (Athanasiadis, 1990).

Plants collected in June bore immature tetrasporangia; mature tetrasporangia were present in July. In France, tetrasporangia were noted in June, August and November; gametophytes were not observed (L'Hardy-Halos, 1968b). Gametophytes occur in Pacific populations (Wollaston, 1972a).

Athanasiadis (1990) placed *A. defectum* and *A. sparsum* Tokida (1932, p. 105), in synonymy with *A. densum* after examination of the relevant types. Hybridization experiments between isolates from Korea (as *A. sparsum*) and a Californian isolate (as *A. defectum*) showed incomplete interfertility, suggesting that these populations were in the process of speciation (Boo & Lee, 1983). Both exhibited *Polysiphonia*-type life histories with irregularities such as the development of monoecious gametophytes and and the formation of mitotic tetrasporangia on females (West & Norris, 1966; Boo & Lee, 1983).

A. densum can be distinguished from *Pterothamnion plumula* by the position of the gland cells, which lie alongside two cells in *A. densum* but only cover one cell in *P. plumula*; likewise in *Scagelia pusilla* (q. v.), gland cells cover part of only one branchlet cell.

Antithamnion villosum (Kützing) Athanasiadis in Maggs & Hommersand, stat. nov.

Holotype: L 939.194.1186. France (Calvados).

Callithamnion cruciatum var. *villosum* Kützing (1861), p. 28.
Antithamnion cruciatum var. *scandinavicum* Athanasiadis (1986), p. 707 (see Athanasiadis, 1990).

Thalli spreading over the substratum by prostrate axes that give rise to erect axes 0.5-2 cm high; prostrate and erect axes branched, 1-1.5 mm wide including whorl-branches, with densely tufted apices, rose-red to bright red in colour, delicate and flaccid in texture.

Main axes enlarging from apical cells 8-10 μm wide to axial cells 90-125 μm diameter and 2-4 diameters in length; *whorl-branches* borne in an opposite pair on each axial cell, sequential pairs arranged at 90° to each other, 960-1100 μm long and 15-35 μm wide, consisting of up to 14 cells, the basal cell isodiametric and unbranched, other cells elongate, 2-6 diameters long; *branchlets* occasionally absent, usually present, formed in an irregularly alternate or irregularly secund pattern on successive cells, mostly 6-10 cells long; apical cells of whorl-branches and branchlets conical, occasionally (in female plants) terminating when young in a hair; *gland cells* usually absent or vestigial, borne on branchlets reduced to 2-4 cells in length, partly covering the adaxial face of 2 cells, very rarely fully-formed, ovoid, 16-18 x 8-12 μm; *indeterminate laterals* formed alternately, replacing whorl-branches at intervals of 3-10 cells, completely suppressing the opposite whorl-branch; further indeterminate axes developing from basal cells of whorl-branches; *rhizoids* 1-3 per cell, 20-36 μm in diameter, multicellular, terminating in multicellular discoid attachment pads; *plastids* initially discoid, becoming filiform in axial cells.

Spermatangial branchlets borne adaxially on up to 4 successive cells of whorl-branches, 1(-2) per cell, 32-100 μm long x 20-28 μm wide, consisting of an axis 5-10 cells long, 5-6.5 μm in diameter, with all cells isodiametric or the basal 1-4 cells elongate, bearing whorls of 2-4 spermatangial mother cells, each with 3-4 ovoid spermatangia measuring 4 x 2.5 μm. *Procarps* formed in series of up to at least 4 in successive segments of fertile branch; *cystocarps* 360-480 μm wide, consisting of 3-4 rounded gonimolobes of different ages, 240-330 μm in diameter, containing numerous angular carposporangia 36-48 μm in diameter. *Tetrasporangia* adaxial, 76-90 x 52-60 μm when mature, formed singly on isodiametric pedicels, or in clusters resulting from elongation and division of the pedicel; tetrasporocytes ovoid, elongate, resembling bisporangia after the first division.

Epiphytic on a variety of algae, epilithic on bedrock and pebbles, also growing on dead shells and on metal and plastic structures, from near extreme low water to 26 m depth at sites with slight to moderate wave-exposure, often exposed to moderate to strong currents.

Widely distributed on south and south-west coasts, recorded from Kent, Hampshire, Dorset, N. and S. Devon, N. Cornwall, Pembroke, Caernavon, Isle of Man, Cheshire.

W. Sweden to N. France (Athanasiadis, 1988, 1990); North Carolina (Kapraun, 1977a, as *A. cruciatum*). European North Atlantic records of *Antithamnion tenuissimum* (Hauck) Schiffner (e. g., L'Hardy-Halos, 1968b), may be based on specimens of *A. villosum* (Athanasiadis, 1990).

Thalli have been collected only from July-October and March, with spermatangia in July and October, procarps and cystocarps in July, August and October, and tetrasporangia

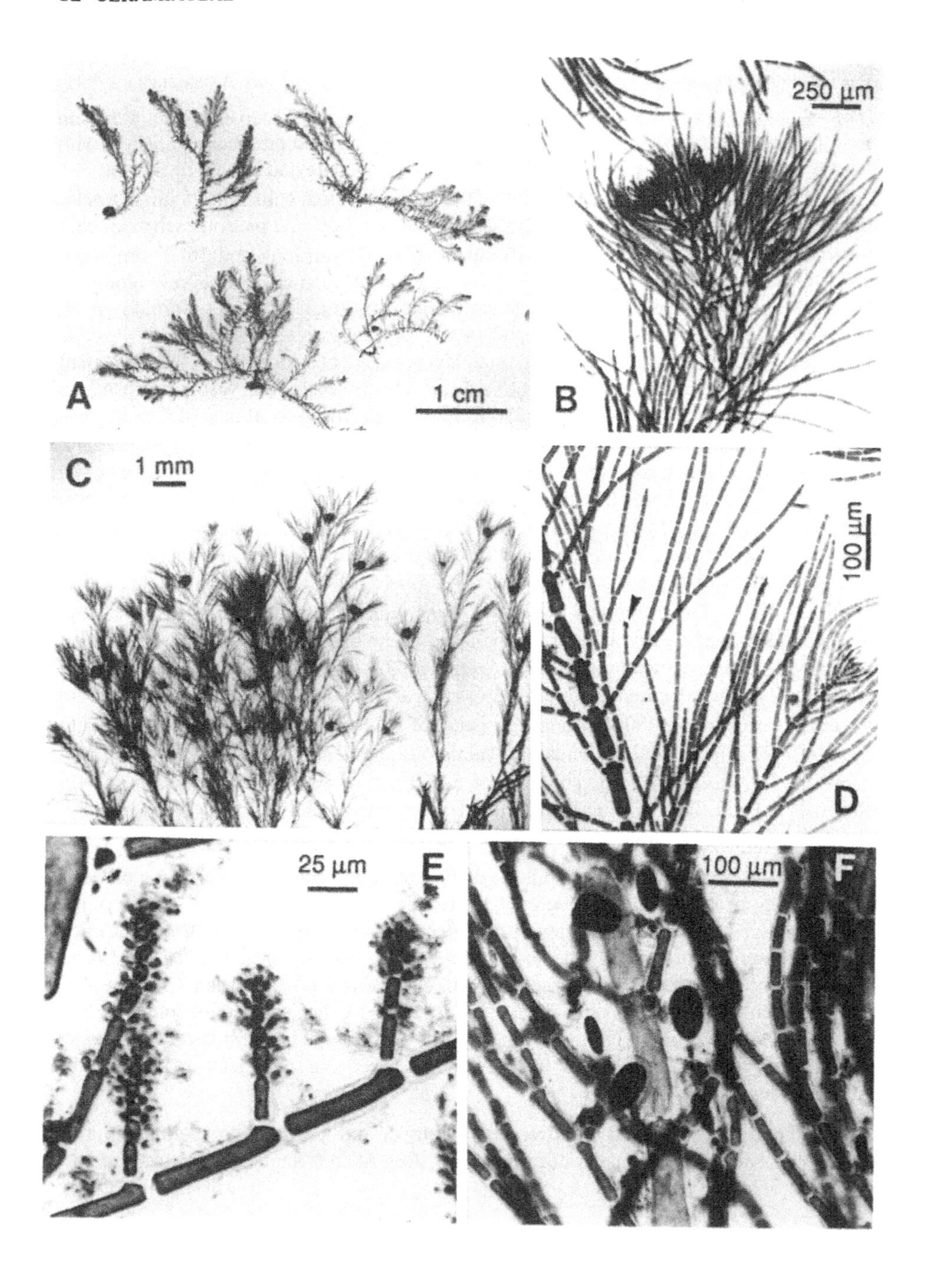
A
1 cm
B
250 μm
C
1 mm
100 μm
D
25 μm
E
100 μm
F

in July-August. Sparse tetrasporangia have been observed on female plants. An isolate from Sweden failed to reproduce in culture, forming only abortive tetrasporangia (Athanasiadis, 1986); mitotic chromosome counts of 55-65 were obtained. North Carolina populations, apparently of *A. villosum*, consist of dioecious gametophytes and tetrasporophytes presumably undergoing a *Polysiphonia*-type life history (Kapraun, 1977a, figs 4-6); chromosome counts of n = c. 24 and $2n$ = c. 48 were obtained by Kapraun (Cole, 1990).

Athanasiadis (1987, p. 6) suggested that *A. cruciatum* var. *scandinavicum* might merit specific status, but poor reproduction in culture hampered hybridization studies. Comparison of numerous reproductive specimens of *A. cruciatum* 'var. *cruciatum*' and *A. cruciatum* 'var. *villosum*' (= var. *scandinavicum*) indicates that these are distinct entities with no morphological overlap.

A. villosum is unlikely to be confused with members of any other genus in the British Isles.

ANTITHAMNIONELLA Lyle

ANTITHAMNIONELLA Lyle (1922), p. 347.

Lectotype species: *A. sarniensis* Lyle (1922), p. 348 [= *A. ternifolia* (J. D. Hooker & Harvey) Lyle (1922), p. 350] (see Kylin, 1956, p. 373).

Thalli uniseriate, ecorticate, differentiated into prostrate and erect indeterminate axes; prostrate axes attached to substratum by rhizoids arising from basal cells of determinate branches (= whorl-branches) and terminated by digitate haptera; determinate branches variable in number, single, opposite or in whorls of 3 or 4, simple or 1-2 times branched; basal cells of determinate branches short or the same length as other cells in the branch, naked or bearing a lateral; gland cells formed laterally, 1-2 per cell on determinate branches; indeterminate axes replacing determinate laterals, rarely borne on basal cells of determinate branches.

Growth of indeterminate axis sinusoidal, by slightly oblique division of apical cells successively to one side and then the other, shifting orientation at sites of future indeterminate branches; branching of determinate branches variable; gland cells cut off laterally, covering the length of the bearing cell, or proximal to a lateral filament; nuclei and plastids as in *Antithamnion*.

Fig. 3. *Antithamnion villosum*

(A) Habit of several plants, showing erect axes arising from creeping axes (Devon, Oct.). (B) Tufted apices (Pembroke, July). (C) Erect axes bearing numerous cystocarps (as A). (D) Paired whorl-branches bearing branchlets and vestigial gland cell (arrow) on reduced branchlet (Dorset, July). (E) Spermatangial branchlets (as D). (F) Pedicellate tetrasporangia (Kent, Aug.).

Gametophytes dioecious. Spermatangial mother cells in whorls on short-celled filaments 1-5 cells long borne on determinate branches; spermatangia 2-4 per mother cell. Procarps 1-3 per fertile filament, consisting of a supporting cell, a reduced, 1-celled sterile group, and a 4-celled, lateral carpogonial branch; postfertilization development as in *Antithamnion*; up to 6 gonimolobes in a cystocarp. Tetrasporangia sessile or pedicellate, borne successively on the upper sides of basal cells (occasionally suprabasal cells) of determinate branches, ovoid to subspherical, tetrahedrally divided.

References: Wollaston (1968, 1972b), Lindstrom & Gabrielson (1989).

The two species of *Antithamnionella* previously recognized in the British Isles, *A. floccosa* and *A. spirographidis*, are both included here; in addition we follow L'Hardy-Halos (1985) in treating *A. ternifolia* (= *A. sarniensis*) as a separate species from *A. spirographidis*.

KEY TO SPECIES

1 Whorl-branches paired, distichous; gland cells absent; tetrasporangia pedicellate........ .... *A. floccosa*
Whorl-branches 2-4 per axial cell; gland cells present; tetrasporangia sessile 2
2 Basal cell of whorl-branch elongate, equal in length to next cell; apical cell of mature whorl-branches with obtuse, rounded tip; usually 2-3 whorl-branches per whorl .... *A. spirographidis*
Basal cell of whorl-branch isodiametric, much shorter than next cell; apical cell of mature whorl-branches pointed; usually 3-4 whorl-branches per whorl .... *A. ternifolia*

Antithamnionella floccosa (O. F. Müller) Whittick (1980), p. 77.

Lectotype: O. F. Müller (1782), pl. 828, fig. 1, based on material from Norway (A. Athanasiadis, pers. comm.).

Conferva floccosa O. F. Müller (1782), pl. 828.
Callithamnion floccosum (O. F. Müller) C. Agardh (1828), p. 158.
Callithamnion pollexfenii Harvey (1844), p. 186.
Pterothamnion floccosum (O. F. Müller) Nägeli (1862), p. 144.
Antithamnion floccosum (O. F. Müller) Kleen (1874), p. 21.
Antithamnion plumula var. floccosum Rosenvinge (1893), p. 789.

Thalli consisting of spreading prostrate axes that become erect at the tips and also give rise to further erect axes; erect axes (2-)5-15.5 cm high, tufted or entwined, much-branched, complanate, 240-600 μm wide including whorl-branches, brownish-red, delicate and flaccid.

Main axes enlarging from apical cells 8 μm wide to axial cells 180-1700 x 30-125 μm, cylindrical but truncated at branch insertions; young laterals borne singly in a secund row on 3 (rarely 4-6) successive axial cells, followed by a row of 3(-6) on the opposite side of the axis; *whorl-branches* developing from 2 of the 3 laterals of each row, distichously arranged when mature (in prostrate axes sometimes sparse or alternate), simple or

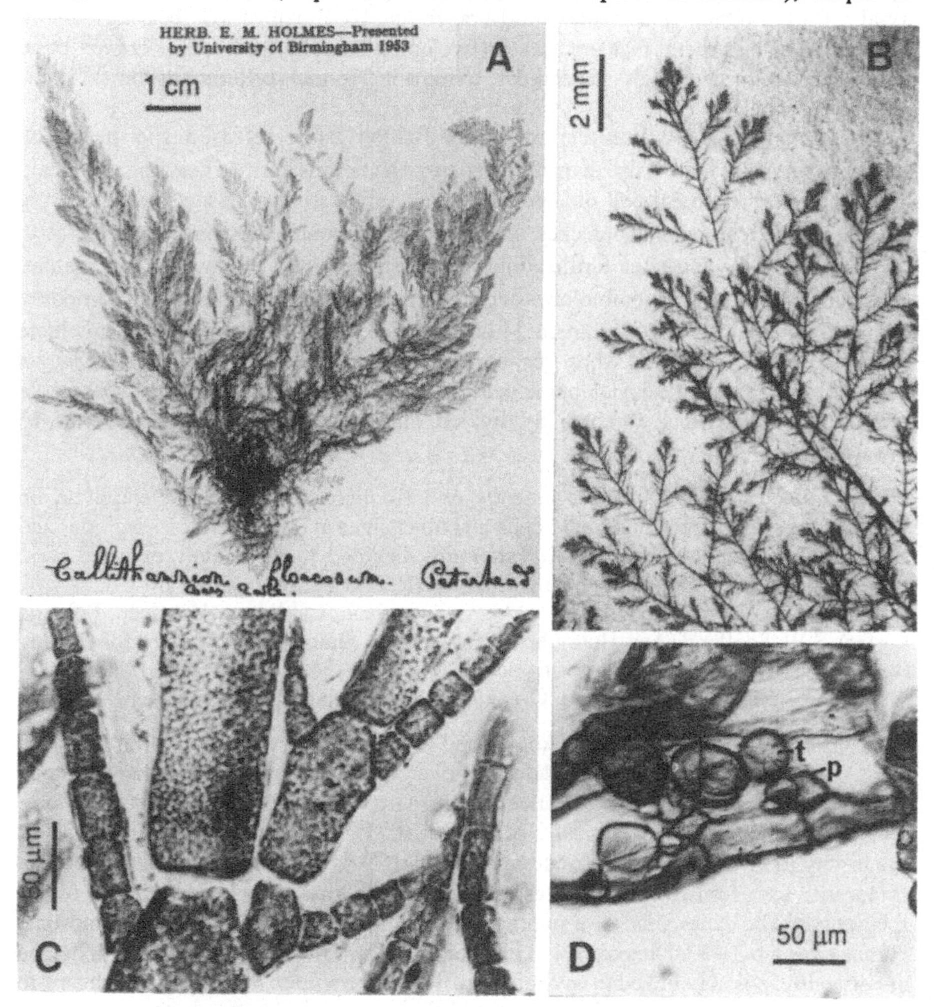

Fig. 4. *Antithamnionella floccosa*

(A) Habit of herbarium specimen (Aberdeen, undated, in BM). (B) Main axes bearing alternate laterals and paired whorl-branches (B-D Aberdeen, Apr., in BM). (C) Main axis bearing lateral axis with paired, unbranched whorl-branches. (D) Tetrasporangia (t) in clusters on pedicels (p).

occasionally with a few lateral branchlets, straight, awl-shaped, 8-12 cells long, 24-32 µm diameter, with the basal cell shorter than others, and the terminal cell small, conical; *gland cells* absent; *indeterminate laterals* developing in plane of whorl-branches every 3(-6) axial cells, not affecting opposite whorl-branch, the basal cell of axis bearing only an abaxial whorl-branch; further erect axes developing laterally from whorl-branch basal cells, *rhizoids* formed singly, unicellular, terminating in multicellular attachment pads; *plastids* discoid.

Spermatangial plants not recognized in British Isles; released spermatia (on trichogynes) spherical, c. 6 µm diameter. *Procarps* numerous, formed near apices of main and lateral axes on basal cell of 2-celled lateral, suppressing opposite lateral, bearing straight 4-celled carpogonial branches, with conical carpogonium c. 8 µm diameter; growth of fertile apex ceasing after fertilization; *cystocarps* 500-600 µm wide when mature, consisting of 3-5 globular gonimolobes of different ages, up to 330 µm diameter, composed of numerous angular carposporangia 24-60 µm wide, surrounded by a loose involucre consisting of 1-2 additional whorl-branches on subtending axial cells. *Tetrasporangia* terminal on 1-8-celled adaxial branchlets or on reduced whorl-branches, solitary or grouped, 1-3 per pedicel; tetrahedrally divided, often appearing cruciate, 60-80 x 36-60 µm when mature.

On barnacles, mussels, rock, concrete and wooden structures, and epiphytic on *Laminaria hyperborea* (Gunnerus) Foslie and other algae at low water of spring tides and submerged on floating structures at extremely sheltered to moderately exposed sites; collected once from 15 m depth on a worm tube (R. Huxley, pers. comm.).

Shetland and N. and E. Scotland (Orkney, Sutherland, Caithness, Aberdeen, Fife and Forfar), S.W. Scotland (Ayr, Bute); N. E. England (Northumberland); record for Galway (see Guiry, 1978) almost certainly erroneous.

Norway and Iceland to Scotland; Greenland to Massachusetts (Athanasiadis, 1990).

Plants collected in March-May, August and October, with cystocarps in May, August and October and tetrasporangia in March, April and August. In Newfoundland, Whittick (1984) found an extreme preponderance of tetrasporophytes; these appeared to perennate whereas gametophytes were annual. Jacobsen et al. (1991) obtained a *Polysiphonia*-type life history in a Norwegian isolate cultured at 8°C, 16 h days; at 14°C gametangia, but not cystocarps, were formed. They suggested that the southern geographical boundary of this species might be determined by a summer lethal temperature of 17°C. The chromosome number was $n = 31$-33. Dimensions of axial cells and whorl-branches show a high degree of variability, possibly linked to environment. Plants from more sheltered habitats tend to have narrower axes.

Lindstrom & Gabrielson (1989) pointed out the close similarity between *A. floccosa* and *Antithamnionella pacifica* (Harvey) Wollaston (1972a, p. 87) [basionym: *Callithamnion floccosum* var. *pacificum* Harvey (1862), p. 176] and the need for experimental investigation of their relationship. Athanasiadis (1990) also discussed the possibility that they are conspecific; Whittick (pers. comm.), however, has found constant differences between these entities.

The parasitic fungus *Olpidiopsis antithamnionis* Whittick & G. R. South was described from *A. floccosa* collected in Newfoundland.

A. floccosa can be distinguished from species of *Antithamnion* by the simple whorl-branches, which differ from the branched whorl-branches of *Antithamnion* spp.

Antithamnionella spirographidis (Schiffner) Wollaston (1968), p. 345.

Holotype: BERL (Wollaston, letter in BM). Isotype: BM. Italy (Trieste), 12 viii 1914, *Schiffner*.

Antithamnion spirographidis Schiffner (1916), p. 137.
Antithamnion tenuissimum Gardner (1927), p.377, non *Antithamnion tenuissimum* (Hauck) Schiffner.

Thalli forming carpets or fringes 0.8-3.5 cm high, spreading extensively over the substratum by prostrate axes that become erect and also give rise secondarily to erect axes, much-branched, complanate or terete depending on the number of whorl-branches formed, 190-480 μm wide including whorl-branches, bright red in colour, very delicate and flaccid.

Main axes enlarging from apical cells 5-7 μm wide to axial cells 45-100 μm wide and 4-8 diameters long, cylindrical or constricted in the middle, with young laterals borne singly in a secund row by (2-)3-5 (typically 4) successive axial cells, then in a row on the opposite side by the next (2-)3-5 axial cells; *whorl-branches* borne in pairs or whorls of 3-4, simple or occasionally with a few lateral branchlets, slightly to strongly tapering, curved adaxially, later lying parallel to axial cells, 8-15 cells long and up to 12 μm wide, the basal cell equal in length to other cells, 2-2.5 diameters long, the terminal cell with a blunt rounded tip, bearing a hair 25-175 μm long when near female apices; *gland cells* frequent, usually one per whorl-branch on the second or third cell, but occasionally several, lens-shaped to almost spherical, about two-thirds the length of bearing cell; *indeterminate laterals* developing in plane of whorl-branches, on alternate sides of main axis, usually on every fourth axial cell, not affecting opposite whorl-branch, the basal cell remaining shorter than the others, the first 1-2(-4) cells typically bearing a single adaxial whorl-branch; further axes developing from basal and occasionally suprabasal cells of whorl-branches and lateral branches; *rhizoids* formed singly or in pairs, multicellular, terminating in loose, branched multicellular attachment pads; *plastids* ribbon-like, becoming filiform in axial cells.

Spermatangial branchlets formed adaxially on whorl-branches, 1-3 per cell, 1-5 cells long, up to 17 μm in length, simple, curved or straight, each cell developing into an isodiametric spermatangial mother cell and forming 1-4 ovoid spermatangia 3-4 x 2.5 μm; released spermatia spherical, 3 μm diameter. *Procarps* formed near apices on basal cell of 2-celled reduced lateral, bearing 4-celled, straight or curved carpogonial branch; *cystocarps* to 330 μm wide when mature, consisting of 3-5 globular gonimolobes of different ages, 60-170 μm in diameter, containing few to numerous angular carposporangia

20-36 μm wide, surrounded by a loose involucre formed by additional branchlets on subtending axial cells, some of which continue growth and replace the fertile apex. *Tetrasporangia* sessile, adaxial, 1-3 per cell, usually developing simultaneously on several whorl-branch cells, ovoid, tetrahedrally divided, sometimes appearing cruciate when mature, 46-60 x 28-40 μm.

Growing on pebbles, various algae, shells, hydroids, sponges and ascidians, and on

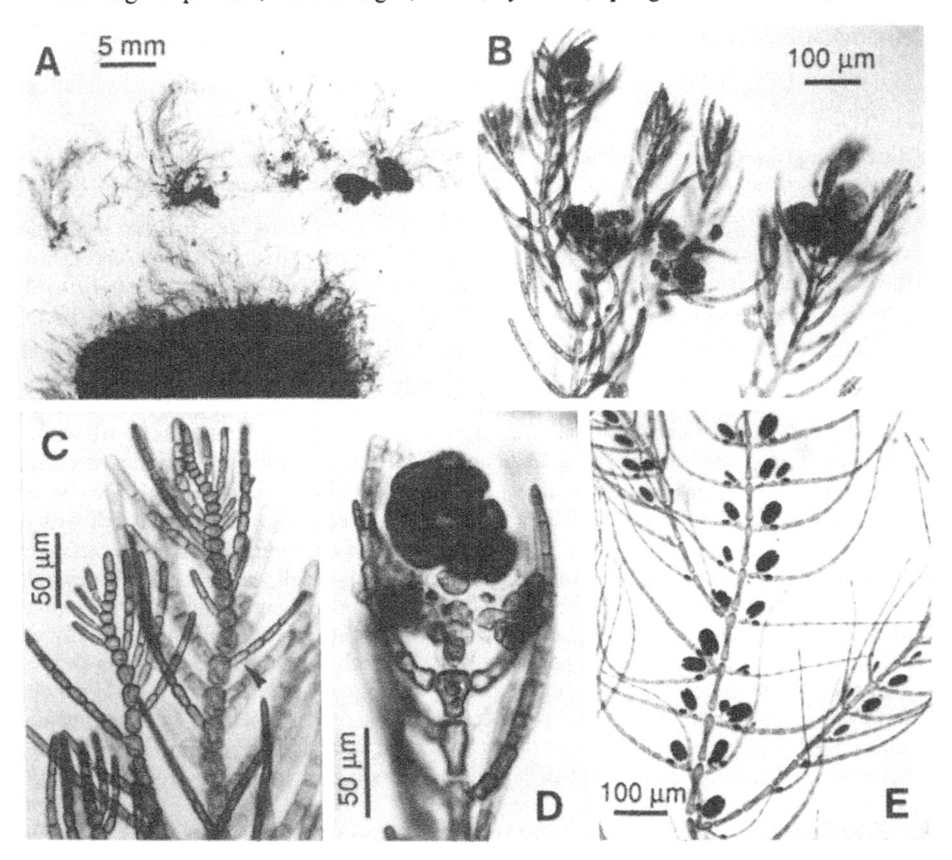

Fig. 5. *Antithamnionella spirographidis*

(A) Habit of extremely delicate thalli on stone, with others removed from substratum (Sutherland, Sep.). (B) Part of female thallus with cystocarps (B-D Devon, Oct.). (C) Apices, showing sinusoidal main axes and development of young whorl-branches; basal cell of whorl-branch (arrow) is equal in length to next cell. (D) Cystocarp with 3 gonimolobes. (E) Tetrasporangia borne directly on unbranched whorl-branches (Argyll, June).

wood and artificial materials such as plastic, submerged on floating pontoons and subtidal at 2-18 m depth at moderately to extremely wave-sheltered sites, tolerant of slightly reduced salinity; most common in marine inlets such as sea-lochs.

Collected at scattered locations on southern and western coasts (Dorset, S. Devon, Pembroke, Argyll; Cork, Down), probably more widespread but distribution difficult to assess owing to previous confusion with *Antithamnionella ternifolia* (as *A. sarniensis*).

British Isles to France; Mediterranean; North Pacific from Alaska to Mexico, Japan, Korea, Australia (Wollaston, 1968; Lindstrom & Gabrielson, 1989).

Plants observed in June-December and March, individuals probably ephemeral. Spermatangia recorded in July and December; procarps and cystocarps in July-December; tetrasporangia in June-July, September, December and March. Tetraspores from plants that appeared in aquaria in Devon gave rise in culture to a 1:1 ratio of male and female gametophytes, and the resultant carpospores grew into tetrasporophytes; females later also formed spermatangia or tetrasporocytes (Drew, 1955). Cultures from France exhibited the same life-history patterns, and vegetative reproduction was very effective (L'Hardy-Halos, 1985; 1986).

Numbers of gland cells vary considerably from occasional single cells to several per whorl-branch; Westbrook (1934) noted that they were more frequent on tetrasporophytes. Branched whorl-branches occur in some gametophytes and tetrasporophytes but not others.

A. spirographidis was first reported from England by Westbrook (1934). The pattern of its subsequent spread in the British Isles has been obscured by Sundene's (1964) proposal that it was conspecific with *A. sarniensis*. L'Hardy-Halos (1985, 1986) later showed that these species are morphologically distinct and do not hybridize in culture. Wollaston (1968) suggested that *A. spirographidis* was introduced into Australia from Europe by shipping as it is associated with dockyards and harbours. Lindstrom & Gabrielson (1989), having placed several Pacific species in synonymy with *A. spirographidis*, considered that the North Pacific was a likely place of origin.

Specimens of *A. spirographidis* with branched whorl-branches resemble *Antithamnion* species; they can be distinguished by the position of gland cells alongside single cells, rather than cupped by 2-3 cells as in *Antithamnion* spp.

Antithamnionella ternifolia (J. D. Hooker & Harvey) Lyle (1922), p. 350.

Lectotype: BM-K (see Ricker, 1987). Chile (Cape Horn).

Callithamnion ternifolium J. D. Hooker & Harvey (1845a), p. 272.
Antithamnionella sarniensis Lyle (1922), p. 348.
Antithamnion sarniensis (Lyle) Feldmann-Mazoyer (1941), p. 269.

Thalli forming carpets or fringes 2-4 cm high, spreading extensively over the substratum by prostrate axes that become erect at the tips and also give rise secondarily to erect axes, much-branched, irregularly terete, about 250 μm in diameter including whorl-branches; deep rose-red in colour, very delicate and flaccid.

Main axes enlarging from apical cells 4-6 μm wide to axial cells 48-53 μm in diameter and 4-5.5 diameters in length, cylindrical; axial cells initially cutting off a single lateral in an irregularly spiral sequence, and when mature bearing 2-4 whorl-branches distally; *whorl-branches* simple or very sparsely branched, 8-19 cells (265-400 μm) long and 7-11 μm wide, straight or curved when young, when mature angled from the basal cell, lying alongside main axis, the basal cell initially 1-1.5 diameters long, later enlarging, other

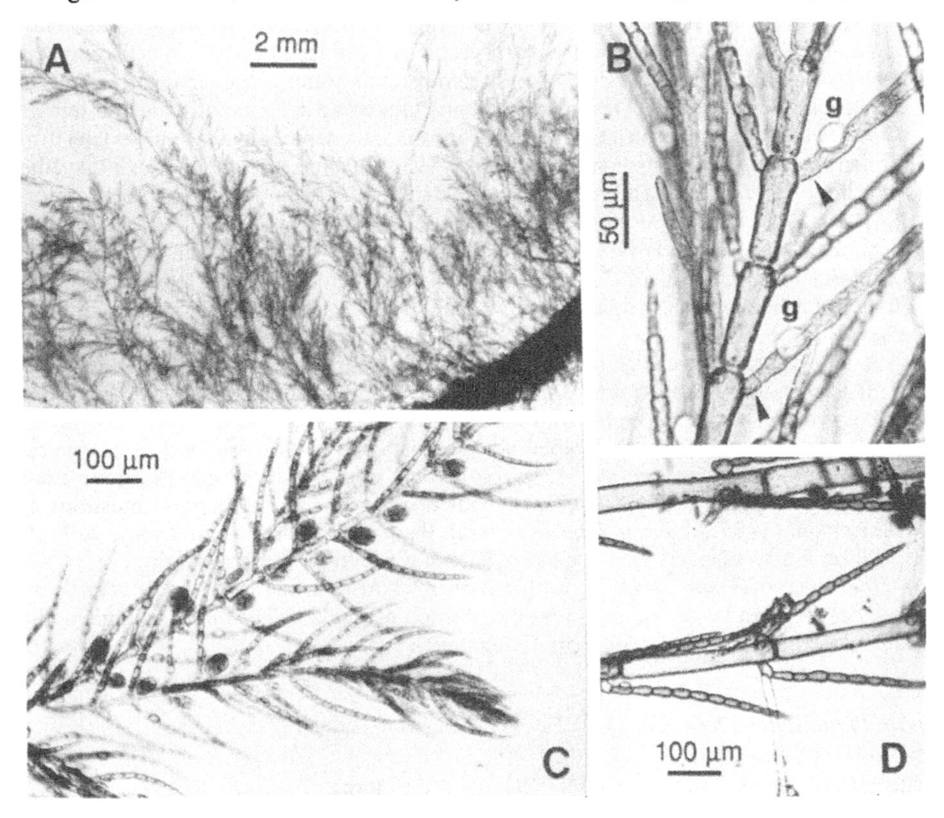

Fig. 6. *Antithamnionella ternifolia*

(A) Habit of thalli forming a fringe on axis of *Gracilaria* sp. (Devon, Nov.). (B) Axis bearing whorl-branches with basal cells (arrows) shorter than suprabasal cells; gland cells (g) are conspicuous (as A). (C) Axes with whorl-branches bearing tetrasporangia (Cornwall, Sep.). (D) Older axis with rhizoid formed by basal cell of sharply-angled whorl-branch (as A).

whorl-branch cells about 2 diameters long, the terminal cell with a blunt rounded tip when young, becoming conical or mucronate; *gland cells* frequent, usually 1 per whorl-branch on the second to fourth cell, but occasionally 2, lens-shaped to ovoid, larger than adjacent whorl-branch cell; *indeterminate laterals* formed at intervals of 3-5 cells, replacing a whorl-branch, with the basal cell typically unbranched and remaining shorter than the others, the second cell bearing a pair of whorl-branches; further lateral indeterminate axes developing from enlarged basal and occasionally suprabasal cells of whorl-branches; *rhizoids* formed singly from basal cells of whorl-branches, multicellular, branched, terminating in multicellular discoid attachment pads; *plastids* ribbon-like, becoming filiform in axial cells.

Gametophytes unknown in British Isles. *Tetrasporangia* sessile, formed singly, initially on the first or second whorl-branch cell, sometimes developing subsequently on other cells, tetrahedrally divided, often appearing cruciate when mature, 42-52 x 28-40 μm.

Growing abundantly on all types of substrata including eelgrass leaves, algae, animals, pebbles and artificial materials, in intertidal pools, on floating pontoons and from near low water to 25 m depth, over a wide range of conditions from extremely wave-sheltered to extremely wave-exposed, tolerant of very strong tidal currents.

Widely distributed on south and west coasts, northwards to Argyll, rarely recorded from east coasts (e.g. Berwick).

British Isles, Netherlands to Portugal; Argentina, Australia, New Zealand and Southern Ocean (Ardré, 1970; Ricker, 1987; Athanasiadis, 1990; Magne, 1991).

Large thalli are found throughout the year and the majority of plants are usually sterile. Gametophytes have not been observed in the British Isles; tetrasporangia recorded for July-November. Cultured tetraspores of a Guernsey isolate (Sundene, 1964) gave rise to males and females that formed cystocarps. More detailed studies of French cultures (Magne, 1986a, 1986b; L'Hardy-Halos, 1986) showed, however, that tetrasporophyte recycling by apomictic tetrasporangia was much more common than a sexual life history. Chromosome number is n = c. 34 (Magne, 1986b).

The earliest known European collection of *A. ternifolia* was made at Plymouth (in BM, 15 viii 1906, *Batters*, as *A. cruciatum* f. *tenuissimum*) (R. Huxley, pers. comm.), and the next in N. France in 1910 (Westbrook, 1930b). It was found in the Channel Isles in 1921 (Lyle, 1922), was common in S. Devon by 1926-1929 (Westbrook, 1930b), and was then reported from Ireland and the Isle of Man (De Valéra, 1939), and W. Scotland (McAllister, Norton & Conway, 1967). When Lyle (1922) described her Channel Isles collections as a new species, *A. sarniensis*, she noted that the only similar plants were from the southern hemisphere, and assumed that the European population was introduced. The similarities between *A. sarniensis*, *A. verticillata* (Suhr) Lyle (1922, p. 349) from South Africa and *A. ternifolia* were pointed out but no distinguishing characters were suggested. Recent studies of *A. verticillata* (Stegenga, 1986) show that it differs markedly in features such as the formation of up to 6 whorl-branches per axial cell. Athanasiadis (1990) suggested that *A. sarniensis* might be conspecific with *A. ternifolia*; recent collections of *A. ternifolia* from the Southern Ocean (Ricker, 1987) are apparently identical to British Isles material. The probable conspecificity of *A. sarniensis* and *A. ternifolia* was obscured until recently by the mistaken view of Sundene (1964) that *A. sarniensis* should be placed in synonymy with *A. spirographidis* (q. v.).

Axial cells vary in length and diameter so that some plants are short and stocky while others are more slender. L'Hardy-Halos (1986) found this variation to be genetically rather than environmentally controlled and suggested that it might be evidence of interspecific hybridization during the evolutionary history of the species.

Specimens of *A. ternifolia* with branched whorl-branches somewhat resemble *Antithamnion* species; they can be distinguished by the position of gland cells alongside single cells, rather than cupped by 2-3 cells as in *Antithamnion* spp.

PTEROTHAMNION Nägeli

PTEROTHAMNION Nägeli (1855), p. 66.

Lectotype species: *P. plumula* (Ellis) Nägeli (1855), p. 66 [see Nägeli (1862), pp. 375, 377].

Thalli uniseriate, erect to 6 cm high, alternately branched in one plane, attached to substratum by multicellular rhizoids issuing from basal cells of lower branches and terminating in digitate haptera; determinate lateral branches (= whorl- branches) opposite, equal or subequal, mostly distichous in plane of branching, 1-2 minor determinate branches transverse, developing later than the first pair, or absent; determinate branches branching adaxially and sometimes also laterally, to 3 orders of filaments, their basal cells similar in length to other branch cells and often provided with an adaxial lateral; gland cells adaxial on determinate branches, shorter than the length of the bearing cell and often proximal to a lateral filament; indeterminate axes replacing determinate branches at intervals of 4-7 axial cells, curving toward apex; adventitious indeterminate axes occasional, formed on basal cells of determinate branches; rhizoidal cortication absent except at base.

Growth of indeterminate axes sinusoidal, by oblique division of apical cells to one side and then the other, shifting orientation at sites of future indeterminate axes; the first initial of a determinate branch usually abaxial, the second adaxial; gland cells cut off laterally, below a lateral filament when present; nuclei and plastids as in *Antithamnion*.

Gametophytes dioecious. Spermatangial mother cells formed in whorls on short-celled axes, 2-3 cells long, borne on determinate branches, each producing up to 4 spermatangia. Procarps initiated 5-7 (-10) segments below apex on indeterminate axes, 1-3 (-4) per fertile axis, consisting of a supporting cell, an unmodified vegetative branch, and a 4-celled lateral carpogonial branch; post-fertilization development as in *Antithamnion*; cystocarp development either suppressing or not suppressing further growth of the fertile axis. Tetrasporangia sessile or pedicellate, borne singly or in clusters on cells of determinate branches, subspherical to pyriform, cruciately divided.

References: L'Hardy-Halos (1970, 1971, as *Antithamnion*), Moe & Silva (1980), Athanasiadis (1985a).

Pterothamnion crispum has recently been reinstated as a separate species (Athanasiadis, 1985a), whereas in previous British checklists it was considered to be a variety of *P. plumula*. Two varieties of *P. plumula* are recognized at present in the British Isles, but one of these, *P. plumula* 'var. *bebbii', is probably based on a specimen of P. crispum* (see Athanasiadis, 1990), and is here treated as a possible synonym of *P. crispum*. The tristichous or tetrastichous form of *P. plumula* usually known by the name 'var. *bebbii*' is keyed out separately below. Athanasiadis (pers. comm.) has pointed out that there are some unresolved taxonomic problems among members of this genus in the British Isles, and that our treatment of *P. crispum* may include two distinct species.

KEY TO SPECIES

1 2 whorl-branches per axial cell in a paired distichous arrangement .... *P. plumula*
  3-4 whorl-branches per axial cell 2
2 All whorl-branches bear branchlets in a secund arrangement......................................
  .... *P. plumula* (tri- or tetrastichous form)
  One pair (the smaller) or both pairs of whorl-branches branched pseudodichotomously
  .... *P. crispum*

Pterothamnion crispum (Ducluzeau) Nägeli (1862), p. 376.

Lectotype: BM-K, ex. Herb. Gray. Unlocalized, undated, *Desvaux**

Ceramium crispum Ducluzeau (1806), p. 47.
Callithamnion plumula var. *crispum* (Ducluzeau) J. Agardh (1851), p. 29.
Antithamnion crispum (Ducluzeau) Thuret ex Le Jolis (1863), p. 112.
?*Callithamnion bebbii* Reinsch (1875), p. 47 (see Athanasiadis, 1990).
Antithamnion plumula var. *crispum* (Ducluzeau) Hauck (1885), p. 73.
Platythamnion crispum (Ducluzeau) J. Feldmann (1937), p. 276.

Thalli consisting of small tufts of erect axes 0.5-1.5 cm high, flabellate, formed of branching complanate axes 330-750 μm wide (including whorl-branches), attached by loose, spreading rhizoidal filaments, rose-pink in colour, delicate and flaccid in texture.

Main axes enlarging from apical cells 7-8 μm wide to axial cells 150-160 μm wide and 1-2 diameters long, cylindrical, with young axial filaments curved towards last-formed indeterminate lateral in a forcipate arrangement; *whorl-branches* paired, distichously arranged, or a second pair of whorl-branches formed at 90° to the first pair, resulting in

* This specimen discovered by Athanasiadis (pers. comm.) takes priority over the neotype designated by Athanasiadis (1985a, fig. 12.).

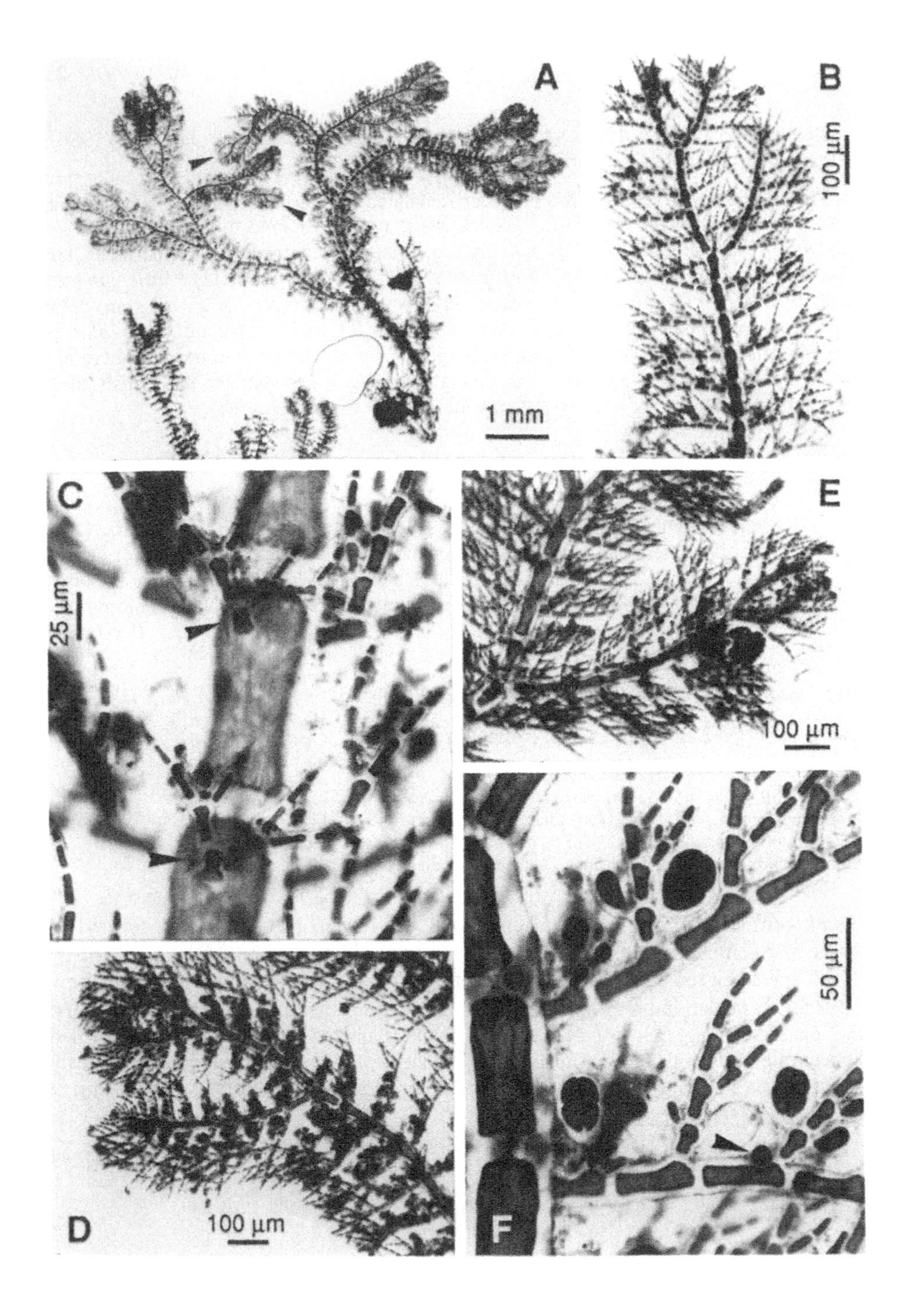
A
1 mm
B
100 µm
C
25 µm
D
100 µm
E
100 µm
F
50 µm

whorls of 4 in a distal position on axial cell, the first pair 130-320 μm long, consisting of 8-9 cells, with the basal cell 1-1.5 diameters long and next few cells 1.5-2.5 diameters long; each cell except terminal 2-4 cells branched according to position on the thallus, pseudodichotomously branched 5-6 times on older axes, sometimes with a second branchlet on the basal cell, or, on younger axes, bearing a secund series of branched abaxial branchlets; second pair of whorl-branches remaining simple, spine-like, or branching pseudodichotomously 3-4 times; *branchlets* spine-like, terminating in conical cell with pointed apical extension, lacking hairs; *gland cells* frequent, adaxial on bearing cell, initially hemispherical, becoming conical, erect, 12-20 μm long x 16-28 high; *indeterminate laterals* replacing a whorl-branch every 4-7 axial cells, on alternate sides, sometimes appearing dichotomous, not affecting opposite whorl-branch but stimulating production of second pair of whorl-branches, the basal cell remaining shorter than the others; further indeterminate laterals also formed occasionally by basal whorl-branch cells; *rhizoids* developing from basal cells of lowermost whorl-branches, 1-2 per cell, 16-36 μm in diameter; *plastids* ribbon-like, becoming filiform in axial cells.

Spermatangial branchlets developing adaxially on whorl-branches, on up to 4 successive cells, or replacing second-order branchlets, obconical or globular, 28-77 μm long x 25-32 μm wide, consisting of an axis of 2-3 cells, the basal cell elongate, often bearing a lateral branchlet, each cell cutting off a whorl or cluster of spermatangial mother cells that bear 3-4 ovoid spermatangia measuring 3-4 x 2.5 μm. *Procarps* formed on basal cells of laterals near secondary apices, up to 3 per axis, only one of which undergoes post-fertilization development, suppressing further axial growth; *cystocarps* consisting of 3-4 rounded gonimolobes of different ages, 120-140 μm in diameter, containing numerous angular carposporangia 14-25 μm in diameter. *Tetrasporangia* formed on first-order or second-order branchlets in place of lateral branchlet, normally sessile, occasionally pedicellate or terminal on 2-celled branchlet, 1-3 per cell, the primary sporocyte adaxial, the next abaxial, pyriform before division, ellipsoid when mature, 36-46 x 28-30 μm.

Epiphytic on a variety of algae, including maerl, and epizoic on ascidians, from near extreme low water to 10 m depth at sites with little wave-exposure, often exposed to moderate to strong currents; tolerant of silt cover.

Recorded from scattered localities on south and south-west coasts: Sussex, Dorset, Devon, S. Cornwall, Glamorgan, Pembroke, Anglesey, Isle of Man; Dublin, Cork, Clare, Galway, Mayo; Channel Isles.

Fig. 7. *Pterothamnion crispum*

(A) Habit of small plant. Note forcipate tips (arrows) and complanate branching (A-F Pembroke, undated). (B) Apex focussed on large complanate pair of whorl-branches that bear secund branchlets. (C) Pairs of small dichotomously branched whorl-branches (arrows) perpendicular to large whorl-branches. (D) Male thallus with abundant spermatangial branchlets. (E) Branch bearing cystocarp. (F) Tetrasporangia sessile on branchlets; gland cell (arrow) is much shorter than bearing cell.

Netherlands to Spain; Mediterranean (Athanasiadis, 1990).

Small regenerating fragments can be found throughout the year. Spermatangia recorded for July and October, procarps and cystocarps in July, August and October, and tetrasporangia in July, September and October. A *Polysiphonia*-type life history was reported for interfertile isolates from the Isle of Man and Guernsey (Sundene, 1975). The Guernsey strain was interfertile with a Mediterranean isolate (Athanasiadis, 1985a), when grown at 15°C under 14 h days. The tetrasporophytic offspring formed non-functional spermatangia in addition to tetrasporangia. Mitotic chromosome counts in gametophytes were of $n = 24 \pm 2$. *P. crispum* isolates did not interbreed with isolates of *P. plumula* (Athanasiadis, 1985a).

Scagelia pusilla can be distinguished from *P. crispum* by the form of the gland cells, which are lens-shaped, with their long axis along the branchlet, whereas gland cells in *P. crispum* are cylindrical or truncated-conical, with the long axis perpendicular to the branchlet.

Pterothamnion plumula (Ellis) Nägeli (1855), p. 66.

Lectotype: Ellis (1768), pl. 18, fig. G. Unlocalized, probably from Sussex.

Conferva plumula Ellis (1768), p. 425.
Ceramium plumula (Ellis) C. Agardh (1817), p. 62.
Callithamnion plumula (Ellis) Lyngbye (1819), p. 127.
Antithamnion plumula (Ellis) Thuret ex Le Jolis (1863), p. 112.

Thalli consisting of tufts of erect axes 3-15 cm high, attached by loose, spreading rhizoidal filaments; erect thalli branched in one plane, flabellate, formed of dichotomously branched complanate axes 1-3 mm wide (including whorl-branches) with forcipate tips, rose-pink or brownish-red in colour, delicate and flaccid in texture.

Main axes enlarging from apical cells 6 µm wide to axial cells 200-260 µm wide and 0.5-1.5 diameters long, cylindrical or with a median constriction, with young axial filaments curved towards the last-formed indeterminate lateral; *whorl-branches* paired and distichous, or borne in whorls of 3 (= tristichous) or 4 (= tetrastichous), the additional 1-2 formed at 90° to the first pair, 600-1800 µm long, consisting of 10-13 cells, with the basal cell 1-1.5 diameters long, and the next few cells 1.5-4 diameters long; each cell except terminal 4-5 cells bearing a branched adaxial branchlet, with the older ones occasionally pseudodichotomously branched; *branchlets* terminating in conical cell with pointed apical extension, lacking hairs; *gland cells* often abundant, adaxial on whorl-branches and branchlets, borne adjacent to a single cell, initially hemispherical, becoming cylindrical or conical, erect, 14-20 µm long x 22-34 high; *indeterminate laterals* formed in plane of whorl-branches every 4-6 axial cells, alternate, sometimes appearing dichotomous, not affecting opposite whorl-branch but stimulating production of second pair of simple or branched whorl-branches; *rhizoids* formed from basal cells of lowermost whorl-branches, 1-2 per cell, 14-48 µm in diameter; *plastids* ribbon-like, becoming filiform in axial cells.

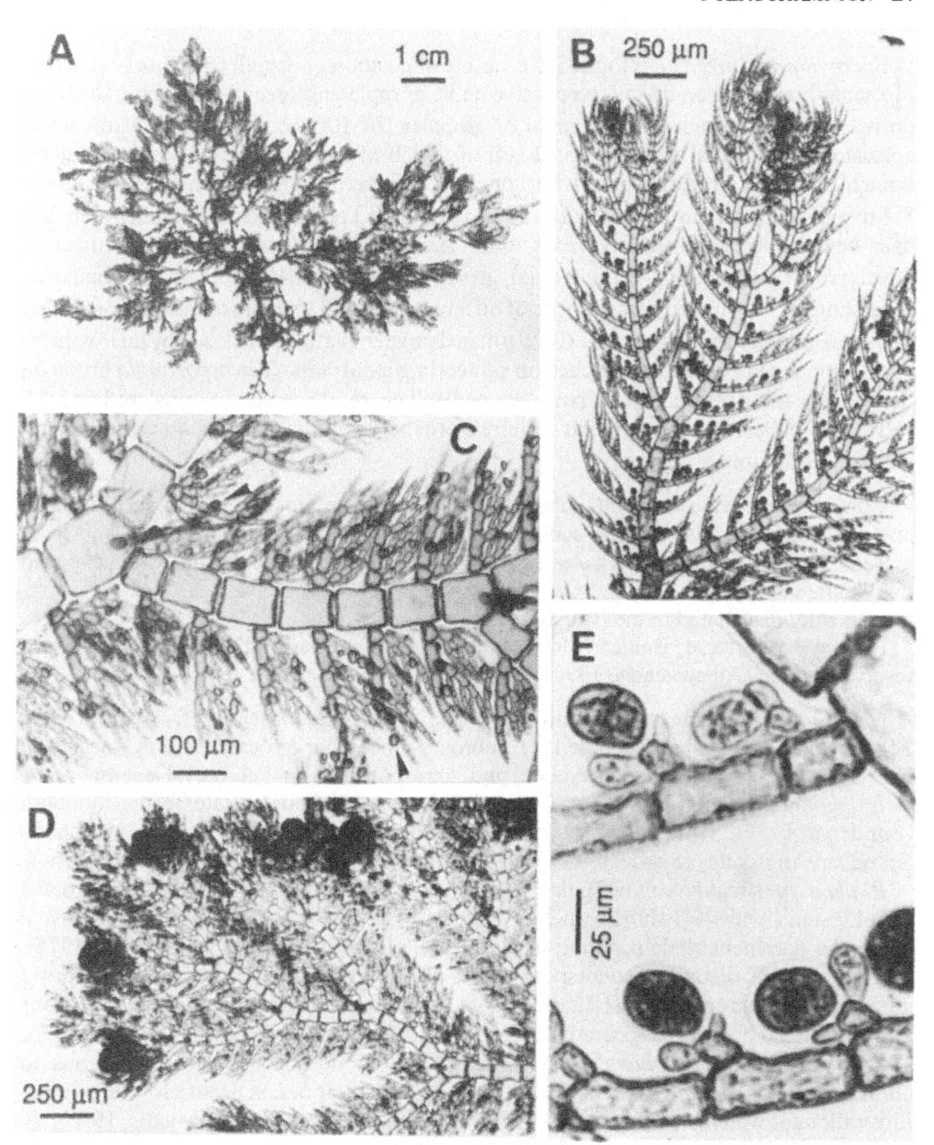

Fig. 8. *Pterothamnion plumula*

(A) Habit (Devon, Nov.). (B) Apex, showing forcipate tips and position of tetrasporangia on paired whorl-branches (as A). (C) Paired whorl-branches with abundant gland cells (arrows) on branchlets (Cornwall, Sep.). (D) Branches bearing cystocarps (as C). (E) Tetrasporangia as A.

Spermatangial branchlets formed in a variety of positions, adaxially on whorl-branches, replacing branchlets on up to 4 successive cells, or replacing second-order branchlets, or on reduced whorl-branches, obconical or globular, 24-100 µm long x 25-44 µm wide, consisting of a central axis, the basal cell of which often bears one or a pair of lateral branchlets, each cell cutting off a whorl or cluster of spermatangial mother cells that bear 3-4 ovoid spermatangia 2.5-3 x 2 µm. *Procarps* formed on basal cells of whorl-branches near lateral apices, up to 3 per axis, only one of which undergoes post-fertilization development, suppressing further axial growth; *cystocarps* 400-500 µm in diameter, consisting of 3-4 rounded gonimolobes of different ages, to 325 µm in diameter, containing numerous angular carposporangia 20-28 µm in diameter, partially enclosed by an involucre of additional pairs of whorl-branches on subtending axial cells. *Tetrasporangia* borne on first- or second-order branchlets, normally pedicellate, rarely sessile, or clustered on each cell of 2-3- celled branchlet, 1-3 per cell, pyriform before division, ellipsoid when mature, 36-46 x 28-30 µm.

Common on a wide variety of substrata, including algae, invertebrates, pebbles, shells and artificial materials, often in sandy or silty habitats, in lower-shore pools and from extreme low water to 20 m depth at sites with little to strong wave-exposure, often exposed to moderate to strong currents.

Generally distributed in the British Isles.

Norway to Portugal; Baltic; ?Mediterranean; southern hemisphere records requiring re-investigation (Athanasiadis, 1990).

Mature plants occur throughout the year; sporelings observed from February-September. Spermatangia recorded for February-December, procarps and cystocarps in February, March and June-December, and tetrasporangia in February-December. A *Polysiphonia*-type life history was reported for a distichous isolate from Plymouth (Sundene, 1959), which was interfertile with several distichous and tetrastichous Scandinavian isolates (Sundene, 1975). Chromosome number is $n = 23$ (Magne, 1964).

P. plumula is highly polymorphic. The morphology known as 'var. *bebbii* (Reinsch) J. Feldmann', with 3-4 whorl-branches, appears to be based on a single-gene recessive allele, the dominant allele resulting in the distichous form (Rueness & Rueness, 1975). Hybrid '*bebbii*'-distichous tetrasporophytes were always distichous, and offspring gametophytes segregated in a 1:1 ratio. More complex results involving subtle variations and transitional phenotypes occurred in crosses between isolates from Oslofjord (Rueness, 1978). The name '*P. plumula* var. *bebbii*'' should not be used for tristichous to tetrastichous *P. plumula*, as it is based on *Callithamnion bebbii* Reinsch, the original illustrations of which probably represent *Pterothamnion crispum* (Athanasiadis, 1990). *P. plumula* var. *spinescens* Strömfelt (1888) appears to be restricted to Norway and the Faroes (Athanasiadis, 1990), and a report from Berwick (Batters, 1902) needs further investigation.

Scagelia pusilla can be distinguished from *P. plumula* by the form of the gland cells, which are lens-shaped, with their long axis along the branchlet, whereas gland cells in *P.*

plumula are cylindrical or truncated-conical, with the long axis perpendicular to the branchlet. In species of *Antithamnion*, gland cells are restricted to special short branchlets, on which they cover 2-3 adjacent cells.

SCAGELIA Wollaston

SCAGELIA Wollaston (1972a), p. 88.

Type species: *S. occidentale* (Kylin) Wollaston (1972a), p. 88.

Thalli uniseriate, ecorticate, prostrate or erect, alternately branched to 5 (-10) cm high, attached by unicellular or multicellular rhizoidal filaments; determinate lateral branches (= whorl-branches) often irregularly branched and unequal in length, formed in whorls of 1-4, usually 3, per axial cell; basal cells of determinate branches similar in length to other branch cells, naked or provided with lateral filaments; gland cells lateral, 1-2 per cell, shorter than the length of the bearing cell and proximal to lateral filaments, when present; indeterminate axes replacing determinate branches, formed alternately at intervals of (1-) 3-4 axial cells, curving towards apex; occasionally arising adventitiously from basal cells of determinate branches.

Growth of indeterminate axis straight to sinusoidal by slightly oblique division of successive apical cells tilted first to one side and then the other at sites of origin of indeterminate branches; determinate branches originating near apex in irregular succession, the first usually abaxial; gland cells cut off laterally; nuclei and plastids as in *Antithamnion.*

Gametophytes dioecious. Spermatangial mother cells formed in whorls on short-celled axes 2-5 cells long borne distally on cells of determinate branches, each producing up to 4 spermatangia. Procarps formed singly in series on axial cells beginning approximately 5 cells below the apex, 2-3 (-4) per fertile axis, consisting of a supporting cell, a terminal unmodified vegetative branch, and a lateral 4-celled carpogonial branch; postfertilization development as in *Antithamnion*; fertile axes elongating after fertilization, often producing a series of cystocarps on the same axis. Tetrasporangia sessile, borne adaxially on cells of determinate branches, ellipsoid, cruciately divided, but the tetraspores often appearing tetrahedrally arranged.

References: Wollaston (1972a), Hansen & Scagel (1981).

Since the segregation of *Scagelia* from *Antithamnion*, its specific and infraspecific taxonomy have been examined in detail. Wynne (1985) showed that the earliest valid name for the circumboreal species complex was *S. pylaisaei*, and listed a large number of synonyms, including several European taxa. Athanasiadis & Rueness (1992) considered, however, that at least two distinct species should be recognized in this complex: *S. pylaisaei* (type locality: Newfoundland), and a sexually reproducing species from Newfoundland and Scandinavia (possibly equivalent to *Callithamnion americanum* Harvey, Athanasiadis, pers. comm.). They also reported three other distinct entities in the North

Atlantic, which might be subspecies or varieties of these two species. One of these, for which the oldest name available was *Callithamnion pusillum* Ruprecht (1851) (Athanasiadis & Rueness, 1992), appears to correspond with Scottish material. Pending definitive nomenclatural work, this is the epithet used here for the species of *Scagelia* found in Scotland. A second entity is known within our geographical area only from Rockall; this has been provisionally identified as *S. pylaisaei.*

KEY TO SPECIES

Paired branchlets present on whorl-branches .... *S. pylaisaei*
Branchlets not paired but irregularly alternate or secund on whorl-branches .... *S. pusilla*

Scagelia pusilla (Ruprecht) Athanasiadis in Maggs & Hommersand, comb. nov.

Lectotype: LE. Novja Zemlja, Arctic Ocean.

Callithamnion pusillum Ruprecht (1851), p. 343.
?*Antithamnion plumula* var. *boreale* Gobi (1878), p. 47.
?*Antithamnion boreale* (Gobi) Kjellman (1883), p. 180.
Antithamnion boreale (Gobi) Kjellman var. *typica* Sundene (1962), p. 15.

Thalli erect, to 1.5 cm high, tufted, alternately branched, complanate, 0.6-1.2 mm wide, rose-pink in colour, with a delicate and flaccid texture.

Main axes enlarging from apical cells 7-8 µm wide to axial cells 80-100 µm diameter and 4-5 diameters in length, cylindrical or slightly constricted in the middle; each young axial cell forming (1-)2(-3) opposite whorl-branches in an irregular or spiral sequence, with successive pairs distichously or decussately arranged; *whorl-branches* straight, up to 12 cells long, tapered, with a pointed apical cell; the basal cell 2-3 diameters in length and other whorl-branch cells elongating to up to 6 diameters, with the first few cells of whorl-branches bearing one branchlet per cell, adaxially, abaxially or laterally; *branchlets* simple, 3-9 cells long, spine-like, with a pointed apical cell; *gland cells* abundant, borne on whorl-branches and branchlets, on up to 3 successive cells, lens-shaped, 12-16 µm diameter, covering less than half, usually much less, of the bearing cell; *indeterminate laterals* alternate, formed at intervals of 4-5 axial cells, with the basal cell of laterals shorter and bearing only a single whorl-branch; *rhizoids* formed by basal cells of whorl-branches, 12-20 µm in diameter, multicellular; *plastids* initially ribbon-like, becoming filiform.

Gametophytes unknown in British Isles. *Tetrasporangia* sessile, formed on whorl-branches and branchlets, in series on successive cells, usually absent on cells bearing gland cells, 1-2 per cell, formed simultaneously or sequentially, resembling bisporangia when immature, regularly cruciately divided when mature, ellipsoid, 60-70 x 48-52 µm.

Epilithic on subtidal bedrock to at least 12 m depth, also reported from the lower intertidal (Clokie & Boney, 1979), at wave-sheltered silty sites with some current exposure, occurring sparsely amongst other ceramiaceous algae.

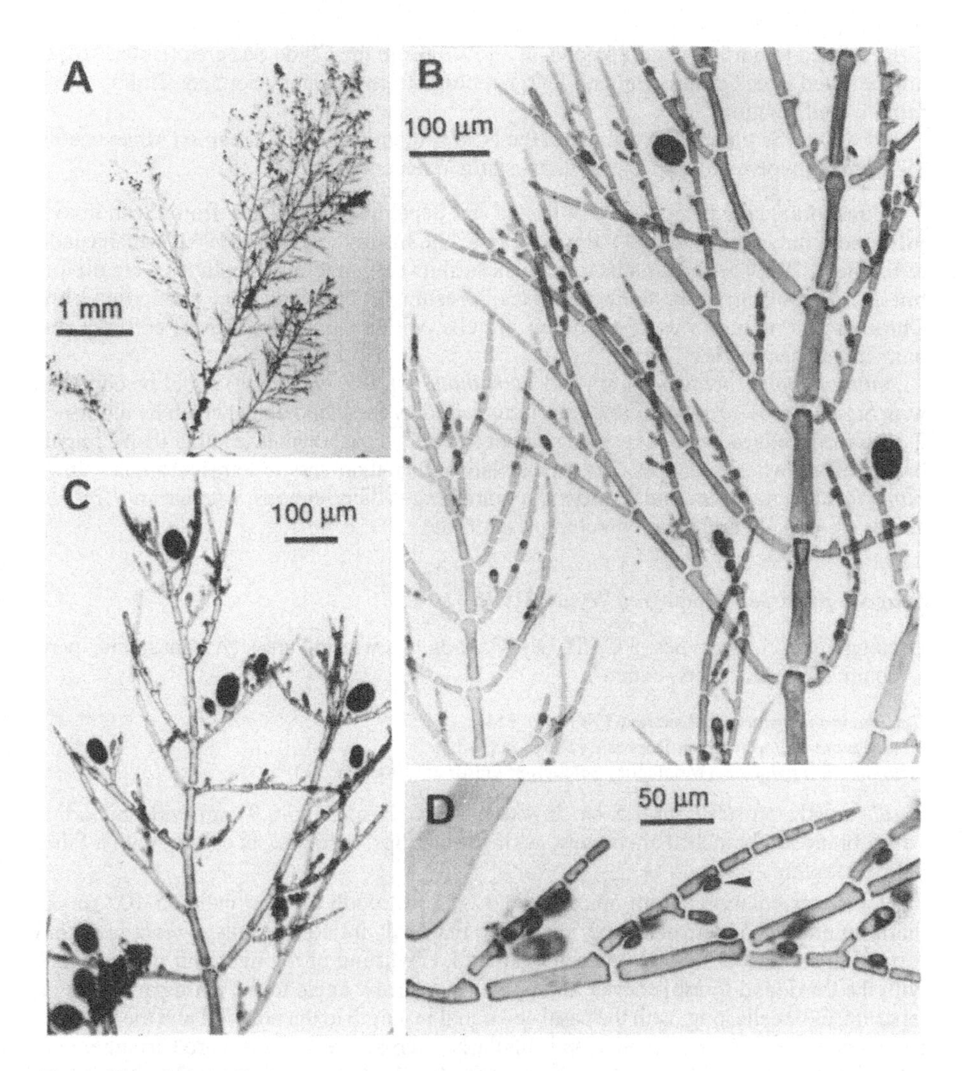

Fig. 9. *Scagelia pusilla*

(A) Part of thallus (A-D Argyll, Sep.). (B) Main axis bearing paired whorl-branches with irregularly arranged branchlets. (C) Tetrasporangia sessile on whorl-branches and branchlets. (D) Gland cells (arrow) and developing tetrasporangia on branchlet cells.

Restricted to northern areas, known at a few sites in the Clyde Sea area (Bute, Argyll) and reported from Orkney (Batters, 1902); records from England (South & Tittley, 1986) require confirmation.

Norway to Scotland. Records from the N.W. Atlantic and N.E. Pacific (Athanasiadis, 1990) may involve other species (Athanasiadis & Rueness, 1992).

Tetrasporangial plants were collected in September. Isolates from Scandinavia followed a direct tetraspore to tetrasporophyte life history (Sundene, 1962; Athanasiadis & Rueness, 1992). Athanasiadis and Rueness showed that tetrasporophytes were diploid but meiosis occurred in tetrasporocytes, presumably preceded by endopolyploidy. Chromosome number was 56-64 in apical cells, with a similar number of chromosome pairs in tetrasporocytes.

Antithamnion cruciatum (q. v.) and *Pterothamnion plumula* (q. v.) could be confused with *S. pusilla*. *A. cruciatum* can be distinguished by the gland cells, which lie alongside 2-3 adjacent cells rather than covering part of one cell, the isodiametric rather than elongate basal cell of whorl-branches, and pedicellate rather than sessile tetrasporangia. In *P. plumula*, apices are sinusoidal, and main branching is dichotomous, whereas in *S. pusilla* apices are straight and main branching is alternate.

Scagelia pylaisaei (Montagne) Wynne (1985), p. 85.

Lectotype: PC. Syntypes: PC, TCD. Canada (Newfoundland) (Athanasiadis, pers. comm.; Whittick, pers. comm.).

Callithamnion pylaisaei Montagne (1837), p. 351.
?*Callithamnion lapponicum* Ruprecht (1851), p. 343.
Antithamnion pylaisaei (Montagne) Setchell & Gardner (1903), p. 342.

Thalli mostly prostrate, to 1.5 cm in extent, complanate, 0.6-0.9 mm wide including whorl-branches, branched alternately, with straight tips, rose-pink in colour, with a fairly delicate texture.

Main axes enlarging from apical cells 12-15 μm wide to axial cells 65-100 μm in diameter and 2.5-5 diameters long, slightly to markedly inflated distally; *whorl-branches* formed in a whorl of 2-3(-4) on each axial cell, consisting of a longer pair in one plane, with the third (and fourth) shorter and borne at a narrow angle to the main pair, straight, tapering, 6-10 cells long, with the basal cell equal in length to the next cell and the terminal cell conical; *branchlets* formed in an irregularly alternate, secund, or paired arrangement, 3-5 cells long, spine-like, with a pointed apical cell, simple or forming 3-celled lateral branchlets; *gland cells* borne on whorl-branches and branchlets, typically 1 or more per whorl-branch, lens-shaped, 16-18 x 10-12 μm, covering about half the bearing cell; *indeterminate laterals* alternate, at intervals of 4-5(-6) axial cells; further indeterminate axes arising from basal cells of whorl-branches; *rhizoids* formed by basal and suprabasal cells of whorl-branches, multicellular, 14-20 μm in diameter.

Reproductive structures unknown in British Isles.

Epizoic on crustose bryozoa, spreading onto creeping algae, at 26-31 m depth at an extremely wave-exposed offshore site.

The only definite collection of this form is from Rockall; reports of *Scagelia* from Scotland are mostly attributable to *S. pusilla* (q. v.).

Specialized reproductive structures are unknown in the British Isles; non-reproductive material was collected in June.

S. pylaisaei sensu lato has a wide distribution in the North Atlantic and North Pacific Oceans (Wynne, 1985). The Rockall plants correspond morphologically to elements of the type material of *Callithamnion lapponicum* from the Murman Sea, which has also been found at Spitzbergen (Athanasiadis, pers. comm.). Since it is not clear at present whether this form should be recognized as a separate species, we have provisionally used the name *S. pylaisaei*.

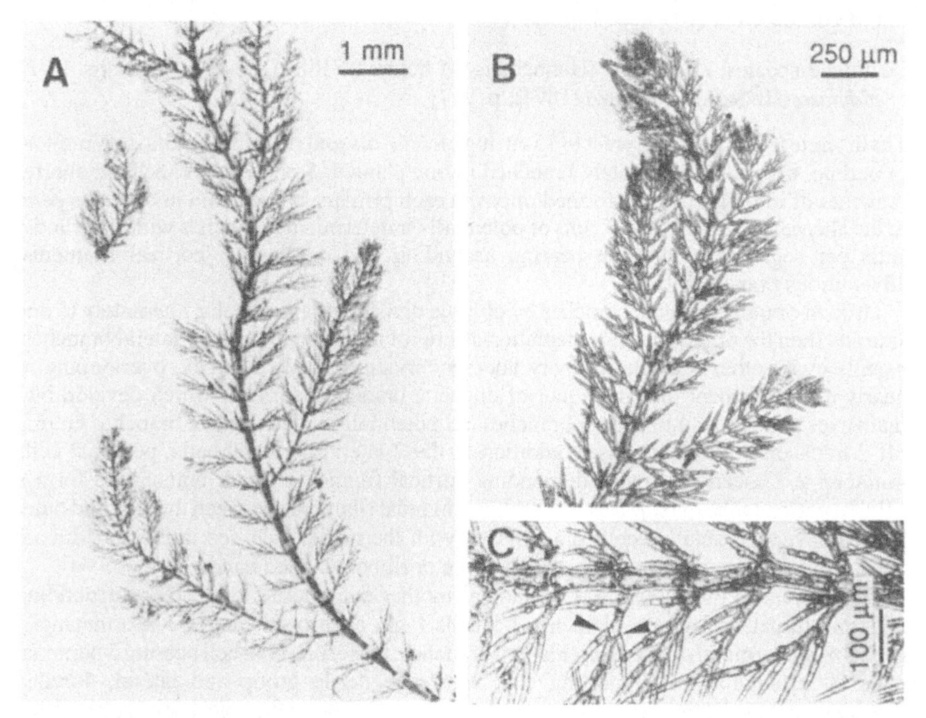

Fig. 10. *Scagelia pylaisaei*

(A) Part of prostrate thallus (A-C Rockall, June). (B) Apex showing alternate branching of major laterals and paired whorl-branches. (C) Paired whorl-branches bearing paired branchlets (arrows) with gland cells.

S. pylaisaei can be distinguished from species of *Antithamnion* by the position of gland cells, which cover part of one cell, rather than lying along 2-3 adjacent cells. In *Pterothamnion plumula*, main branching is dichotomous, whereas in *S. pylaisaei* the main branching pattern is alternate.

Tribe PTILOTEAE Cramer (1863), p. 106.

PLUMARIA Schmitz, nom. cons.

PLUMARIA Schmitz (1896), p. 5.

Lectotype species: *P. elegans* (Bonnemaison) Schmitz (1889), p. 450, typ. cons. [= *P. plumosa* (Hudson) O. Kuntze (1891), p. 911].

Thalli terete to compressed, erect to 15 cm high from a discoid rhizoidal holdfast, pyramidal in outline, irregularly alternately branched in one plane to 5 orders of branching; shorter branches of indefinite length formed opposite each primary branch, and in opposite pairs in the alternate segments; axial cells of potentially indeterminate branches with 4 periaxial cells per segment, with each bearing ascending and descending cortical filaments; adventitious branches absent.

Growth of indeterminate branches by oblique division of apical cells, alternately to one side and then the other, shifting orientation at sites of initiation of primary lateral branches, usually every other segment; primary laterals curving towards the apex, overtopping it; nearly every segment forming a pair of opposite branches, some of which develop into main axes and others into shorter branches; all potentially indeterminate branches cutting off 2 transverse periaxial cells in addition to the 2 lateral periaxial cells; periaxial cells initiating 2-3 ascending and 2 descending cortical filaments which branch and form a cellular cortex to the outside and descending rhizoidal filaments between the axis and outer cortical layers; vegetative cells uninucleate, with the nuclei enlarged in elongated axial cells; plastids parietal, dissected into bead-like or ribbon-shaped bands.

Gametophytes dioecious. Spermatangial mother cells borne in clusters surrounding cells in ultimate uniseriate filaments 4-5 cells long, each producing 2-3 spermatangia. Procarps subterminal on the shorter lateral branches; the fertile axial cell bearing 3 periaxial cells: a fertile abaxial supporting cell bearing a sterile group and lateral, 4-celled carpogonial branch, and 2 transverse vegetative periaxial cells; the apical cell, sterile group and vegetative periaxial cells all producing short, monosiphonous filaments terminated by a hair cell; postfertilization stages largely unknown, the cystocarp probably developing as in *Ptilota*; cystocarp consisting of 3-5 gonimolobes of different ages surrounded by monosiphonous involucral filaments formed from periaxial cells immediately below the fertile segment. Tetrasporangia terminal on ultimate branchlets, rarely lateral, pyriform, tetrahedrally divided. Parasporangia initiated from terminal cells on ultimate branchlets, forming globular to irregularly lobed clusters of parasporangia.

References: Suneson (1938), Drew (1939), L'Hardy-Halos (1970, 1971).

The type and only British species of *Plumaria* was previously known as *P. elegans*, but Irvine & Dixon (1982) showed that *Fucus plumosus* Hudson is an earlier name for this species.

Plumaria plumosa (Hudson) O. Kuntze (1891), p. 911.

Lectotype: Dillenius in Ray (1724), pl. 2, fig. 5 (see Irvine & Dixon, 1982).

Fucus plumosus Hudson (1762), p. 473.
Ptilota plumosa var. *tenuissima* C. Agardh (1822), p. 386.
Ptilota elegans Bonnemaison (1828), p. 22.
Ptilota sericea sensu Harvey (1848), pl. 191, non *Fucus sericeus* S.G.Gmelin (1768), p. 149.
Plumaria elegans (Bonnemaison) Schmitz (1889), p. 450.

Thalli growing in dense tufts 4-13 cm high from discoid holdfast 0.5 cm in diameter, composed of densely matted rhizoidal filaments; erect axes complanate with an irregularly pyramidal or rounded outline, consisting of a single main axis below, 0.5-0.9 mm in diameter, irregularly alternately branched to 5 orders of major branching, with the main axes also bearing regular series of shorter branches, dull brownish- or purplish-red in colour, with a soft and flaccid texture.

Main axes growing from apical cells 12 µm wide; young axial filament regularly sinuous, with axial cells increasing to 4 diameters long, covered by continuous cortication of angular cells 8-20 µm in diameter; older axes heavily corticated by 2-5 layers of cortical cells, clothed with simple incurved adventitious branchlets 20-24 µm in diameter and with descending rhizoidal filaments between axial filament and cortical cells; *indeterminate laterals* formed in pairs of unequal length on each axial cell, long ones borne on alternate sides at intervals of 2-3 axial cells, short ones formed by the next 1-2 axial cells on the same side as the long one, the long lateral on one side paired with the second of a group of 3-4 short laterals of increasing lengths on the opposite side, all laterals strongly incurved, corticate, bearing a further order of indeterminate laterals near apices of long laterals; *determinate laterals* borne in a paired distichous arrangement on indeterminate laterals, increasing in length distally, with sparse cortication; *branchlets* paired, simple, angled forwards, tapering from 22-36 µm at the base to 14-16 µm, composed of 6-10 isodiametric cells with a domed apical cell; *plastids* discoid to ribbon-like.

Spermatangial branchlets borne on ultimate branchlets, developing on all sides of terminal 4-5 cells, simple or branched, each spermatangial mother cell bearing 2-3 subspherical spermatangia. *Procarps* formed singly near apices of determinate branches, appearing to be lateral on basal cell of a branchlet; *cystocarps* about 180 µm in diameter, consisting of 3-5 spherical or subspherical gonimolobes of different ages, up to 140 µm

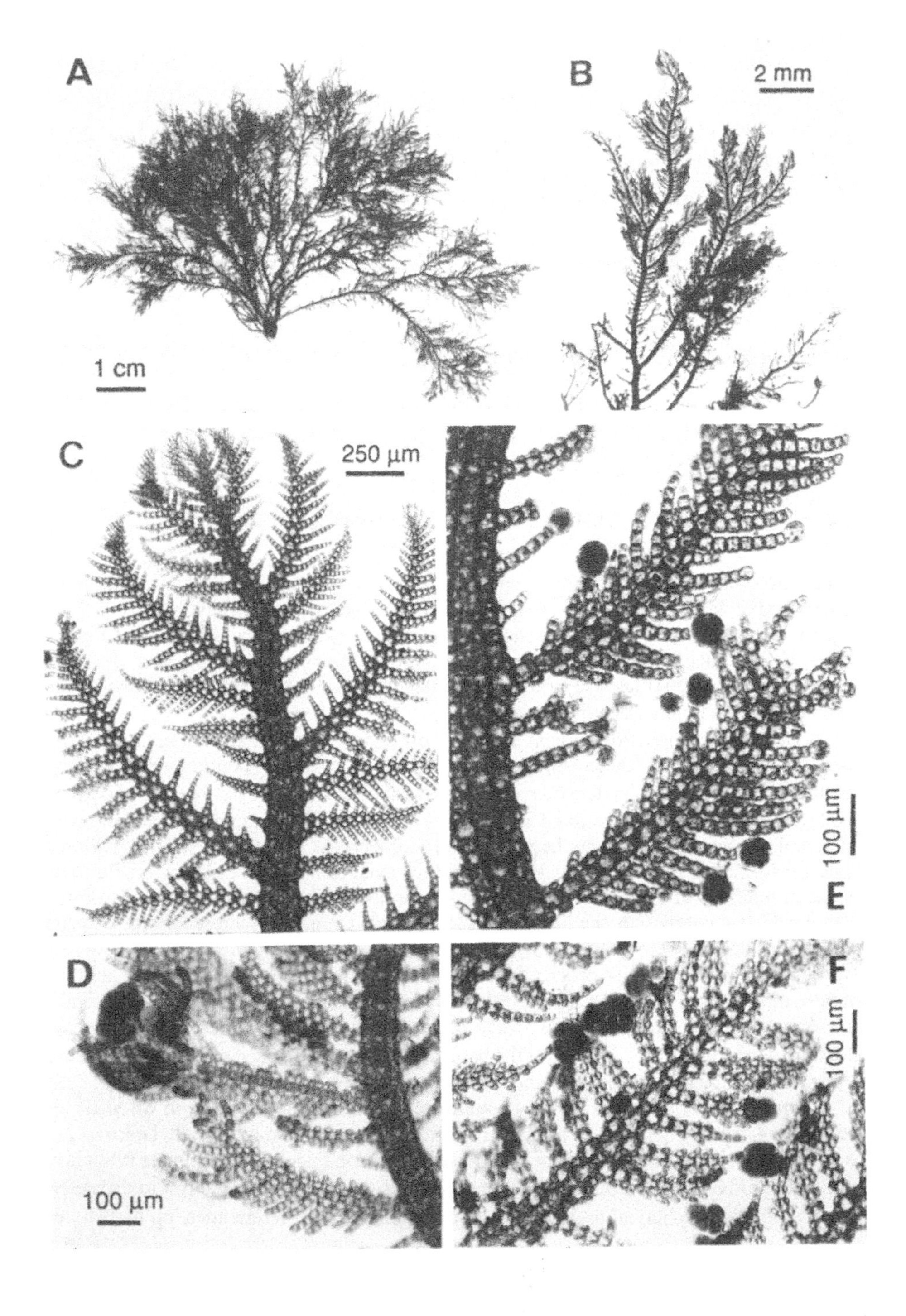
A
B
2 mm
1 cm
C
250 µm
100 µm
E
D
F
100 µm
100 µm

in diameter, containing relatively few angular carposporangia 24-40 μm in diameter, partially surrounded by simple involucral filaments formed by cells below cystocarp. *Tetrasporangia* borne on ultimate branchlets, usually terminal, single, often on 1-celled branchlets, rarely lateral near branchlet apices, pyriform before division, spherical when mature, 52-66 μm in diameter. *Parasporangial clusters* borne terminally on ultimate branchlets, initiated as pyriform cells that divide obliquely, globular or lobed when mature, containing 5-16 angular parasporangia 20-32 μm wide, enclosed in a thick envelope.

Normally epilithic on bedrock, occasionally epiphytic on small algae, rarely on *Laminaria hyperborea* stipes, typically on vertical and overhanging shaded rock faces and in caves, less often in pools from mid-shore to extreme low water, occasionally subtidal to 1 m on bedrock, regenerating plants recorded to 10 m on mobile substrata, reported from 19 m in S. Cornwall (Price et al. 1981), at moderately sheltered to moderately wave-exposed sites, not found in very sheltered inlets.

Generally distributed in the British Isles (Norton, 1985).

Arctic Norway and Iceland to Portugal; Newfoundland to New Jersey (South & Tittley, 1986).

Large thalli occur throughout the year. Spermatangia recorded for August and December, procarps and cystocarps for March, April, August, September, November and December; tetrasporangia in January, April, July and September-December; parasporangia were found throughout the year at St Andrews, Scotland (Blackler, 1974). Drew (1939) showed that gametophytes are haploid (n = 31), tetrasporophytes are diploid, and parasporangial plants are triploid. Gametangial and tetrasporangial plants appear to undergo a *Polysiphonia*-type life history, while parasporangial plants reproduce only by mitotic paraspores. Triploid plants appear to have a wider geographical range, occurring further north than gametophytes and tetrasporophytes. This suggestion is consistent with the presence in Norway and Newfoundland only of parasporangia, paraspores from which gave rise to several generations of parasporangial plants in culture (Rueness, 1968; Whittick, 1977). Tetrasporangia formed by parasporangial plants appeared to be non-functional. Plants growing in low-light situations such as caves show less regular branching than those in more open situations; a regular branching pattern is restored by culturing shade forms in brighter light (Norton, 1975).

Robust thalli of *Plumaria plumosa* resemble *Ptilota gunneri* (q. v.), but can be distinguished by the branching pattern at main apices: whereas *Plumaria plumosa* forms

---

Fig. 11. *Plumaria plumosa*

(A) Habit (Devon, Nov.). (B) Part of thallus showing complanate branching (Pembroke, Dec.). (C) Apex of main axis, showing paired laterals on every cell, and monosiphonous last-order branchlets (Donegal, May). (D) Cystocarp with loose involucral filaments (as A). (E) Tetrasporangia terminal on monosiphonous last-order branchlets (as B). (F) Parasporangial clusters borne in same position as tetrasporangia (as B).

a lateral branch on every axial cell, in *Ptilota gunneri* the axial filament bears a branch only every 2-3 cells. In *Plumaria plumosa*, last-order branchlets are normally uniseriate and non-corticated, with rounded apices, but in *Ptilota gunneri* they are several cells wide, resembling small blades, with a pointed conical apical cell.

PTILOTA C. Agardh, nom. cons.

PTILOTA C. Agardh (1817), pp. xix, 39.

Type species: *Ptilota gunneri* Silva, Maggs & L. Irvine, nom. nov. [= *Fucus ptilotus* Gunnerus (1772), p. 135], typ. cons. prop.*

Thalli compressed, branched in one plane, decumbent or erect to 35 (-60) cm high from a discoid rhizoidal holdfast; consisting of 3-4 orders of irregularly alternate to subdichotomous primary branches and shorter branches opposite the primary branches or in opposite pairs at every node, giving the thallus a feathery appearance; opposite branches mostly similar, potentially indeterminate, or dissimilar, one potentially indeterminate and the other determinate, leaflet-like with serrated margins, or sometimes wanting; axis corticated by a small-celled cortex, up to 7 cell layers thick, composed of subspherical inner and angular outer cells; rhizoidal filaments between axial cells and cortex; adventitious branches, when present, originating from outer cortical cells.

Growth of indeterminate branches by oblique division of apical cells, alternately to one side and then the other, shifting orientation at sites of primary lateral initials, usually every other axial segment; primary laterals curving towards the apex, overtopping it, developing into indeterminate branches, some of which form primary indeterminate axes; almost every segment ultimately forming a pair of opposite branch initials which either develop similarly into potential indeterminate branches, or one of the pair producing a determinate branch or only forming cortex; transverse periaxial cells absent in vegetative axes; each lateral periaxial cell typically producing 2 ascending and 3 descending cortical filaments that exhibit quadri-, tri-, and dichotomous branching to form a cellular cortex; rhizoidal cortication formed secondarily; cells uninucleate; plastids discoid to ribbon-like.

Gametophytes dioecious. Spermatangial mother cells borne in clusters around cells of uniseriate filaments 3-6 cells long, 2-3 spermatangia per spermatangial cluster. Procarps subterminal on long or short indeterminate branches or subterminal on axes regenerated on serrations of determinate branches, the fertile axial cell bearing 4 periaxial cells, an abaxial supporting cell with a sterile group and a lateral, 4-celled carpogonial branch, two lateral periaxial cells and an adaxial periaxial cell; the apical cell, sterile group and

* This type has been proposed because *Ptilota plumosa* (Hudson) C. Agardh (1817), p. 39 is based on a species of *Plumaria*.

vegetative periaxial cells all producing short monosiphonous filaments terminated by hair cells; auxiliary cells cut off distally from the supporting cell after fertilization, the fertilized carpogonium cutting off a terminal cell and fusing directly with the auxiliary cell, depositing a derivative of the fertilization nucleus; auxiliary cell dividing into a foot cell containing the haploid nucleus, which may divide several times, and a gonimoblast initial containing the diploid nucleus; gonimoblast initial cutting off a distal and 2-3 lateral gonimolobe initials which elongate and produce globose clusters of carposporangia terminally that mature successively; fusion cell incorporating the central cell, supporting cell, foot cell and primary gonimoblast cell; gonimoblasts surrounded by involucral branches formed from lateral and transverse periaxial cells in the segments just below. Tetrasporangia borne terminally in clusters on ultimate, uniseriate filaments, naked or surrounded by sterile filaments, pyriform, tetrahedrally divided.

References: Kylin (1923), Masuda & Sasaki (1990).

The only member of this genus in the British Isles is currently known as *Ptilota plumosa* (Hudson) C. Agardh, but this name is based on a specimen of *Plumaria* (Irvine & Dixon, 1982). An alternate name is based on *Fucus ptilotus* Gunnerus (1772), p. 135, but the tautonym *Ptilota ptilota* cannot be used, so a new name, *Ptilota gunneri*, has been proposed.

Ptilota gunneri Silva, Maggs & L. Irvine in Maggs & Hommersand, nom. nov.

Lectotype: TR, Herb. Gunnerus 1045.1. Unlocalized (presumably Norway), undated.

Fucus ptilotus Gunnerus (1772), p. 135, pl. II, fig. 15.
Fucus plumosus sensu Hudson (1778), p. 587, non *Fucus plumosus* Hudson (1762), p. 473.

Thalli forming tufts 7-30 cm high from small solid discoid holdfast; each erect axis complanate, with an irregularly flabellate or rounded outline, consisting of a single main axis below, 0.9-1.2 mm in diameter, irregularly alternately branched to 4 orders of branching; main axes compressed, bearing regular, feather-like series of shorter branches; bright red to brownish-red in colour, with robust, flexible main axes and soft branchlets.

Main axes growing from apical cells 10 μm wide, regularly sinuous when young, the axial cells increasing to 4 diameters in length, with cortication continuous from near apices, consisting of angular cells 6-12 μm in diameter; older axes heavily corticated by 4-7 layers of cells, with narrow rhizoidal filaments growing downwards between axial filament and cortical cells; *indeterminate laterals* borne on every alternate axial cell in pairs, long ones on alternate sides paired with opposite short ones, all strongly incurved, overtopping main apex, some developing into major axes equal in diameter to main axes, all bearing pairs of long and short indeterminate laterals in a distichous arrangement; *determinate laterals* borne on indeterminate laterals in pairs of unequal length, increasing in length distally, bearing pairs of equal branchlets from every cell; *branchlets* curved, becoming winged

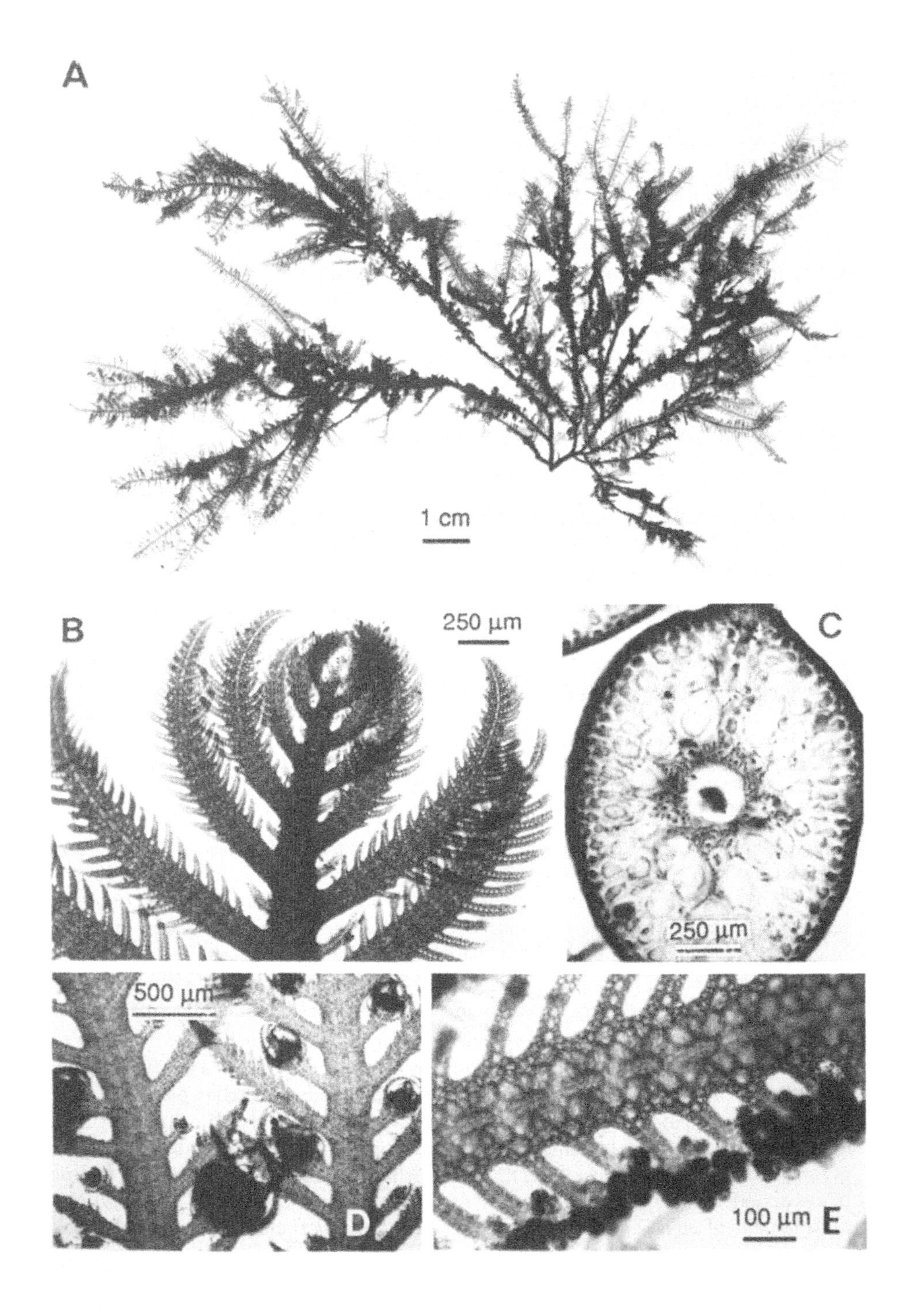
A
1 cm
B
250 µm
C
250 µm
500 µm
D
100 µm
E

and blade-like due to cortication, tapering from 36-44 μm at the base to the conical pointed apical cell; *plastids* discoid to ribbon-like.

Spermatangial branchlets borne on uniseriate ultimate branchlets 3-6 cells long that develop in pairs from winged determinate branchlets, giving rise to short branchlets from all sides, these bearing spermatangial mother cells with 2-3 subspherical spermatangia 6 μm in diameter. *Procarps* formed singly near apices of determinate branches, borne in a cluster with 2-3 carpogonium-like structures terminating in hairs; *cystocarps* about 180 μm in diameter, consisting of 2-4 spherical or subspherical gonimolobes of different ages, up to 140 μm in diameter, containing angular carposporangia 40-56 μm in diameter, enclosed by simple or branched corticate involucral filaments borne on subtending axial cells. *Tetrasporangia* formed around apical portions of penultimate determinate branches, terminally on wing-like ultimate branchlets and in dense clusters on branched 2-4 celled uniseriate filaments, pyriform before division, spherical when mature, 52-64 μm in diameter.

Normally epiphytic on *Laminaria hyperborea* stipes and holdfasts, rarely epilithic on bedrock, small plants occasional on mobile substrata such as maerl and pebbles, from extreme low water to 10 m, rarely to 27 m depth, at moderately to extremely wave-exposed sites. At St Abbs, Berwick, *P. gunneri* was most abundant at 1-2 m depth, where it occupied the basal part of kelp stipes, whereas at 6 m and 10 m depth it grew above the middle of the stipes; it was absent at 12 m depth (Whittick, 1983).

Northern coasts of the British Isles, southwards to Isle of Man and Anglesey on west coasts, and York on east coasts; widely distributed in Ireland except on S.E. coasts (Norton, 1985).

Arctic Norway, Spitzbergen and ?Greenland to England and Denmark (South & Tittley, 1986). Records from eastern Canada are based on *Ptilota serrata* Kützing (see South, 1984), which can be distinguished from *P. gunneri* because short laterals are spine-like and long laterals are dentate, whereas all laterals in *P. gunneri* are feathery (see Masuda & Sasaki, 1990, figs 2 and 9).

Large thalli occur throughout the year. At St Abbs, biomass was maximal in June and July, followed by a dramatic loss in November (Whittick, 1983). Spermatangia recorded for May; cystocarps and tetrasporangia occurred throughout the year at St Andrews, Scotland (Blackler, 1974).

There is little morphological variation.

See under *Plumaria plumosa* for distinctions between it and *Ptilota gunneri*.

Fig. 12. *Ptilota gunneri*

(A) Habit (A-E Donegal, May). (B) Apex of main axis, showing paired long and short laterals on every alternate axial cell, and paired wing-like branchlets. (C) T.S. through main axis, with rhizoidal filaments surrounding axial cell. (D) Branches bearing cystocarps of different ages. (E) Clusters of tetrasporangia on branchlets.

Tribe CERAMIEAE (Dumortier) Schmitz (1889), p. 106.

CERAMIUM Roth, nom. cons.

CERAMIUM Roth (1797), p. 146, nom. cons.

Type species: *C. virgatum* Roth (1797), p. 146, typ. cons. [? = *C. nodulosum* (Lightfoot) Ducluzeau (1806), p. 61].

Thalli cylindrical or slightly compressed, corticated at the nodes, or cortication intermittent to continuous, erect to 30 cm high or partly to wholly prostrate, variously attached to substratum by unicellular or multicellular, branched rhizoids which often terminate in digitate pads; branching alternate-distichous to pseudodichotomous, with straight, curved or involute tips; adventitious branches rare to frequent, restricted to plane of branching or borne on all sides.

Growth monopodial by oblique division of apical cells followed by elongation of axial cell; lateral branches originating abaxially at more or less regular intervals by steeply oblique division of apical cell followed by oblique division of subterminal cell, and with the laterals developing as strongly as the main axis (pseudodichotomous branching) or less strongly developed (alternate branching); periaxial cells 3 to 10, the first abaxial, the rest cut off in alternate (rhodomelacean) sequence; periaxial cells typically cutting off 2 anterior and 1-3 (usually 2) posterior cortical initials, or lacking posterior initials; cortical filaments from anterior initials ascending and those from posterior initials usually descending; cortex 1-3 cells thick, continuous over axial cell, or discontinuous, forming bands at the nodes, either loose, or adhering tightly to axial cell as it enlarges; covered by a soft or firm to cartilaginous outer membrane; deciduous hairs, spines, or darkly staining 'gland' cells present or absent; axial cell initially rectangular, broader than long, elongating to 1-many times as long as broad, sometimes continuing to elongate after cortical growth has ceased; adventitious branches originating from periaxial or cortical cells.

Gametophytes dioecious, sometimes monoecious or mixed phases present. Spermatangia formed in extensive superficial sori, protruding through surface membrane; 1-3 spermatangia per spermatangial mother cell. Procarps formed abaxially in rows from the first periaxial cell, consisting of a sterile group that resembles a vegetative cortical filament and a 4-celled carpogonial branch; a second, lateral sterile group rare; auxiliary cell cut off distally from the supporting cell after fertilization; fertilized carpogonium cutting off a terminal capping cell and fusing directly with the auxiliary cell, depositing a derivative of the fertilization nucleus; auxiliary cell dividing into a residual auxiliary cell (foot cell) containing the haploid nucleus, which may divide, and the gonimoblast initial containing the diploid nucleus and a conspicuous protein body; gonimoblasts abaxial, composed of one to several gonimolobes that usually mature in succession, naked, or surrounded by 1-several involucral branchlets; most gonimoblast cells maturing into carposporangia; foot cell, supporting cell and axial cell united into a central fusion cell, separated from the primary gonimoblast cell by a broad pit connection. Tetrasporangia initially formed from periaxial cells, later often from cortical cells as well, immersed to partly exposed or naked, tetrahedrally to irregularly cruciately divided.

References: Kylin (1923), Feldmann-Mazoyer (1941), Dixon (1960a, 1960b), Hommersand (1963), Itono (1977), Womersley (1978).

Ceramium species are unusually morphologically variable. The cortical spines formed by a few species are diagnostic, and this group has been well studied (Dixon, 1960a), but the non-spiny species are in a state of taxonomic chaos in the British Isles and elsewhere. In an attempt to clarify the systematics of the non-spiny members of this genus in the British Isles, a plastid DNA-based approach has been taken (Ward, 1992; Ward & Maggs, unpublished), and some of the results have been incorporated in the following treatment. Initial results showed that certain taxonomic characters previously considered significant, such as the details of cortical filament development, are of little practical value in delimiting British Isles species, except in dividing the genus into species that are potentially fully corticated and those that always have ecorticate internodes. On the other hand, the most important character, which has generally been dismissed, is the branching pattern, in particular the intervals between branches; the degree of incurvature of the apices is also significant. Thallus texture and colour are useful, but require some comparative experience. We concur with Womersley (1978) that gametangial reproductive structures are of little value in species delimitation.

Nomenclature is also a major problem. Thirteen non-spiny *Ceramium* species are listed for the British Isles by South & Tittley (1986), who also included 6 further species in synonymy with *C. rubrum*, which is an illegitimate name. There appear to be four British Isles representatives of the potentially fully corticate '*C. rubrum*' group, for which we have used the following provisional names: *C. pallidum*, *C. botryocarpum* (a species that has been ignored recently), *C. nodulosum* (a valid name for the species usually known as *C. rubrum*), and *C. secundatum*, which has frequently been referred to '*C. pedicellatum* Duby'. In addition, we have recently found *C. circinatum* Kützing (1842, p.733) at a few sites in S. England. This primarily Mediterranean species, which is up to 12 cm high, can readily be distinguished from our other *Ceramium* species by the presence of 10 periaxial cells. It is illustrated by Harvey (1850, pl. 276), as *C. decurrens*. Most British Isles records of other members of the '*C. rubrum*' group appear to be based on collections of one or more of these four species, or spiny species, as follows:

C. arborescens J. Agardh: *C. pallidum*;
C. crouanianum J. Agardh: *C. pallidum* and *C. nodulosum*;
C. derbesii Solander ex Kützing: *C. nodulosum* and *C. ciliatum*;
C. fruticulosum J. Agardh: *C. pallidum* and *C. nodulosum*;
C. tenue J. Agardh: *C. pallidum, C. nodulosum and C. gaditanum*;
C. vimineum Petersen: *C. secundatum* and *C. gaditanum*.

South & Tittley (1986) listed 7 species of partially corticated *Ceramium* for the British Isles. Of these, records of *C. codii* and *C. fastigiatum* are transferred here to *C. cimbricum*, following suggestions made by Rueness (1992); *C. deslongchampii* and *C. flaccidum* are unchanged, and *C. tenuissimum* records are transferred to *C. diaphanum*. *C. strictum* sensu Harvey is a distinctive species but this name is illegitimate and a valid alternative name remains to be determined. *C. siliquosum* is one of the names available for *C. diaphanum* sensu Harvey, and has been used here pending further study. The nomenclatural situation of this group is extremely confused, partly because type material of *Conferva diaphana* Lightfoot is heterogeneous, including at least three species.

KEY TO SPECIES

1 Spines present on cortical bands near apices 2
  Spines absent 5
2 Spines single-celled .... *C. echionotum*
  Spines multicellular 3
3 Each spine multicellular at the base, merging with the cortical band .... *C. shuttleworthianum*
  Spines consisting of a single row of cells, clearly delimited from cortical band 4
4 Spines whorled, very conspicuous, basal cell of spine 1.5 diameters long ... *C. ciliatum*
  Spines borne singly, often sparse and inconspicuous, basal cell of spine 1 diameter long .... *C. gaditanum*
5 Main axes of mature thalli entirely corticate, younger axes and juvenile thalli entirely corticate or with cortical bands of variable widths and irregular or dentate upper and lower borders (see Fig. 22G) 6
  Mature axes incompletely corticate, cortical bands more-or-less even in width, with straight upper and lower borders, separated by ecorticate internodes (see Fig. 16B) 9
6 Axes branching every 4-8(-10) segments, mostly at intervals of 5-7 segments; branching pattern appearing regular .... *C. pallidum*
  Axes branching at intervals of (6-)10-18 segments; branching pattern appearing irregular 7
7 All axes brittle and cartilaginous; periaxial cells 7-9, typically 8; subtidal and deep lower-shore lagoons .... *C. secundatum*
  Younger axes soft and flexible; periaxial cells 6-7; intertidal and sublittoral fringe 8
8 Forming turfs due to development of extensive prostrate axes; adventitious branching often abundant; periaxial cells typically 6 .... *C. botryocarpum*
  Growing as individual thalli without prostrate axes, adventitious branching usually inconspicuous, periaxial cells typically 6-7 in individual thalli .... *C. nodulosum*
9 Rhizoids unicellular; apices obviously alternately branched every 5-6 segments .... *C. flaccidum*
  Rhizoids multicellular; apices pseudodichotomously branched 10
10 Periaxial cells 10 .... *C. circinatum*
  Periaxial cells 4-8 11

11 Non-reproductive apices strongly inrolled 12
Non-reproductive apices straight or incurved but not strongly inrolled 14
12 Cortical bands collar-like, protruding markedly from internodal axial cells (see Fig. 17D); tetrasporangia formed singly or in small groups on nodes........ *C. diaphanum*
Cortical bands level with internodal axial cells, not protruding except when reproductive (see Fig. 16B); tetrasporangia formed in rings around nodes.............. 13
13 Periaxial cells 6-7; branch intervals 6-12 segments; axes 120-200 μm in diameter ....*C. strictum* sensu Harvey
Periaxial cells 8; branch intervals 4-7 segments; axes 200-500 μm in diameter ....*C. siliquosum*
14 Dull greyish-purple in colour; texture cartilaginous; tetrasporangia entirely naked; common in muddy pools and on damp intertidal rock faces ....*C. deslongchampii*
Rose-pink to bright red in colour, texture fragile and delicate; tetrasporangia immersed or partially cupped by cortical filaments; subtidal and in deep sheltered pools .... *C. cimbricum*

Ceramium botryocarpum Griffiths ex Harvey (1848), pl. 215.

Lectotype: TCD. Syntypes: TCD; BM. Devon (Preston Rocks, Paignton), June 1846, *Griffiths*.

Ceramium lanciferum var. β *monstruosum* Kützing (1847), p .33.

Thalli 3-12 cm high, consisting of turfs or irregularly shaped tufts of one to several erect axes attached by a dense mass of multicellular rhizoidal filaments; base of main axes decumbent and attached by a felt-like rhizoidal mass, giving rise to numerous lateral branches that become erect; axes 0.2-0.4 mm in diameter basally, branching irregularly or regularly dichotomously, often in one plane, dull purplish-red in colour bleaching to yellow or greyish-brown; young axes fairly soft but not delicate, becoming rather cartilaginous when older.

Main axes with forcipate to strongly inrolled apices; axial segments increasing to 1.0-1.5 diameters long, either remaining cylindrical or becoming markedly constricted at intervals due to pyriform shape of axial cells; laterals developing at intervals of 14-18 segments; *adventitious branches* common to abundant; *nodes* consisting of 6 or 7 periaxial cells that give rise to a complete cortex; axes conspicuously banded when mature due to a thinner covering of cortical filaments over internodes; *internodes* 15 μm wide in young axes, visible only when stained; *plastids* irregularly plate-like in cortical cells.

Gametophytes dioecious. *Spermatangial sori* continuous around axes. *Cystocarps* usually consisting of a single gonimolobe, 250-300 μm in diameter, containing numerous angular carposporangia 25-40 μm in diameter, surrounded by 3-5 involucral branchlets. *Tetrasporangia* completely immersed within the cortex or partially exposed, 50-70 μm in diameter.

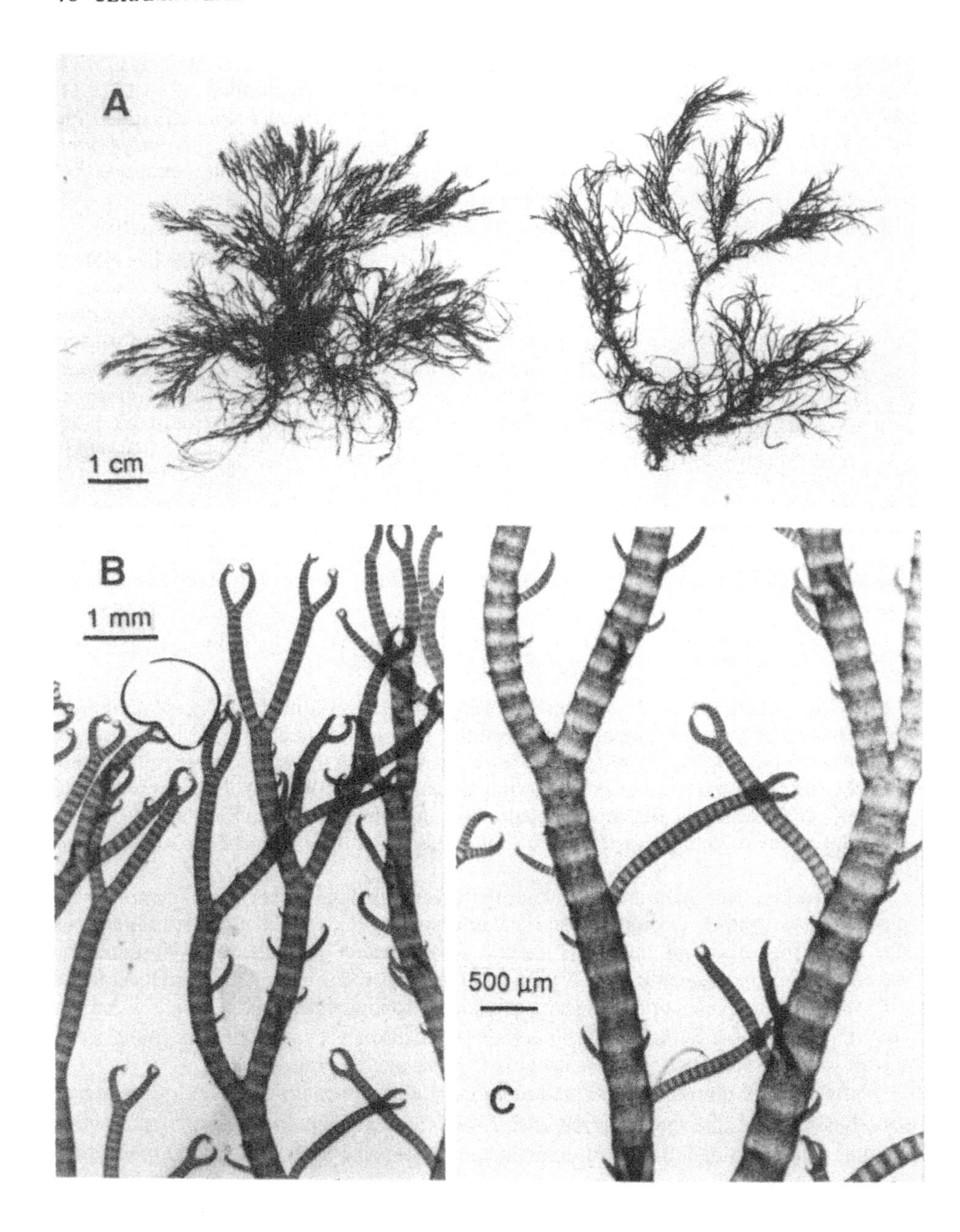
A
1 cm
B
1 mm
500 µm
C

Epiphytic on perennial algae, such as *Cladostephus verticillatus* and *Laurencia pinnatifida*, also growing on silty rock, in upper-shore pools and on open rock around the level of low water of neap tides, in habitats where large brown algae are absent, at moderately sheltered sites.

Probably widely distributed in the British Isles (Batters, 1902); recently observed in Devon, Fife, Cork, Clare, Dublin and Channel Isles.

France (Brittany); wider distribution unknown.

Large thalli have been observed between March and September, with spermatangia, cystocarps and tetrasporangia throughout this period.

This species occurs in two rather dissimilar forms, related in part to habitat. Plants growing on open, often vertical, rock have wide axes with abundant adventitious branching and are rather pale in colour. Populations in upper-shore pools have narrower, darker axes, and a more complanate branching pattern with regular dichotomies.

Ceramium ciliatum (Ellis) Ducluzeau (1806), p. 64.

Lectotype: Ellis (1768), pl. 18, fig. H. Unlocalized (?England).

Conferva ciliata Ellis (1768), p. 425.
Ceramium diaphanum var. *ciliatum* (Ellis) Duby (1830), p. 967.
Echinoceras ciliatum (Ellis) Kützing (1842), p. 736.

Thalli densely tufted to loosely branched, 2.5-7(-15) cm high, consisting of interwoven prostrate axes giving rise to groups of erect axes, more or less entangled or matted together by rhizoids at the base; axes repeatedly pseudodichotomously branched, rose-pink in colour, fairly rigid and brittle in texture.

Main axes with forcipate, strongly inrolled apices; apical cells 16-20 μm wide including walls 4 μm thick; axial cells initially rounded, becoming cylindrical and increasing in length to 660-960 μm, increasing only slightly in diameter with age; mature axes 175-310 μm in diameter, with segments 2-4.5 diameters long, branching pseudodichotomously at intervals of 10-14 axial cells; *adventitious branches* sparse to numerous, often in pairs, on older axes; *nodes* consisting of 6-7 periaxial cells that give rise to ascending and descending cortical filaments decreasing from 24 μm to 8-10 μm in diameter; descending filaments longer than ascending, forming 2-3 cortical layers over periaxial cells; *spines* formed on nodal bands near apices in whorls of 5-6, non-pigmented, pointed, 100-125 μm long x 26-36 μm wide, 3-celled, the basal cell being 1.5-3 diameters long; *internodes* of naked

Fig. 13. *Ceramium botryocarpum*

(A) Habit of two plants, showing turf-like growth form and abundant adventitious branching (Dublin, Aug.). (B) Young axes, with numerous developing adventitious branches and strongly inrolled apices (B-C Devon, Apr.). (C) Mature axes, showing conspicuous banded appearance due to differential cortical development over nodes and internodes.

axial cells usually separating cortical bands, variable in length; *rhizoids* formed by cortical cells in whorls around nodes, simple, multicellular, 16-32 μm wide, terminating in multicellular discoid pads; *plastids* ribbon-like in axial cells, becoming filiform and gradually breaking down.

Gametophytes dioecious; spermatangia not noted. *Cystocarps* consisting of 2-5 (-8) globular gonimolobes 360-410 μm in diameter, containing numerous angular carposporangia 24-44 μm in diameter, enclosed in a membrane to 20 μm thick, surrounded when

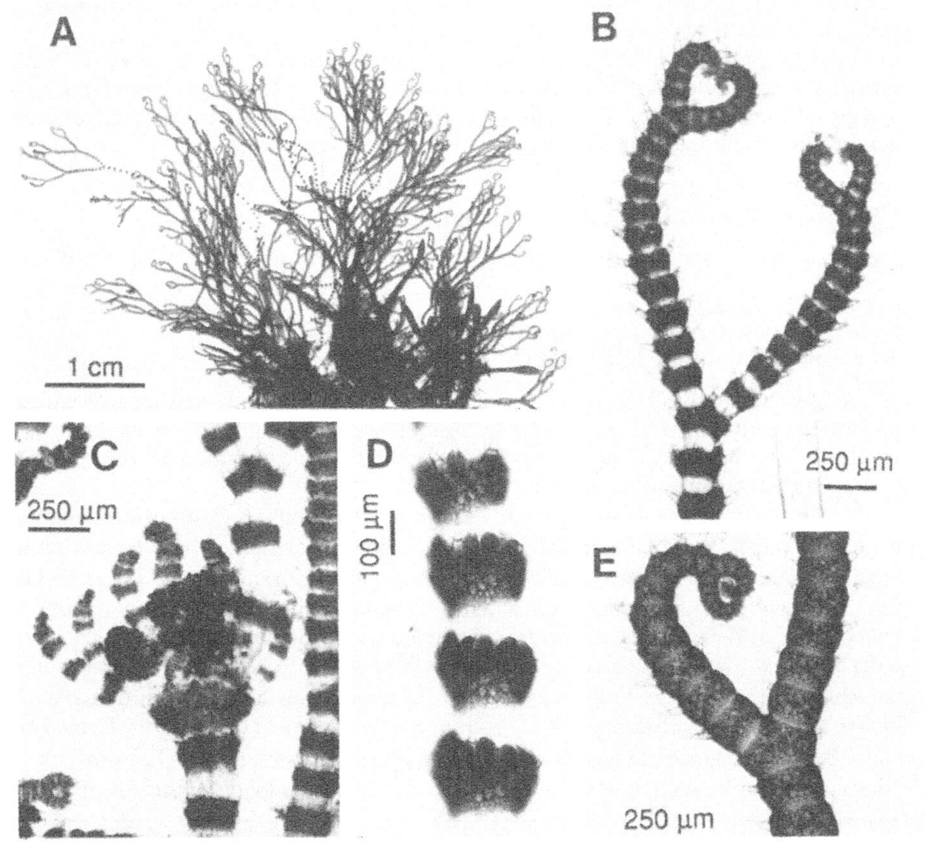

Fig. 14. *Ceramium ciliatum*

(A) Habit (Cork, Nov.). (B) Young axes with inrolled apices, bearing multicellular spines on cortical bands; internodes are ecorticate (B-D Dublin, May). (C) Cystocarp with several gonimolobes surrounded by involucral branches. (D) Whorls of immersed tetrasporangia in each cortical band. (E) Incurved apices of fully-corticate thallus (as A).

mature by a whorl of 6 simple, inrolled, involucral branchlets formed by subtending axial cell. *Tetrasporangia* borne on younger axes, in whorls of (1-)5-12 around the nodes, initially entirely covered by cortical cells, when mature protruding apically from upward-growing cortical filaments, ellipsoid, 64-70 x 42-56 μm, including the thick walls.

Epilithic on bedrock, epiphytic on various algae and occasionally epizoic on limpets, in pools and damp places in the mid- to lower intertidal and subtidal to 10 m depth, often on mobile substrata such as maerl, at extremely sheltered to moderately wave-exposed sites, sometimes with exposure to strong tidal currents.

Widely distributed on south and west coasts, northwards to Orkney (records for Shetland erroneous), rarely recorded from E. Scotland and N. E. England; widely distributed in Ireland; Channel Isles.

Norway, British Isles and France; Azores, Canary Is., Cape Verde Is. (South & Tittley, 1986; Price et al., 1986).

Plants have been noted only from February-November; large thalli are often non-reproductive. Spermatangia not apparently recorded in the British Isles; cystocarps recorded in May and June; tetrasporangia in March, May-July and September-November.

Dixon (1962a) commented that *C. ciliatum* sensu lato is a particularly polymorphic species. While most British specimens have 3-celled spines, some have 2- or 4-celled spines; Mediterranean populations form spines of 3-5 cells.

Ceramium cimbricum H. Petersen in Rosenvinge (1923-24), p. 378.*

Lectotype: C 8843. Denmark (Egerslev Røn, Limfjord), 27 vii 1920, *H. Petersen* (J. Rueness, pers. comm.)

Ceramium fastigiatum Harvey (1834), p. 303, non *Ceramium fastigiatum* Roth (1800a), p. 463.
Ceramium fastigiramosum Boo & I. K. Lee (1985), p. 223.

Thalli creeping or erect, 0.5-8(-12) cm high, consisting of prostrate axes sometimes giving rise to dense hemispherical or irregularly shaped tufts of erect axes 0.3 mm in diameter, repeatedly pseudodichotomously branched; rose-pink to bright red in colour, with larger axes distinctly banded to the naked eye, turgid and very fragile when young, becoming more rigid when older.

Prostrate axes 30-200 μm in diameter, attached by rhizoids, becoming erect apically; *erect axes* with straight or incurved to slightly inrolled apices; young axial cells transversely ellipsoid in surface view, cylindrical when mature, occasionally becoming

---

* *Ceramothamnion codii* H. Richards (1901) from Bermuda, transferred to *Ceramium* by G. Mazoyer (1938, p. 324), and *Ceramothamnion adriaticum* Schiller (1911) from the Adriatic, may be older names for this species.

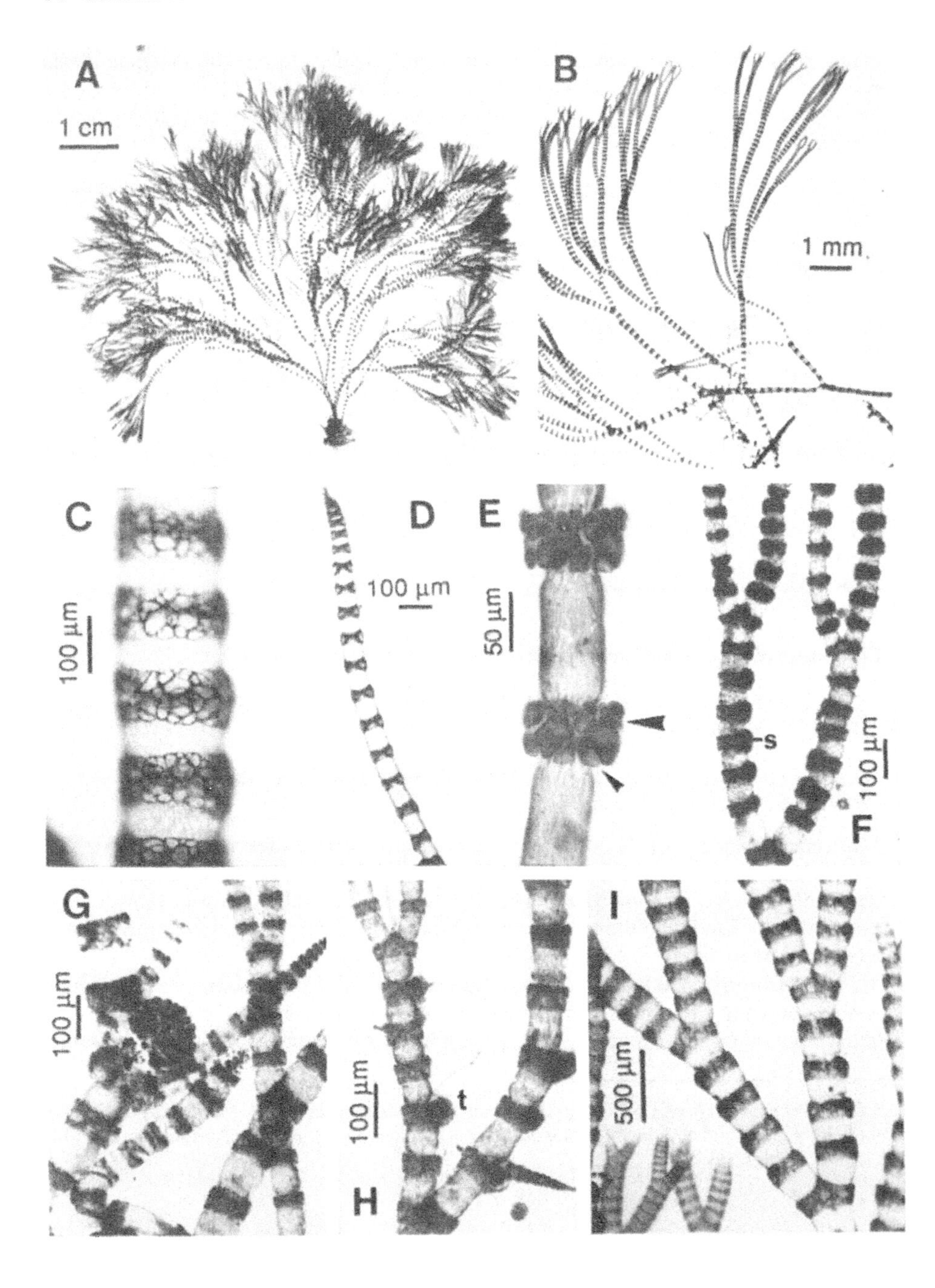
A
1 cm
B
1 mm
C
100 µm
D
100 µm
E
50 µm
F
s
100 µm
G
100 µm
H
t
100 µm
I
500 µm

swollen and pyriform, 30-300 µm in diameter; axes incompletely corticate, branching pseudodichotomously at intervals of 5-30 axial cells, with branch angles of <45°; *adventitious branches* sparse in erect thalli, abundant in prostrate axes, sometimes from every node, developing into numerous erect axes; *nodes* consisting of 4-6 periaxial cells that either remain as a single whorl of cells or each give rise to 1-2 ascending filaments one to several cells long (sometimes also to 1-2 descending filaments), composed of angular cortical cells; nodes sometimes giving rise to smaller cells that form rosettes around periaxial and inner cortical cells; cortical bands protruding only slightly beyond outer walls of internodal axial cells, 100-150 µm long, the upper and lower margins remaining sharply defined; *internodes* 2-5 diameters long, distinctly pigmented; *rhizoids* non-pigmented, multicellular, 20-30 µm in diameter; *plastids* reticulate in cortical cells, ribbon-like to filiform in axial cells.

Gametophytes dioecious. *Spermatangial sori* initially formed on adaxial sides of young cortical bands, later entirely covering them; spermatangia ellipsoid, 4 x 3 µm. *Cystocarps* consisting of 1-3 globular gonimolobes of different ages, c. 250 µm in diameter, containing numerous angular carposporangia 35-50 µm in diameter, when mature subtended by a whorl of 1-4 simple straight involucral branchlets. *Tetrasporangia* formed on large axes in whorls near the upper end of each cortical band, entirely immersed in cortex, developing on narrow axes in small groups on one side of the node only, protruding laterally and partially covered by cup-like ascending cortical filaments, spherical to ellipsoid, 50-60 x 45-50 µm including the thick walls.

Growing on a wide variety of substrata, including bedrock and pebbles, large and small algae, maerl, ascidians, hydroids and dead shells, from extreme low water to 26 m depth, found at greater depths on exposed shores than at sheltered sites, particularly common from 3-10 m depth at sites sheltered from wave action with slight to strong tidal currents, such as in Scottish sealochs.

Widely distributed on south and west coasts, northwards to Shetland; S.E. Scotland; probably generally distributed in the British Isles.

Fig. 15. *Ceramium cimbricum*

(A) Habit of large plant with wide axes (Down, May). (B) Narrow prostrate axes bearing relatively broad erect axes with slightly incurved apices (Donegal, Dec.). (C) Broad axis with well-developed cortical bands (as A). (D) Narrow axis with narrow, little-protruding cortical bands separated by long internodes (as B). (E) Detail of cortication, showing 2-3 -celled ascending filaments (large arrow) and single-celled descending filaments (small arrow) (as B). (F) Male axes with spermatangial sori (s) on adaxial faces of cortical bands (F-H Donegal, May). (G) Cystocarp surrounded by involucral branches. (H) Tetrasporangia (t) borne singly, protruding from cortical bands of narrow axes. (I) Whorls of immersed tetrasporangia in each cortical band of thallus with wide axes (as A).

Norway to N. France; widely distributed in N.W. Atlantic (South & Tittley, 1986); Mediterranean; ?North Pacific (Boo & Lee, 1985).

Creeping thalli found throughout the year, forming larger erect axes in May-September. Spermatangia recorded in May-June and September; cystocarps in June-September; tetrasporangia in May-September. The majority of field-collected plants are non-reproductive or tetrasporangial; mixed-phase tetrasporangial/female plants have been observed. A *Polysiphonia*-type life history was obtained in Norwegian isolates, but some mixed-phase reproduction also occurred (Rueness, 1992). Yabu et al. (1981) also observed a *Polysiphonia*-type life history in cultures from Japan attributed to this species, and obtained chromosome numbers of $n = 28$.

This species shows a wide range of variation in axial diameter, tetrasporangial position and branching frequency. Very narrow axes with only 4-5 periaxial cells may entirely lack cortication; this form appears to have been reported from the British Isles as *C. codii* (Dixon, 1958). Rueness (1992) observed this type of morphology in cultures grown under low irradiances. Slightly more corticate thalli, in which tetrasporangia protrude markedly from the narrow axes, have been known as *C. cimbricum*. Thalli with wider axes and immersed tetrasporangia were described as *C. fastigiatum*. Erect thalli are fairly regularly branched, usually every 5-10 segments, but branching intervals in prostrate thalli can be much longer.

Ceramium deslongchampii Chauvin ex Duby (1830), p. 967.

Lectotype: PC. France (Normandy), Chauvin exsiccatum 'Algues de la Normandie', no. 83.

Ceramium agardhianum Griffiths ex Harvey (1841), p. 99.
Gongroceras deslongchampii (Chauvin ex Duby) Kützing (1842), p. 735.
Gongroceras strictum Kützing (1842), p. 735.
Ceramium strictum (Kützing) Harvey (1849b), p. 164, non *Ceramium strictum* Roth (1806), p.130.

Thalli 2.5-8(-12) cm high, hemispherical or irregular in shape, consisting of prostrate axes giving rise to numerous densely tufted erect axes 0.1-0.4 mm in diameter, repeatedly pseudodichotomously branched; dark brown or greyish brown in colour, older axes distinctly banded to the naked eye, fairly rigid and cartilaginous in texture.

Prostrate axes c. 200 μm in diameter, attached by rhizoids, becoming erect apically; *erect axes* with straight or incurved but not tightly inrolled apices; axial cells diamond-shaped in surface view in young axes, cylindrical or pyriform when mature; axes increasing from 60-80 μm near tips to 125-300 μm in diameter when mature, rarely to 400 μm if tetrasporangial, incompletely corticate, with segments to 2 diameters long; branching pseudodichotomous at intervals of 6-17 axial cells, with branch angles of c. 60° between young branches, wider in older axes; *adventitious branches* formed abundantly by prostrate axes, developing into erect axes, sparse to numerous on older erect axes, initially more-or-less perpendicular to axis, later curving inwards; *nodes* consisting of 5 periaxial

Fig. 16. *Ceramium deslongchampii*

(A) Habit (A-D Glamorgan, June). (B) Detail of cortication, with cortical bands level with ecorticate internodes, rather than forming a protruding collar. (C) Sparsely branched tetrasporangial axes showing constant width of cortical bands. (D) Naked tetrasporangia protruding from cortical bands (arrow). (E) Cystocarp surrounded by sparse straight involucral branches (Pembroke, May).

cells that each give rise to 2 ascending and 2 descending filaments of angular cortical cells decreasing to 8-10 μm in diameter; upper and lower margins of cortical bands remaining sharply defined; *internodes* distinctly pigmented, with a clear wall 20 μm thick; *rhizoids* non-pigmented, multicellular, 30-60 μm in diameter; *plastids* plate-like in cortical cells, irregularly reticulate in axial cells, becoming filiform.

Gametophytes dioecious. *Spermatangial sori* conspicuous, entirely covering the last 2-4 orders of branching; spermatangia c. 6 μm in diameter; released spermatia 4 μm in diameter. *Cystocarps* consisting of 1-3 globular gonimolobes of different ages, 200-250 μm in diameter, containing numerous angular carposporangia 30-45 μm in diameter, usually subtended when mature by a whorl of 1-4 simple straight involucral branchlets. *Tetrasporangia* formed initially in a ring around the middle of each node, later in two rings or becoming crowded and irregular, naked, protruding to about half their diameter, spherical to ellipsoid, 50-65 x 50-55 μm including the thick walls.

Epilithic on bedrock and pebbles and epiphytic on a wide range of small algae and fucoids in shaded crevices and under fucoids, with other turf-forming algae, at mid- to upper-shore levels; also epiphytic, especially on *Corallina* spp. and *Polysiphonia nigrescens,* in pools and damp places in mid- to lower intertidal, rarely subtidal, at moderately wave-exposed to moderately sheltered sites and at extremely wave-sheltered sites with exposure to tidal currents; frequently found on seawalls and slipways, becoming abundant in these habitats and probably tolerant of low salinity.

Widely distributed in the British Isles, reported northwards to Orkney (Batters, 1902); recently collected on east, south and west coasts of England, in Wales, and in Ireland (Cork, Down, Waterford), particularly common from Lincoln to Kent.

Reported from Norway and Iceland to Spain, and from Newfoundland to New York (South & Tittley, 1986); there are definite records from France (BM), Denmark (Rosenvinge, 1923-24), Helgoland (Kornmann & Sahling, 1978) and Netherlands (Stegenga & Mol, 1983).

Plants have been collected throughout the year; spermatangia were found in March, May and July; cystocarps recorded in March and May; tetrasporangia in March-August. Some populations seem to consist largely of small, non-reproductive plants.

This species shows relatively little morphological variation and is one of the most easily recognizable members of this genus in the British Isles.

Ceramium diaphanum (Lightfoot) Roth (1806), p. 154.

Lectotype: BM-K (Figs 17A-B). Unlocalized (?Scotland), undated.

Conferva diaphana Lightfoot (1777), p. 996.
?*Conferva nodulosa* Hudson (1778), p. 600, non *Conferva nodulosa* Lightfoot (1777), p. 994.
Ceramium diaphanum var. *tenuissimum* Roth (1806), p. 156.
Gongroceras tenuissimum Kützing (1842), p. 736.
Ceramium nodosum sensu Griffiths & Harvey ex Harvey (1847), pl. 90.

Ceramium tenuissimum (Roth) J. Agardh (1851), p. 120, non *Ceramium tenuissimum* Bonnemaison (1828), p. 132.

Thalli forming unattached or loosely attached masses up to 20 cm in diameter, consisting of numerous erect axes ≤0.2 mm in diameter, repeatedly pseudodichotomously branched, rose-pink to bright red in colour, with a fairly tough and rigid texture.

Main axes with tightly inrolled apices, the outer faces of which are conspicuously dentate; axial cells diamond-shaped in surface view when young, cylindrical or occasionally slightly pyriform when mature, increasing to 125-200 μm in diameter, remaining incompletely corticate; branching pseudodichotomous at intervals of 8-13 segments, with branch angles >60°; *adventitious branches* frequent in young and old axes; *nodes* consisting of 6-7 periaxial cells that each give rise to 2 ascending and 2 descending filaments of angular cortical cells, the ascending cortical filaments decreasing to 6-8 μm in diameter and the descending filaments decreasing to 8-12 μm in diameter; ovoid to reniform gland cells 12-20 μm in diameter sometimes abundant; cortical bands swollen, collar-like, protruding beyond outer walls of internodal axial cells, 100-125 μm long, the upper and lower margins remaining sharply defined; *internodes* little pigmented, 4-8 diameters long; *rhizoids* multicellular, non-pigmented, 20-25 μm in diameter; *plastids* filiform in axial cells.

Gametophytic material not available. *Tetrasporangia* borne singly or in small groups on adaxial sides of nodes, protruding laterally and partially covered by cup-like ascending cortical filaments, spherical to ellipsoid, 50-60 x 45-50 μm including the thick walls.

Epiphytic on algae and *Zostera* rhizomes, also growing on muddy sand, from extreme low water to c. 3 m, at moderately to extremely wave-sheltered sites with little or no current exposure; apparently uncommon except at a few sites.

South coast of England; widely distributed in Ireland (Harvey, 1847); reported to be widespread in Scotland and on east coasts (Batters, 1902), but literature records difficult to assess owing to taxonomic confusion.

Reported (as *C. nodosum*) from Norway to Spain; USA (Long Island) (South & Tittley, 1986).

This species appears to be an early summer ephemeral, with decaying thalli observed from August onwards. Gametophytes are extremely rare in the populations we have examined. Spermatangia have been reported in August; cystocarps in August-September; and tetrasporangia have been observed in May and July-September. Rueness (1978) found a *Polysiphonia*-type life history in an isolate from S. Norway attributed to this species; it failed to hybridize with isolates identified as '*Ceramium strictum*' from Norway and New Jersey.

Relatively little morphological variation has been observed.

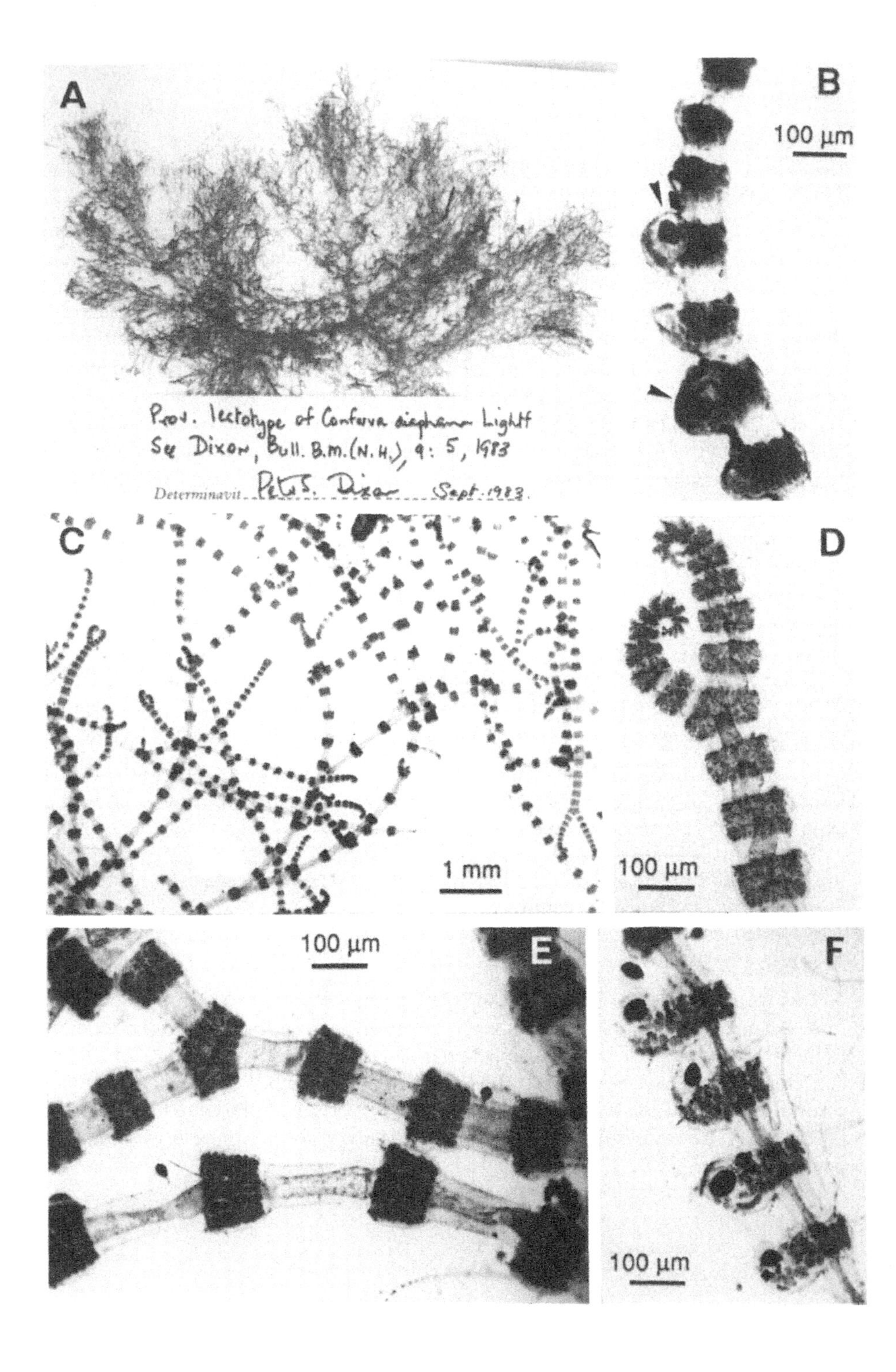
A
B
100 µm
Prov. lectotype of Conferva diaphana Lightf
See Dixon, Bull. B.M.(N.H.), 9: 5, 1983
Determinavit Peter S. Dixon Sept. 1983
C
1 mm
D
100 µm
E
100 µm
F
100 µm

Ceramium echionotum J. Agardh (1844), p. 27.

Lectotype: LD 20792. Devon (Torquay), September 1837, *Griffiths*.

Thalli densely tufted to loosely branched, 1.5-15 cm high, consisting of interwoven prostrate axes giving rise to groups of erect axes, more-or-less entangled or matted together by rhizoids at the base, usually repeatedly pseudodichotomously branched, less often with a distinct main axis bearing irregularly alternate laterals; apices complanate, spreading, composed of overlapping fans of regularly pseudodichotomously branched axes; dull purplish-red in colour, fairly rigid in texture when fresh.

Main axes with strongly inrolled forcipate apices, enlarging from apical cells 16-18 µm wide, including the thick walls; axial cells initially rounded, becoming cylindrical and increasing in length to 150-670 µm, the axes increasing only slightly in diameter with age, 240-310 µm in diameter when mature, slightly to markedly constricted at nodes, with segments 0.7-2 diameters long, completely or incompletely corticated; branching pseudodichotomous at intervals of 10-15 axial cells; *adventitious branches* sparse to numerous, often in pairs, on older axes; *nodes* consisting of 8 periaxial cells that give rise to ascending and descending cortical filaments decreasing from 24-33 µm to 6 µm in diameter; cortical bands continuous or covering only about half of axial cells due to axial cell elongation, sometimes giving rise to sparse cortication of narrow ascending filaments; *spines* developing near apices on only the outer faces of nodes or in complete whorls of 5-7, angled in different directions, non-pigmented, pointed, unicellular, 50-110 µm long x 8-12 µm wide, also formed adventitiously from cortex of older axes, to 40 µm in length, abundant; *rhizoids* formed by groups of cortical cells, multicellular, branched, 16-28 µm wide, terminating in multicellular discoid pads; *plastids* ribbon-like and reticulate.

Gametophytes dioecious. *Spermatangial sori* continuous over cortex of younger axes; spermatangia teardrop-shaped, 6 µm x 4 µm, releasing spherical spermatia 3-4 µm in diameter. *Cystocarps* consisting of 1-3 globular gonimolobes 215-335 µm in diameter, containing numerous angular carposporangia 44-60 µm in diameter, enclosed in a membrane 12 µm thick, surrounded when mature by a whorl of 6-7 simple, inrolled, involucral branchlets from subtending axial cell. *Tetrasporangia* formed on younger axes, 1-2 per segment, on outer face of axis, initially entirely covered by cortical cells that may bear numerous spines, causing nodal swelling when mature, mostly covered by ascending

---

Fig. 17. *Ceramium diaphanum*

(A) Lectotype in BM (no locality or date; width = 13 cm). (B) Tetrasporangia (arrows) borne singly, protuberant, covered by cortical filaments (as A). (C) Matted irregularly branched axes (C-F Donegal, July). (D) Inrolled apices with dentate outer faces. (E) Older axes, with collar-like cortical bands composed of equal ascending and descending narrow cortical filaments. (F) Tetrasporangia borne singly, protruding beyond cortical bands and covered by cortical filaments.

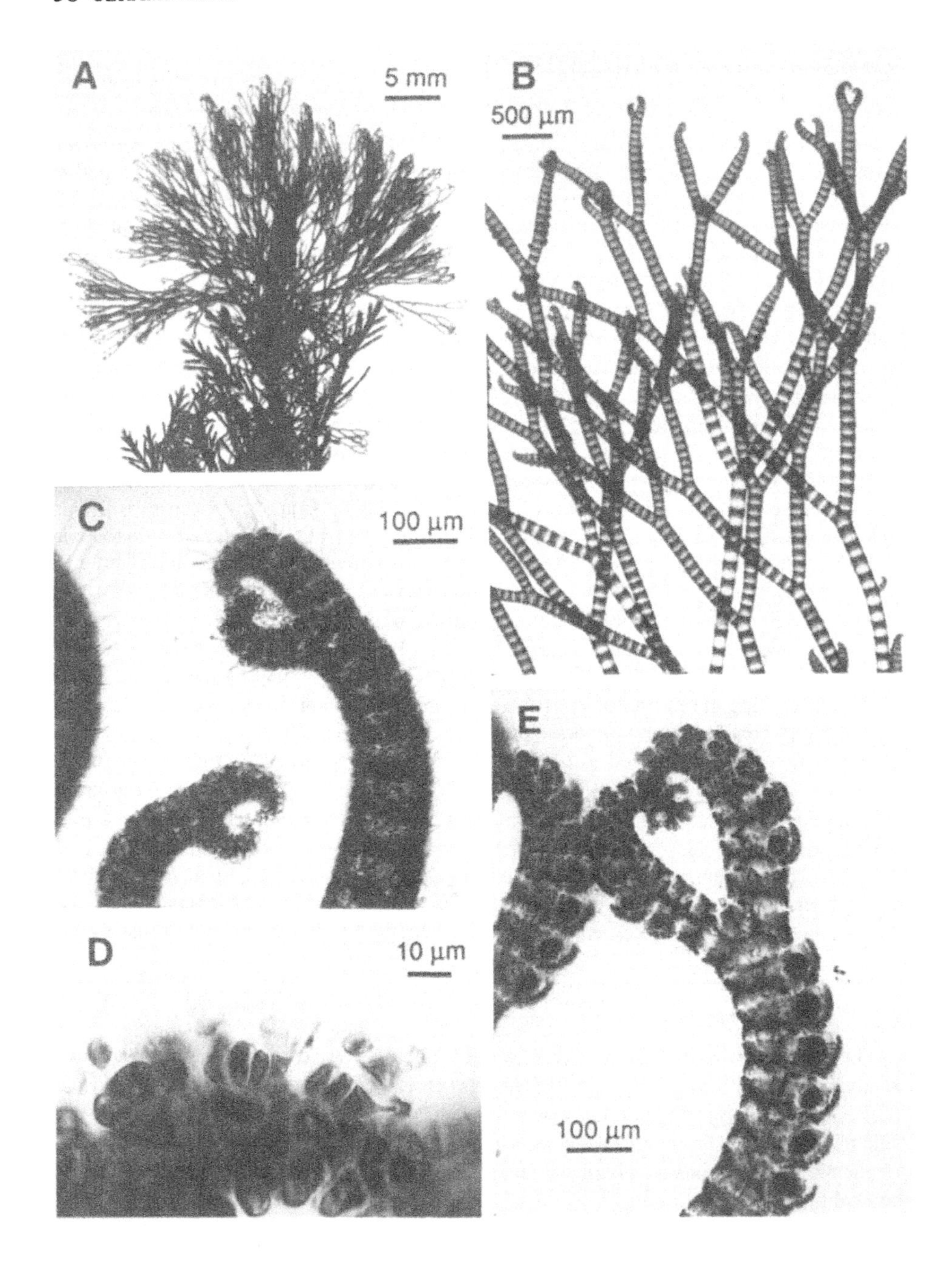
A
5 mm
B
500 µm
C
100 µm
D
10 µm
E
100 µm

cortical filaments, spherical to subspherical, 64-80 μm in diameter, including the thick walls.

Epiphytic on various algae and epilithic on bedrock, in pools and damp places in the mid- to lower intertidal and subtidal to 10 m depth, often on mobile substrata such as shell gravel, at extremely sheltered to moderately wave-exposed sites, sometimes with exposure to strong tidal currents.

Widely distributed on south and west coasts, northwards to Ross & Cromarty and Orkney, rarely recorded from E. Scotland and N. E. England; widely distributed in Ireland; Channel Isles.

Norway to Canaries and Azores (Ardré, 1970; South & Tittley, 1986).

Plants probably perennate by basal fragments, which have been recorded from March-December; large mature thalli occur from June-October. Spermatangia recorded in February and August; cystocarps in July-October; tetrasporangia in June-October.

C. echionotum shows considerable variation in habit, from small, turf-forming plants on open rock, to loose, open plants growing in pools, but vegetative and reproductive morphology are relatively constant.

Ceramium flaccidum (Kützing) Ardissone (1871), p. 40.

Lectotype: L 940.265.55. Syntypes: L; BM; TCD (see Womersley, 1978). Clare (Kilkee), *Harvey*.

Hormoceras flaccidum Kützing (1862), p. 21, pl. 69 a-d.

Ceramium gracillimum sensu Harvey (1834), p. 303, non *Ceramium gracillimum* (Kützing) Zanardini (1847), p. 223.

Thalli 2-12 cm high, attached by prostrate axes that forming spreading or irregularly shaped tufts of erect axes <0.2 mm in diameter, consisting of numerous main axes bearing alternate arrangements of densely branched laterals, obviously complanate when young; bright red in colour; older axes appearing banded to the naked eye, with an extremely flaccid, diaphanous texture.

Main axes with incurved apices, c. 50 μm in diameter when young; axial cells initially diamond-shaped in surface view, cylindrical when mature, 120 μm in diameter; axes incompletely corticate; branching distinctly alternate every 5-6 segments, with incurved

---

Fig. 18. *Ceramium echionotum*

(A) Habit, epiphytic on *Corallina* sp. (Devon, Nov.). (B) Pseudodichotomously branched young axes with incurved apices (Donegal, July). (C) Apices showing abundant unicellular spines (Donegal, Nov.). (D) Carpogonial branches on outer face of young axis (Clare, July). (E) Tetrasporangia protruding on adaxial sides of axes, covered by cortical filaments (spines not visible) (Donegal, Dec.).

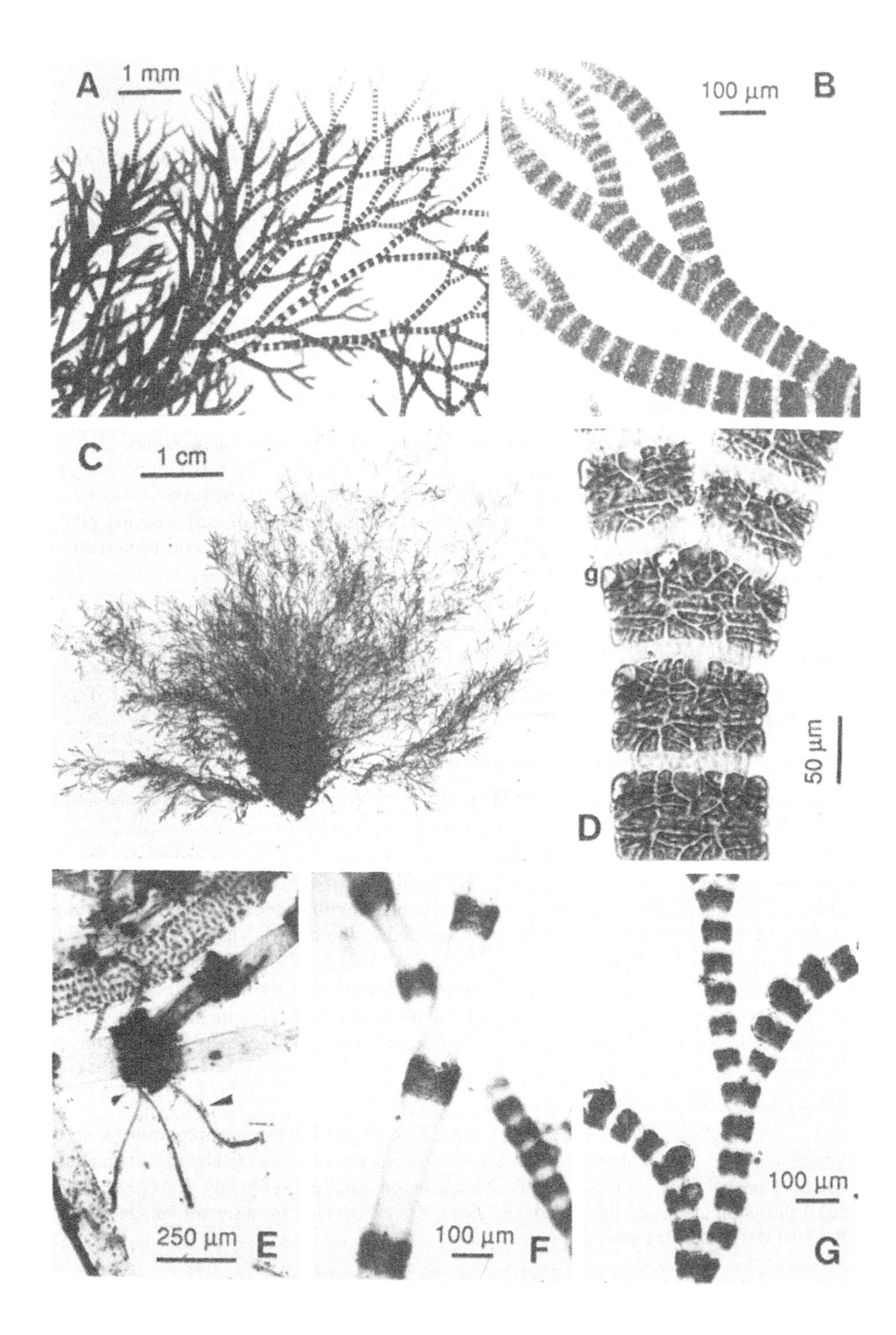
A
1 mm
100 µm
B
C
1 cm
g
50 µm
D
250 µm
E
100 µm
F
100 µm
G

young laterals, branched again to several orders in the same pattern at an angle of c. 30°; *adventitious branches* sparse; *nodes* consisting of 6 periaxial cells that each give rise to 2 ascending cortical filaments and 1 descending filament, the first cell of which cuts off a few more cells basipetally and either remains as wide as periaxial cell, or divides transversely, obscuring the pattern of cortical filaments; angular gland cells 15-25 µm frequent in ascending filaments; young cortical bands level with outer walls of internodal axial cells, later protruding slightly, 5-8 cells long, the upper and lower margins remaining sharply defined; *internodes* 8-12 diameters long, little pigmented; *rhizoids* numerous, unicellular, non-pigmented, 20-25 µm in diameter; *plastids* reticulate in cortical cells, ribbon-like to filiform in axial cells.

Gametangial material not available. *Tetrasporangia* formed on branchlets of last 2-3 orders of branching, causing these to become recurved as they mature, 1-3 per cortical band, usually only on adaxial side of the node, protruding laterally and partially covered by cup-like ascending cortical filaments, spherical, 40-80 µm in diameter including walls.

Epiphytic on a wide variety of algae, and epilithic on bedrock and pebbles, on damp surfaces near extreme low water and in pools up to mid-tide level, occasionally on strongly wave-exposed shores, more frequently at moderately to extremely sheltered sites, particularly those with some tidal currents; occurring sporadically but usually in abundance.

South coast of England, from Cornwall eastwards to Sussex, and in Norfolk; Cork, Clare; Channel Isles.

England to Canaries and Cape Verde; Mediterranean; S. Australia; probably cosmopolitan in cold temperate to tropical seas (Ardré, 1970; Womersley, 1978).

Small plants have been collected in April and May; rapid growth results in the presence of large plants in June-November, but these very rarely bear reproductive structures. Populations probably reproduce largely by fragmentation. Spermatangia recorded for September (Buffham, 1888); cystocarps in September (Harvey, 1848); and tetrasporangia in July and September. Chromosome number is $n = 42$ (Rao et al., 1978).

In the British Isles, this species shows little variation in morphology other than in overall size. Elsewhere, it appears to be more variable, particularly in the presence or absence of specialized hairs (Womersley, 1978).

Fig. 19. *Ceramium flaccidum*

(A) Part of young plant with short internodes and dense branching (Devon, Nov.). (B) Apex of main axis with alternate branches (as A). (C) Habit of extremely delicate mature plant (Dorset, July). (D) Detail of cortication, showing single, branched, descending cortical filaments cut off by periaxial cells, and abundant gland cells (g) (as A). (E) Unicellular rhizoids (arrow) (Channel Isles, Sep.). (F) Young and older axes showing increase in length of internodes (as C). (G) Tetrasporangia borne singly, protruding adaxially (as E).

Ceramium gaditanum (Clemente) Cremades in Cremades & Pérez-Cirera (1990), p. 489.

Lectotype: MA-Algae 2115 (Cremades, 1993, fig. 3). Spain (?Cadiz) (see Cremades & Pérez-Cirera, 1990, and Cremades, 1993).

Conferva gaditana Clemente (1807), p. 322.
Ceramium flabelligerum J. Agardh (1844), p. 27.

Thalli loosely to densely tufted, 2.5-15 cm high, consisting of prostrate axes giving rise to groups of erect axes, either with dominant main axes bearing irregularly alternate major laterals, or composed of numerous much-branched, equal-sized axes, occasionally entangled or matted together by rhizoids, dull brownish-red in colour and fairly cartilaginous.

Main axes with forcipate but not strongly inrolled apices, enlarging from apical cells 12-16 μm wide including the thick walls, increasing greatly in diameter with age, 170-350 μm in diameter when mature, constricted at nodes, usually completely corticate but appearing banded due to varying density of cortical cells; branching pseudodichotomous at intervals of 8-12 segments; *adventitious branches* sparse to numerous, often in pairs, with basal constrictions; *nodes* consisting of 5-6 periaxial cells that give rise to ascending and descending cortical filaments of angular cells 6-14 μm in diameter, frequently forming rounded gland cells 8-14 μm wide; *spines* developing near apices, typically one per segment, in a row on the outer face of the axis, also formed adventitiously in irregular whorls in older axes, slightly pigmented, conical, 36-44 μm long x 18-28 μm wide, 3-4 cells long, the basal cell 0.7-1 diameter long and other cells isodiametric, with a pointed terminal cell 1-2 diameters long; *internodes* occasionally present between cortical bands; *rhizoids* arising from groups of cortical cells, multicellular, branched, 16-32 μm wide, terminating in multicellular discoid pads; *plastids* discoid in cortical cells, becoming ribbon-like and reticulate in axial cells.

Gametophytes dioecious. *Spermatangial sori* continuous on younger axes; spermatangia ellipsoid, 5-6.5 x 5 μm. *Cystocarps* consisting of 2-4 globular or slightly lobed gonimolobes 216-336 μm in diameter, containing numerous angular carposporangia 26-40 μm in diameter, surrounded by a membrane 16 μm thick, partially enclosed when mature by a whorl of 2-5 simple or branched involucral branchlets from subtending axial cells. *Tetrasporangia* borne on younger axes in whorls of 3-7, covered by cortical cells, causing swelling of nodes as they enlarge, spherical to ellipsoid, 48-68 μm long x 40-60 μm wide, including the thick walls.

Epiphytic on various algae and epilithic on bedrock, in pools and damp places in mid- to lower intertidal, never subtidal, at moderately wave-exposed sites.

Widely distributed except in northern Scotland, occurring northwards to Ross & Cromarty and E. Sutherland; records from Ireland sparse but widely distributed; Channel Isles.

British Isles, Netherlands to Spain; Canaries, ?Azores (Price et al., 1986; South & Tittley, 1986).

Fig. 20. *Ceramium gaditanum*

(A) Habit (Pembroke, Dec.). (B) Part of plant showing irregular branching and slightly incurved apices (as A). (C) Apex of male thallus showing sparse, relatively inconspicuous, multicellular spines (arrow) (Pembroke, undated). (D) Cortex of fully corticated axis (as C). (E) Tetrasporangia borne in whorls, partly exposed (as A).

Large thalli occur throughout the year. Spermatangia recorded in July; cystocarps in May-October and December; tetrasporangia in January-March and May-December. An isolate from Sutherland derived from tetraspores gave rise in culture to gametophytes that formed cystocarps after fertilization (Edwards, 1973). Dixon (1965) reported that this species does not appear to form reproductive structures near its northern limits, yet the majority of young plants in these areas were derived from spores.

Ceramium nodulosum (Lightfoot) Ducluzeau (1806), p. 61.

Lectotype: BM-K. ? Hampshire (labelled 'Milford'), undated.

Conferva nodulosa Lightfoot (1777), p. 994, non *Conferva nodulosa* Hudson (1778), p. 600.
Conferva rubra Hudson (1762), p. 486.*
Ceramium rubrum (Hudson) C. Agardh (1811), p. 17.

Thalli 3-30 cm high, consisting of irregularly shaped tufts of one to several erect axes attached by a dense mass of multicellular rhizoidal filaments; base of main axes decumbent and attached by a felt-like rhizoidal mass, giving rise to numerous lateral branches that become erect, branching extremely irregular, either irregularly dichotomous or with dominant main axes, not normally complanate, often very bushy; axes 0.5-1 mm in diameter basally, sometimes swollen or nodular with internodal constrictions when older, dull brownish-red in colour; young axes fairly soft but not delicate, rarely flaccid but not fragile, usually becoming cartilaginous when older.

Main axes with straight or forcipate to slightly inrolled apices; axial segments increasing from 0.7-1.0 to 1.5 diameters long; axes either remaining cylindrical or becoming markedly constricted at intervals due to pyriform shape of axial cells, typically fully corticate, rarely incompletely corticate; laterals developing at irregular intervals of (6-)10-18 segments; *adventitious branches* infrequent to common in sterile plants, often abundant in fertile thalli, especially females; *nodes* consisting of 6, 6-7 or 7 periaxial cells, each giving rise

Fig. 21. *Ceramium nodulosum*

(A) Habit of typical large plant (A-C Down, Nov.). (B) T.S. of axis with 7 periaxial cells. (C) Tips of fully corticated plant showing irregular branching intervals (e.g. 9-12 segments) and slightly incurved apices. (D) Axes with cortical bands separated by ecorticate internodes. Note irregular lower margins of cortical bands (arrow) resulting from secondary growth of cortical filaments (Down, June). (E) Tetrasporangia completely immersed in fully corticated axes (Devon, Nov.). (F) Cystocarps, in this example lacking involucral branches (as A).

* *Conferva rubra* Hudson is a superfluous name because Hudson cited *Fucus cartilagineus* Linnaeus in synonymy (see Dixon, 1983); we have selected a lectotype in BM.

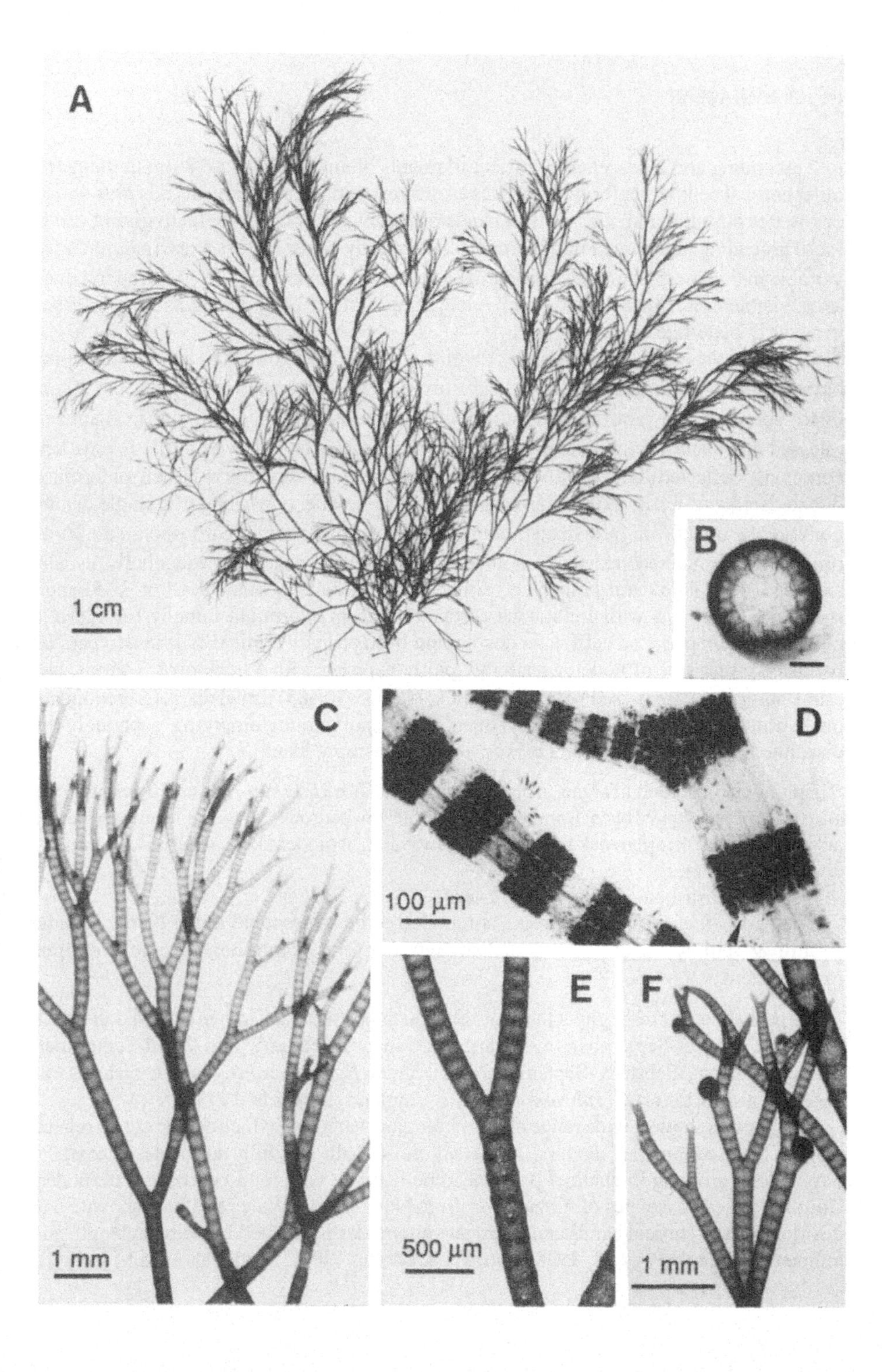
A
1 cm
B
C
1 mm
D
100 µm
E
500 µm
F
1 mm

to 2 ascending and 2 descending cortical filaments of angular cells 8-18 μm in diameter; outer cortical cells either forming rosettes around expanded inner cortical cells near nodes, or rosettes obscure or absent; cortical bands of incompletely corticate thalli giving rise to loose ascending and descending filaments that entirely cover mature axes; mature cortex variable in thickness, sometimes obscuring articulations; *internodes* 10 μm wide in young axes, visible only when stained, rarely elongating up to 0.3-0.8 diameters long; *plastids* irregularly plate-like in cortical cells.

Gametophytes dioecious. *Spermatangial sori* initiated on adaxial side of upper branches, later continuous around axis or forming extensive raised pale patches to 20 μm thick; spermatangia protruding through outer membrane, ellipsoid, 6 μm in diameter; released spermatia 4.5 μm in diameter. *Cystocarps* either lateral, with fertile female axis commonly deflected about 45° away from them, sometimes exposing axial cell, or terminal due to deciduous fertile axis; gonimoblasts globose, sessile, consisting of a single central gonimolobe c. 400 μm in diameter, containing numerous angular carposporangia 30-45 μm in diameter, the secondary gonimolobes suppressed or developing sequentially, usually surrounded by involucral branchlets, either 3-5 long banded branchlets, or 3(-5) short spine-like branchlets with continuous cortication. *Tetrasporangia* initially formed in a nodal ring from periaxial cells, later developing from primary cortical cells in two parallel bands on either side of node or scattered, partly exposed with a thick outer wall or, less often, completely immersed within the cortex, (40-)55-70(-85) μm in diameter, secondary ones usually smaller than primary ones, each sporangium emptying separately by dissolution of outer wall leaving behind a pore and empty locule.

Epiphytic on perennial algae, such as *Fucus* spp., *Corallina* spp., *Mastocarpus stellatus* and *Laminaria hyperborea* fronds, and epilithic on bedrock, in pools from mid-shore downwards, and on open rock near extreme low water, at moderately sheltered to extremely wave-exposed sites.

Generally distributed in the British Isles.

The '*Ceramium rubrum* complex' is reported to be widespread in the North Atlantic (South & Tittley, 1986), but the distribution of its component species requires reassessment.

Large thalli occur throughout the year. Spermatangia recorded in January, March-April, June and August-September; cystocarps in January-February and April-September; tetrasporangia in February-September. A regular *Polysiphonia*-type life history was obtained in isolates of '*C. rubrum*' from N.E. England (Edwards, 1973).

This species shows a wide range of morphological variation, which appears to be related in part to environmental factors. Whereas most thalli are fully corticate, occasional populations growing in shallow pools become banded, with long ecorticate internodes. Culture studies on isolates of '*C. rubrum*' from Nova Scotia, Wales and Italy showed that development of cortical bands and ecorticate internodes is affected by both daylength and temperature (Garbary *et al.*, 1978; Cormaci & Motta, 1987). Under short (8 h) days the isolates were fully corticate, whereas under long (16 h) days, axial cell elongation resulted in ecorticate internodes. These findings raised doubts about the taxonomic distinctions between *C. rubrum* and other species such as *C. rubriforme* Kylin and *C. areschougii* Kylin. (It should be noted, however, that some of these cultures may have been of another *Ceramium* species.) Whether cortication develops primarily upwards or downwards is variable in Irish material, although the direction of growth has previously been used as an important feature in species identification (e. g. Newton, 1931). Curvature of the apices also varies: in general, apices are forcipate rather than inrolled, but some thalli appear to have slightly inrolled apices. The morphological limits of this species require further investigation.

Ceramium pallidum (Nägeli ex Kützing) Maggs & Hommersand, comb. nov.

Lectotype: L 940.264.427. Syntypes: L, BM, etc. Devon (Torquay), 1845, *Nägeli.*

Trichoceras pallidum Nägeli ex Kützing (1849), p. 680.
Ceramium pennatum J. Agardh (1851), p. 136, non *Ceramium pennatum* (Hudson) Roth (1800a), p. 171.
Ceramium armoricum P. Dixon & H. Parkes (1968), p. 83.

Thalli 3-12 cm high, consisting of fan-shaped or cylindrical tufts of several erect axes attached by a dense mass of multicellular rhizoidal filaments, with base of main axes decumbent and attached by felt-like rhizoids; erect axes 0.3-0.7 mm in diameter at the base, either with dominant, zig-zag, main axes bearing laterals at regular intervals, or branching pseudodichotomously, with branch angles of 45-55°; branching complanate and flabellate, or spiral around main axes, or bushy and pyramidal; first-order laterals branched regularly pseudodichotomously or alternately to at least 5 orders of branching; adventitious branches usually inconspicuous; dull brownish-red in colour, bleaching to yellow-brown, varying in texture from extremely flaccid and delicate to fairly robust and brittle, older axes becoming cartilaginous.

Main axes with strongly inrolled apices; axial segments increasing from 0.8 to 1-2 diameters long, either remaining cylindrical or becoming markedly constricted at intervals due to pyriform shape of axial cells, forming laterals at intervals of 4-8(-10) segments, typically 6-7 segments, either fully corticate or with ecorticate internodes; *adventitious branches* formed with variable frequency on young axes, usually less than one per node, initially inrolled; *nodes* consisting of 6-7 periaxial cells, each of which gives rise to 2 ascending and 2 descending cortical filaments of angular cells 6-12 μm in diameter, which may extend rapidly to form a continuous cortex, with rosette patterns around inner cortical cells near nodes, variable in thickness when mature, or remain separated by ecorticate internodes, resuming growth in mature axes, which are entirely covered by loose ascending and descending filaments; *internodes* 10 μm wide in young axes, sometimes elongating to up to 0.7-1.7 diameters in length, transparent and non-pigmented; *plastids* irregularly plate-like to reticulate in cortical cells, filiform in axial cells.

Gametophytes dioecious or monoecious. *Spermatangial sori* continuous over or

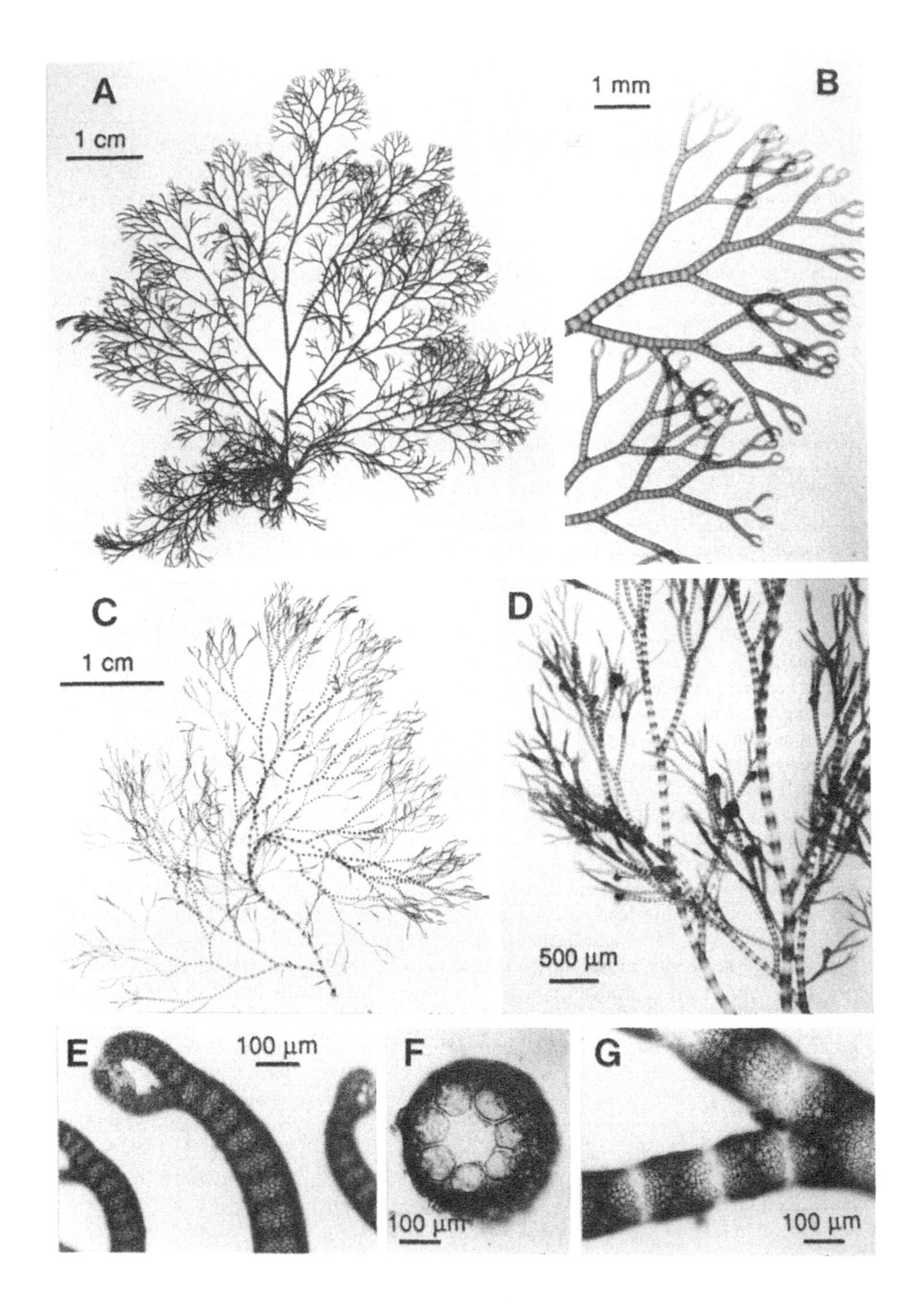
A
1 cm
1 mm
B
C
1 cm
D
500 µm
E
100 µm
F
100 µm
G
100 µm

forming extensive pale patches on young axes; spermatangia ellipsoid, 2-3 µm in diameter. *Cystocarps* consisting of 2-3 globular or slightly lobed gonimolobes 400-600 µm in diameter, containing numerous angular carposporangia 30-45 µm in diameter, partially enclosed when mature by a whorl of 2-5 straight or incurved, simple or branched involucral branchlets with distinct ecorticate internodes. *Tetrasporangia* borne in whorls at each node on younger axes, immersed in cortex, formed in wide bands in mature axes, causing swelling of nodes as they enlarge, spherical when mature, 65-90 µm in diameter.

Epiphytic on larger perennial algae, such as *Fucus* spp., *Chorda filum*, *Himanthalia elongata*, *Mastocarpus stellatus*, and *Laminaria hyperborea* fronds, in upper- and mid-shore pools and near extreme low water, also growing on permanently submerged floating structures such as marina pontoons, at moderately to extremely wave-exposed sites, and at sites sheltered from wave action but exposed to strong currents, frequently at the edges of tidal rapids.

Widespread and common on all coasts of the British Isles.

This species appears to be widespread in the North Atlantic, known under various synonyms, and probably also occurs in the Mediterranean, but further assessment of its distribution is required.

Large thalli occur throughout the year. Spermatangia recorded in January-May, July-October and December; cystocarps throughout the year; tetrasporangia in January-May and July-November.

This species shows an extraordinary range of morphological variation, which appears to be related in part to environmental factors. On open coasts, thalli tend to have dominant main axes bearing alternate or spiral arrangements of short laterals; in tidal rapids and at sheltered sites plants are usually dichotomously branched. Cortical development varies from fully corticate to banded with long ecorticate internodes; in the latter form, additional cortication is either largely ascending or descending. The characteristic inrolled apices occur only on vegetative and male axes; mature tetrasporangial axes straighten out at the tips, and cystocarpic axes branch rather less regularly.

Ceramium secundatum Lyngbye (1819), p.119.

Lectotype: C. Faroes (Hoyviig), 27 vi 1817, *Lyngbye* (J. Rueness, pers. comm.).

Fig. 22. *Ceramium pallidum*

(A) Habit of plant growing in moderate wave exposure (Down, Nov.). (B) Part of same fully corticated plant, showing regularly alternate branching every 6-9 segments (as A). (C) Habit of plant from wave-sheltered site (Donegal, May). (D) Axes with ecorticate internodes from sheltered site (D-G as A). (E) Incurved apices of fully-corticate thallus. (F) T.S. of axis with 7 periaxial cells. (G) Cortical bands showing irregular upper and lower margins, separated by ecorticate internodes.

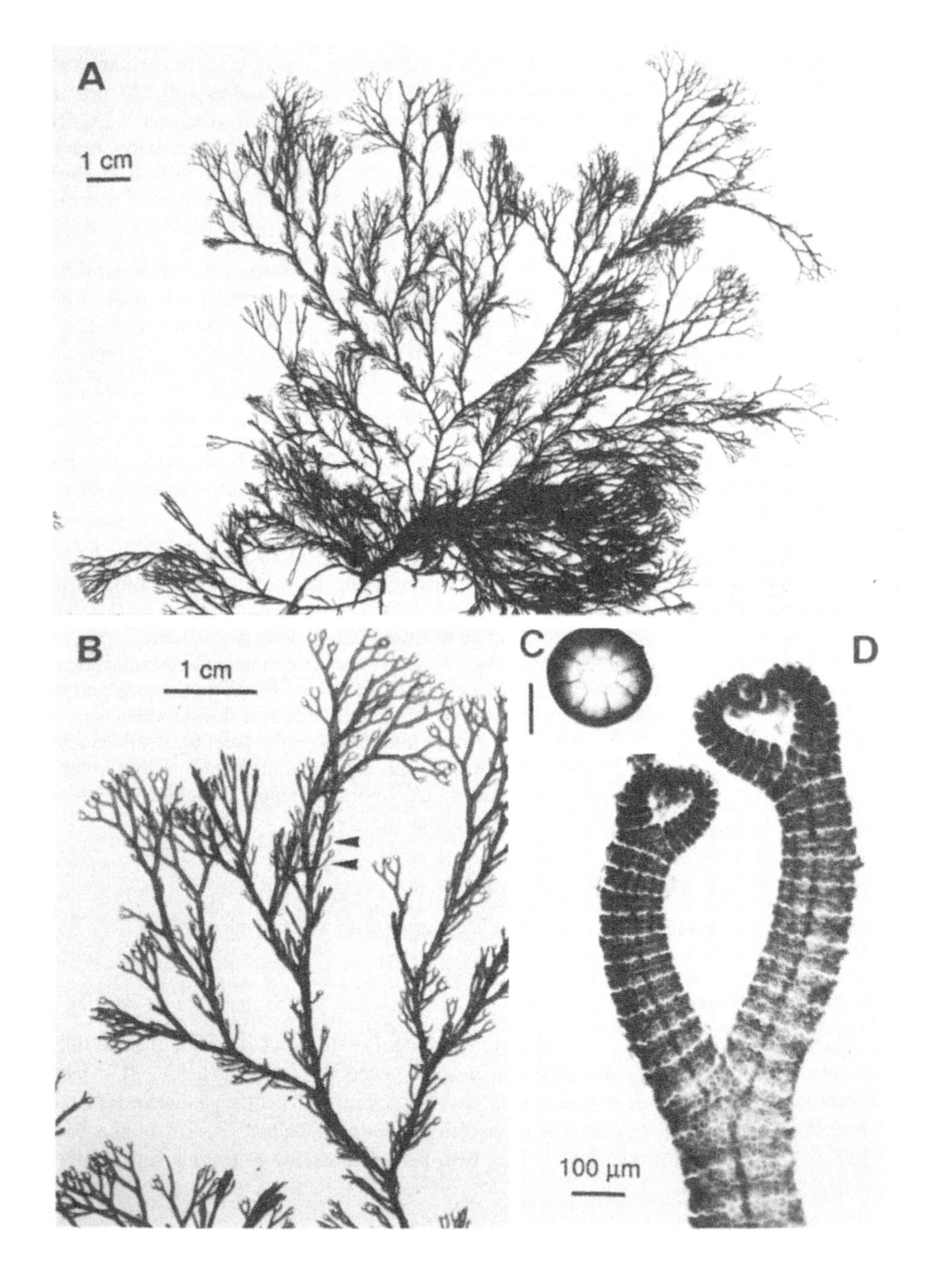
A
1 cm
B
1 cm
C
D
100 µm

Thalli 5-14 cm high, consisting of fan-shaped or cylindrical tufts of several erect axes attached by a dense mass of multicellular rhizoidal filaments c. 25 µm in diameter; erect axes 0.4-1.0 mm in diameter at the base, initially branching pseudodichotomously at 45° angles but often appearing to have a distinct main axis due to unequal development of the branches, usually complanate and flabellate initially; adventitious branches often formed later near apices and increasing in number during maturation of the thallus, sometimes eventually clothing and partly obscuring main axes, dull brownish-red in colour, brittle when fresh, becoming cartilaginous when older.

Main axes with strongly inrolled apices, consisting of segments increasing from 0.8 to 1-2 diameters long, usually becoming markedly constricted between nodes, completely corticated, appearing banded when young because periaxial and large inner cortical cells are visible through outer cortex; lateral branches pseudodichotomous, formed at intervals of 10-15 segments; *adventitious branches* developing abundantly on young axes, often several per node in a whorl, initially tightly inrolled, later branching and increasing to up to 500 µm in diameter; *nodes* consisting of 7-8(-9) periaxial cells, giving rise to ascending and descending cortical filaments of angular cells 8-30 µm in diameter; gland cells not observed; outer cortical cells arranged when mature into rosette-like or reticulate patterns around inner cortical cells near nodes and elongate over internodes; *internodes* 4-6 µm wide, only visible in young axes when stained, occasionally wider in rapidly-growing branches; *plastids* plate-like to reticulate in cortical cells.

Gametophytes dioecious. *Spermatangial sori* forming a continuous cover or extensive pale patches on young axes; spermatangia ellipsoid, 6 x 3-4 µm. *Cystocarps* consisting of 2-3 globular or slightly lobed gonimolobes 350-600 µm in diameter, containing numerous angular carposporangia 30-55 µm in diameter, partially enclosed when mature by a whorl of 1-5 straight or incurved, simple or branched involucral branchlets from subtending 1(-2) axial cells, often banded with distinct ecorticate internodes. *Tetrasporangia* borne in whorls at each node on younger axes, covered by cortical cells, causing swelling of nodes as they enlarge, spherical, 50-60 µm in diameter when mature.

Epiphytic on larger red algae, such as *Gracilaria* spp. and *Polysiphonia elongata*, less often on smaller perennial species, and on *Laminaria hyperborea* fronds, common in submerged positions on artificial materials such as marina pontoons and mooring buoys, occasional to frequent in deep sandy lower-shore pools, large populations occurring from 1 m to at least 11 m depth in sandy areas, such as *Zostera marina* beds, at wave-sheltered to moderately wave-exposed sites, exposed to currents at the sheltered sites.

Fig. 23. *Ceramium secundatum*

(A) Habit of typical plant with fan-like branching (Donegal, Feb.). (B) Young axes showing fan-like branching and abundant adventitious branches (arrows) (as A). (C) T.S. of axis with 8 periaxial cells (Cork, Apr.). (D) Tips of plant showing inrolled apices and narrow cortical gaps after staining (as C).

Recorded from Hampshire, Dorset, Devon, Cornwall, Pembroke; Cork, Kerry, Clare, Mayo, Donegal, Antrim and Down; probably widely distributed.

Wider distribution cannot be assessed at present.

Large thalli occur throughout the year. Spermatangia recorded in February-March, June, September and December; cystocarps in March, April, June-September and December; tetrasporangia in February-April, July-September and December.

The characteristic inrolled apices occur only on vegetative and male axes; mature tetrasporangial axes straighten out at the tips, and cystocarpic axes branch less regularly.

Ceramium shuttleworthianum (Kützing) Silva (1959), p. 64.

Lectotype: L 940.265.127. Ireland (see Dixon, 1960b).

Acanthoceras shuttleworthianum Kützing (1842), p. 739.
Ceramium ciliatum β *acanthonotum* Carmichael ex Harvey in W. J. Hooker (1833), p. 336.
Ceramium acanthonotum (Carmichael ex Harvey in W. J. Hooker) J. Agardh (1844), p. 2.
Ceramium acanthonotum var. *coronata* Kleen (1874), p. 19.

Thalli densely tufted, 0.5-4 cm high, sometimes forming extensive turfs, consisting of interwoven prostrate axes giving rise to much-branched, entangled erect axes that reattach by secondary holdfasts, bright red to reddish-brown in colour, soft but not flaccid.

Main axes with forcipate but not strongly inrolled apices, enlarging from apical cells 14-16 μm wide, including the thick walls, to cylindrical axial cells 100-145 μm wide and 1.3-5 diameters long; axes incompletely corticate, enclosed in a tough acid-resistant wall; branching regularly pseudodichotomous at intervals of 6-10 segments; *adventitious branches* frequent, particularly in damaged thalli; *nodes* consisting of 4 periaxial cells that give rise to ascending and descending cortical filaments 3-8 cells long, decreasing in diameter from 20 to 6 μm, frequently forming gland cells (Dixon, 1960b); cortical bands reaching a maximum width of 85-120 μm; *spines* formed on cortical bands near apices, typically one per band in a row on the outer face of the axis, alternating in position from one side of a median line to the other, less often in whorls of 2-4, also formed adventitiously on mature cortical cells, pigmented, conical, sometimes curved adaxially or abaxially, 48-60 μm long x 36-40 wide, the base multicellular, merging into the cortical band, narrowing to terminal row of 2-3 cells; *rhizoids* formed by periaxial cells and groups of cortical cells, multicellular, branched, 22-54 μm wide, terminating in multicellular discoid pads; *plastids* discoid in cortical cells, becoming ribbon-like and reticulate in axial cells.

Gametophytes dioecious. *Spermatangial sori* developing on younger axes, almost

---

Fig. 24. *Ceramium shuttleworthianum*

(A) Habit (Antrim, June). (B) Detail of thallus (Pembroke, Dec.). (C) Apices, with conspicuous multicellular spines merging basally into cortical bands (Antrim, Nov.). (D) Cystocarps with sparse involucral branches (as B). (E) Whorls of tetrasporangia partially protruding from each cortical band (as C).

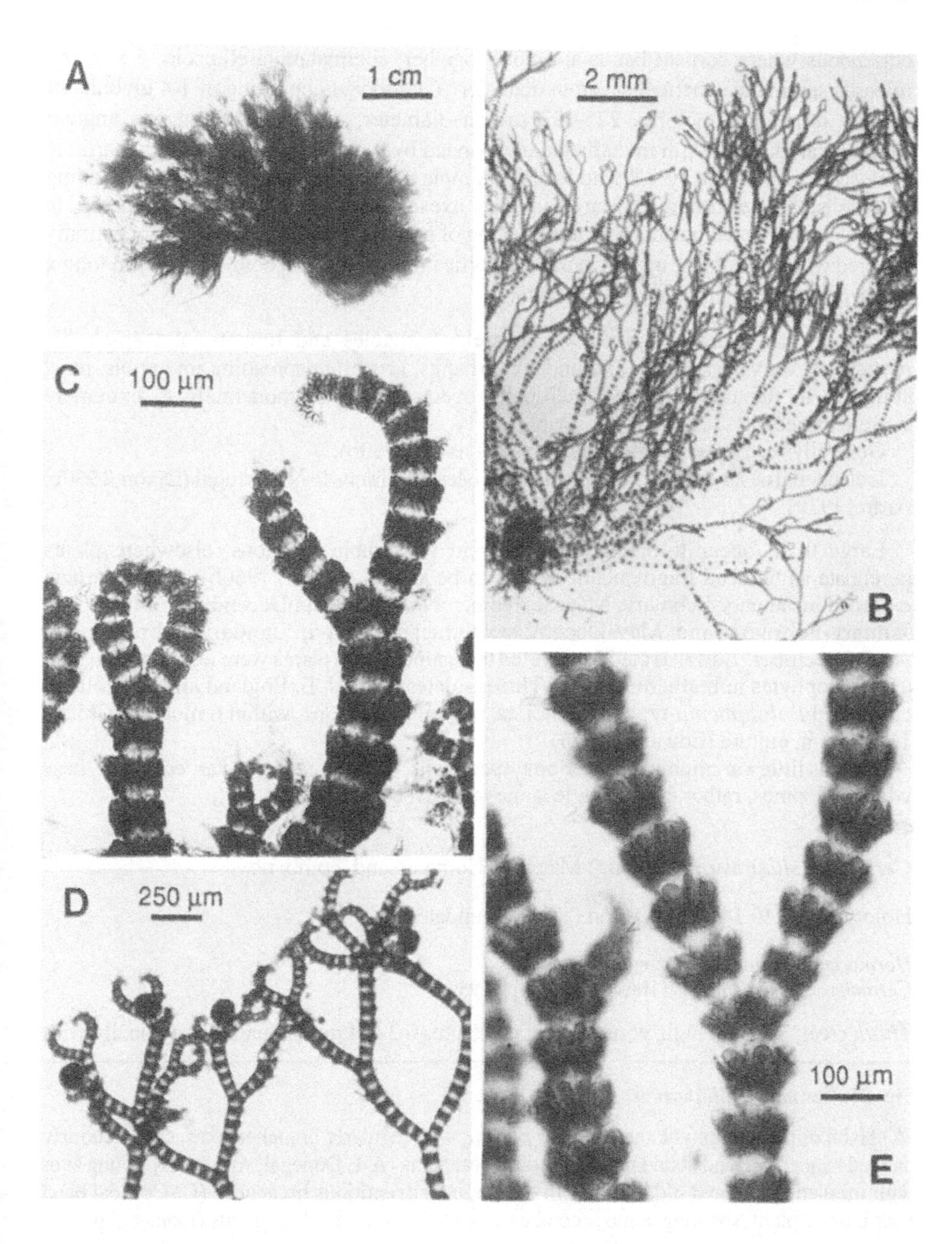
A
1 cm
2 mm
B
C
100 µm
D
250 µm
100 µm
E

continuous where cortical bands are close together; spermatangia ellipsoid, 4 x 3 µm; released spermatia spherical, 3 µm in diameter. *Cystocarps* consisting of 1-4 globular or slightly lobed gonimolobes 225-332 µm in diameter, containing numerous angular carposporangia 36-48 µm in diameter, surrounded by a membrane 16 µm thick, partially enclosed when mature by 1-2 whorls of 2-4 simple involucral branchlets from subtending axial cells. *Tetrasporangia* borne on younger axes in whorls of 1-8, initially immersed in cortical bands, causing distortion and swelling of nodes as they enlarge, and only partially covered when mature by upward-growing cortical filaments, ellipsoid, 50-100 µm long x 20-48 µm wide, including the thick walls.

Epilithic on bedrock, epizoic on mussels, barnacles and limpets, and epiphytic, particularly on vertical faces and under overhangs, less often spreading into pools, most abundant in the upper to mid-intertidal, never subtidal, at moderately to extremely wave-exposed sites.

Generally distributed in the British Isles (Dixon, 1960b).

Iceland, Faroes, N. and S. Norway, British Isles, N. France to N. Portugal (Dixon, 1960b, Ardré, 1970).

Large thalli occur throughout the year in favourable locations, elsewhere plants perennate as reduced fragments or appear to be annual (Dixon, 1960b). Spermatangia recorded in January-February, May-September, November and December; cystocarps in January-February and May-December; tetrasporangia in January-February and April-December. Edwards (1973) reported that gametangial plants were less common than tetrasporophytes in northern Britain. Three isolates from N. E. England and E. Scotland completed *Polysiphonia*-type life histories, without deviations, within 6 months at 15°C, 16 h days in culture (Edwards, 1973).

There is little variation in habit. Some specimens, corresponding to var. *coronata*, bear whorls of spines, rather than a single spine, on each cortical band.

Ceramium siliquosum (Kützing) Maggs & Hommersand, comb. nov.

Holotype: L 940.142.282. Devon (Torbay), undated.

Hormoceras siliquosum Kützing (1847), p. 35.
Ceramium diaphanum sensu Harvey (1848), pl. 193.

Thalli erect, 3-18 cm high, consisting of erect axes 0.2-0.4 mm in diameter, typically with

---

Fig. 25. *Ceramium siliquosum*

(A) Habit of plant epiphytic on *Ahnfeltia plicata*, with regularly branched axes, conspicuously banded major axes, and abundant adventitious branching (A-B Donegal, Aug.). (B) Young axes with inrolled apices and older axis with developing adventitious branches. (C) Cortical band near base of plant, showing some secondary elongation of cortical filaments (Dorset, Apr.).

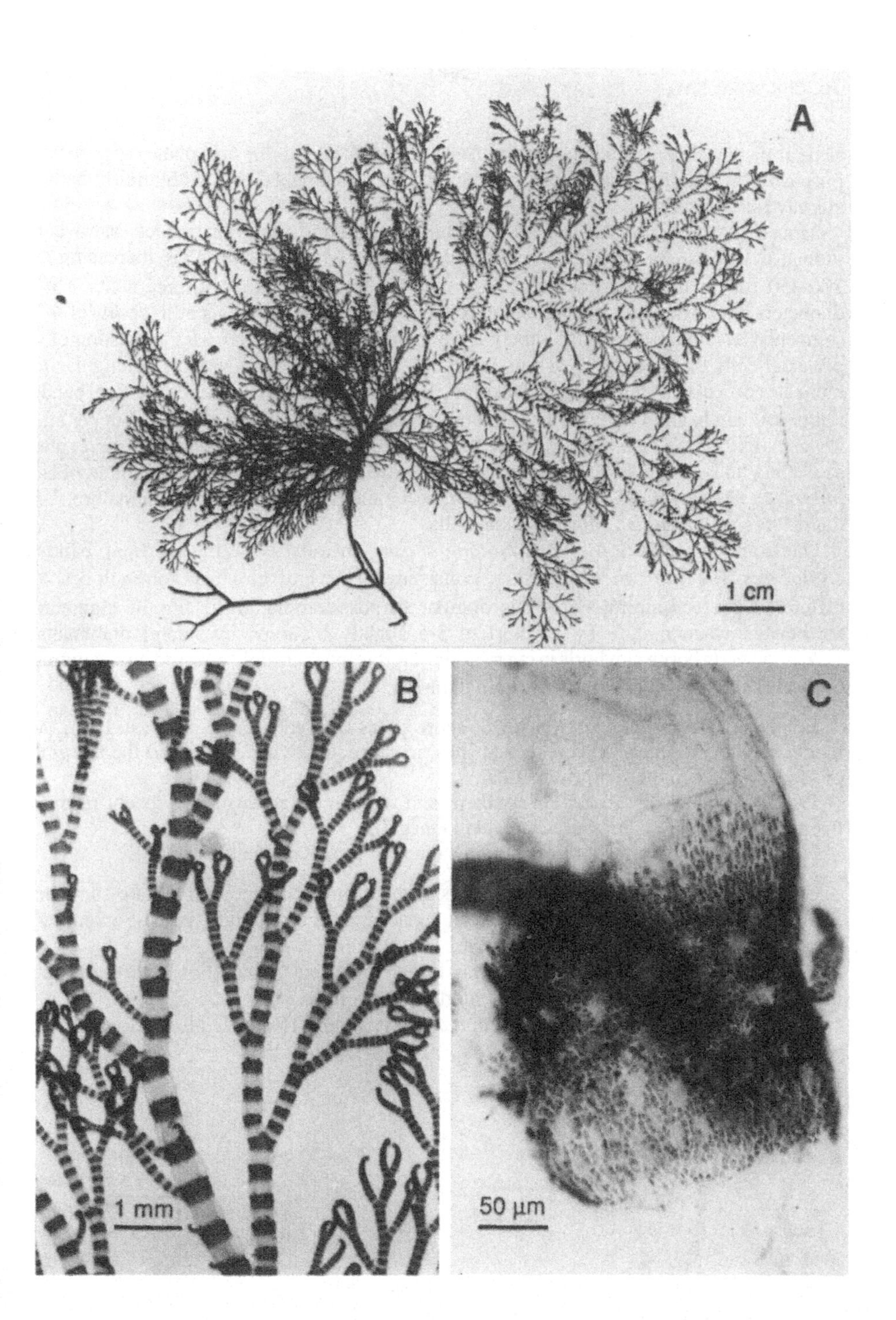
A
1 cm
B
1 mm
C
50 µm

several distinct main axes bearing regularly alternate laterals in one plane, rose-red to purplish in colour, the older axes distinctly banded to the naked eye, with a fairly harsh, slightly rigid texture.

Main axes with tightly inrolled apices; axial cells diamond-shaped in surface view when young, cylindrical or occasionally slightly pyriform when mature; axes increasing to 200-450 μm in diameter and remaining incompletely corticate, with segments 1.5-3 diameters long when mature; branching regularly pseudodichotomous at intervals of 4-7 segments; *adventitious branches* abundant on young and old axes; *nodes* consisting of 8 periaxial cells that each give rise to ascending and descending filaments of angular to elongate cortical cells that form rosette arrangements over the periaxial cells; cortical bands not protruding beyond outer walls of internodal axial cells, c. 300 μm long, the upper and lower margins remaining sharply defined except near the base of main axes where irregular ascending and descending secondary cortical growth may almost cover the internodes; *internodes* strongly pigmented when young, becoming little-pigmented when mature, 3-6 diameters long; *plastids* filiform in axial cells.

Gametophytes dioecious. *Spermatangial sori* entirely covering cortical bands. *Cystocarps* 300-450 μm in diameter, consisting of several globular gonimolobes of different ages containing numerous angular carposporangia to 45 μm in diameter, surrounded when mature by a whorl of 3-5 slightly incurved involucral branchlets. *Tetrasporangia* borne in whorls in every node of fertile axes, entirely covered by ascending cortical filaments, spherical, c. 60 μm in diameter.

Epiphytic on smaller algae in lower-shore pools and subtidal to at least 12 m, at moderately to extremely wave-exposed sites, most frequently found around the margins of sandy patches in the shallow subtidal.

Probably widely distributed on southern and western coasts (we have recent records from Dorset, Devon, Cornwall, Clare and Donegal).

N. France, N. Spain; wider distribution cannot be assessed at present.

Subtidal populations of this species appear to be most conspicuous in late summer. Spermatangia have been observed in April; cystocarps in April and August-September; and tetrasporangia in August-September.

This species apparently shows relatively little morphological variation, except that the internodes are unusually short in some intertidal collections.

Many collections attributed to *C. diaphanum*, including Harvey's illustrations (1848, pl. 193), are actually of this species.

Ceramium strictum sensu Harvey (1849b), p. 163.*

Typical specimen: TCD. Kerry (Dingle), undated, *Harvey*.

Non *Ceramium strictum* Roth (1806), p. 130, nec *Ceramium strictum* (Kützing) Harvey (1849b), p. 163.

Thalli erect, 1-12 cm high, consisting of entangled axes giving rise to dense hemispherical to cylindrical tufts of erect axes 0.2 mm in diameter, repeatedly pseudodichotomously branched; younger axes dull purple in colour, bleaching to yellowish; older axes less pigmented, distinctly banded to the naked eye, with a fairly harsh, slightly rigid texture.

Main axes with tightly inrolled apices, the outer faces of which are slightly dentate or smooth; axial cells diamond-shaped in surface view when young, cylindrical or occasionally slightly pyriform when mature, increasing to 120-200 µm in diameter, incompletely corticate; branching pseudodichotomous at intervals of 6-12 segments, branch angles 30-45°; *adventitious branches* abundant on young and old axes; *nodes* consisting of 6-7 periaxial cells that each give rise to 2 ascending and 2 descending filaments of angular cortical cells that cover periaxial cells when mature; ascending and descending filaments equal in length, decreasing to 4-10 µm in diameter; gland cells not observed; cortical bands protruding only slightly beyond outer walls of internodal axial cells, 100-120 µm long, the upper and lower margins remaining sharply defined; *internodes* little pigmented when mature, 3-6 diameters long; *rhizoids* numerous, multicellular, non-pigmented, 15-40 µm in diameter; *plastids* filiform in axial cells.

Gametophytes dioecious. *Spermatangial sori* entirely covering cortical bands, spermatangia ellipsoid, 4 x 2.5 µm; released spermatia spherical, 3 µm in diameter. *Cystocarps* consisting of 3-5 globular gonimolobes of different ages, up to 180 µm in diameter, containing numerous angular carposporangia 20-35 µm in diameter, surrounded when mature by a whorl of 4-6 simple or branched incurved involucral branchlets. *Tetrasporangia* borne in whorls in every node of fertile axes, entirely covered by ascending cortical filaments, spherical to ellipsoid, 45-70 x 40-70 µm.

Epiphytic on smaller algae such as *Polysiphonia atlantica*, epizoic on mussels and other molluscs, in upper-shore pools and in open damp situations near low water of neap tides, at moderately wave-sheltered to extremely wave-exposed sites, sometimes abundant in these habitats.

Dorset, Devon, Cornwall, Pembroke; Kerry, Clare, Galway, Dublin; probably widely distributed on southern coasts but literature records difficult to assess owing to taxonomic confusion.

Reported from Norway to France and Azores and from the western North Atlantic (South & Tittley, 1986), but this requires reassessment.

---

* A valid name has not yet been determined for this species.

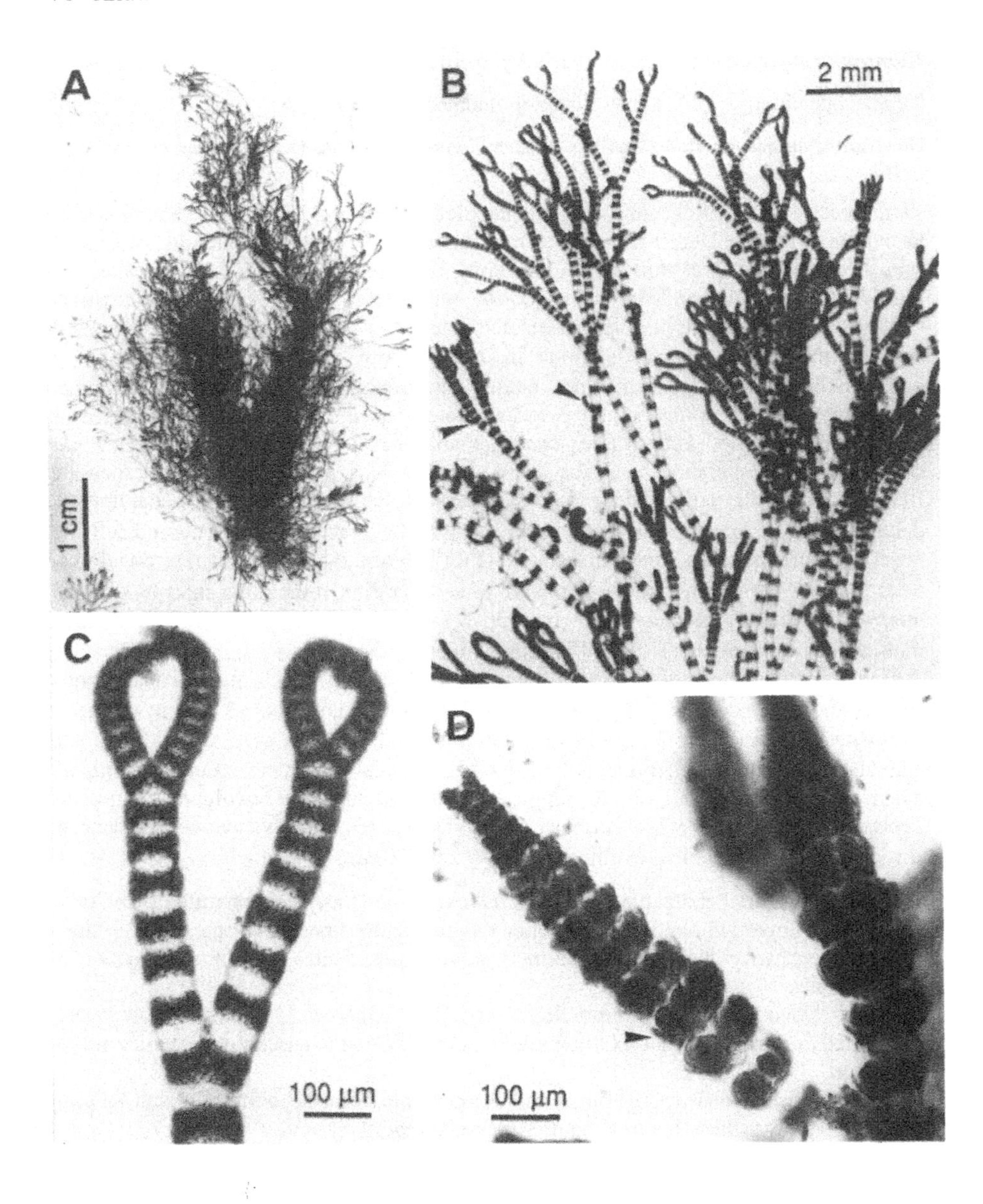
A
B
2 mm
1 cm
C
100 µm
D
100 µm

Spermatangia have been observed in April, June and July; cystocarps in July; and tetrasporangia in July. Rueness (1973, 1978) found *Polysiphonia*-type life histories in isolates from S. Norway and from New Jersey that were both attributed to this species but were non-interfertile. Futher studies are required to determine whether either of these entities is conspecific with *C. strictum* sensu Harvey from the British Isles. A parasporangial life history was also observed in a Norwegian isolate (Rueness, 1973), but parasporangia have not been noted in the British Isles.

Relatively little morphological variation has been observed. Tetrasporangial axes show little elongation of axial cells, and thus appear to be almost fully corticate, with cortical bands separated by very narrow internodes. Harvey (1850) described the apices as straight or slightly hooked, whereas we have found that growing tips are strongly incurved; it is possible that Harvey made his observations on axes with developing tetrasporangia, which are usually straight.

Ceramium strictum (Kützing) Harvey is based on *Gongroceras strictum* Kützing (1842), p. 735 (= *Ceramium deslongchampii).* Feldmann-Mazoyer (1941, p. 302) designated a Harvey specimen in PC as type, but this neotype has no status because the holotype is in L. As yet, we have not been able to determine whether there is an existing valid name for *C. strictum* sensu Harvey.

MICROCLADIA Greville

MICROCLADIA Greville (1830), pp. 1, 99.

Type species: *M. glandulosa* (Solander ex Turner) Greville (1830), p. 99.

Thalli subcylindrical or compressed, corticated throughout, erect to 40 cm high, decumbent, or prostrate, attached to solid substratum by a fibrous holdfast, or epiphytic, attached by rhizoids penetrating into host tissue; primary axes straight or zig-zag, alternately branched in one plane, branch angles narrow to wide; adventitious branches present or absent, often developing in plane of the alternate branches.

Growth monopodial, by oblique division of apical cells, lateral branches sometimes overtopping the thallus apex; first periaxial cell cut off abaxially, the rest in alternating (rhodomelacean) sequence; each periaxial cell bearing two acropetal and two basipetal cortical initials that develop cortical filaments primarily in acropetal direction; cortex 1-4 cells thick, the surface layer continuous, composed of angular cells, usually including some terminal, darkly staining 'gland' cells, covered by a soft or firm, cartilaginous outer

Fig. 26. *Ceramium strictum* sensu Harvey (1851), pl. 334.

(A) Habit (A-D Kerry, July). (B) Part of tetrasporangial thallus branched pseudodichotomously and with adventitious branches (arrows). (C) Incurved apices, with cortical bands protruding very little from ecorticate internodes. (D) Tetrasporangial branchlets, with whorls of tetrasporangia in each node, covered by ascending cortical filaments (arrow).

membrane; adventitious branches, when present, usually originating from surface cortical cells; cells uninucleate with enlarged nuclei in axial cells; plastids parietal, reticulate and ribbon-like.

Gametophytes dioecious. Spermatangia formed in superficial sori, originating from surface cortical cells and protruding through outer membrane. Procarps abaxial, on first periaxial cells in upper branches, developing as in *Ceramium* before and after fertilization; gonimoblasts abaxial, usually surrounded by 3-7 involucral branches; terminal and lateral gonimolobes maturing successively, or the lateral gonimolobes reduced. Tetrasporangia immersed, arising initially from periaxial cells, later from cortical cells, or originating from cortical cells only and exserted, tetrahedrally or irregularly cruciately divided.

References: Feldmann-Mazoyer (1941), Hommersand (1963), Stegenga (1986), Gonzales & Goff (1989).

Microcladia is distinguished from *Ceramium* primarily by its consistent alternate-distichous branching, forwardly-directed cortical filaments, and continuous surface layer of relatively small, angular cells. As presently constituted, *Microcladia* appears to be heterogeneous.

One species in the British Isles.

Microcladia glandulosa (Solander ex Turner) Greville (1830), p. 99.

Lectotype: BM-K. Unlocalized, undated.

Fucus glandulosus Solander ex Turner (1802), p. 82, pl. 38.
Delesseria glandulosa (Solander ex Turner) C. Agardh (1822), p. 182.

Thalli attached by small discoid holdfast composed of matted rhizoidal filaments, consisting of small groups of erect or spreading decumbent axes that reattach to the substratum and develop tufts of new axes, 3-10 cm high, complanate, with an irregularly flabellate or rounded outline, composd of compressed axes increasing in width upwards from about 0.3 mm near the holdfast to a maximum of 0.8-1.5 mm, narrower towards apices; main axes zig-zag, repeatedly alternate-distichously or subdichotomously branched to 6 orders, bearing short, irregularly alternate, incurved branches; branch angles narrow to wide; laterals rarely overtopping main axes; adventitious branchlets rare in non-reproductive thalli, common in tetrasporangial plants; bright red to brownish-red in colour, main axes becoming cartilaginous.

Main axes growing from dome-shaped apical cells 16-20 μm in width; young apices forcipate; axial cells in mature axes rounded in surface view, oval in T.S., about 250 μm wide; branching pseudodichotomous, formed at intervals of 6-10 axial cells; *periaxial cells* 6-8, the first abaxial, the rest irregularly alternate, with the second either to right or left of the first, and the last adaxial, expanding to same size as the first, remaining recognizable in mature axes in T.S., each giving rise to 4 ascending cortical filaments, those forming the 'wings' more branched than the others; *cortication* soon complete, consisting of one or more layers of large inner cortical cells and an outer cortex of angular cells 20-38 μm in diameter, with abundant gland cells, through which the axial cells are visible, all enclosed in an acid-resistant wall 6 μm thick; *rhizoids* formed from groups of cortical cells, multicellular, 10-40 μm wide, terminating in multicellular discoid attachment pads, coalesced into solid secondary holdfasts; *plastids* elongate, becoming ribbon-like and reticulate.

Spermatangia formed in irregular patches on young axes, developing from outer cortical cells that give rise to spermatangial mother cells bearing spermatangia 3 μm in diameter (Buffham, 1896). *Cystocarps* sessile on abaxial margins of upper branchlets, enclosed by 2-3 simple involucral branchlets [details sparse as material is lacking]. *Tetrasporangia* initiated in abaxial rows on upper branchlets, singly or in pairs from sides of first periaxial cells, spreading around axis, then arising from periaxial and cortical cells, abundant on adventitious branchlets, immersed, completely covered by cortex, spherical when mature, 40-50 μm in diameter, divided tetrahedrally or irregularly cruciately.

Epilithic on bedrock in deep shaded lower-shore pools, epiphytic on larger algae and on bedrock from extreme low water to 10 m depth, at moderately to extremely wave-sheltered sites, with slight to strong exposure to tidal currents.

South coast of England: Sussex, Hampshire, Devon, Cornwall, including Isles of Scilly; widely distributed in Ireland, recorded from Dublin, Wicklow, Cork, Mayo; Channel Isles. Records from eastern England dubious.

England to Morocco; Mediterranean (Ardré, 1970).

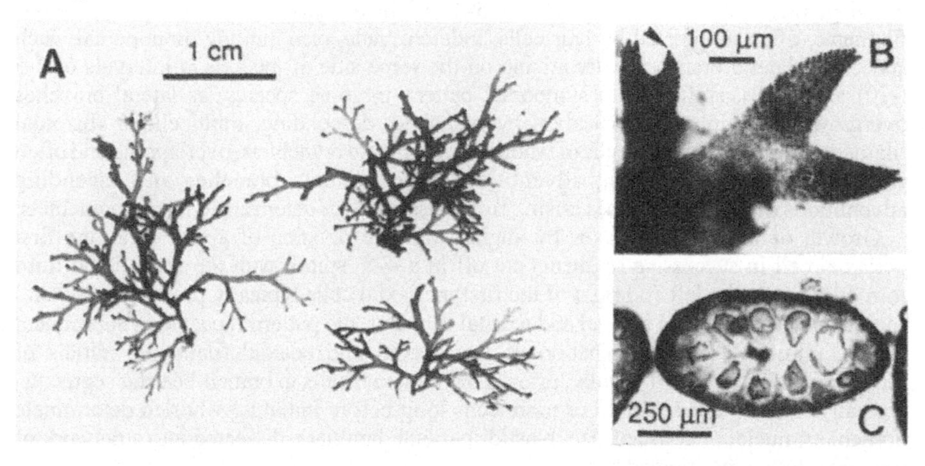

Fig. 27. *Microcladia glandulosa*

(A) Habit of 3 plants (A-C Cornwall, July). (B) Apices growing from conspicuous single apical cells (arrow). (C) T.S. of axis, showing axial cell surrounded by periaxial cells.

Large thalli occur throughout the year. Spermatangia recorded in August; cystocarps in April, May, July and September-November; tetrasporangia in June, August-October and December.

Thalli with narrow branches may resemble *Ceramium* species, from which they can be distinguished by the compressed rather than cylindrical axes. *Callophyllis cristata* (see Vol. 1, 2A), is occasionally confused with *M. glandulosa*, but it is multiaxial whereas *M. glandulosa* has conspicuous single apical cells.

Tribe CROUANIEAE Schmitz (1889), p. 451.

CROUANIA J. Agardh

CROUANIA J. Agardh (1842), p. 83.

Type species: *Crouania attenuata* (C. Agardh) J. Agardh (1842), p. 83.

Thalli growing in tufts, erect, up to 7 cm high, attached by basal rhizoids, or differentiated into prostrate and erect axes; erect axes terete, some with annular constrictions, each axial cell bearing a whorl of 3 determinate lateral branchlets (whorl-branches) embedded in a mucilaginous matrix and composed of 5-8 orders of quadri-, tri-, and dichotomous filaments, often terminated by hair cells; indeterminate axes initially monopodial, each successive lateral branch usually arising on the same side of the axis at intervals of 2-5 (-10) axial cells, shifting to a sympodial pattern in some species, as lateral branches overgrow the main axis; cortication by branched, descending, multicellular rhizoidal filaments issuing from basal cells of whorled determinate branchlets, overlapping and often attaching to the substratum; adventitious indeterminate branches and ascending adventitious rhizoids sometimes arising from basal cells of determinate lateral branchlets.

Growth of indeterminate axes by slightly oblique division of apical cells, the first periaxial cell in successive segments cut off in a 40° spiral, with the second and third formed 120° to the left and right of the first; periaxial cells typically producing 2 lateral filaments followed by an abaxial and a distal filament, the pattern repeated in subsequent orders, ultimately with elimination of the abaxial and adaxial filaments; initials of indeterminate axes cut off distally, prior to the periaxial cells in branch-bearing segments, typically forming chains 10-15 or more cells long before initiating whorled determinate branchlets; nuclei of constant size, plastids parietal, laminate, dissected into a network of fine, longitudinally oriented bands.

Gametophytes dioecious. Spermatangial mother cells terminal, solitary or in clusters on whorled determinate branchlets, each bearing clusters of 2-3 spermatangia. Procarps initiated near apices on main axes or lateral branches which cease further growth, usually

1 per fertile segment and consisting of a supporting cell (fertile periaxial cell) and a 4-celled, curved carpogonial branch; sterile groups absent; auxiliary cell cut off distally from the supporting cell after fertilization and containing a large proteinaceous body; the fertilized carpogonium cutting off a terminal cell and a latero-posterior connecting cell that fuses with a process from the auxiliary cell; auxiliary cell dividing into a foot cell containing a haploid and a diploid nucleus, and a gonimoblast initial containing a diploid nucleus and protein body; gonimoblast initial elongating and cutting off a distal primary gonimoblobe initial and 1-2 (-3) lateral gonimolobe initials which develop successively into globose clusters of carposporangia; fusion cell absent, the pit connections broadening between the axial cell, supporting cell, foot cell and gonimoblast initial; cystocarps terminal on branches that may be displaced laterally by new branches formed lower down; carposporangia liberated through a pore in the mucilaginous sheath. Tetrasporangia nearly spherical, sessile, produced successively on the upper sides of basal cells of whorled determinate branches, 1-several (-10) per cell in upper parts of the plant, tetrahedrally divided.

References: Feldmann-Mazoyer (1941); Wollaston (1968); Coomans & Hommersand (1990).

One species in the British Isles.

Crouania attenuata (C. Agardh) J. Agardh (1842), p. 83.

Lectotype: LD 20263 (see Dixon, 1962a). France (Brittany), *Bonnemaison*.

Mesogloia attenuata C. Agardh (1824), p. 51.
Crouania bispora P. Crouan & H. Crouan (1848), p. 374.

Thalli consisting of spreading prostrate axes giving rise to dense tufts of erect axes 1-5 cm high, tapering from 0.3-0.4 mm in diameter to 0.15 mm near apices; young axes cylindrical, becoming annular when older, branched irregularly alternate-distichously to irregularly spirally, to 3 orders of branching; apices dark red in colour; older axes red-brown, soft and mucilaginous but not slimy in texture.

Main axes growing from apical cells 10 µm in diameter; axial cells cylindrical, increasing to 24-56 µm in diameter and 3-4 diameters in length, each bearing a whorl of 3 whorl-branches, successive whorls becoming separated by elongation of axial cells; *whorl-branches* consisting of a basal cell 12-14 µm diameter, 1-1.3 diameters long, bearing 4 branchlets terminally in a quadrichotomous arrangement, each of these branching quadrichotomously from the first cell, successive orders of branching trichotomous, then dichotomous; ultimate branchlets 3-4 cells long with cylindrical terminal cells 4-5 µm in diameter that have bluntly conical apices, often terminating in hairs; *indeterminate laterals* arising from young axial cells, in a position distal to the insertion of the whorl-branchlets, at intervals of 4-10 segments; *rhizoids* formed by basal cells of whorl-branches, multicellular, 8-25 µm in diameter, growing around axes and attaching to substratum; *plastids* elongate to filiform in axial cells.

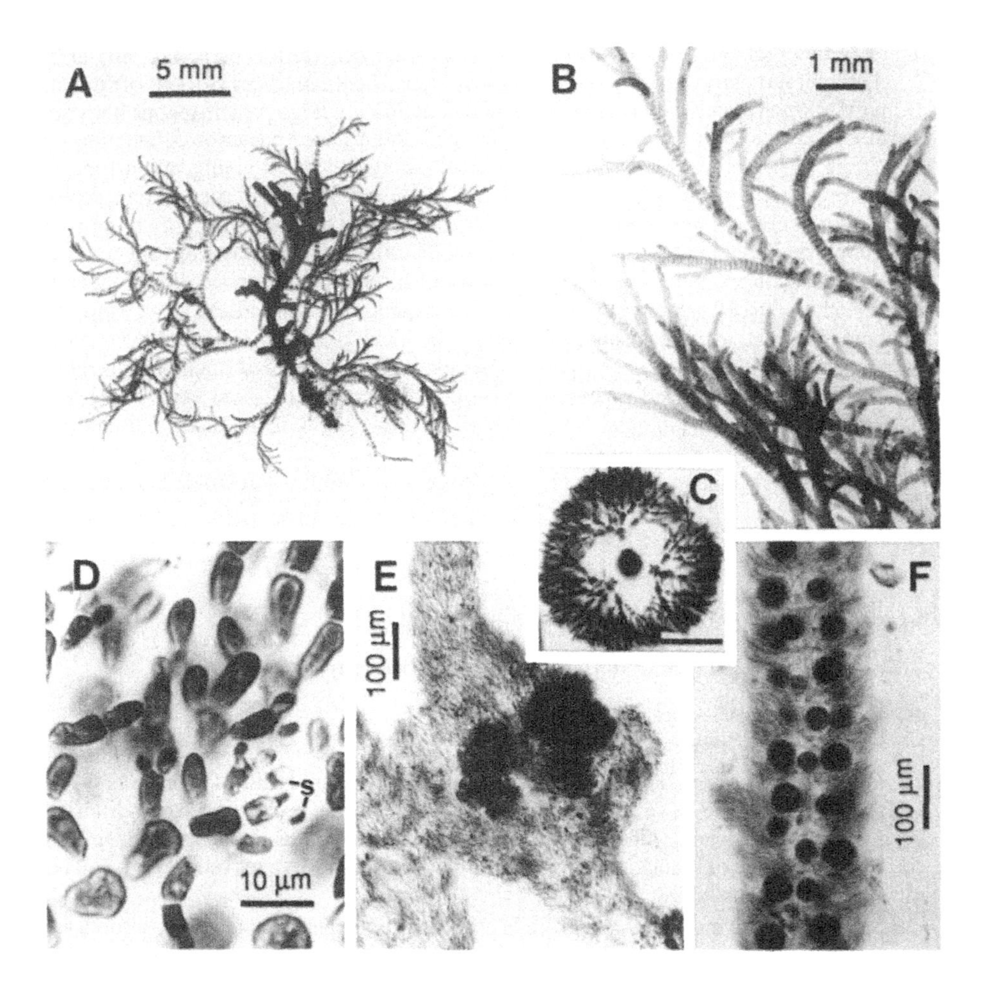

Fig. 28. *Crouania attenuata*

(A) Habit, epiphytic on *Laurencia truncata* (A-C Cork, Nov.). (B) Tips of thallus, showing alternate branching from main axes; axial cells visible between whorls of whorl-branches in older axes. (C) T.S. of axis, showing axial cell bearing whorl of 3 much-branched filaments. (D) Outer cortical filaments terminating in spermatangial mother cells bearing spermatangia (s) (Channel Isles, Sep.). (E) Cystocarp immersed in swollen axis (as D). (F) Tetrasporangia in whorls (Cork, Aug.).

Spermatangia formed on mature axes, the spermatangial mother cells developing in groups of 2-3 on terminal cells of whorl-branches, elongate, 4 x 2.5 µm, each cutting off 3 ellipsoid spermatangia measuring 3 x 2 µm. *Procarps* formed near apices of lateral branches; *cystocarps* developing singly on branch, 430 µm in diameter, consisting of 3-4 rounded gonimolobes of different ages, 150-210 µm in diameter, containing numerous rounded carposporangia 32-56 µm in diameter. *Tetrasporangia* formed on young and mature axes, on the upper side of the basal cell of whorl-branches, initially one per cell, 1-2 developing later, pyriform before division, spherical when mature, tetrahedrally divided, 45-64 µm in diameter, with walls 4 µm thick.

Epiphytic on various algae in mid- to lower-shore pools on wave-exposed coasts and from extreme low water to 5-8 m depth at wave-sheltered sites, with little to moderate exposure to currents; very local in distribution but sometimes growing in abundance.

South coasts of Devon and Cornwall, N. Devon (Lundy); Wexford, Cork; Channel Isles. The few records from outside these areas require confirmation.

British Isles to Canaries; Mediterranean; West Indies; Red Sea; Indian Ocean (Ardré, 1970; Athanasiadis, 1987).

Plants have been collected in April and July-November, and apparently become reproductive only from August onwards. Spermatangia, procarps and cystocarps found in September; tetrasporangia in August and September. Bisporangia have not been found in the British Isles, although they occur in this species elsewhere (see Feldmann-Mazoyer, 1941).

There is little morphological variation.

C. attenuata could be confused with several other species that have cylindrical mucilaginous axes. *C. attenuata* is always epiphytic, whereas *Schmitzia neapolitana* (Berthold) Lagerheim ex Silva and *Platoma bairdii* (Farlow) Kuckuck (Vol. 1, 1) are found only on pebbles and shells; in both latter species the outermost whorl-branch cells are closely united into a continuous cortex in contrast to the shaggy imbricated whorl-branches of *C. attenuata*. *Calosiphonia vermicularis* (J. Agardh) Schmitz (Vol. 1, 1) also forms a continuous outer cortex, and has whorls of four branches rather than three. *Dudresnaya verticillata* (Withering) Le Jolis (Vol. 1, 2A) is commonly found epiphytic like *C. attenuata*, but it is up to 20 cm in length, forms whorls of four whorl-branches, and has abundant rhizoidal filaments obscuring the axial filament.

Tribe CALLITHAMNIEAE Schmitz (1889), p. 450.

AGLAOTHAMNION Feldmann-Mazoyer

AGLAOTHAMNION Feldmann-Mazoyer (1941), p. 451.

Type species: *A. furcellariae* (J. Agardh) Feldmann-Mazoyer (1941), p. 451 [= *A. byssoides* (Arnott ex Harvey in W. J. Hooker) L'Hardy-Halos & Rueness (1990), p. 352].

Thalli uniseriate, erect to 8 (-15) cm high, attached by a primary rhizoidal disc and commonly also by prostrate filaments bearing unicellular or multicellular rhizoids and secondary erect axes; axes spirally, alternate-distichously, or pseudodichotomously branched, ecorticate or corticated by descending rhizoids; determinate lateral branches alternately branched or the branching secund, sometimes overtopping the axis; indeterminate branches developing from tips of determinate branches and then resembling main axes; hyaline, deciduous hairs present or absent.

Growth of indeterminate axes by oblique division of uninucleate apical cells, each axial cell cutting off a single lateral initial from its distal end and elongating below; growth of determinate lateral filaments by oblique or transverse divisions of apical cells; rhizoidal filaments, when present, originating singly or in groups of 2-3, mostly from basal cells of lateral branches, growing downwards externally or within the cell wall matrix, potentially contributing to basal attachment and sometimes bearing adventitious branchlets; cells uninucleate; plastids elongate or ribbon-like at maturity.

Gametophytes dioecious or monoecious, sometimes with mixed phases present. Spermatangial filaments small-celled, clustered, usually 2 or more per cell on adaxial surface of branchlet, sometimes encircling it; spermatangia hyaline in gelatinous matrix, 1-several per spermatangial mother cell; procarp formed near apex of potentially indeterminate axis, consisting of opposite pair of periaxial cells situated at right angles to lateral vegetative filament, typically with one periaxial cell functioning as the supporting cell of a horizontal, 4-celled carpogonial branch in L-shaped, U-shaped or zig-zag arrangement; sterile groups absent. Spermatia evidently attaching first to a hair cell and then to the trichogyne; supporting cell and opposite periaxial cell both usually enlarging after fertilization and cutting off auxiliary cells; fertilized carpogonium usually dividing vertically into two cells, each of which cuts off a connecting cell that fuses with one of the auxiliary cells; connecting cell nucleus entering auxiliary cell, dividing and cutting off a persistent, residual cell [= connecting cell in the literature]; auxiliary cell dividing by incomplete septum into foot cell containing a haploid nucleus, which may divide, and a terminal gonimoblast initial that produces 1 terminal and 1 or more lateral gonimolobe initials; gonimolobes conical, spherical or irregular in shape, enclosed in a common outer envelope; gonimoblast cells mostly maturing into carposporangia. Tetrasporangia usually sessile, 1-2 (-3) per cell at the distal ends of cells of ultimate branchlets, ellipsoid, ovoid or subspherical, tetrahedrally or irregularly divided; bisporangia or parasporangia present in some species.

References: Feldmann-Mazoyer (1941), L'Hardy-Halos & Rueness (1990).

The genus *Aglaothamnion* was erected by Feldmann-Mazoyer (1941) for uninucleate species of Ceramieae that had previously been included in *Callithamnion*. The principal features characterizing *Aglaothamnion* were uninucleate vegetative cells, zig-zag or U-shaped carpogonial branches, and lobed gonimolobes. The segregation of *Aglaothamnion* from *Callithamnion* was not universally accepted, particularly by British workers (e.g. Dixon & Price, 1981). All characters listed by Feldmann-Mazoyer for *Aglaothamnion*, other than the number of nuclei, have also been found in species of *Callithamnion*, and are variable within *Aglaothamnion* (L'Hardy-Halos & Rueness, 1990). However, numbers of nuclei appear to be of phylogenetic significance, and we consider that placement of uninucleate and multinucleate species within the same genus is not justified; we therefore recognize the genus *Aglaothamnion* as distinct from *Callithamnion*.

In the most recent checklist for the British Isles, South & Tittley (1986) listed under *Callithamnion* five uninucleate species (i.e. *Aglaothamnion* species). Of these, we have confirmed the presence in the British Isles of *A. byssoides*, *A. hookeri*, *A. roseum* and *A. sepositum*; the species previously listed as *C. decompositum* is treated here as *A. priceanum*. During the course of the present research, six more *Aglaothamnion* species have been added to the flora of the British Isles. British records of *A. bipinnatum*, *A. gallicum* and *A. tripinnatum* were previously included under *C. hookeri*, while *A. diaphanum*, *A. feldmanniae* and *A. pseudobyssoides* had not been found in the British Isles. *C. scopulorum* has recently been found to be a species distinct from *A. hookeri* (Rueness & L'Hardy-Halos, pers. comm.).

The principal morphological features separating species of *Aglaothamnion* include vegetative, spermatangial, cystocarpic and sporangial characters. Identification of single specimens can be difficult, so if possible a number of thalli should be collected. Nuclei are clearly visible in live material of multinucleate species (see Fig. 44F). Formalin-preserved specimens and herbarium material soaked in formalin-seawater should be bleached in bright light, and stained lightly with aniline blue or with haematoxylin or acetocarmine nuclear stains. *A. hookeri* is extremely variable in morphology and it therefore appears several times in the following key.

KEY TO SPECIES

1 Branching alternate-distichous, obviously in one plane in at least part of thallus; apex of main axis easily identifiable, level with or overtopping its lateral branches 2
  Branching spiral; apex of main axis surrounded by and concealed by overtopping lateral branches 9
2 Basal cell of lateral branch bears **abaxial** branchlet (at least a few examples present on every specimen, see Fig. 32B)..3

Basal cell of lateral branch bears **adaxial** branchlet or lacks branchlet 4

3 Axes ecorticate; tetrasporangia ovoid, thin-walled, often borne in pairs on same cell; gametangial plants rare, unknown in British Isles.......................... ... *A. feldmanniae*

Main axes corticated, often densely; tetrasporangia subspherical, thick-walled, solitary; gametangial plants common .... *A. gallicum*

4 Basal cell (and often also second cell) of lateral branch bears adaxial branchlet (at least a few examples present on every specimen, see Fig. 31B)...................................... 5

First cell (and usually also next 3-5 cells) of lateral branch lacks branchlet 8

5 Parasporangia are the only reproductive structures, forming globose clusters, often on short adaxial branchlets .... *A. priceanum*

Parasporangia, if present, formed in addition to other reproductive structures........... 6

6 Bisporangia present (gametophytes extremely rare), thalli 2-10 mm high .... *A. diaphanum*

Bisporangia absent, gametophytes and tetrasporophytes common, thalli 1.5-7 cm high 7

7 Thalli monoecious, tetrasporangia often mixed with spermatangia; irregularly lobed parasporangium-like masses terminal on main and lateral axes .... *A. bipinnatum*

Thalli dioecious, gametangia not mixed with tetrasporangia; parasporangium-like cells borne in zig-zag pattern terminally on lateral branches .... *A. tripinnatum*

8 Thalli dioecious, spermatangial branchlets hemispherical, cushion-like; gonimoblasts rounded or irregular; tetrasporangia globose, thick-walled; intertidal only .... *A. hookeri* (including *scopulorum* form)

Thalli monoecious, spermatangial branchlets small, erect; gonimoblasts cordate [heart-shaped]; tetrasporangia ovoid, thin-walled; subtidal .... *A. bipinnatum*

9 Main axes corticated at base (see Fig. 33F) 10

Adherent cortication absent; although loose rhizoidal filaments may occur 15

10 Bisporangia present, other reproductive structures absent; cortication sparse, thalli delicate and flaccid; subtidal or damp lower shore ....*A. byssoides*

Bisporangia absent; axes strongly corticated, thalli robust 11

11 Basal cell of most first-order lateral branches bears an abaxial branchlet 12

First few cells of all first-order lateral branches lack branchlets 14

12 First-order laterals irregularly branched, often lacking branchlets on first few cells, never branched from every cell .... *A. hookeri*

First-order laterals regularly branched, most bearing a branchlet on every cell 13

13 Thalli very robust, main axes 0.4-5 mm diameter, obscured by dense spongy covering of lateral and adventitious branchlets; epilithic and epizoic on open intertidal rock, rarely on *Fucus* spp. .... *A. sepositum*

Thalli soft, main axes 0.4 mm diameter, clearly visible between lateral branches; epiphytic on smaller algae on lower shore and in pools .... *A. gallicum*

14 Ultimate branchlets strongly incurved; spermatangial branchlets erect, discrete; gonimoblasts spherical; tetrasporangia ellipsoid .... *A. roseum*

Ultimate branchlets nearly straight; spermatangial branchlets forming large spreading cushions; gonimoblasts irregularly rounded or conical; tetrasporangia globose .... *A. hookeri*

15 Main axes to 150 µm diameter; spermatangial branchlets discrete, 1-3 per cell; gonimoblasts deeply lobed, lobes pointed; damp intertidal and subtidal ... *A. byssoides*

Main axes to 75 µm diameter; spermatangial branchlets (1)2-4 per cell, in cushions; gonimoblasts spherical to slightly conical, not lobed; subtidal only........................... .... *A.. pseudobyssoides*

Aglaothamnion bipinnatum (P. Crouan & H. Crouan) Feldmann-Mazoyer (1941), p. 192.

Lectotype: CO (see Dixon & Price, 1981). France (rade de Brest), August 1850, *Crouan*.

Callithamnion bipinnatum P. Crouan & H. Crouan (1867), p. 136.

Thalli consisting of single or grouped erect axes, attached by discoid filamentous holdfast and prostrate filaments that can give rise to further erect axes; erect axes 5-13(-18) mm high and up to 23 mm wide, complanate, with single principal axis and an oval to irregularly flabellate outline, wider than high in well-developed plants, rose-pink in colour, delicate and flaccid.

Main axes enlarging from apical cells 14-16 µm wide to axial cells 130-215 µm in diameter; cells near the base 0.3-1 diameter long, cylindrical, in the middle of the plant up to 2-3 diameters long, with a median constriction; *first-order laterals* formed by oblique division of the third cell from the apex, one per axial cell, becoming level with main apical cells, borne in a regularly alternate-distichous pattern, the first two cells of each first-order lateral bearing an adaxial branchlet, the third cell an abaxial branchlet, followed by an alternate-distichous series of branchlets; *second-order laterals* generally unbranched for the first 1-2 cells, then bearing branchlets in an alternate or secund adaxial pattern, a further order of alternate branching occurring in larger plants; *parasporangium-like structures* developing terminally on branches, laterally in place of tetrasporangia, or around intercalary branchlet cells, initiated as enlarged pyriform cells, later consisting of clusters, 160-265 µm long x 240-300 µm wide, of irregularly branched chains of lightly-staining uninucleate cells up to 40 x 32 µm, surrounded by a thick wall; *rhizoidal filaments* 6-25 µm in diameter formed on main axes from basal cells of lateral branches, growing downwards within the walls as a sparse adherent cortex and merging with the holdfast, also formed on lateral axes in larger plants; *plastids* discoid to ribbon-like.

Gametophytes monoecious. *Spermatangial branchlets* 1(-2) per cell, erect, 24-32 µm long x 18-20 µm wide, consisting of 1-2 cells bearing a cluster of spermatangial mother cells, each forming 1-3 spermatangia measuring 7 x 4 µm, with a median constriction around the nucleus. *Carpogonial branches* U-shaped; *cystocarps* consisting of 3 or more gonimolobes of different ages, developing gonimoblasts distinctly cordate (heart-shaped), mature gonimoblasts cordate to triangular, up to 270 µm long x 240 µm wide, composed of angular to rounded carposporangia 20-28 µm in diameter. *Tetrasporangia* borne on

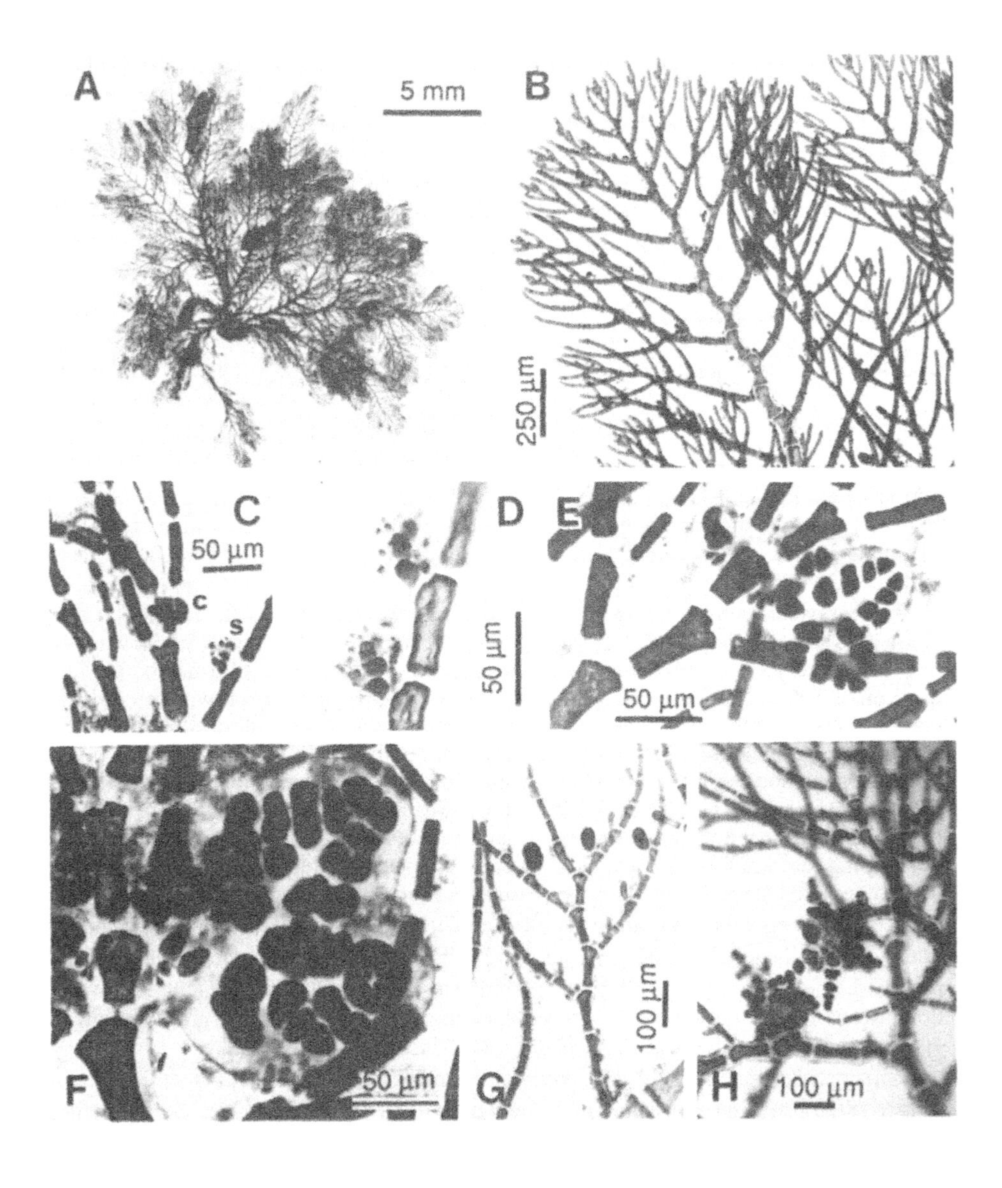
A
5 mm
B
250 µm
C
50 µm
c
s
D
E
50 µm
50 µm
F
50 µm
G
100 µm
H
100 µm

last two orders of branching, in adaxial series of up to 5 on successive cells, mostly sessile, some terminal on short branchlets or pedicellate, 1-3 of different ages per cell, pyriform to ovoid before division, ovoid when mature, with a length: width ratio of 1.3-2.0, 52-60 μm long x 28-42 μm wide, the walls 4-8 μm thick.

Epiphytic on blades of *Laminaria hyperborea* and on smaller algae, and epizoic on ascidians and hydroids from 1-15 m depth at very sheltered to moderately wave-exposed sites, most commonly near mouths of sea-lochs.

West coasts of Scotland (Argyll, Inverness, Ross & Cromarty, Sutherland) and Ireland (Clare, Galway).

Norway to N. France (Rueness & Rueness, 1980) and Atlantic Spain (Gallardo et al., 1985).

Mature thalli have been collected in April-November, the majority of them bearing a mixture of reproductive structures. Some plants were predominantly gametangial and also bore parasporangium-like structures, others were tetrasporophytes bearing spermatangia, occasional unfertilized carpogonia, and parasporangia. Spermatangia, procarps and cystocarps were present in April, May, July, and September-November; tetrasporangia in April-July and September-November; and parasporangium-like structures in April, May, July and September-November. The parasporangium-like cells are of unknown function as spores do not appear to be released from them. They resemble parasporangial clusters in other *Aglaothamnion* species (e.g. *A. priceanum*) and may be vestigial. Isolates of *A. bipinnatum* from Norway were monoecious and underwent a *Polysiphonia*-type life history (Rueness & Rueness, 1980). Some apparently non-functional gametangia developed on tetrasporophytes, and abortive tetrasporangia were formed on gametophytes.

Little morphological variation has been observed in this species, except that the first 1-3 cells of some first-order laterals may be unbranched, lacking the characteristic pattern of 2 adaxial branchlets followed by an abaxial one.

Dixon & Price (1981) considered that the status of *A. bipinnatum* required clarification with regard to *A. hookeri*. In Norway, *A. bipinnatum* is distinct and recognizable (Rueness & Rueness, 1980); we have found that British Isles material is readily distinguishable from related species.

Fig. 29. *Aglaothamnion bipinnatum*

(A) Habit of large plant (Sutherland, Sep.). (B) Apex of same plant, with basal few cells of each first-order lateral either unbranched or bearing an adaxial branchlet on basal and suprabasal cells (as A). (C) Carpogonial branch (c) and spermatangial branchlet (s) on same thallus (C-F Inverness, July). (D) Spermatangial branchlets. (E) Young, cordate (heart-shaped) gonimoblast. (F) Mature cordate gonimoblast. (G) Sessile tetrasporangia (as A). (H) Much-branched parasporangium-like clusters (as C).

Aglaothamnion byssoides (Arnott ex Harvey in W. J. Hooker) L'Hardy-Halos & Rueness (1990), p. 352.*

Lectotype: TCD (L'Hardy-Halos & Rueness, 1990, figs 2-4, 19b). Cornwall (Whitsand Bay), 1829, *Arnott*.

Callithamnion byssoides Arnott ex Harvey in W. J. Hooker (1833), p. 342.
Callithamnion arnottii Trévisan (1845), p. 77 (see Dixon & Price, 1981).
Callithamnion furcellariae J. Agardh (1851), p. 37 (see L'Hardy-Halos & Rueness, 1990).
Aglaothamnion furcellariae (J. Agardh) Feldmann-Mazoyer (1941), p. 451.

Thalli 0.5-3(-7) cm high, attached by prostrate or endophytic filaments that give rise to new axes; erect axes growing singly or in tufts, either with a distinct naked main axis below or composed of dense tufts of numerous erect axes, with a cylindrical to almost spherical outline, rose-pink to brownish-pink in colour, very delicate and flaccid.

Main axes enlarging from apical cells 10-14 μm in diameter; mature axial cells 75-150 μm in diameter and 1.5-2 diameters long, increasing in length upwards to 5 diameters; *first-order laterals* developing from the third or fourth cell behind apex, one per axial cell, surrounding and overtopping main apex, borne in an irregular spiral divergence, rarely distichous in places, often becoming equal in length to main axis, usually branched from every cell in a irregular spiral arrangement; *second-order laterals* bearing branchlets in an irregular spiral, branchlets usually curved inwards, less often spine-like, hairs sometimes present on apical cells of gametophytes and bisporophytes; *rhizoidal filaments* formed by some axial cells and basal cells of main laterals, 12-32 μm wide, growing downwards loosely but not developing into an adherent cortex; *plastids* discoid to ribbon-like.

Gametophytes dioecious. *Spermatangial branchlets* borne adaxially on last three orders of branching, (1-)2-3 per cell, erect, straight or slightly curved adaxially, e.g. 36-52 x 16-20 μm, 3-5 cells long, simple or with up to 2 single-celled branches, all cells becoming spermatangial mother cells, each forming 1-3 ovoid spermatangia 6.5-8.5 x 3-4 μm;

Fig. 30. *Aglaothamnion byssoides*

(A) Habit of small, robust intertidal plant (Channel Isles, Sep.). (B) Numerous delicate main axes with spirally arranged laterals (Galway, Feb.). (C) Developing and mature erect spermatangial branchlets, 1-3 per cell. Note single central nucleus in each vegetative cell (stained with haematoxylin) (C-E Devon, Nov.). (D) Sessile ellipsoid tetrasporangia, 1-2 per cell. (E) Cystocarp with deeply lobed gonimolobes. (F) Pedicellate bisporangia and released bispore (arrow) (Donegal, Dec.).

---

* The older name *Ceramium tenuissimum* Bonnemaison (1828), p. 132, is typified by a tetrasporophyte that resembles *A. byssoides* and *A. pseudobyssoides* but cannot be identified with certainty (L'Hardy-Halos & Rueness, 1990).

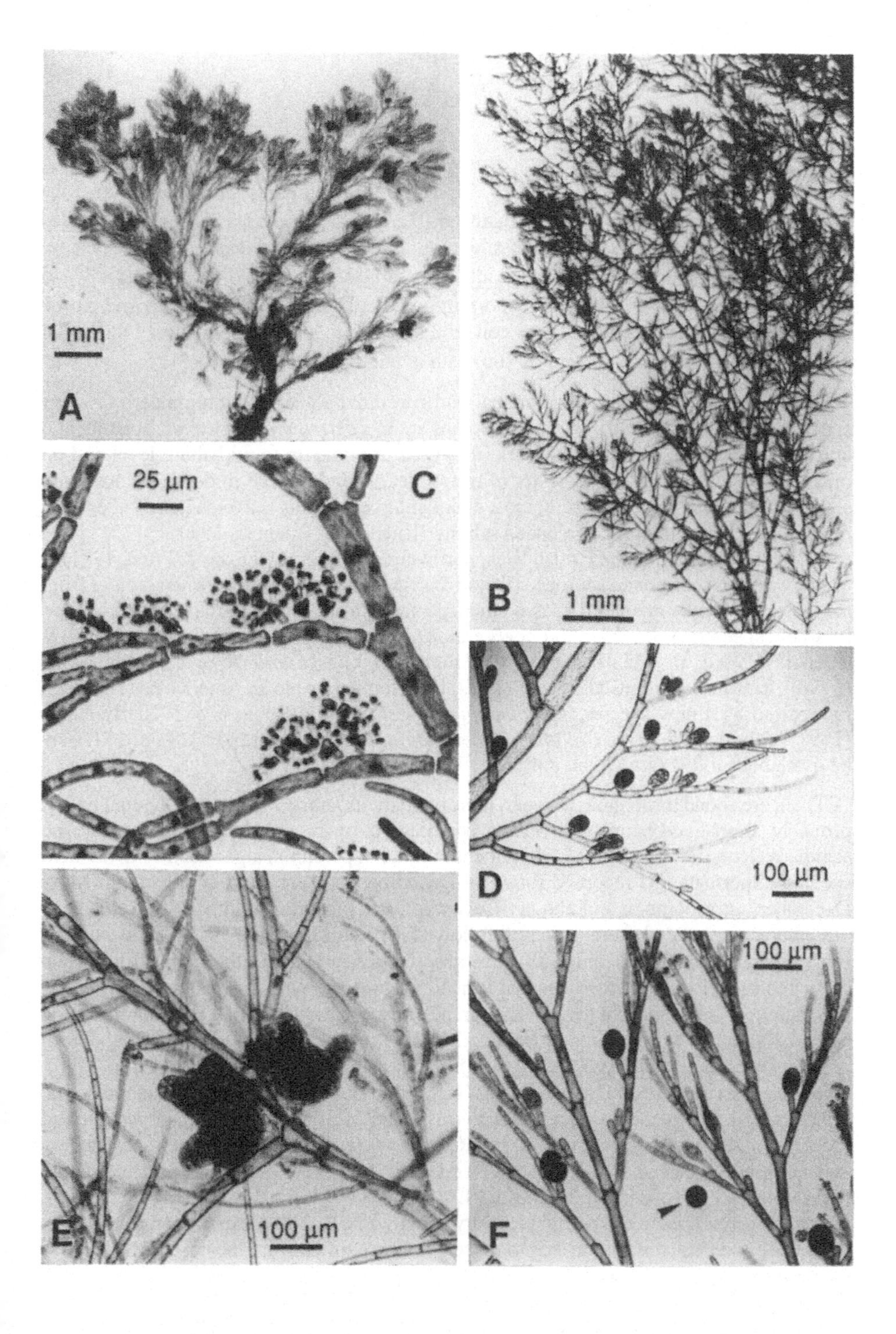
1 mm
A
25 µm
C
B
1 mm
100 µm
D
100 µm
E
100 µm
F

carpogonial branches zig-zag; *cystocarps* with several gonimolobes up to 130 µm long x 92 µm wide, slightly lobed to deeply dissected into segments, containing numerous angular carposporangia 16-30 µm in diameter. *Tetrasporangia* adaxial, in series of up to 4 on successive cells, 1-2 per cell, sessile, initially cylindrical, elongate before division, ellipsoid when mature, with a length: width ratio of (1.2-)1.3-1.7, 50-60 µm long x 30-45 µm wide, the walls 2-4 µm thick; *bisporangia* pedicellate or sessile, often borne on the same cell as a lateral branch, 1-2 per cell, ellipsoid, obliquely divided, often with partial tetrahedral furrowing, 65-80 x 45-55 µm, with a length: width ratio of 1.3-1.5.

Epiphytic on a wide variety of annual and perennial algae, including maerl, *Codium* species, blades of *Laminaria* spp., also epiphytic on *Zostera marina* leaves, epizoic, and epilithic on pebbles, rarely on bedrock, intertidal in pools up to mid-shore level and on open damp surfaces up to low water of neap tides, subtidal to 18 m depth, at sites with slight to moderate wave exposure, most abundant at extremely wave-sheltered sites with moderate currents; tolerant of reduced salinity (Rueness & Rueness, 1980).

Widely distributed in the British Isles, northwards to Orkney (Dixon & Price, 1981).

Norway to Canaries and Morocco (Dixon & Price, 1981). Records of *C. byssoides* from the western Atlantic are based on two other species. Caribbean plants have been assigned to a new species, *Aglaothamnion boergesenii* (Aponte & Ballantine) L'Hardy-Halos & Rueness (1990), p. 362 [basionym *Callithamnion boergesenii* Aponte & Ballantine (1990)]. *C. byssoides* sensu Edwards (1969) from temperate North America has also been recognized as a new species, *Aglaothamnion westbrookiae* Rueness & L'Hardy-Halos (1991) [although the name *Callithamnion arachnoideum* C. Agardh (1828), p. 181, may be available for this (see Dixon & Price, 1981, for typification)].

Thalli are found throughout the year, less abundant in January-March; individual fronds probably short-lived, some individuals perennating by prostrate filaments epizoic on ascidians (Rueness & Rueness, 1980) or endophytic in perennial algae such as *Gracilaria* species. Spermatangia recorded for April-December, procarps and cystocarps in May-December, tetrasporangia in February-December, and bisporangia in May-September.

Cultures inoculated from bispores followed two different life histories in culture. Bispores from Donegal (11 xii 1988) gave rise to tetrasporophytes, the spores from which grew into dioecious gametophytes that formed cystocarps; carpospores formed a further generation of tetrasporophytes. Tetrasporangia were meiotic, with 27-30 chromosome pairs, comparable with counts in field-collected tetrasporophytes (n = 29-34) and with previous reports (Harris, 1962: n = 28-33). In a second isolate (Donegal, 19 vii 1990), bispores gave rise to several generations of bisporophytes without forming tetrasporangia. Field-collected bisporangia were mitotic, with about 66 chromosomes. Norwegian and French isolates exhibited a regular *Polysiphonia*-type life history, except that a few tetrasporangia occurred on male plants (Rueness & Rueness, 1980), and were completely interfertile (L'Hardy-Halos & Rueness, 1990).

Gametophytes and tetrasporophytes vary greatly in branching pattern and robustness, from thalli with a distinct main axis to those which are tufted from the base, but are never corticate. Bisporophytes are corticate, and often have a distinct main axis surrounded by a regular arrangement of branches. These can be very difficult to distinguish from bisporangial *Seirospora interrupta* (q. v. for differences).

Aglaothamnion diaphanum L'Hardy-Halos & Maggs (1991), p. 468.

Holotype: PC (Fig. 31A); Isotypes: PC, BM, O, GALW, US; Donegal (St John's Point), 28 viii 1990, *Maggs*.

Thalli consisting of one to a few erect axes attached by small discoid filamentous holdfast and secondary creeping axes, 2-10 mm high, branched in one plane, with a single principal axis and linear to flabellate outline, pale pink in colour, with a delicate texture.

Main axes enlarging from apical cells 5 µm wide and 1-3 diameters long to axial cells 50-90 µm in diameter; cells near the base 1-2 diameters long, with a slight to marked median constriction, up to 3 diameters long in the middle of the plant; *first-order laterals* formed by oblique division of the third cell from the apex, one per axial cell, becoming level with main apical cells, borne in a regularly alternate-distichous pattern, the basal cell forming an adaxial branchlet, the next cell an abaxial one, in an alternate-distichous series; *second-order laterals* and one to two further orders of branching also branched in this pattern; *rhizoidal filaments* 16-22 µm in diameter formed by some axial cells near the holdfast, branching, growing downwards and attaching to the substratum but not developing into an adherent cortex; *plastids* ribbon-like, arranged longitudinally.

Plants dioecious but gametangia apparently non-functional; spermatangia found in the British Isles only on bisporangial plants. *Spermatangial branchlets* adaxial on lateral branches, 1-3 per cell, consisting of 3-4 cuboid to obconical spermatangial mother cells bearing 1-2 spermatangia, 8.5-10.5 µm long and 4-4.5 µm wide, markedly constricted around the central nucleus. *Carpogonial branches* U-shaped. *Bisporangia* developing adaxially in series of up to 4 on consecutive cells of lateral branchlets, usually sessile and borne singly, occasionally terminal on short branchlets in pairs, ovoid, 42-58 µm long x 22-36 µm wide, containing two uninucleate spores separated by an oblique curved wall.

Usually epiphytic on larger algae growing on bedrock, rarely epilithic on pebbles, subtidal from 7-25 m depth at moderately wave-exposed sites.

Known in the British Isles from only a few collections from Cornwall (Isles of Scilly), Donegal and Kerry.

British Isles to Atlantic France (Brittany).

Plants have been collected only in July-November, with male and female gametangia in August, and developing and mature bisporangia in July-September and November. Bispores of an Irish isolate gave rise in culture to further bisporophytes, some of which formed spermatangia in addition to bisporangia (L'Hardy-Halos & Maggs, 1991). Bisporangia were meiotic, with c. 32 pairs of chromosomes, and germinating spores were

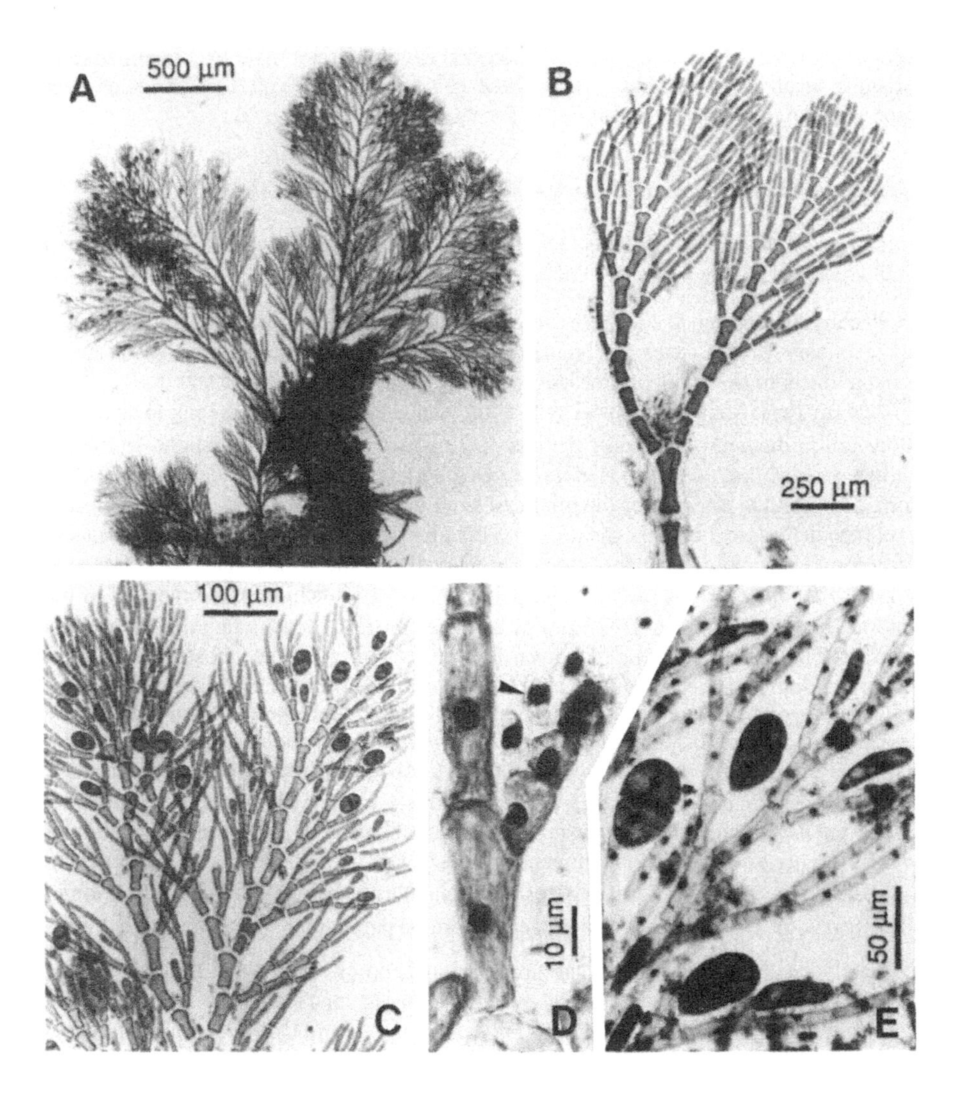
A
500 µm
B
250 µm
100 µm
C
10 µm
D
50 µm
E

haploid. The diploid chromosome number was apparently restored during sporangium development.

This species shows little variation except in the degree of development of first-order laterals. In some thalli, occasional laterals become equal in length to the main axis, producing a flabellate rather than linear frond outline. Hairs are abundant on female plants but absent on bisporophytes.

A. diaphanum could be confused with very small non-reproductive thalli of some growth forms of *Compsothamnion gracillimum* (q. v.), but *Compsothamnion* species are multinucleate in contrast to *Aglaothamnion*, which is uninucleate; *A. diaphanum* thalli are reproductive from a very small size and are usually easily recognizable by the bisporangia.

Aglaothamnion feldmanniae Halos (1965b), p. 126.

Holotype: PC (M.-Th. H. R638). France (Baie de Morlaix), *Halos*.

Callithamnion feldmanniae (Halos) South & Tittley (1986), p. 50.

Thalli consisting of erect axes growing singly or in tufts from small filamentous holdfasts and prostrate axes, 2-5 cm high and up to 6 cm wide, complanate, with distinct main axes and a rounded or flabellate overall shape, rose-pink, soft and flaccid.

Main axes enlarging from apical cells 10 μm in diameter; mature axial cells 68-104 μm in diameter, cylindrical, 1.5-3 diameters long near the holdfast, in the middle of the plant 3 diameters long, with a median constriction; *first-order laterals* developing 4-5 cells behind apex, not overtopping apical cell, borne in a regularly alternate-distichous series, equal in length to the main axis and up to 50 μm in diameter, the basal cell bearing an abaxial branchlet, followed by an alternate-distichous arrangement of narrow-angled branchlets; *second-order laterals* bearing a further similar pattern of branching in some thalli, ultimate branchlets all approximately equal in length, giving each branch a pennate outline; *rhizoidal filaments* formed by axial cells near holdfast, branched, consisting of cylindrical to bead-like cells 15-75 μm in diameter, growing downwards without becoming adherent to the main axis, attaching to the substratum and giving rise to further erect axes; *plastids* irregularly ribbon-like in young and older cells.

Gametophytes unknown in British Isles. *Tetrasporangia* formed in adaxial series of

---

Fig. 31. *Aglaothamnion diaphanum*

(A) Holotype (Donegal, Aug.). (B) Young thallus, showing branchlets on every cell of laterals, adaxially on the basal cell and then alternately (Cornwall, Feb.). (C) Apex of mature bisporangial plant (as B). (D) Spermatangial branchlet bearing spermatangia with central nuclei (arrow), formed on bisporophyte in culture from Donegal. Note single central nucleus in each vegetative cell (stained with haematoxylin). (E) Developing and mature sessile bisporangia (as A).

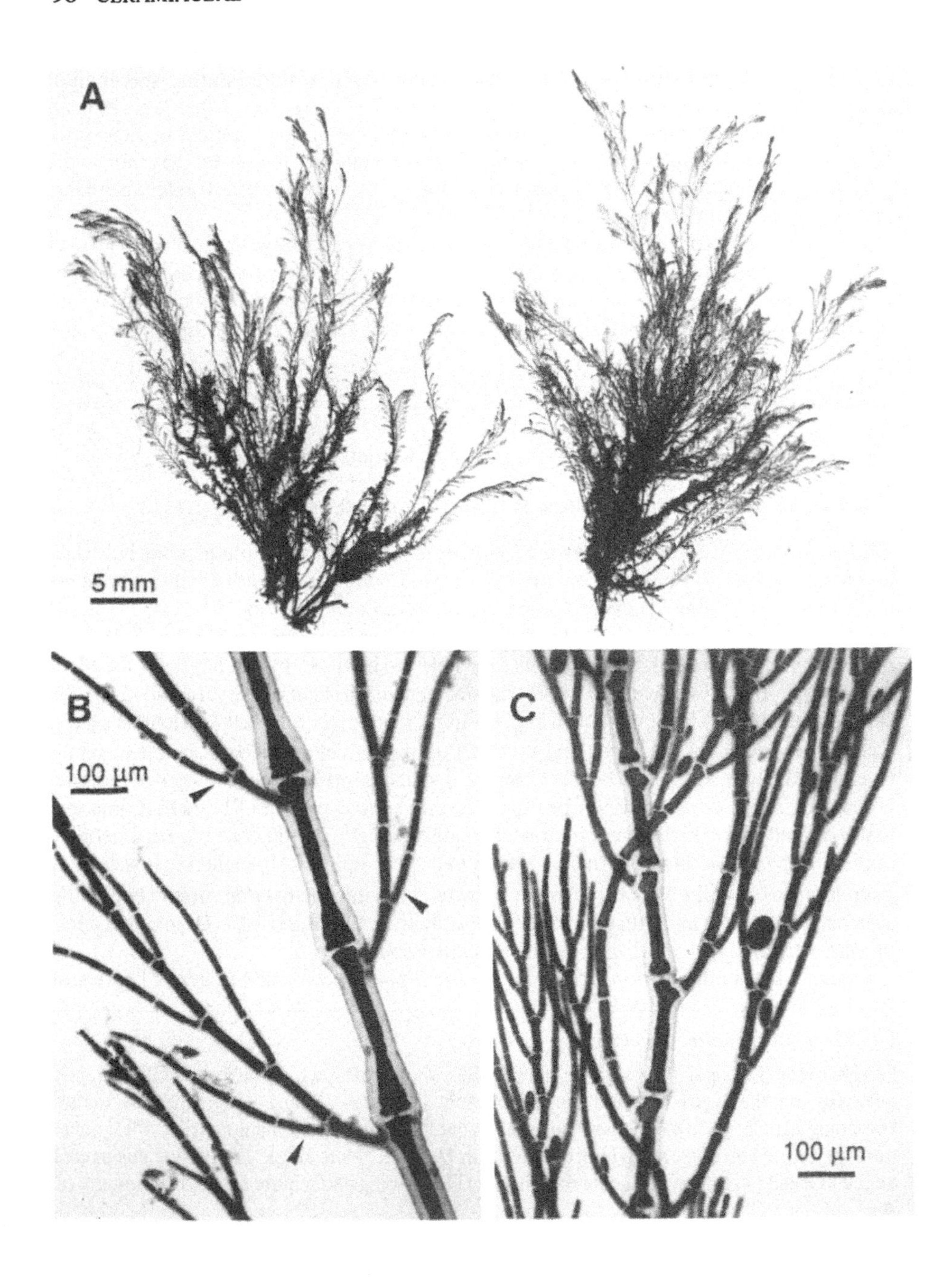
A
5 mm
B
100 µm
C
100 µm

which only 1-2 are mature, sessile, 1-2 per cell, ovoid, 50-55 µm long x 32-40 µm wide, with a length: width ratio of 1.3-1.6 and walls 3-4 µm thick.

On stones and dead shells, occasionally epiphytic, at extreme low water to 14 m depth in wave-sheltered bays and inlets, with moderate to strong exposure to tidal currents.
South coast of England from Dorset to S. Cornwall; Pembroke.
England to Atlantic France (Brittany).

Mature overwintering thalli were found in February, young plants appearing in June, becoming fertile in July, and persisting until at least October; tetrasporangia recorded in February, July and September- October. In Brittany, thalli perennate by persistent creeping axes, gametangia occur in July and August, and tetrasporangia in July-January (Halos, 1965b). In French isolates, carpospores gave rise to tetrasporophytes (L'Hardy-Halos & Rueness, 1990).

The proportion of branches bearing the characteristic abaxial branches on the basal cell is less in lax, little-branched thalli than in densely-branched plants. Some thalli have largely unbranched secondary laterals, but all specimens show at least a few abaxial branchlets.

L'Hardy-Halos & Rueness (1990) and Rueness & L'Hardy-Halos (1991) investigated the interbreeding of *A. feldmanniae* isolates with various North Atlantic *Aglaothamnion* species. *A. feldmanniae* was distinct from, but closely related to, a new species from temperate North America, *Aglaothamnion westbrookiae* Rueness & L'Hardy-Halos. This species occurs in Helgoland and Norway and may represent a recent introduction. As yet, it has not been found in the British Isles; it can be distinguished from *A. feldmanniae* by the lack of branchlets on the basal few cells of lateral branches.

Compsothamnion gracillimum (q. v.) resembles *A. feldmanniae* in habit, but the branchlet borne on the basal cell of a lateral is always adaxial rather than abaxial.

Aglaothamnion gallicum (Nägeli) Halos ex Ardré (1970), p. 174.

Lectotype: SBR. France (rade de Brest). 'Algues marines du Finistère', no. 154. *Crouan.*

Maschalosporium gallicum Nägeli (1862), p. 371.
Aglaothamnion brodiaei sensu Feldmann-Mazoyer (1941), p. 452.

Thalli 0.6-5 cm high and 0.2-4 cm broad, with a cylindrical, pyramidal or flabellate outline, consisting of erect axes attached singly by discoid holdfast 370-850 µm diameter that may give rise secondarily to new axes from margins; main axes distinct below, 0.1-0.3 mm in

---

Fig. 32. *Aglaothamnion feldmanniae*

(A) Habit of 2 plants, showing long main axes with regular feather-like arrangement of short laterals (Pembroke, Feb.). (B) Laterals with abaxial branchlets on basal cells (arrows) (Pembroke, undated). (C) Sessile tetrasporangia (as B).

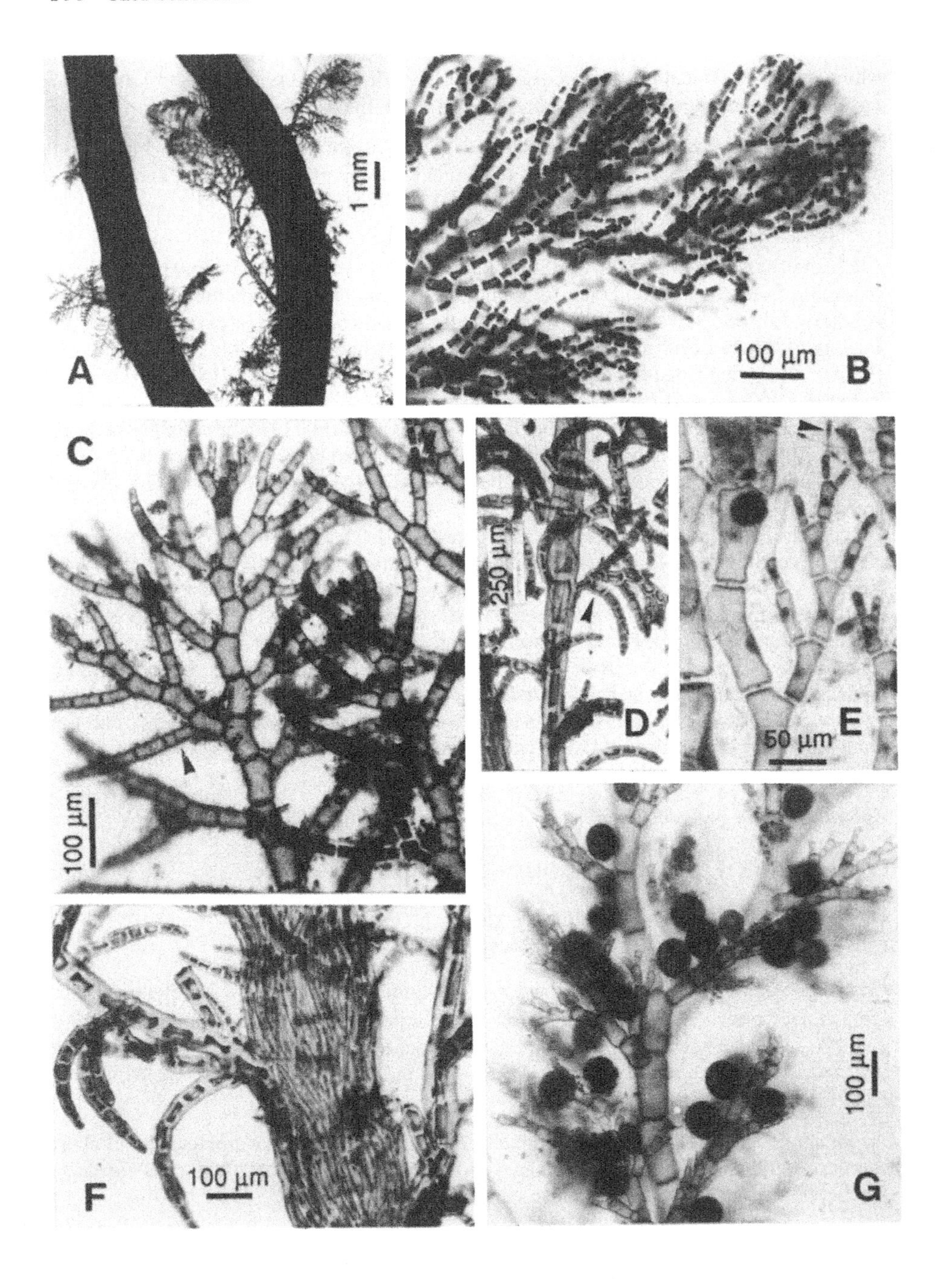
1 mm
A
100 µm
B
C
100 µm
250 µm
D
50 µm
E
100 µm
F
100 µm
G

diameter, often becoming indistinguishable amongst laterals, brownish-pink in colour, flaccid but not delicate.

Main axes growing from apical cells 10-12 µm wide; axial cells rapidly increasing in diameter so ratio of apical cell diameter to diameter of tenth cell is 1:3.5-3.8; *first-order laterals* developing from the third to fourth axial cell, one per cell, level with or overtopping and concealing main apex, borne in a regularly alternate-distichous to irregular 1/3 spiral pattern, often equal in length to main axis, usually branched from every cell in an irregularly spiral or alternate-distichous arrangement, with an abaxial branchlet on the basal cell, or basal 1-2 cells unbranched; *second-order laterals* bearing branchlets in the same pattern, lowermost ones usually simple, others branched alternate-distichously, twisted to lie perpendicular to the previous order of branching, curved inwards and tapering markedly; hairs sometimes borne abundantly on apical cells; *rhizoidal filaments* 16-30 µm in diameter developing from basal cells of lateral branches, growing downwards within the walls, branching abundantly and forming a dense cortex on main axis and first- and second-order laterals, also giving rise to sparse adventitious branchlets and contributing to holdfast tissue; *plastids* discoid in young cells, ribbon-like to filiform in axial cells.

Gametophytes dioecious. *Spermatangial branchlets* 1-2 per cell, curved forwards and lying along branch, aggregated when mature into a continuous layer 20-26 µm thick, covered with spermatangia measuring 6.5 x 4 µm. *Cystocarps* consisting of several globular or bilobed gonimoblasts, 70-200 µm x 65-175 µm, often containing <25, but occasionally >50, rounded carposporangia 25-50 µm in diameter. Tetrasporangia borne adaxially, or occasionally laterally, in series of 1-3 on successive cells of last two orders of branching, sessile, 1 per cell, sometimes replaced by a small adaxial branchlet, pyriform before division, spherical to ovoid when mature, length: width ratio 1.0-1.3, 60-88 µm long x 56-74 µm wide, with bilayered walls 7-8 µm thick.

Epiphytic on a variety of perennial algae, including *Corallina* sp., *Chondrus crispus* Stackhouse and *Mastocarpus stellatus* (Stackhouse) Guiry, in lower-shore pools and near extreme low water, on moderately to extremely wave-exposed shores.

South and south-west coasts (Devon, Pembroke; Clare, Donegal; Channel Isles), possibly more widespread, but recent literature records are lacking as this species has been considered conspecific with *A. hookeri*.

Fig. 33. *Aglaothamnion gallicum*

(A) Habit of small plants epiphytic on *Chondrus crispus* (Torquay, Devon, Oct.). (B) Apices of spirally branched thalli (Pembroke, undated). (C) Apex of thallus with complanate branching, showing branchlet on basal cell of some laterals (arrow) (as A). (D) Branch of Griffiths specimen in TCD, showing branchlet on basal cell of some laterals (arrow) (Torquay, May). (E) Branchlets with numerous terminal hairs (arrow), stained with haematoxylin to show single nucleus in each cell (Donegal, Sept.). (F) Major axis with heavy cortication (as D). (G) Sessile spherical tetrasporangia (as E).

Sweden, Denmark, British Isles and France; Mediterranean (Feldmann-Mazoyer, 1941).

Young thalli were noted in February; populations are most conspicuous in late summer and autumn. Spermatangia recorded for September and October; procarps, cystocarps and tetrasporangia in May, September and October.

This species shows wide variation in overall size and shape and in density of branching. Some plants have spiral branching throughout, whereas others are partially or entirely distichous.

The lectotype of *Callithamnion brodiaei* Harvey in W. J. Hooker (1833, p. 340), is from Forres, Scotland (see Dixon & Price, 1981). The protologue also included specimens from Torquay that were not identical to the Scottish material but "resemble the Scottish specimens so closely" that Harvey was "afraid to describe them as distinct". The lectotype of *C. brodiaei* is a specimen of *A. hookeri* (L'Hardy-Halos, pers. comm.) and thus Dixon & Price (1981) were justified in placing it in synonymy with *A. hookeri*, although Price (1978) had previously noted that "'*brodiaei*' may conceal a good taxon". For authors such as Rosenvinge (1923-24) and Feldmann-Mazoyer (1941), the concept of *A. brodiaei* was apparently based on the Torquay plants. Recently, however, L'Hardy-Halos and Rueness (pers. comm.) have found that *Maschalosporium gallicum* Nägeli, which was based on a Crouan specimen of *C. brodiaei*, is a valid name for this species.

Although Rueness & Rueness (1982) found that an isolate from Bergen, Norway, identified as *Callithamnion brodiaei*, resembled *A. hookeri* in culture and was interfertile with it, the appropriate morphological features were not compared and the material may have been misidentified. L'Hardy-Halos & Rueness (pers. comm.) have now shown that an isolate of *A. gallicum* from the type locality was non-interfertile with *A. hookeri*. Harris (1966) reported that the chromosome number of plants identified as *Callithamnion brodiaei* was $n = 29\text{-}30$, in contrast to that of *A. hookeri*, which was $n = 33$.

Aglaothamnion hookeri (Dillwyn) Maggs & Hommersand, comb. nov.*

Lectotype: TCD. Moray (Cawsie) (see Dixon & Price, 1981).

Conferva hookeri Dillwyn (1809), pl. 106.
?*Callithamnion polyspermum* C. Agardh (1828), p. 169 (see Dixon & Price, 1981).
Callithamnion brodiaei Harvey in W. J. Hooker (1833), p. 340 (see entry under *A. gallicum*).
Callithamnion grevillei Harvey in W. J. Hooker (1833), p. 345 (see Dixon & Price, 1981).
Callithamnion spinosum Harvey in W. J. Hooker (1833), p. 345 (see Dixon & Price, 1981).
Aglaothamnion brodiaei (Harvey in W. J. Hooker) Feldmann-Mazoyer (1941), p. 452 (see entry under *A. gallicum*).

Thalli consisting of erect axes attached singly or in small groups by discoid holdfasts; some

* Although this name was used by Feldmann (1954, p.104), his new combination was invalid as the basionym was not included.

axes becoming prostrate and developing secondary holdfasts in mat-forming thalli; erect thalli 0.3-7 cm high and 0.5-9 cm broad, with a distinct naked main axis below, 0.1-0.5 mm in diameter, often becoming indistinguishable amongst laterals, and a pyramidal, triangular, flabellate or linear outline, depending on the degree of development of first-order laterals, brownish-red to purplish in colour, flaccid to fairly robust.

Main axes growing from apical cells 16 µm wide; axial cells increasing in width gradually so the ratio of apical cell diameter to diameter of tenth cell is 1:1.1-2; *first-order laterals* developing from the third to eighth axial cell, one per cell, borne in a regularly alternate-distichous to 1/4 spiral divergence; apices remaining shorter than the conspicuous main apex when distichously arranged, or overtopping and concealing main apex in spirally-branched plants, when mature often becoming equal in length to main axis; basal cell and usually next 3-5 cells naked, other cells branched irregularly spirally or alternate-distichously; *second-order laterals* and up to 3 further orders of laterals branching in the same pattern; ultimate laterals cylindrical with rounded apices; *rhizoidal filaments* 16-30 µm in diameter developing from basal cells of lateral branches, sparse to very dense, developing close to apices of main and lateral axes when dense, forming an adherent cortex and giving rise to adventitious branchlets; *plastids* discoid to polygonal in young cells, ribbon-like in axial cells.

Gametophytes dioecious, although a few spermatangial branchlets may occur on females. *Spermatangial branchlets* formed on last 2-3 orders of branching, 1(-2) per cell, curved forwards and lying along branch, 2-3 cells long, much-branched, forming a hemispherical tuft, later aggregating into a continuous layer covered with spermatangia 6.5-9 x 4-4.5 µm. *Carpogonial branches* zig-zag; *cystocarps* consisting of several typically globular, sometimes irregularly conical or slightly lobed, gonimolobes, 100-240 µm x 90-170 µm, containing 10-40 angular carposporangia 35-50 µm in diameter. *Tetrasporangia* sessile, borne on last two orders of branching, 1 per cell, adaxial or abaxial, sometimes replaced by a small adaxial branchlet, pyriform before division, spherical to subspherical when mature, 52-78 µm long x 48-68 µm wide, with a length: width ratio of 1.0-1.3, and bilayered walls 8-11 µm thick.

A. hookeri occupies a very wide range of intertidal habitats (see Price, 1978). Near extreme low water, and in lower-shore pools on moderately to extremely wave-sheltered shores, sometimes exposed to tidal currents, it is typically epiphytic on a wide variety of algae. It also grows on open rock and epiphytically in the mid-shore of moderately to very wave-exposed shores, and the *scopulorum* form is found in shaded situations among other turf-forming algae, epizoic and epiphytic in the mid- to upper intertidal (Westbrook, 1927). Cultured isolates obtained from upper-shore plants, probably the *scopulorum* form, were tolerant of reduced salinity (Edwards, 1979).

Widespread, probably present throughout the British Isles (Dixon & Price, 1981).

N. Norway to N. France (Rueness & Rueness, 1982). Its wider distribution in Europe is difficult to assess owing to problems in circumscription, while reports from N. E. North America (e.g. Whittick, 1981) may involve another species (Whittick, pers. comm.).

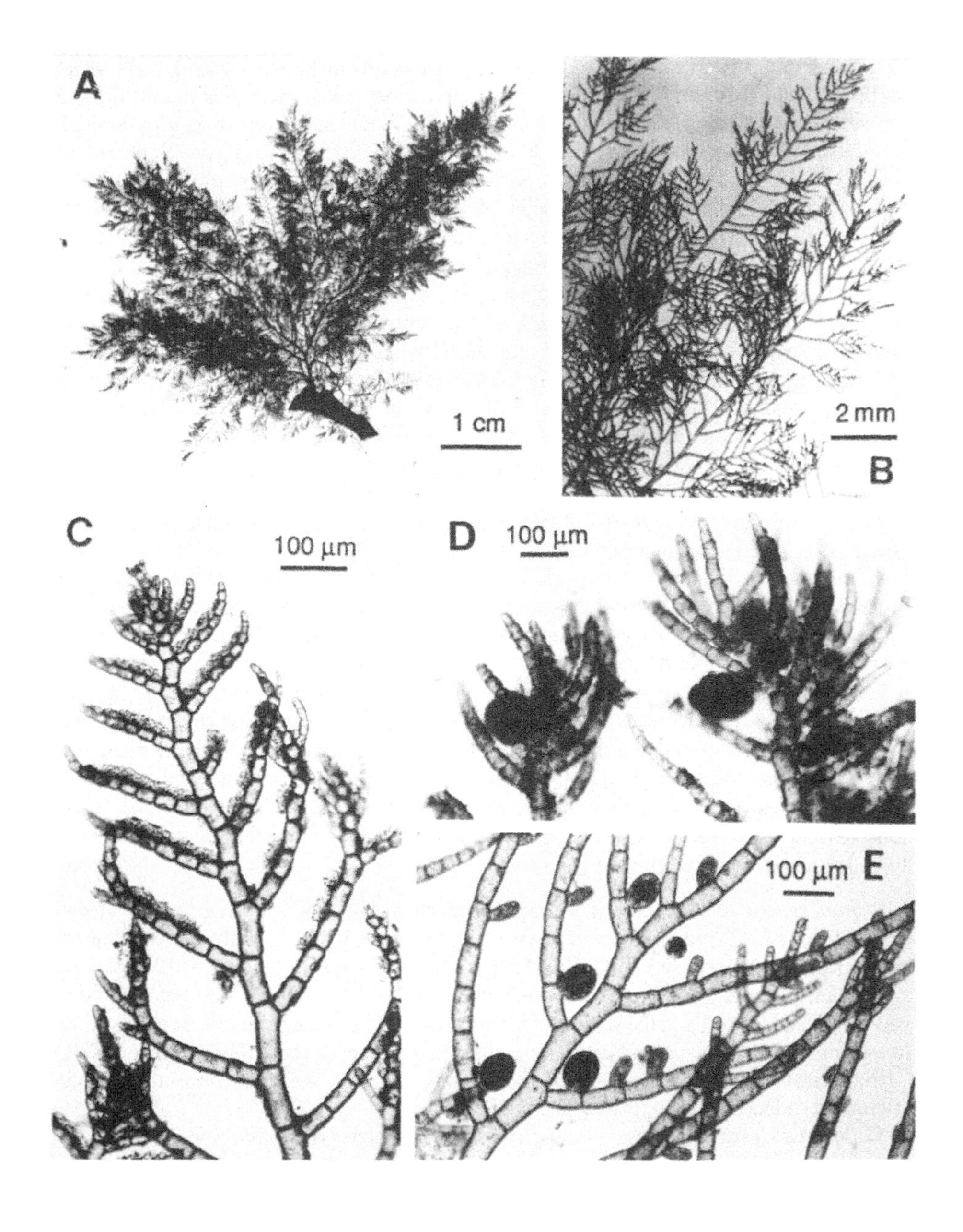
A
1 cm
2 mm
B
C
100 µm
D
100 µm
100 µm
E

Mature thalli are found throughout the year, but individual fronds are probably ephemeral. At Durham, England, plants (probably the *scopulorum* form) were denuded and reduced in size in winter (Edwards, 1979), whereas at Plymouth, the *scopulorum* form reproduced in spring and early summer and died back in mid-summer (Westbrook, 1927). Spermatangia have been recorded for January-July and September to November; procarps and cystocarps in January, February and April-November; tetrasporangia throughout the year.

Edwards (1979) found that although gametangial structures were rare in field populations, 13 interfertile isolates from N.E. England, E. Scotland, Devon and Cornwall underwent a *Polysiphonia*-type life history in culture. Complete life histories were obtained in 3-6 months, with optimum growth at 20°C; there were no photoperiodic effects on growth or reproduction. Chromosome number in British material is $n = 33$ (Harris, 1966). Parasporangia, the only reproductive structures in some Scandinavian and Danish populations (Rosenvinge, 1923-24; Rueness & Rueness, 1978), have not been found in the British Isles.

This species shows a very wide variation in overall size and shape and in pattern and density of branching. There is often a strong dimorphism between female plants, which usually show partly or entirely spiral branching and are strongly corticated, and males and tetrasporophytes which tend to have distichous branching throughout. This difference was also apparent in several European isolates (Rueness & Rueness, 1982). Female plants probably grow more slowly, as they are frequently heavily epiphytized. Various forms previously regarded as separate species (see Price, 1978) were placed in synonymy with *A. hookeri* by Dixon & Price (1981); further critical biosystematic studies are needed. The dwarf *scopulorum* form, based on *Callithamnion scopulorum* C. Agardh from the Faroes, has some distinct morphological characters (Westbrook, 1927), including a lack of cortication, and has recently been shown to be a separate species non-interfertile with *A. hookeri* (Rueness & L'Hardy-Halos, pers. comm.).

Aglaothamnion priceanum Maggs, Guiry & Rueness (1991), p. 344.

Holotype: BM (Fig. 35A). Mayo (Clare Is.), 24 vi 1990, *Maggs*.

Thalli consisting of small groups of erect axes attached by discoid filamentous holdfasts or endophytic rhizoidal filaments; holdfasts forming prostrate filaments that give rise to

---

Fig. 34. *Aglaothamnion hookeri*

(A) Habit of large epiphytic plant from near extreme low tide level at a moderately wave-sheltered site (Down, Nov.). (B) Major axes bearing laterals in one plane; first few cells of each lateral lack branchlets (B-D Down, Jan.). (C) Apex of male thallus with spermatangial branchlets forming a continuous layer on branchlets. (D) Apex of spirally branched female plant with spherical cystocarps containing few large carposporangia. (E) Sessile subspherical tetrasporangia (as A).

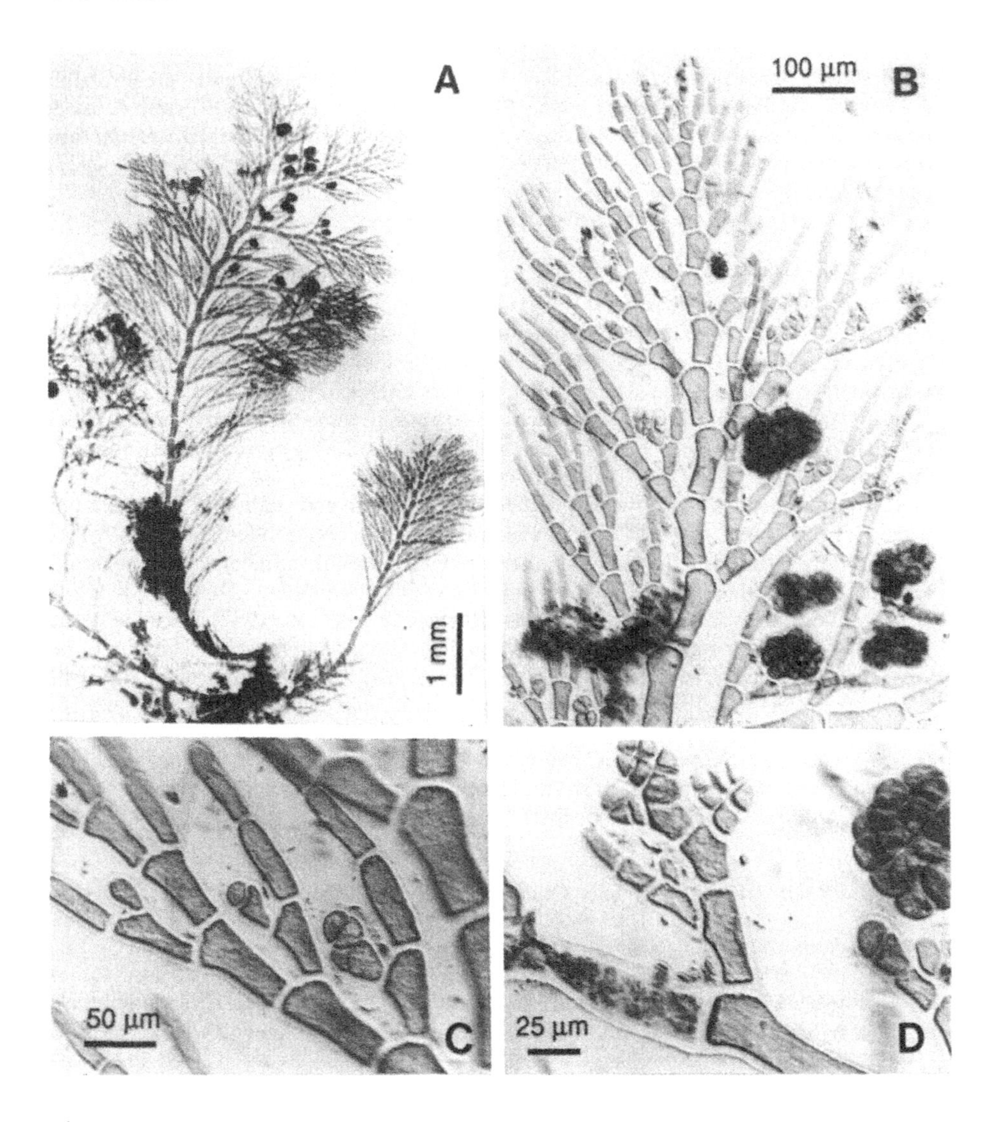

Fig. 35. *Aglaothamnion priceanum*

(A) Holotype (A-D Mayo, June). (B) Apex of mature parasporangial plant showing adaxial branchlets on first 2 cells of laterals, followed by alternate branching. (C) Developing parasporangial clusters. (D) Developing and mature compound parasporangial clusters.

further erect axes; erect axes 0.2-0.8 (-2) cm high, complanate, with a single principal axis, and a linear to irregularly flabellate outline, deep pink in colour, soft but not flaccid.

Main axes enlarging from apical cells 14-16 μm wide, including the thick walls; axial cells 60-200 μm in diameter, 0.8-1 diameter long near the holdfast, in the middle of the plant up to 2-3 diameters long, constricted centrally, with walls 10-35 μm thick; *first-order laterals* formed by oblique division of the third cell from the apex, apices nearly level with main apical cells, borne in a regularly alternate-distichous pattern, the first two cells bearing an adaxial branchlet, the third cell an abaxial one, followed by an alternate-distichous series; *second-order laterals* and 1-2 further orders of branching generally repeating the same pattern, although occasional higher-order branches bear series of up to 3 secund adaxial branchlets; *rhizoidal filaments* 10-18 μm in diameter formed in larger thalli from basal cells of lateral branches, growing downwards within the walls as a sparse, adherent cortex; *plastids* ribbon-like, arranged longitudinally; axial cells containing numerous small crystals.

Gametangial and tetrasporangial plants unknown in the British Isles. *Parasporangial clusters* developing adaxially on second to fourth-order branchlets, sometimes in series on adjacent cells, or terminally on adaxial or abaxial branchlets, initiated as enlarged pyriform cells resembling tetrasporocytes, dividing obliquely to form two cells, both of which or only the distal cell divide further, globular when mature, slightly or strongly lobed depending on the pattern of cell divisions during development, 53-250 μm long x 40-170 μm wide, containing 4-40 or more uninucleate, angular to rounded parasporangia, 25-45 x 20-35 μm, surrounded by a membrane 6-7 μm thick.

On small perennial algae and blades of *Laminaria hyperborea* growing on bedrock in depths of 5-30 m at moderately to extremely wave-exposed sites.

Collected rarely in the British Isles, from islands off northern and western coasts: Shetland (Price, 1978, fig. 15A, B), Inverness, Argyll; Mayo.

Norway (J. Rueness, pers. comm.) and Faroes to N. W. France (Maggs et al., 1991).

Plants collected in June-September, possibly perennating as endophytic rhizoidal filaments; parasporangia present in June, August and September. Paraspores from Ireland and the Faroes released in culture gave rise to several generations of parasporophytes without forming any other reproductive structures (Maggs et al., 1991). Chromosome counts of >80 suggested that parasporangial plants may be triploid. Reports of gametangia and carposporophytes in this species (Halos, 1965a, 1965b; Dixon & Price, 1981) appear to be based on specimens of other species (Maggs et al., 1991).

Aglaothamnion pseudobyssoides (P. Crouan & H. Crouan) Halos (1965b), p. 117.

Lectotype: CO (see L'Hardy-Halos & Rueness, 1990, figs 10 and 11). France (Kervallon, rade de Brest), *Crouan*.

Callithamnion pseudobyssoides P. Crouan & H. Crouan (1867), p. 136.

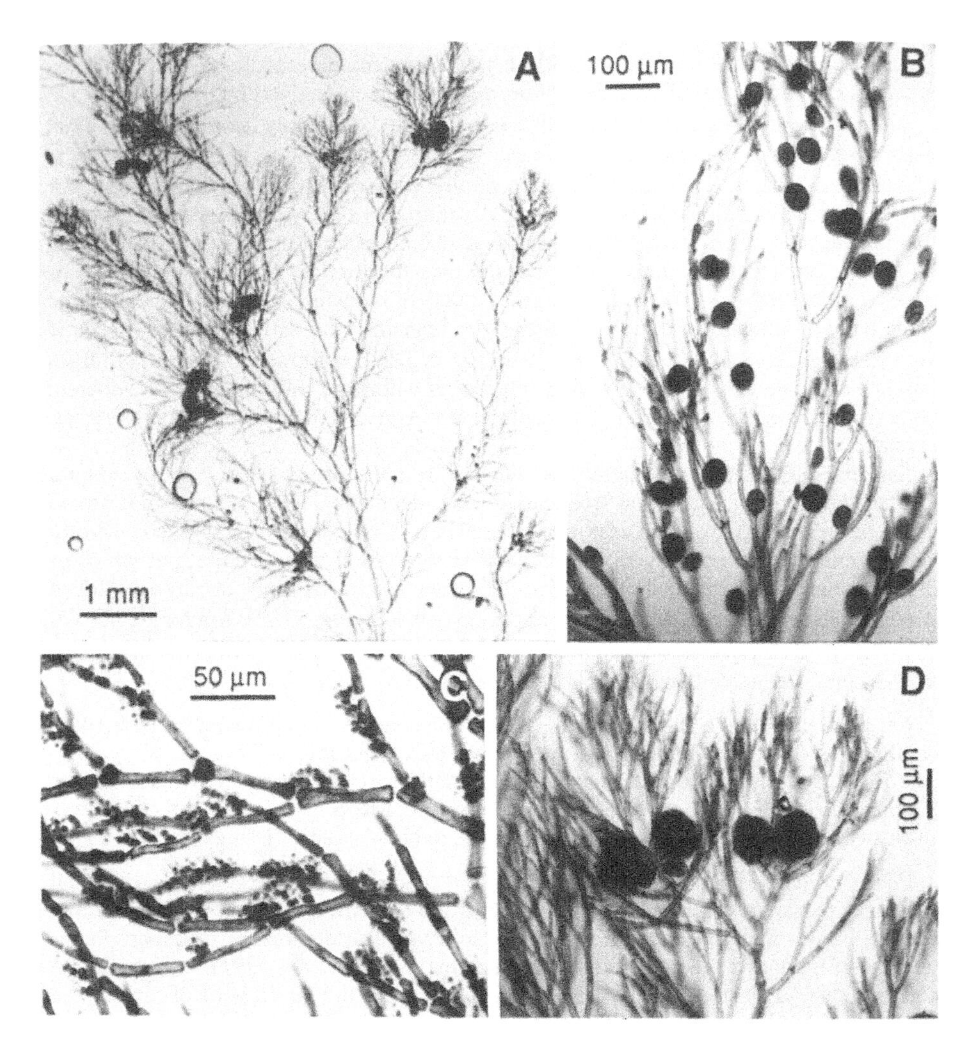

Fig. 36. *Aglaothamnion pseudobyssoides*

(A) Habit of female plant with delicate main axes and spirally arranged branches (Cornwall, July). (B) Sessile tetrasporangia (B-D Dorset, July). (C) Erect spermatangial branchlets, 1-3 per cell. (D) Cystocarps with globular gonimolobes.

Thalli consisting of erect axes growing in tufts or singly from prostrate or endophytic filaments, 0.3-3 cm high, with distinct naked main axes below, and a cylindrical, rounded or obconical outline, rose-pink in colour, very delicate and flaccid.

Main axes enlarging from apical cells 10 µm in diameter; mature axial cells 14-44 µm in diameter, 2.5-3 diameters long near the holdfast, increasing in size upwards to 75 µm wide, up to 7 diameters long, with a median constriction; *first-order laterals* formed from the third cell behind apex, surrounding and overtopping main apex, arranged when mature in a 1/3-1/4 spiral divergence, short below, increasing in length upwards, some equal to main axis; *second-order laterals* formed in this pattern in larger plants only; all major laterals clothed with a spiral arrangement of incurved, alternately branched branchlets, hairs formed abundantly on apical cells of gametophytes and sporophytes; *rhizoidal filaments* formed by some axial cells and basal cells of main laterals, growing downwards loosely but not forming an adherent cortex; plastids discoid to ribbon-like.

Gametophytes dioecious. *Spermatangial branchlets* borne adaxially on last 3 orders of branching, (1-)2-4 per cell, coalescing into hemispherical cushions 32-40 µm in diameter or continuous; branchlets simple or branched, consisting of 2-6 spermatangial mother cells, each forming several ovoid spermatangia 4.5-5 x 2-2.5 µm. *Carpogonial branches* zig-zag; *cystocarps* composed of 1-4 gonimolobes, usually two pairs of different ages, gonimolobes initially conical, spherical to ovoid when mature, 100-160 µm diameter, enclosed in a 3 µm-thick membrane, containing angular carposporangia 20-36 µm in diameter. *Tetrasporangia* developing adaxially at upper end of branchlet cell, often formed on same cell as a lateral cell or branchlet, appearing axillary, 1-2 per cell, sessile, ovoid, with a length: width ratio of 1.2-1.4, 48-60 µm long x 36-48 µm wide, the walls 2 µm thick.

On pebbles and epiphytic on larger red algae, in depths of 4-14 m, at sites with slight to moderate wave exposure, sometimes abundant on silty shell and pebble bottoms.

Known in the British Isles only from a few localities in Dorset, S. Devon and N. Cornwall.

Netherlands (Boddeke, 1958, figs 4, 14; Stegenga & Mol, 1983, pl. 84, figs 1-7, both as *Callithamnion byssoides*), Atlantic France, Spain and Portugal (L'Hardy-Halos & Rueness, 1990).

Observed in the British Isles only in June-August, probably perennating as creeping or endophytic filaments; spermatangia present in July-August, procarps and cystocarps in June-August, and tetrasporangia in July-August.

Thalli with little-developed lateral axes have a cylindrical outline whereas plants with well-developed laterals are wider.

A. pseudobyssoides was placed in synonymy with *A. byssoides* by Miranda (1932), but Halos (1965b) and L'Hardy-Halos & Rueness (1990) showed that *A. pseudobyssoides* could be distinguished from *A. byssoides* (as *A. furcellariae*) by several morphological features, and that isolates were intersterile. Suggestions by Rueness & Rueness (1980) that *A. pseudobyssoides* might be closely related to American *A. byssoides* sensu Edwards (1969) [=*Aglaothamnion westbrookiae* Rueness & L'Hardy-Halos (1991)] were refuted by L'Hardy-Halos & Rueness (1990).

Callithamnion corymbosum (q. v.) can resemble *A. pseudobyssoides*, but has corticated main axes and multinucleate vegetative cells, in contrast to *A. pseudobyssoides*, which is ecorticate, with uninucleate cells.

Aglaothamnion roseum (Roth) Maggs & L'Hardy-Halos (1993), p. 522.

Lectotype: LD 18466. France (Bayonne).

Ceramium roseum Roth (1798), p. 47.
Conferva rosea (Roth) J. E. Smith (1802), pl. 966.
Callithamnion roseum (Roth) Lyngbye (1819), p. 126.

Thalli consisting of erect axes growing singly or in groups from discoid holdfast formed by aggregation of corticating rhizoidal filaments; further erect axes arising secondarily from holdfast, 0.7-8 cm high, with a distinct naked main axis below, often indistinguishable above amongst laterals, the outline pyramidal or irregularly rounded, pinkish-purple to brownish-purple in colour, flaccid but not extremely delicate.

Main axes enlarging from apical cells 11 µm in diameter to 150-360 µm in diameter; axial cells increasing in size upwards to 5 axial diameters long, with a median constriction; *first-order branches* developing from the third or fourth cell behind apex; laterals often corymbose, borne when mature in an irregularly distichous or 1/3-1/4 spiral arrangement, short below, increasing in length upwards, some becoming equal in length to main axis and developing dense cortication, basal 1-7 cells unbranched, then every cell bearing a branch in a distichous or spiral pattern; branching often distichous below and becoming spiral towards apices; *second-order branches* typically branched alternate-distichously after a few naked cells, bearing long, inwardly-curved branchlets that may bear a few further branchlets near apices, branchlet cells cylindrical to barrel-shaped, occasionally bearing hairs; *rhizoidal filaments* 12-32 µm wide, branched, forming dense adherent cortication, obscuring axial cells, but not forming adventitious branchlets; *plastids* discoid.

Gametophytes dioecious. *Spermatangial branchlets* borne adaxially on last 2 orders of branching, 1(-2) per cell, 35-50 µm long x 30-50 µm wide, erect, curved adaxially, initially

---

Fig. 37. *Aglaothamnion roseum*

(A) Habit of robust plant (Cork, Dec.). (B) Detail of branching, showing long, slightly curved last-order branchlets and sessile tetrasporangia (B-C Devon, Nov.). (C) Apex of main axis stained with haematoxylin showing single nucleus in each cell (arrows). (D) Cortication developing within wall matrix of main axis (Pembroke, Dec.). (E) Developing (d) and mature (m) spermatangial branchlets, 1 per cell (Cornwall, May). (F) Cystocarps with globular gonimolobes (as D).

conical, becoming almost spherical, 4-5 cells in length, the basal one often large and pigmented, each cell forming several lateral branches, all cells of which, except the basal one, develop into spermatangial mother cells and bear several spermatangia, 6-7 x 2.5-4 µm. *Carpogonial branches* zig-zag; *cystocarps* with 1-4 gonimolobes, often two pairs of

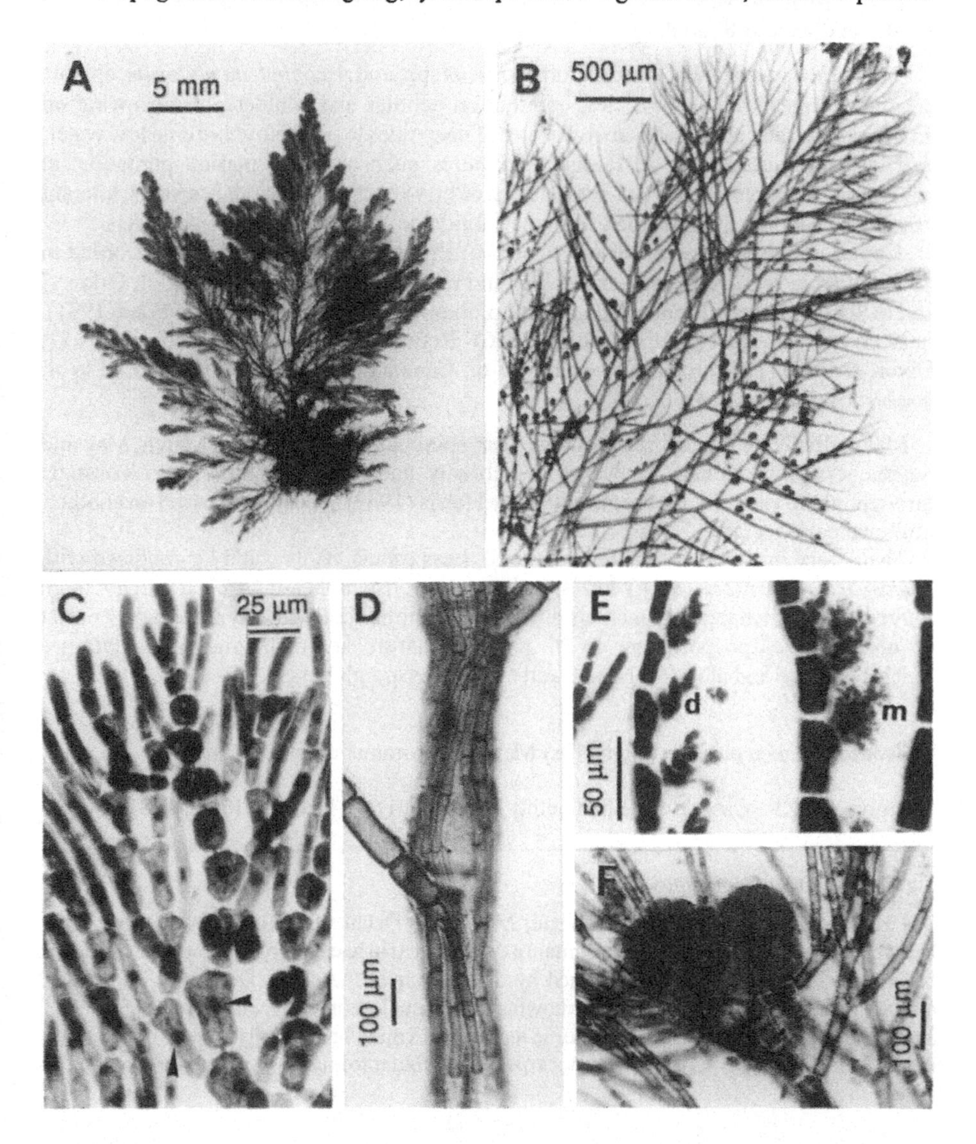

different ages, gonimolobes spherical to ovoid when mature, 160-230 µm long x 125-200 µm wide, containing rounded carposporangia 32-50 µm in diameter. *Tetrasporangia* formed in adaxial series of 1-7 on branchlets, 1 per cell, sessile, pyriform before division, ellipsoid when mature, 64-80 µm long x 52-58 µm wide, with a length: width ratio of 1.2-1.6 and walls 5-8 µm thick.

Epiphytic on a variety of algae including *Fucus* spp. and *Ascophyllum nodosum*, epizoic on invertebrates such as barnacles, epilithic on pebbles and boulders, also growing on artificial substrata, intertidal from low water of neap tides to just below extreme low water, and typically submerged on floating structures such as yacht marina pontoons, at moderately to extremely wave-sheltered sites, often with exposure to tidal currents; tolerant of reduced salinity; also recorded from the subtidal to 20 m (Dixon & Price, 1981).

Common on south and south-west coasts of England, Wales and Ireland; recorded in Scotland from scattered localities on west and north coasts (Ayr, Bute, Argyll, Orkney) and in Shetland, recently collected in Ayr, "probably widespread" (Dixon & Price, 1981).

Norway to Portugal (South & Tittley, 1986). Reports from eastern USA (Taylor, 1957; Dixon & Price, 1981) require re-examination; Canadian records were transferred to *A. hookeri* by South (1984).

Mature thalli are found throughout the year; spermatangia recorded in March, May and September; procarps and cystocarps in February and April-December; tetrasporangia throughout the year. Chromosome counts by Harris (1962) appear to be based on another, multinucleate, species.

Thalli vary in overall size and robustness. Less robust plants could be confused with *Seirospora interrupta* (q. v.) but are always purplish rather than clear pink. In *S. interrupta*, spermatangial branchlets are erect and little-branched and cystocarps consist of chains of carposporangia. In *A. roseum*, mature spermatangial branchlets are much-branched and almost spherical and cystocarps are globose.

Aglaothamnion sepositum (Gunnerus) Maggs & Hommersand, comb. nov.

Lectotype: OXF. Caernavon (Lhanfaethly, Anglesey) (see Dixon & Price, 1981).

Fig. 38. *Aglaothamnion sepositum*

(A) Habit of thalli on mussels (Donegal, May). (B) Detail of thalli, showing very dense branching (Antrim, Feb.). (C) Apex of main axis squashed in haematoxylin showing main apical cell (arrow) surrounded and overtopped by dense branches (Antrim, Dec.). (D) Corticated major axis squashed in haematoxylin showing single nucleus in each cell (arrows) (Cornwall, June). (E) Spermatangial branchlets borne densely on young branches; last-order branches are up to 11 cells in length (as C). (F) Cystocarps with globular to slightly conical gonimolobes (as B).

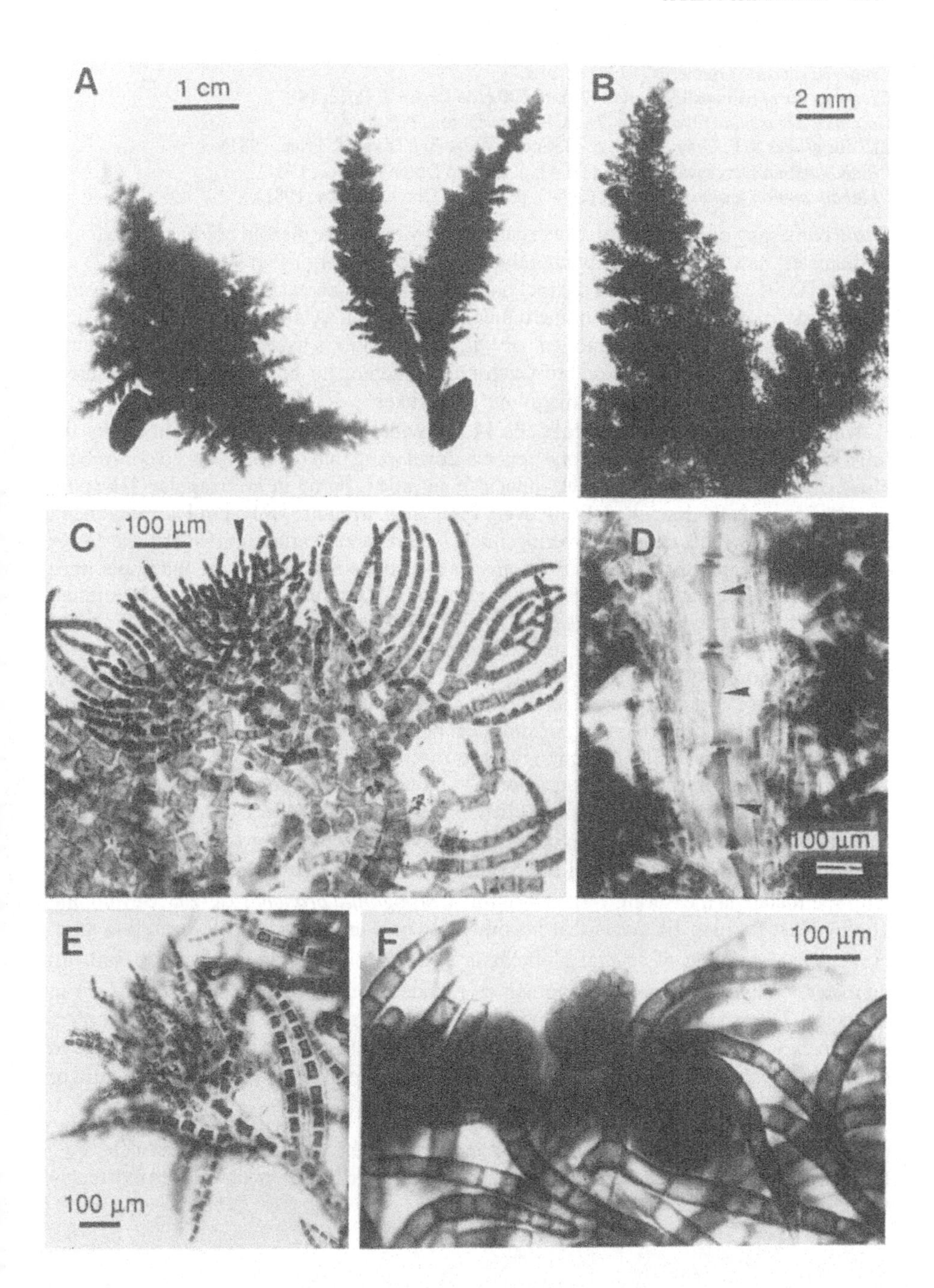
A
1 cm
B
2 mm
C
100 µm
D
100 µm
E
100 µm
F
100 µm

Conferva seposita Gunnerus (1772), p. 116.
Conferva purpurascens Hudson (1778), p. 600 (see Dixon & Price, 1981).
Conferva arbuscula Dillwyn (1807), pl. 85 (see Dixon, 1960c).
Ellisius glaber S. F. Gray (1821), p. 333, nom. illeg. (see Dixon & Price, 1981).
Phlebothamnion faroense Kützing (1864), p. 30 (see Dixon & Price, 1981).
Phlebothamnion scoticum Kützing (1864), p. 30 (see Dixon & Price, 1981).

Thalli consisting of groups of erect axes attached by extensive discoid holdfast up to 5 mm in diameter; new axes arising abundantly from holdfast margins; erect axes 2-8(-13) cm high and 0.7-5 cm broad, with a distinct naked main axis, decreasing in diameter upwards from a maximum 0.3-1.5 mm; outline typically pyramidal, as lateral branches decrease in length upwards, less often cylindrical or with the main axis naked below and branched above; colour deep purplish or brownish-red, bleaching to brown or straw-coloured; texture spongy and becoming cartilaginous in old axes.

Main axes enlarging from apical cells 14 µm wide; *first-order laterals* cut off by the third to fourth cell from the apex, one per cell, developing into densely-packed corymbose clusters of incurved filaments that conceal main apex, borne in an irregular 1/4 spiral divergence, usually branched from every cell in a irregular 1/4 spiral arrangement; *second-order laterals* densely covering main laterals and bearing branchlets, the lower ones of limited growth, branched in an alternate-distichous arrangement, the upper ones indeterminate, bearing a spiral arrangement of branchlets that are branched alternate-distichously; *ultimate laterals* (2-)7-16 cells long, curved inwards, cylindrical with bluntly conical apical cells, sometimes bearing terminal hairs; *rhizoidal filaments* developing from lower cells of lateral branches just behind apices, much-branched, forming a dense cortex obscuring axial cells and cells of first- and second-order laterals, and giving rise to a dense covering of simple or branched adventitious branchlets that may be removed by abrasion in old plants; *plastids* discoid in young cells.

Gametophytes dioecious, or occasionally monoecious. *Spermatangial branchlets* formed in the dense corymbose apical clusters, on branches of last 2 orders, 1-3 per cell, formed on adaxial and lateral faces of branch, angled forwards, 4-5 cells long, much-branched, aggregated when mature into a continuous layer 25 µm thick, covered with spermatangia measuring 7-8.5 x 4-6 µm. *Carpogonial branches* zig-zag; developing gonimoblast forming filaments that become entwined around adjacent vegetative axes; *cystocarps* consisting of several globular or slightly lobed gonimoblasts, 240-340 µm diameter, containing numerous angular or rounded carposporangia 36-52(-75) µm in diameter. *Tetrasporangia* borne in long adaxial series on successive cells of last two orders of branching, sessile, 1-2 per cell, becoming spherical prior to division, spherical to subspherical when mature, 65-75 µm long x 60-68 µm wide, with a length: width ratio of 1.0-1.1(-1.3) and walls 6 µm thick.

Epilithic on bedrock, epizoic on mussels and limpets, less often epiphytic on *Fucus* spp., typically on open rock, less often in pools, in the mid-intertidal of moderately to extremely wave-exposed shores.

Widely distributed on northern coasts (Dixon & Price, 1981), most abundant in Scotland and Ireland, uncommon in Devon and Cornwall; rare in the Channel Isles (J. H. Price, pers. comm.).

Norway, Faroes and Iceland (Dixon & Price, 1981); N. France (J. H. Price, pers. comm.).

Large thalli are present throughout the year. During winter and early spring, they are straggly and lack the characteristic corymbose clusters of laterals (Dixon & Price, 1981). Spermatangia have been recorded for February-October; procarps and cystocarps in February-October; and tetrasporangia in January-October. Chromosome number in meiotic tetrasporocytes from Antrim was determined as $n = 32\text{-}35$.

Tetrasporangial specimens are more luxuriantly branched than females, which have sparser, more open branching, and spermatangial plants are intermediate between them; hairs are more abundant on females (Dixon & Price, 1981). Thalli are monoecious or dioecious, apparently varying between populations (Dixon & Price, 1981).

Callithamnion granulatum (q. v.), which may closely resemble *A. sepositum* in habit, can be distinguished from it by the multinucleate vegetative cells and the length of the ultimate laterals, which are typically 3-4 cells in *C. granulatum* and 7-9 cells in *A. sepositum* (Dixon & Price, 1981). *C. granulatum* is generally iridescent, whereas *A. sepositum* is never iridescent.

Aglaothamnion tripinnatum (C. Agardh) Feldmann-Mazoyer (1941), p. 464.

Holotype: LD 18975 (see Dixon & Price, 1981). Atlantic coasts (?France), *Grateloup*.

Callithamnion tripinnatum C. Agardh (1828), p. 168.
Phlebothamnion tripinnatum (C. Agardh) Kützing (1849), p. 654.

Thalli consisting of erect axes attached in dense tufts to solid discoid holdfast composed of densely interwoven filaments, and also arising from secondary prostrate filaments; erect axes 1.5-5 cm high and 1.8-6 cm wide, bushy, pyramidal to rounded, wider than high when well-developed, with the main axis often difficult to distinguish amongst first-order laterals that equal it in length and width; plants are bright red in colour, with tough main axes and flaccid branches.

Main axes enlarging from apical cells 14-16 μm wide to 145-360 μm; cells near the base 0.3 axial diameters long and cylindrical, in the middle of the axis 2-3 diameters long, with a median constriction; *first-order laterals* developing by oblique division of the third cell from the apex, remaining shorter than main apex, borne in a regularly alternate-distichous pattern, the basal cell, and sometimes also the second cell, sometimes forming a curved adaxial branchlet, the next few cells often naked, followed by alternate branching from every cell; *second-order laterals* lying at an acute angle to branches, with a pennate outline, and forming branchlets in the same pattern; *parasporangium-like structures* formed terminally on branches in branched zig-zag rows, composed of clavate to pyriform, deeply staining cells 56-72 x 28-36 μm, surrounded by a thick wall; *rhizoidal filaments* 15-32 μm in diameter developing from basal cells of lateral branches and growing

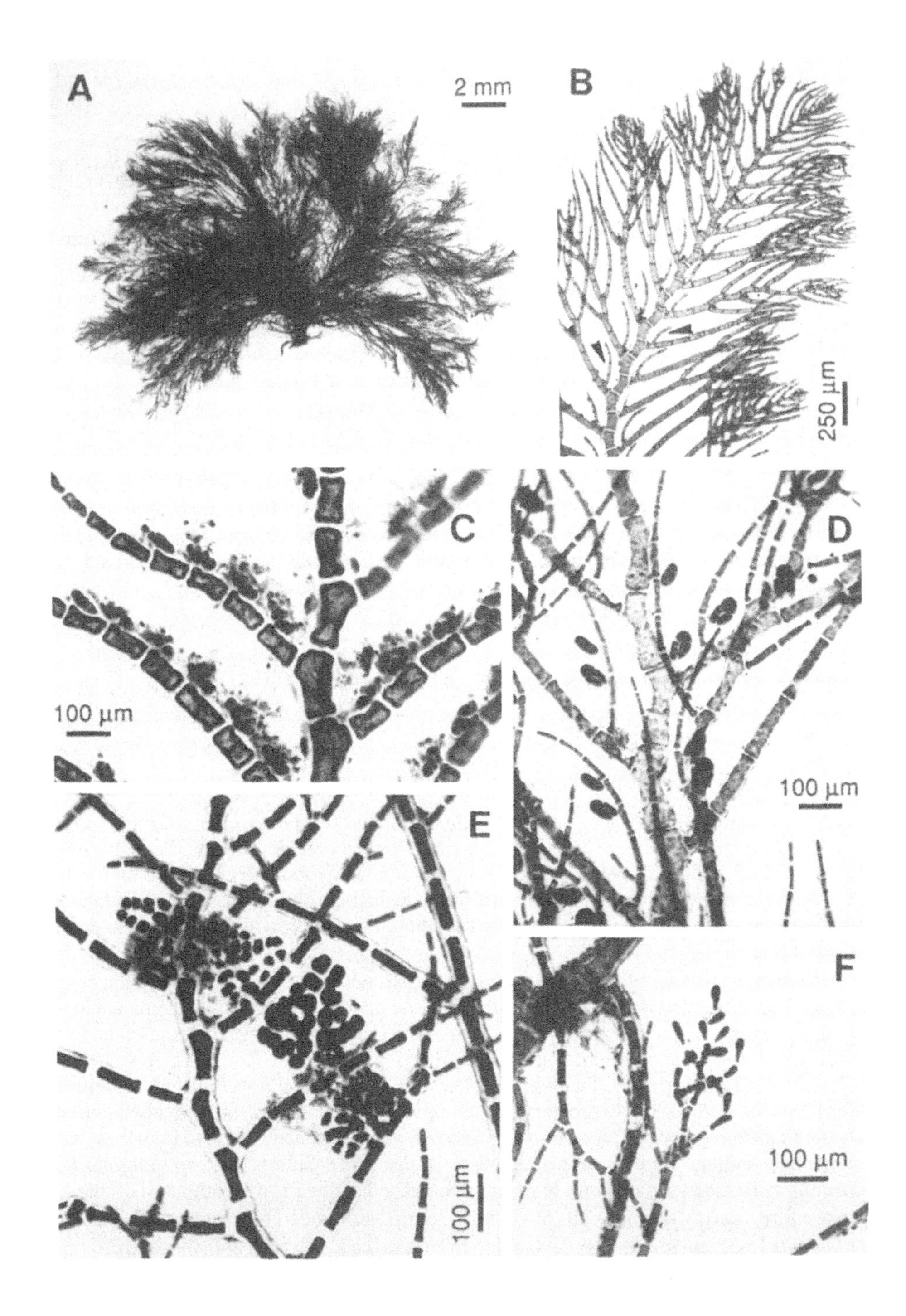
A
2 mm
B
250 µm
C
100 µm
D
100 µm
E
100 µm
F
100 µm

downwards within the walls to form a dense, adherent cortex on main and lower lateral axes, giving rise to sparse adventitious branchlets; *plastids* discoid in young cells, ribbon-like to filiform in axial cells.

Gametophytes dioecious. *Spermatangial branchlets* 1-2 per cell, curved forwards and becoming cushion-like when mature, 20-24 µm long x 20 µm wide, consisting of a cluster of spermatangial cells, sometimes with a single sterile basal cell, each of which forms 1-3 spermatangia measuring 6.5 x 3 µm. *Carpogonial branches* U-shaped; *cystocarps* 240-370 µm in diameter, consisting of 2-3 gonimolobes of different ages, gonimoblasts distinctly cordate (heart-shaped) when developing, and cordate, pyramidal or conical, rarely globular, when mature, up to 225 µm long x 185 µm wide, composed of rounded carposporangia 24-32 µm in diameter. *Tetrasporangia* borne in adaxial series of up to 8 on successive cells, sessile, terminal on short branchlets or pedicellate, 1-3 of different ages per cell, ovoid when mature, with a length: width ratio of 1.3-1.4, 54-60 µm long x 42-44 µm wide, the walls 3-4 µm thick.

Epilithic on bedrock and pebbles, epiphytic on a variety of algae in mid-shore pools and sublittoral from extreme low water to 10 m depth, at moderately to extremely wave-sheltered sites, usually with exposure to moderate to strong tidal currents, often very abundant in these habitats.

South and south-west coasts (Dorset, S. Devon, S. Cornwall, Pembroke; Cork, Clare, Galway).

British Isles to S. Portugal (Albufeira, 23 iii 1989, *Maggs*); ?Sénégal (Price et al., 1986); Mediterranean.

Mature thalli can be found throughout the year, perennating by creeping bases, the individual erect axes probably short-lived; numerous sporelings are usually present among reproductive plants. Spermatangia recorded for February, March, July and October; procarps and cystocarps in Febuary, July and October; tetrasporangia in February-April, June-August, October and December; and parasporangium-like structures present in March, July and October. An isolate from Pembroke (Milford Haven, 28 iii 1990) showed a *Polysiphonia*-type life history. Carpospores from field-collected plants gave rise to tetrasporophytes that became fertile in 4-6 weeks. The characteristic feature of an adaxial branchlet from the basal cell of lateral branches was maintained in culture, and tetrasporangia were formed laterally or in clusters on pedicels. Parasporangium-like cells developed on female plants in culture, but no spores were released from them and they do

---

Fig. 39. *Aglaothamnion tripinnatum*

(A) Habit (A-B Cork, Dec.). (B) Apex of main axis with adaxial branchlets (arrows) on basal cell of laterals. (C) Spermatangial branchlets borne singly on each cell of branches (C-F Pembroke, Oct.). (D) Ellipsoid tetrasporangia sessile on adaxial branchlets. (E) Cystocarps with cordate gonimoblasts. (F) Zig-zag arrangement of parasporangium-like swollen cells.

not appear to function as propagules, despite their resemblance to parasporangia. Chromosome number in dividing tetrasporocytes was approximately $n = 30\text{-}32$.

Plants vary in their regularity of branching. Whereas some thalli form adaxial branchlets from the basal cell of every lateral branch, in other individuals these are formed only sparsely. More robust plants tend to show more regular branching but there is variation even between plants of a single collection.

A. tripinnatum was first reported from the British Isles (Roundstone, Galway) by Harvey (1847, pl. 77), but there were very few subsequent records, and Dixon & Price (1981) considered that its status required clarification with regard to *A. hookeri* and *A. bipinnatum*. Our observations on several populations of *A. tripinnatum* have confirmed that this species is readily distinguishable from *A. hookeri* at sites where they grow together.

Non-reproductive thalli of *A. tripinnatum* could be confused with *Compsothamnion gracillimum* (q. v.), but they have corticate main axes, in contrast to the ecorticate axes of *Compsothamnion* species.

SEIROSPORA Harvey

SEIROSPORA Harvey (1846), pl. 21.

Type species: *S. griffithsiana* Harvey (1846), pl. 21 [= *S. interrupta* (J. E. Smith) Schmitz (1893b), p. 281].

Thalli uniseriate, erect to 12 cm high, attached by a discoid rhizoidal holdfast and consisting of 3-4 orders of indeterminate axes and 2-3 orders of determinate lateral filaments; axes spirally, alternate-distichously or pseudodichotomously branched, extensively corticated or corticated only at the base by descending rhizoidal filaments; determinate lateral branchlets alternately to pseudodichotomously branched, sometimes overtopping the axis; deciduous hyaline hairs present or absent.

Growth of indeterminate axis by oblique division of uninucleate apical cell; indeterminate axes, determinate lateral filaments and rhizoids developing as in *Aglaothamnion*; vegetative cells uninucleate (rarely binucleate), plastids elongate or ribbon-like at maturity.

Gametophytes dioecious or monoecious; asexual reproduction by means of bisporangia or seirosporangia. Spermatangial filaments developing as in *Aglaothamnion*. Procarp formed near apex of potentially indeterminate axis as in *Aglaothamnion*; carpogonial branch L-shaped, the first three cells horizontal and the carpogonium vertical. Early postfertilization stages imperfectly known, probably occurring as in *Aglaothamnion*; cystocarps typically formed in opposite pairs; carposporangia borne in branched, moniliform chains that resemble the seirosporangia. Tetrasporangia solitary, sessile or pedicellate from upper ends of cells of the last 3 orders of filaments, ellipsoid, tetrahedrally divided; bisporangia ellipsoid, binucleate, diploid, developing in the same position as the tetrasporangia; seirosporangia formed sequentially in terminal branched moniliform chains by direct transformation of cells of the ultimate branchlets, subspherical to ellipsoid; seirospores released individually through a lateral fissure or terminally by dissolution of apical walls.

References: Feldmann-Mazoyer (1941), Dixon (1971), Plattner & Nichols (1977).

Dixon (1964) showed that *Seirospora griffithsiana* Harvey was a later name for the species originally described as *Callithamnion versicolor* β *seirospermum* Harvey. Dixon (1971) noted that there appeared to be no clear-cut differences between *Seirospora seirosperma* and *S. interrupta*, the other species of *Seirospora* reported from the British Isles, and concluded that the two taxa should be merged. Problems with typification and the identity of *Conferva interrupta* J. E. Smith, the older name, prevented further progress at this time. *C. interrupta* was based on bisporangial collections from Brighton. Some of these bisporangia show partial second zonate divisions, which is diagnostic of *Seirospora*. Accordingly, *S. interrupta* is the correct name for the single species of *Seirospora* found in the British Isles.

Seirospora interrupta (J. E. Smith) Schmitz (1893b), p. 281.

Lectotype: BM-K. Syntypes: BM-K; LD 18029. Sussex (Brighton), 28 vii 1807, *Borrer*.

Conferva interrupta J. E. Smith (1808), pl. 1838.
Callithamnion interruptum (J. E. Smith) C. Agardh (1828), p. 174.
Callithamnion versicolor β *seirospermum* Harvey (1834), p. 302 (see Dixon, 1964).
Callithamnion seirospermum (Harvey) Harvey (1841), p. 113.
Seirospora griffithsiana Harvey (1846), pl. 21 (see Dixon, 1964).
Callithamnion hormocarpum Holmes (1873), p. 1.
Callithamnion byssoides f. *seirosporifera* Holmes & Batters (1891), p. 98.
Seirospora seirosperma (Harvey) P. Dixon (1964), p. 65.

Thalli erect, consisting of axes growing singly from endophytic filaments or discoid filamentous holdfast, 2-12 cm high, pyramidal or irregularly rounded, with a single main axis below, 0.1-0.4 mm in diameter, the main axis either distinct throughout or indistinguishable above amongst major laterals, rose-pink to bright red in colour, delicate and flaccid.

Main axes enlarging from apical cells 8-12 µm in diameter to axial cells 3 diameters long, with median constrictions; *first-order laterals* developing from the third or fourth cell behind apex, one per cell, in a 1/4-1/8 spiral arrangement, decreasing in length upwards, often regularly, bearing a branch on every cell; *second-order laterals* distichous, spiral towards the apices, increasing in length upwards, bearing a lateral on every cell in an approximately 1/4 spiral; *third-order laterals* incurved, forming dense clusters, typically branched alternate-distichously, bearing incurved branchlets with cylindrical cells and bluntly rounded apices; terminal hairs present only on female plants; *rhizoidal*

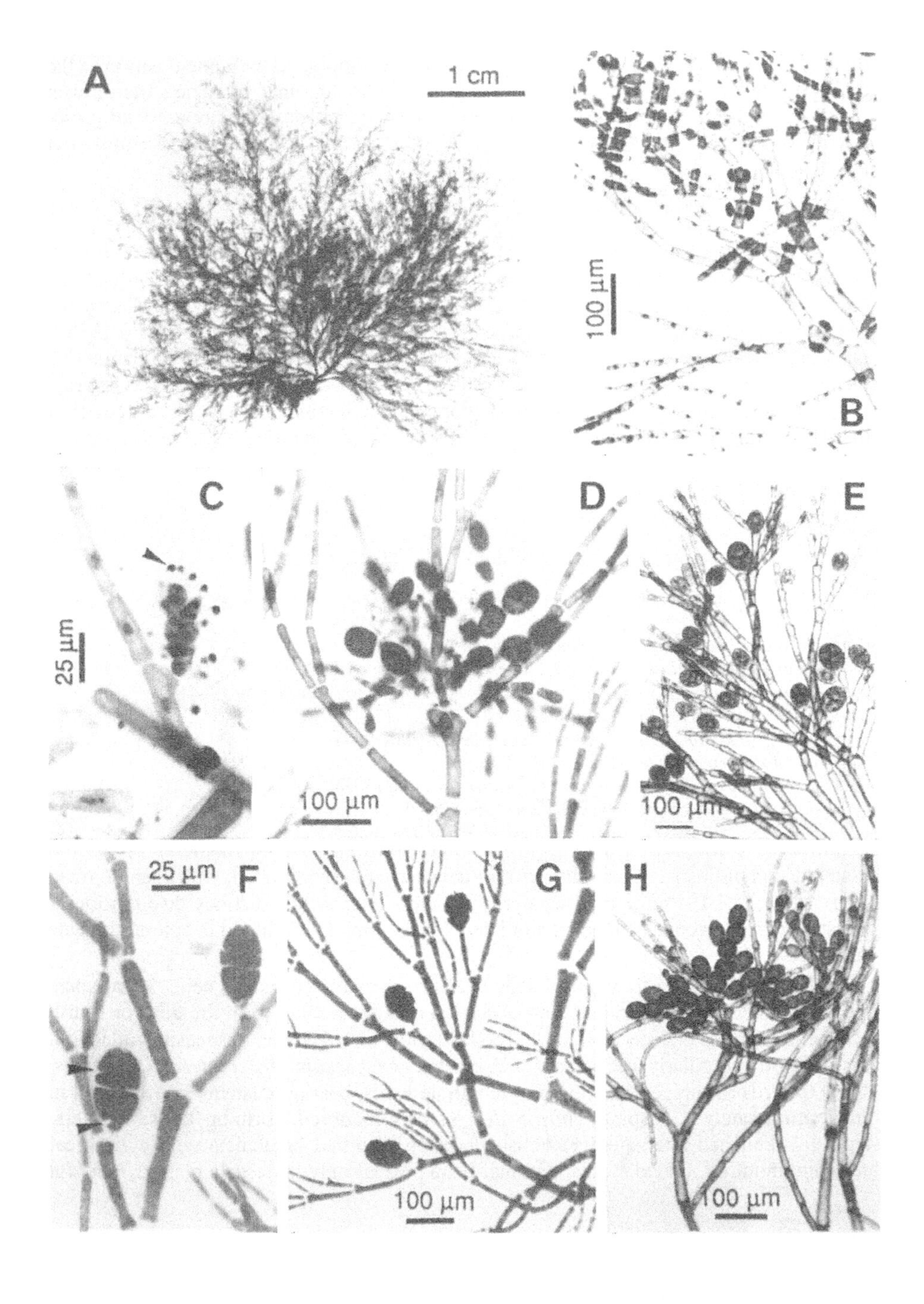
A
1 cm
B
100 µm
C
25 µm
D
100 µm
E
100 µm
F
25 µm
G
100 µm
H
100 µm

filaments 8-60 μm wide, branched, forming a thick cortex on main axes and first-order laterals but lacking adventitious branchlets; *plastids* ribbon-like.

Gametophytes dioecious. *Spermatangial branchlets* borne adaxially on last two orders of branching, 1(-2) per cell, 32-58 μm long x 16-28 μm wide, often terminal on an elongate, pigmented, sterile cell, erect, slightly curved adaxially, 2-5 cells in length, basal 1-3 cells forming 1-3-celled lateral branches, all cells developing into spermatangial mother cells bearing 2-4 ellipsoid spermatangia, 6.5-8 x 5-6.5 μm. *Carpogonial branches* L-shaped, ephemeral if unfertilized but periaxial cells persistent and occasionally giving rise to rhizoidal filaments; *cystocarps* developing as several radiating, repeatedly branched filaments about 7 μm wide, of cells 2 diameters long, maturing into branched chains of spherical to ellipsoid carposporangia 44-56 μm long x 40-48 μm wide, each chain enclosed in a separate mucilage sheath, 270-465 μm in total diameter. *Tetrasporangia* formed adaxially on last 2-3 orders of branching of tetrasporophytes, on up to 4 adjacent cells, 1 per cell, often in addition to a branchlet, sessile or rarely on an elongate pedicel, ellipsoid when mature, with a length: width ratio of 1.2-1.5, 60-70 μm long x 44-52 μm wide, tetrahedrally divided. *Bisporangia* borne on separate plants, adaxial, often axillary to branchlet, sessile or pedicellate on an elongate cell, initially curved and an elongate, ellipsoid when mature, with a length: width ratio of 1.4-1.8, 64-80 x 40-52 μm, divided by an oblique or transverse wall, frequently with irregular second divisions cutting off a small cell basally and partial zonate furrowing of one or both sporangia. *Seirosporangia* formed on separate plants, developing near apices of laterals, arising by modification of vegetative filaments that branch several times, each cell becoming spherical to ellipsoid, 32-48 μm long x 44 μm wide, including the thick walls, occasionally partially cleaved tetrahedrally; seirosporangial plants commonly also bearing apparently non-functional, partially cleaved, sessile or pedicellate tetrasporangia, replaced very rarely by sessile or pedicellate polysporangia about 56 μm long x 44 μm wide, containing about 12 polysporangia 18-23 μm long.

Epilithic on pebbles, epizoic on live and dead molluscs and worm tubes, and epiphytic on a wide range of large and small algae; rarely intertidal in lower-shore pools and on shaded vertical faces in moderately wave-exposed locations, commonly from extreme low

---

Fig. 40. *Seirospora interrupta*

(A) Habit, with main axes densely clothed by tufted branchlets (A-B Devon, Nov.). (B) Axes of female plant with conspicuous paired periaxial cells, stained with haematoxylin to show single nucleus in each cell. (C) Spermatangial branchlet, formed by male thallus that developed from cultured tetraspores (Galway, Nov.), with apical nuclei in spermatangia (arrow). (D) Cystocarp consisting of loose chains of developing and mature carposporangia (Pembroke, Oct.). (E) Tetrasporangia (Galway, Jan.). (F) Bisporangia, with partial zonate cleaving (arrows) (Galway, Sept.). (G) Partially cleaved polysporangia (Inverness, June). (H) Seirosporangia in cultured plant (Donegal, Dec.).

water to 20 m depth, at moderately to extremely wave-sheltered sites, often with exposure to tidal currents.

Widely distributed in the British Isles, northwards to Shetland, although there are few records for eastern coasts.

Norway to Portugal; Mediterranean; Nova Scotia and Massachusetts to New Jersey (Feldmann-Mazoyer, 1941; Bird & Johnson, 1984; South & Tittley, 1986).

Large plants are present in the field throughout the year and probably represent overlapping ephemeral generations, since seirosporangial isolates from Nova Scotia and Massachusetts reproduced within 3 weeks in culture at 15°C (Bird & Johnson, 1984; Plattner & Nichols, 1977). Gametophytes and tetrasporophytes make up a very small proportion of populations, commonly occurring as single plants amongst numerous seirosporangial plants. Tetrasporangia occur sparsely on females. Spermatangia have been recorded for February and September-October; procarps and cystocarps for February, April, July, August and October-November; tetrasporangia on tetrasporophytes for June, August, October and November; bisporangia in February, March, July-September and November-December; seirosporangia in February, March and May-December, mixed with tetrasporangia in February-March, August and November, and mixed with polysporangia on a plant from Inverness (17 vi 1988).

Tetraspores from a plant collected in Galway (24 vi 1989) gave rise in culture to dioecious gametophytes, spermatangia maturing earlier than procarps. Cystocarps formed by females grown in the presence of males released carpospores that grew into a further tetrasporangial generation. A bisporangial isolate from Donegal (5 v 1989) formed several generations of bisporophytes at intervals of 2-6 weeks, bisporangia often developing when plants were only 2.5 mm high. Bisporangia were mitotic and diploid, with approximately 50-55 chromosomes. They contained only two nuclei, but some showed various patterns of partial second division: zonate, tetrahedral or polysporangial. A seirosporangial isolate from Donegal (11 xii 88) formed only seirosporangia in culture.

Plants vary considerably in overall shape, those with well-developed first-order laterals lacking a distinct main axis, producing a rounded or lobed rather than regularly pyramidal outline.

Seirosporangial plants of *S. interrupta* are easily identifiable. Gametangial and tetrasporangial plants of *S. interrupta* may resemble *Aglaothamnion byssoides* (q. v.), but can be identified by the adherent cortication on main axes, which is lacking in *A. byssoides* gametophytes and tetrasporophytes. Bisporangial thalli of *A. byssoides* are corticate, however, and it can be difficult to distinguish bisporophytes of *S. interrupta* and *A. byssoides*. Bisporangia in *S. interrupta* form partial second divisions parallel to the first division (partially zonate) whereas in *A. byssoides* they are perpendicular (partially cruciate). *S. interrupta* could be misidentified as *A. roseum*, which has corticate main axes. Male plants can be distinguished by the form of the spermatangial branchlets, which are narrow and little-branched in *S. interrupta*, and ovoid and much-branched in *A. roseum*; cystocarps of *S. interrupta* form separate chains of carposporangia in contrast to the compact globular clusters of *A. roseum*. Tetrasporangia in *S. interrupta* are commonly borne on a cell that also bears a branchlet, whereas in *A. roseum* they never occur with a branchlet. *S. interrupta* differs from *Callithamnion corymbosum* (q. v.), which is also corticate, in its uninucleate cells in contrast to the conspicuously multinucleate cells of *C. corymbosum*.

CALLITHAMNION Lyngbye

CALLITHAMNION Lyngbye (1819), p. 123.

Lectotype species: *C. corymbosum* (J. E. Smith) Lyngbye (1819), p. 125.

Thalli uniseriate, consisting of 1-several erect main axes up to 25 (-40) cm high, attached by a discoid rhizoidal holdfast and bearing 3 (-4) orders of indeterminate branches and 1-3 orders of determinate branchlets; axes spirally, alternate-distichously or pseudo-dichotomously branched, and sparsely to densely corticated by descending rhizoidal filaments that often contribute rhizoids to the basal disc; determinate lateral branchlets spirally, alternately to subdichotomously branched, sometimes overtopping the axis.

Growth of indeterminate axes and determinate filaments by oblique and transverse divisions of multinucleate apical cells; each axial cell cutting off a single lateral initial from its distal end and elongating below; rhizoids initiated singly or in clusters from basal cells of lateral filaments, growing downwards and covering the main axes; cells multinucleate; plastids discoid to ribbon-like.

Gametophytes dioecious or monoecious. Spermatangial filaments clustered, on adaxial surface of cells of branchlets or encircling them; procarps formed near the tips of potentially indeterminate axes, developing as in *Aglaothamnion*; carpogonial branches L-shaped, the first three cells horizontal and the fourth (the carpogonium) vertical. Postfertilization stages taking place as in *Aglaothamnion*; cystocarps typically consisting of 1-4 globular or heart-shaped gonimolobes. Tetrasporangia sessile, 1-3 per cell, adaxial or lateral on cells of determinate branchlets, ellipsoid to subspherical, tetrahedrally divided.

References: Oltmanns (1922), Rosenvinge (1923-24), L'Hardy-Halos (1971), Itono (1977), Dixon & Price (1981).

The multinucleate vegetative cells that distinguish *Callithamnion* species from *Aglaothamnion* can be seen in live or fixed material, as described in the notes for *Aglaothamnion*. Members of this genus in the British Isles are readily discriminated by vegetative features such as the branching pattern.

KEY TO SPECIES

1 Thalli delicate and flaccid; main axes <0.2(-0.3) mm in diameter; axial cells 6-12 diameters long, visible through sparse cortical filaments; adventitious branchlets absent .... *C. corymbosum*

Thalli robust; main axes 0.2-3.0 mm in diameter; axial cells ≤6 diameters long, obscured by dense cortex forming sparse to abundant adventitious branchlets 2
2 Apical cells of main axes distinct, not surrounded by laterals; ultimate branchlets short, spine-like, alternate-distichous .... *C. tetricum*
Apical cells of main axes surrounded and concealed by dense clusters of laterals 3
3 Ultimate branchlets 1-4(-6) cells long; terminal cell only slightly smaller than sub-terminal cell .... *C. granulatum*
Ultimate branchlets ≥6 cells long; terminal cell minute, conical .... *C. tetragonum*

Callithamnion corymbosum (J. E. Smith) Lyngbye (1819), p. 125.

Lectotype: BM-K. Sussex (Brighton), July 1811, *Borrer* (see Dixon & Price, 1981).

Conferva corymbosa J. E. Smith (1811), pl. 2352.
Ceramium versicolor C. Agardh (1824), p. 140 (see Dixon & Price, 1981).

Thalli consisting of single erect axis attached by discoid holdfast up to 500 μm diameter, formed by aggregation of corticating rhizoidal filaments, 0.6-5.5 cm high, with distinct naked main axes below, indistinguishable above amongst laterals, obconical or irregularly rounded in outline, orange-pink to deep red in colour, flaccid and delicate.

Main axes enlarging from apical cells 8-10 μm wide to 40-300 μm in diameter; axial cells 1.5-3 diameters long near holdfast, increasing in length upwards to 6-12 diameters long, with a median constriction; *first-order laterals* developing from the third or fourth cell behind apex, with corymbose apices, irregularly distichous below, otherwise borne in a 1/3-1/5 spiral arrangement, increasing in length upwards, some equal in length and almost equal in diameter to main axis, branching appearing pseudodichotomous below; *second-order laterals* borne on every cell in a spiral pattern, branched similarly themselves, ultimate branchlets usually short, one to a few cells in length, usually bearing numerous hairs; *rhizoidal filaments* branching, corticating axial cells to a varying degree but lacking adventitious branchlets; *plastids* discoid to ribbon-like.

Gametophytes dioecious. *Spermatangial branchlets* borne adaxially on last 2-3 orders of branching, (1-)2-4 per cell, all of which coalesce into a single hemispherical cushion 40-64 μm long x 18-24 μm thick, each component cell developing into a spermatangial mother cell bearing several spermatangia, 6.5-8 x 2.5-3 μm. *Carpogonial branches* L-shaped; *cystocarps* with 1-4 gonimolobes, often 2 pairs of different ages, gonimolobes

Fig. 41. *Callithamnion corymbosum*

(A) Habit (A-B Galway, Feb.). (B) Detail of thallus with pseudodichotomously branched major axes (arrow). (C) Axes stained with haematoxylin showing numerous nuclei in each cell (arrows) (Down, Sept.). (D) Cystocarp consisting of several globular gonimolobes (Inverness, July). (E) Tetrasporangia, many appearing axillary to branchlets (arrow) (as A).

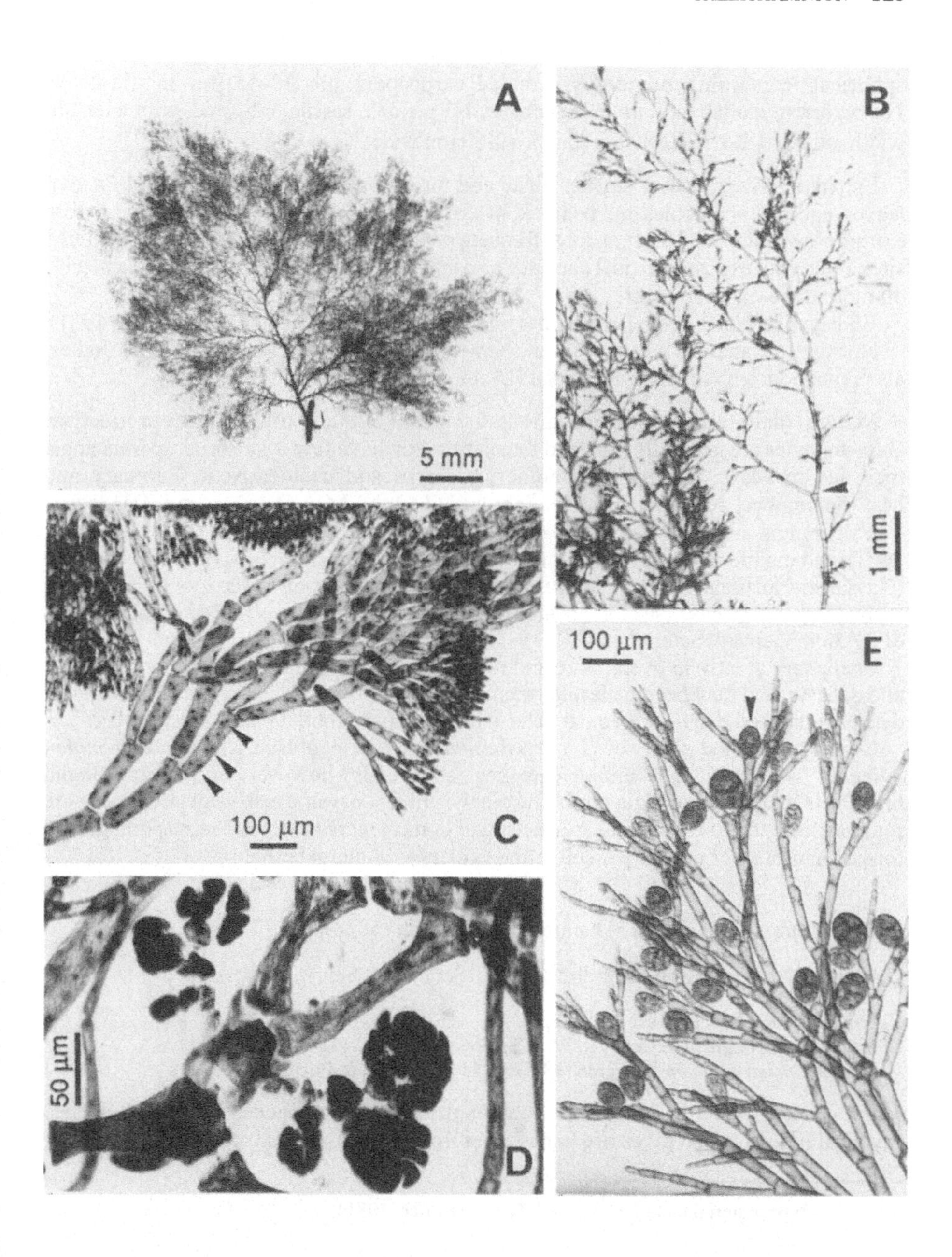
A
5 mm
B
1 mm
C
100 µm
D
50 µm
E
100 µm

spherical, containing numerous rounded carposporangia 30-45 µm in diameter. *Tetrasporangia* often axillary to branchlets, 1-3 per cell, sessile, ellipsoid, with a length: width ratio of 1.2-1.5, 60-80 µm long x 40-60 µm wide.

Epiphytic on a range of smaller algae and maerl, on *Laminaria* fronds and *Zostera* leaves, epilithic on pebbles and bedrock, occasionally on marina pontoons, subtidal from extreme low water to 12(-22) m, most frequent at moderately to extremely wave-sheltered sites, often with exposure to tidal currents; also intertidal in lower-shore pools and channels on more wave-exposed coasts

Widely distributed in the British Isles, recorded from all coasts (Dixon & Price, 1981).

Norway to Canaries; Mediterranean; Newfoundland to New Jersey and West Indies; also reported from Australia and Japan (Dixon & Price, 1981).

Mature thalli are found throughout the year, but are often non-reproductive. Gametophytes are generally rare, and females usually have few cystocarps; spermatangia recorded in May, July and September; procarps and cystocarps in February and May-September; tetrasporangia in February, March, May-October and December. Tetrasporangia and cystocarps occasionally occur on the same thallus (Dixon & Price, 1981), and the life history of isolates from Helgoland included mixed-phase reproduction (Hassinger-Huizinga, 1952). Various chromosome numbers of about $n = 30$ have been reported in this species (see Dixon & Price, 1981), and we have obtained similar counts of 27-30 in Galway material.

Thalli vary greatly in overall size and robustness. Male gametophytes are often small and delicate, and may become fertile when only 0.5 cm in length. Intertidal plants appear densely branched due to short axial cells, whereas some subtidal plants are very lax.

Densely branched plants of *C. corymbosum* could be confused with *Aglaothamnion roseum*. *A. roseum* has uninucleate vegetative cells, however, and spermatangial branchlets and tetrasporangia are borne singly. In *C. corymbosum,* vegetative cells are multinucleate, there are several spermatangial branchlets per cell, and tetrasporangia are formed in groups of up to 3 per cell, often axillary to a lateral branch.

Callithamnion granulatum (Ducluzeau) C. Agardh (1828), p. 177.

Neotype: BM-K, ex Herb. Agardh.* France (Sette, now Sète), *J. Agardh*.

Ceramium granulatum Ducluzeau (1806), p. 72.
Callithamnion spongiosum Harvey in W. J. Hooker (1833), p. 346.
Callithamnion harveyanum J. Agardh (1841), p. 45 (see Dixon & Price, 1981).

Thalli erect, growing in dense tufts from tangled holdfast composed of much-branched rhizoidal filaments that gives rise to new axes from the margins; erect axes 3-15 cm high,

* Original type material cannot be located (Dixon & Price, 1981).

with a distinct main axis below, soon becoming indistinguishable amongst laterals, typically with a pyramidal outline, dark brownish-red in colour, bleaching to pale brown or straw-coloured, with a marked bluish iridescence in all or part of the thallus, spongy and soft when out of water, delicate in water; old axes becoming cartilaginous.

Main axes enlarging from apical cells 10 µm wide to a maximum of 0.4-3 mm in diameter; *first-order laterals* arising from the second to third cell from the apex, one per cell, developing into densely-packed corymbose clusters that conceal main apex, borne in an irregularly spiral pattern; *second-order laterals* indeterminate, often irregularly distichously branched below, clothed with determinate third-order laterals that branch pseudodichotomously 2-9 times; *ultimate laterals* 1-4(-6) cells long, straight, spine-like with conical apical cells and often bearing terminal hairs; *rhizoidal filaments* much-branched, forming a thick cortex and obscuring axial cells and cells of first- and second-order laterals, also giving rise to a dense covering of simple or pseudodichotomously branched adventitious branchlets; *plastids* discoid in young cells, becoming irregularly elongate in older cells.

Gametophytes dioecious or occasionally monoecious. *Spermatangial branchlets* formed on determinate laterals, one to several per cell, much-branched, all branchlets on each cell coalescing into hemispherical or almost globular cushions 45-60 µm in diameter, covered with spermatangia measuring 6.5-8 x 4-6 µm, each cluster remaining distinct. *Carpogonial branches* L-shaped; *cystocarps* consisting of several globular gonimolobes, often 2 pairs of different ages, 260-330 µm in diameter, containing numerous angular or rounded carposporangia 25-52(-70) µm in diameter. *Tetrasporangia* formed on determinate laterals, often on the same cells as vegetative branches, sessile, 1 per cell, pyriform before division, subspherical to ellipsoid when mature, length: width ratio 1.1-1.6, 60-76 µm long x 48-70 µm wide, with walls 4-6 µm thick.

Epiphytic, especially on *Corallina* spp., epilithic on bedrock, and epizoic on mussels and limpets, usually in pools, rarely emersed in shaded or wet places in the lower intertidal of moderately to extremely wave-exposed shores.

Widely distributed in the British Isles but rare on northern coasts (Dixon & Price, 1981). Shetland to Morocco, ?W. Africa; Mediterranean (Dixon & Price, 1981).

Thalli can persist for 2-3 years; perennating plants become straggly as they lose lateral apices, which regenerate in the next growing season (Price, 1978). Spermatangia have been recorded for June and August-October; procarps and cystocarps in May, August and September; and tetrasporangia in February, April-June and August-October. Chromosome counts of 28-33 were obtained for *C. granulatum* (as '*Callithamnion purpurascens*') by Harris (1962).

Harvey in W. J. Hooker (1833) described *Callithamnion spongiosum* from Ireland, but indicated that it might be conspecific with *C. granulatum* from southern Europe. J. Agardh (1851) placed *C. spongiosum* in synonymy with *C. granulatum*, and Harris (1966) concurred with this opinion (although he used the name *C. purpurascens* for *C. granulatum*). Cotton (1912), however, found that at Clare Island, Mayo, two distinct forms

500 µm
A
1 cm
B
C
100 µm
D
50 µm
E
100 µm
100 µm
F

of *C. granulatum* could be distinguished by compactness of branching and habitat differences. The less compact, larger form that grew in damp shady places corresponded to Harvey's *C. spongiosum*. Dixon & Price (1981) considered that *C. spongiosum* differed from *C. granulatum* principally in the lengths of the ultimate branchlets, which were 1-2 and 3-4 cells long respectively. They reported that *C. spongiosum* occurred rarely, usually sporadically among *C. granulatum*, and might be in the process of speciating from it. In view of the variation in length of ultimate branchlets, even within a single specimen, and in the absence of any reproductive differences or studies of interfertility, we do not believe that species status for *C. spongiosum* is justifiable at present.

Distinctions between *Aglaothamnion sepositum* and *C. granulatum*, which may be closely similar in habit, are given under *A. sepositum*. Compact thalli of *Dasya hutchinsiae* (q. v.) and *Halurus equisetifolius* (q. v.) have superficial resemblances to *C. granulatum*, but in *Dasya* the main axes are polysiphonous, and in *Halurus* whorl-branches are borne in whorls around main axes.

Callithamnion tetragonum (Withering) S. F. Gray (1821), p. 329.

Lectotype: GL. Dorset (Portland), *Velley & Stackhouse* (see Dixon & Price, 1981).

Conferva tetragona Withering (1796), p. 405.
Ceramium brachiatum Bonnemaison (1828), p. 136, nom. illeg. (see Dixon & Price, 1981).
Callithamnion fruticulosum J. Agardh (1841), p. 46 (see Dixon & Price, 1981).

Thalli erect, growing singly from discoid holdfast 0.3-1.2 mm in diameter, 2-11 cm high, consisting of a single main axis bearing laterals decreasing in length upwards, giving an irregularly pyramidal outline, bright pink to brownish-red in colour; main axes robust and flexible, the branchlets turgid.

Main axes enlarging from apical cells 8 μm in diameter to a maximum of 225-550 (1000) μm; axial cells 0.3-1 diameter long near holdfast, elongating upwards to 2-6 diameters long; *first-order laterals* developing from third cell behind apex, forming a dense cluster surrounding and overtopping apex, borne initially in an alternate-distichous arrangement, later in a 1/3-1/4 spiral (see Rosenvinge, 1923-24, for details), bearing a branchlet on every cell, the first one abaxial, then in an alternate-distichous series, replaced distally with a 1/3-1/4 spiral; *ultimate laterals* 6-10 cells long, straight or curved, either tapering from the base and composed of cylindrical cells about 40 μm diameter, or awl-shaped with barrel-shaped cells up to 165 μm wide, terminating in a minute conical cell; *rhizoidal filaments* arising from axial cells and basal cells of laterals, circling around the axial cell

---

Fig. 42. *Callithamnion granulatum*

(A) Habit (Donegal, May). (B) Detail of thallus with dense branches (B-C Antrim, June). (C) Last-order branches, less than 6 cells in length (Down, Sept.). (D) Spermatangial branchlets forming globose cushions on branches (Clare, June). (E) Cystocarp with globular gonimoblasts (Caernavon, undated). (F) Globose tetrasporangia (as C).

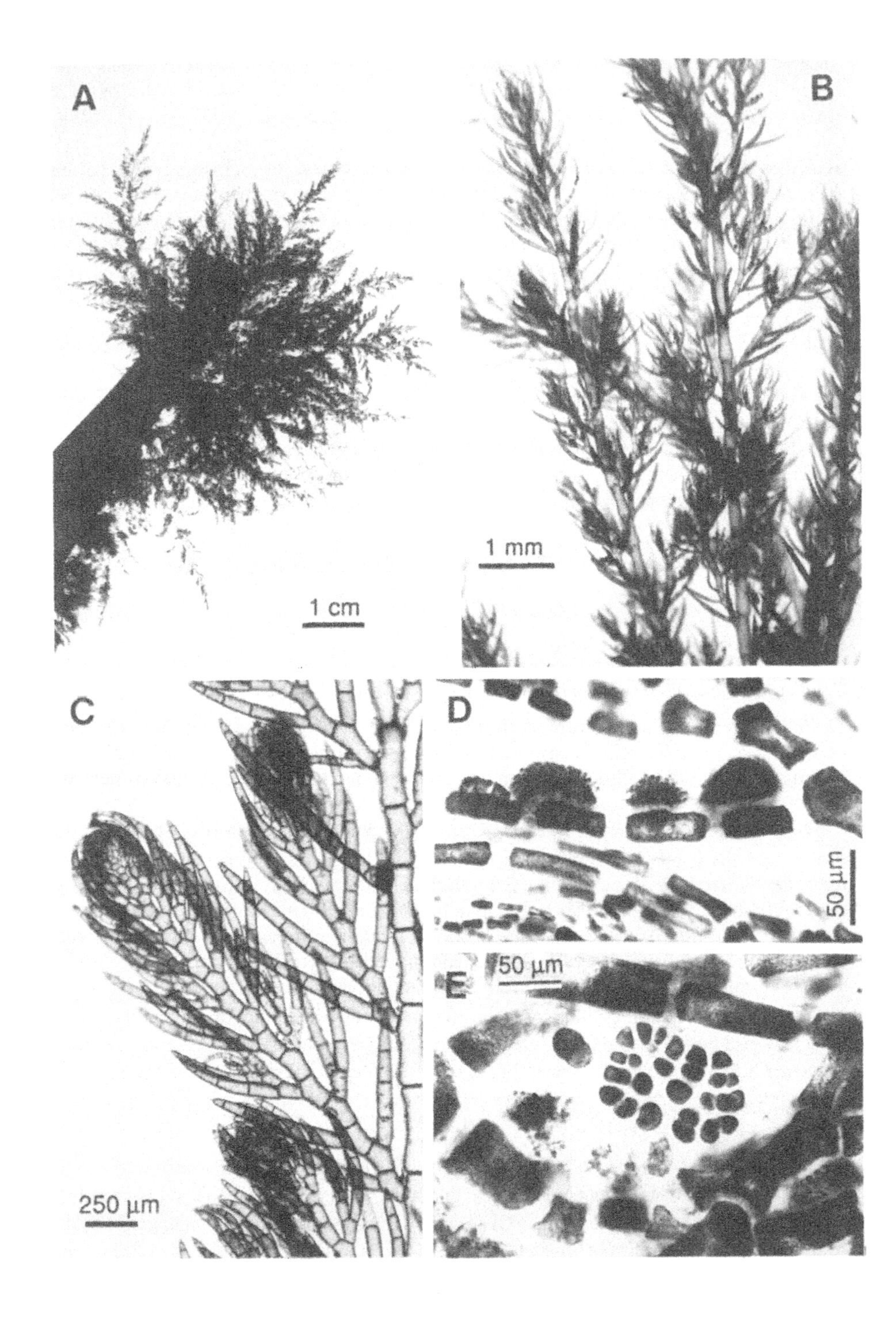
A
B
1 cm
1 mm
C
D
E
50 µm
50 µm
250 µm

before growing downwards within the walls, forming sparse to dense cortication and giving rise occasionally to sparsely branched adventitious branchlets; *plastids* irregularly discoid at apices, elongating to irregularly constricted ribbons.

Gametophytes monoecious. *Spermatangial branchlets* formed adaxially, 1-2 per cell, much-branched, all branchlets on each cell coalescing into a compact hemispherical to hemi-ovoid cushion 36-73 µm diameter and 32-40 µm high, individual cushions remaining distinct, covered with spermatangia measuring 6.5-8.5 x 4-5 µm; released spermatia 3-4.5 µm diameter. *Carpogonial branches* L-shaped; *cystocarps* developing as pairs of regularly cordate (heart-shaped) gonimolobes one cell thick, mature gonimolobes regularly or irregularly cordate, 160-240 µm long x 180-280 µm wide and 2-3 cells thick, containing numerous angular carposporangia 28-64 µm in diameter, surrounded by a membrane 3-6 µm thick; further pairs of cordate gonimolobes developing secondarily. *Tetrasporangia* formed adaxially on last two orders of branching, sessile, 1-3 per cell, at different stages of maturity, ellipsoid, length: width ratio 1.2-1.5, 52-92 µm long x 40-62 µm wide, with walls 3-5 µm thick.

Epiphytic on a wide range of algae, rarely epilithic, in pools and on open surfaces in the lower intertidal of wave-exposed shores; epiphytic on fronds, less often on holdfasts and stipes of *Laminaria hyperborea*, rarely on other algae, from low water to at least 15 m, at sites with either strong wave or current exposure.

Generally distributed in the British Isles (Dixon & Price, 1981).

Iceland and Norway to Portugal; Mediterranean; ?Canaries, Cape Verde; eastern USA (Maine to New Jersey) (Taylor, 1957; Dixon & Price, 1981).

Large thalli can be found throughout the year, but are non-reproductive in early spring. Spermatangia recorded for April and June-September, procarps and cystocarps in April-October, and tetrasporangia in April-December. Plants with mixed cystocarps and tetrasporangia, originally noted by Buffham (1884), are more common in the British Isles than those with only gametangia or tetrasporangia. A Swedish isolate underwent a *Polysiphonia*-type life history in culture for several generations (Rueness & Rueness, 1985), then gave rise to a few mixed-phase plants. Gametangia formed by tetrasporophytes were non-functional, while tetraspores released by gametophytes failed to grow. Mixed-phase reproduction continued for several generations. The haploid chromosome number is in the range 28-33 (Harris, 1962).

The ultimate laterals vary from wide and awl-shaped to narrow and tapering, the latter

---

Fig. 43. *Callithamnion tetragonum*

(A) Habit on blade tip of *Laminaria hyperborea* (Clare, June). (B) Detail of thallus with open branching (Donegal, Aug.). (C) Regularly alternate branches tapering to minute terminal cells (Devon, Nov.). (D) Spermatangial branchlets forming hemispherical cushions on branches (D-E Cornwall, July). (E) Cystocarp with cordate gonimoblast.

form diagnostic of *C. tetragonum* var. *brachiatum* (Bonnemaison) J. Agardh, which is now considered to be a growth form (Dixon & Price, 1981). *Callithamnion baileyi* Harvey (1853, p. 231) from eastern North America was placed in synonymy with *C. tetragonum* by Rosenvinge (1923-24), although their conspecificity has not been universally accepted (e. g., by Taylor, 1957). Isolates of *C. baileyi* (Whittick & West, 1979) followed a life history in culture similar to that of Swedish *C. tetragonum*.

In its typical habitat, epiphytic on *Laminaria hyperborea*, *C. tetragonum* is unlikely to be misidentified. Epilithic intertidal plants can be distinguished from lax plants of *Aglaothamnion sepositum* (q. v.) by the hemispherical spermatangial cushions, cordate gonimolobes and ovoid tetrasporangia in contrast to the continuous spermatangial layer, spherical gonimolobes and spherical tetrasporangia of *A. sepositum*.

Callithamnion tetricum (Dillwyn) S. F. Gray (1821), p. 324.

Lectotype: WELT, Herb. Sylvanus Thompson A4502. Glamorgan (Swansea) (see Dixon & Price, 1981).

Conferva tetrica Dillwyn (1806), pl. 81.

Thalli erect, forming dense tufts from tangled holdfasts up to 10 mm in diameter, composed of much-branched rhizoidal filaments, that give rise to further young axes from the margins; erect axes (1-)4-25 cm high, shaggy and irregularly shaped, resembling frayed rope, with a distinct main axis below, soon becoming indistinguishable amongst laterals, decreasing in diameter upwards from a maximum of 1-3 mm, dull purplish-red or brownish-red in colour, bleaching to yellow, with a marked blue or purple iridescence, coarse and rigid in texture.

Main axes with a prominent apical cell; *first-order laterals* borne in an irregularly spiral or irregularly distichous arrangement, similarly branched themselves; *second-order laterals* unbranched for first few cells, then bearing an alternate-distichous series of branches, these branching alternate-distichously for 1-2 further orders of branching, all laterals acute, those of each order of branching shorter than the axis bearing them; *rhizoidal filaments* much-branched, tangled, forming a thick, loosely adherent cortex, obscuring axial cells and cells of first- and second-order laterals, and giving rise to a dense covering of adventitious branchlets that are simple or branched, recurved on lower axes, straight nearer apices, and borne at an acute angle; *plastids* irregularly discoid.

Gametophytes dioecious. *Spermatangial branchlets* formed on short, recurved, ultimate branches, 1-3 per cell, much-branched, coalescing into a continuous layer about

---

Fig. 44. *Callithamnion tetricum*

(A) Habit (A-F Pembroke, Dec.). (B) Detail of thallus. (C) Spermatangial branchlets forming a continuous irregular layer on last-order branch. (D) Regularly alternate last-order branches. (E) Cystocarp with irregularly globular gonimoblasts. (F) Branch (live), with multiple nuclei visible in each cell (arrows) and sparse, sessile tetrasporangia.

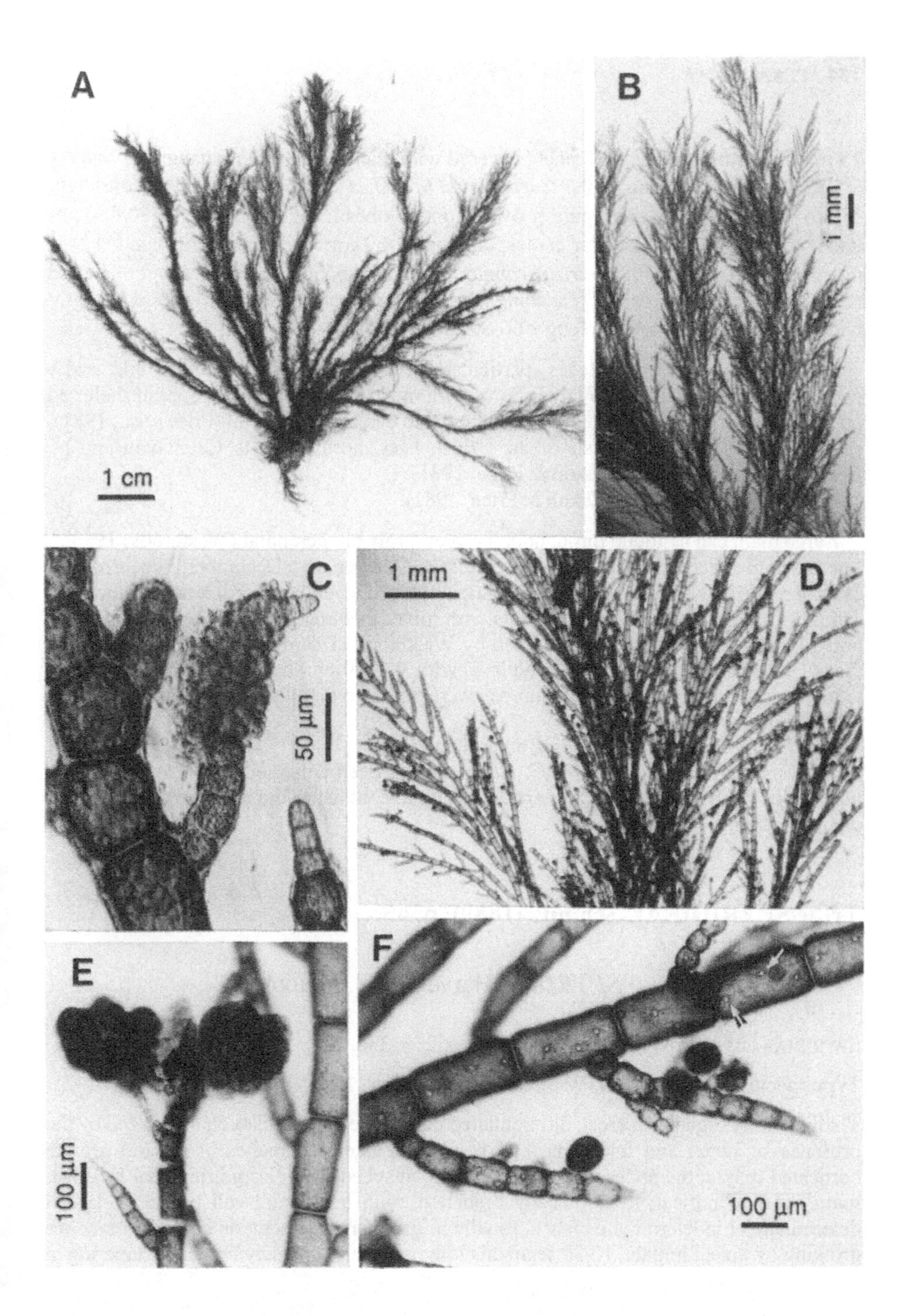
A
1 cm
B
1 mm
C
50 µm
D
1 mm
E
100 µm
F
100 µm

25 µm thick surrounding the branch, covered with thick-walled spermatangia measuring 6.5-8 x 3-4 µm, excluding walls. *Carpogonial branches* L-shaped; *cystocarps* consisting of two pairs of globular to irregularly ovoid gonimolobes 150-240 µm long x 150-180 µm wide, containing 15-50 angular carposporangia 42-60 µm in diameter, surrounded by a membrane 10 µm thick. *Tetrasporangia* formed adaxially and laterally on curved branchlets of last two orders of branching, sessile, 1-2 per cell, spherical to ellipsoid, length: width ratio 1.0-1.3, 77-90 µm long x 66-90 µm wide, with walls 6 µm thick.

Epilithic on emersed surfaces, particularly common on vertical rocks and under overhangs, in the lower intertidal of shores with moderate wave-exposure, or in sheltered positions on very wave-exposed shores; subtidal to 6 m in S. Cornwall (Price *et al.*, 1981).

South and south-west coasts of the British Isles, northwards to Caernavon and N. Ireland, eastwards to Kent (Dixon & Price, 1981).

British Isles to Morocco (Dixon & Price, 1981).

Large thalli can be found throughout the year, probably perennial (Westbrook, 1930a). Spermatangia recorded for April-June, August-October and December; procarps and cystocarps in February, April, June-August and December; and tetrasporangia in January-October and December. Plants with mixed cystocarps and tetrasporangia were noted by Buffham (1884) and discussed by Westbrook (1930a). Harris (1962) reported that *C. tetricum* appeared to be polyploid, with 90-100 chromosomes. We found that tetrasporocytes in material from Pembroke were mitotic rather than meiotic, and contained >70 chromosomes.

C. tetricum shows little variation, other than in size.

Some herbarium specimens of *C. tetricum* have been confused with *Pleonosporium borreri* (q. v.), but *P. borreri* axes are ecorticate in contrast to the well-developed cortex of *C. tetricum*.

Tribe SPYRIDIEAE Schmitz (1889), p. 451.

SPYRIDIA Harvey in W. J. Hooker

SPYRIDIA Harvey in W. J. Hooker (1833), pp. 259, 336.

Type species: *S. filamentosa* (Wulfen) Harvey in W. J. Hooker (1833), p. 337.

Thalli radially organized, erect, differentiated into indeterminate axes characterized by the presence of nodal and internodal cell bands and lateral branches of limited growth corticated only at the nodes, solitary or in tufts, attached by a discoid rhizoidal holdfast; main axes cylindrical, monopodially branched, with each axial cell bearing a lateral determinate or indeterminate branch, usually in spiral arrangement; determinate branches growing by apical initials, 10-30 segments long, simple or sparsely branched, tapering at the base and apex, each segment corticated at the nodes by a band of 6-8 periaxial cells and limited cortex, separated by internodal regions in which the elongated axial cells are exposed; adventitious determinate branches present in some species, opposite or whorled; indeterminate branches replacing determinate branches at intervals along the axis, each segment surrounded by 11-16 nodal periaxial cells and usually twice as many internodal cells; descending rhizoids issuing from nodal and internodal cells and corticating the axis lightly to heavily; adventitious indeterminate branches variable in number, originating from periaxial and cortical cells.

Growth by slightly oblique division of the apical cell with the high side giving rise to determinate and indeterminate lateral branches; periaxial cells cut off in alternate (rhodomelacean) sequence beginning on either side of the branch initial, each nodal periaxial cell cutting off two cells from its lower end which elongate and attach by secondary pit connections to periaxial cells in the segment below; rhizoidal cells frequently forming conjunctor cells from their lower ends which link to periaxial cells and rhizoidal cells forming secondary pit connections; cells uninucleate; plastids ribbon-like to filiform.

Gametophytes dioecious. Spermatangial parent cells produced from cortical cells and forming a continuous layer in the mid-region of determinate branches, each typically bearing 3 spermatangia covered by the outer surface layer; spermatia released *en masse*, each spermatium spherical, covered by 6-8 evenly distributed appendages. Procarps produced on indeterminate lateral branches of limited growth, 3-5 in alternate segments; each fertile segment segment bearing 3 periaxial cells, one of which produces a single, horizontally curved, 4-celled carpogonial branch; sterile groups absent; the supporting cell and each of the other two periaxial cells cutting off auxiliary cells after fertilization; the fertilized carpogonium cutting off 2-3 connecting cells each of which potentially fuses with a process from one of the auxiliary cells, and gives rise to 2 (rarely 3) gonimoblasts in the same segment; auxiliary cell dividing into a foot cell containing a haploid and a diploid nucleus and a gonimoblast initial containing a diploid nucleus and a protein body; gonimoblast initial elongating, cutting off a distal primary gonimolobe initial followed by 2-3 lateral gonimolobe initials; gonimoblasts surrounded by a pericarp composed of ternately to dichotomously branched filaments linked by secondary pit connections forming a loose network around the gonimoblasts; primary gonimolobe cells enlarging, becoming multinucleate and linked by secondary pit connections to inner pericarp cells; gonimolobes loosely branched, with most cells maturing into carposporangia; one compound cystocarp maturing on a fertile branch, without an ostiole. Tetrasporangia borne on lower cells of determinate branches, sessile, mostly adaxial, modified from periaxial cells or replacing them; tetrasporocytes clavate, becoming spherical, and dividing tetrahedrally.

References: Hommersand (1963), Womersley & Cartledge (1975), West & Calumpong (1989), Broadwater, Scott & West (1991).

One species in the British Isles.

Spyridia filamentosa (Wulfen) Harvey in W. J. Hooker (1833), p. 337.

Type material: Vienna? Type locality: Adriatic Sea.

Fucus filamentosus Wulfen (1803), p. 64.
Hutchinsia filamentosa (Wulfen) C. Agardh (1824), p. 159.
Ceramium filamentosum (Wulfen) C. Agardh (1828), p. 141.

Thalli erect, growing in dense tufts from solid discoid holdfast 0.4-1 cm in diameter, 4-18 cm high, with an irregularly rounded outline, consisting of simple main axes below, irregularly radially branched above, to 3-4 orders of branching; all axes cylindrical, the main axes tapering gradually upwards from 1 mm in diameter near the base, the lateral branches to 1 mm wide, clothed with curved pigmented filaments when young, dark red in colour, bleaching to yellowish or almost white; young axes soft and flaccid, older axes more cartilaginous in texture.

Main axes enlarging from apical cells 20 µm in diameter including walls 6 µm thick; axial cells initially 0.1 diameters long, gradually increasing to 0.5 diameters long, entirely corticated, each bearing a determinate or indeterminate lateral in a spiral pattern; *determinate laterals* 1.5-2 mm long, cylindrical with attenuate base and apex, 45-50 µm in diameter; basal 1-3 cells isodiametric, other cells elongating to 2-4 diameters long, with a small conical terminal cell, each cell except those at the extreme base and apex bearing a band 12-20 µm wide of about 6 periaxial cells; *indeterminate laterals* developing in place of determinate branches at irregular intervals and giving rise to branches indistinguishable from main axes, also arising adventitiously from periaxial and rhizoidal cells; *nodes* consisting of a band of 14 isodiametric periaxial cells that elongate to 36-60 x 8-20 µm; *internodes* corticated by a band of 27 cells, each elongating to 44-80 µm long x 5-12 µm wide; *rhizoidal filaments* later growing downwards between original cortical cells, forming a dense covering of 2-3 layers on older axes; *plastids* ribbon-like to filiform in determinate branches.

Spermatangia formed from cortical cells on determinate branches in a sleeve 250-500 µm long x 90-150 µm wide over (3-)5-6 elongate cells near the base of the branch, ellipsoid, 8 µm long x 6 µm wide. *Cystocarps* formed singly on reduced branches, globular, bilobed or trilobed, about 500 µm in diameter, consisting of 1-3 gonimolobes of different ages, when mature bearing branched chains of rounded carposporangia 50-80 µm in diameter, enclosed in a loose non-ostiolate pericarp. *Tetrasporangia* borne on determinate branches, developing in whorls of 3-5 at the distal end of first few elongate cells, clavate before division, spherical when mature, 76-96 µm in diameter.

Epilithic on pebbles and bedrock, epiphytic on maerl and other algae, in silty or sandy habitats from near extreme low water to 13 m depth, at moderately to extremely wave-sheltered sites exposed to slight to strong currents.

South and west coasts of the British Isles, eastwards to Kent, recorded from scattered localities northwards to Anglesey and Wigtown; Cork to Donegal, Dublin; Channel Isles.

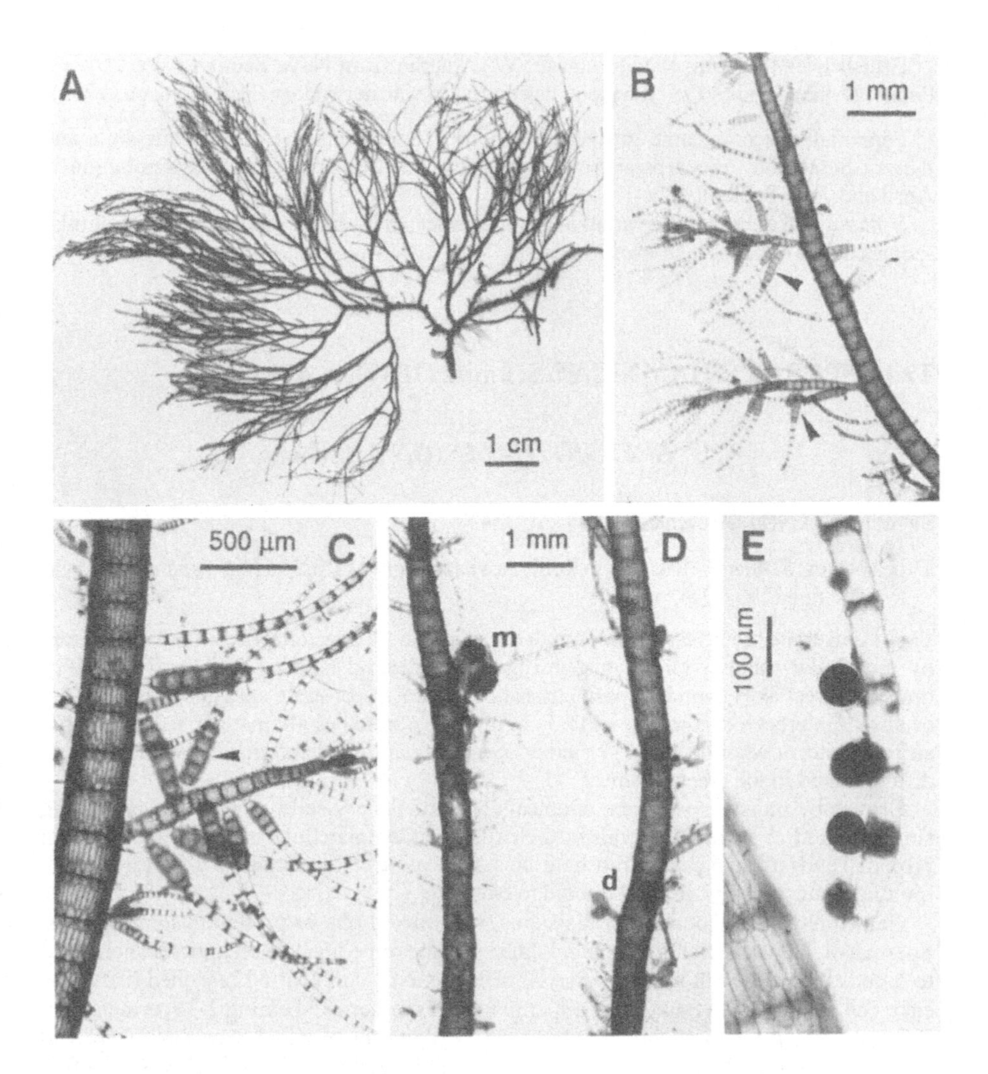

Fig. 45. *Spyridia filamentosa*

(A) Habit (Wigtown, Sep.). (B) Detail of thallus, with spermatangial sori on last-order branches (arrows) (B-E Dorset, June). (C) Spermatangial sori (arrow) forming collars arround last-order branches; main axes are fully corticated and banded whereas last-order branches are partially ecorticate. (D) Developing (d) and mature (m) cystocarps. (E) Last-order branch with tetra- sporangia formed at nodes.

British Isles to Spain; Mediterranean; W. Atlantic from Nova Scotia to West Indies; Pacific (Athanasiadis, 1987); widely distributed in warmer waters (Hommersand, 1963).

Perennial; apices resume growth in January. Spermatangia recorded for June and August-September; procarps and cystocarps for June-September; and tetrasporangia in April and June-September.

S. filamentosa shows little variation in appearance. It is readily identifiable as no similar species occur in the British Isles.

Tribe SPERMOTHAMNIEAE Schmitz (1889), p. 451.

SPERMOTHAMNION Areschoug

SPERMOTHAMNION Areschoug (1847), p. 334.

Type species: *S. turneri* (Mertens in Roth) Areschoug (1847), p. 335 [= *S. repens* (Dillwyn) Rosenvinge (1923-24), p. 298].

Thalli uniseriate, ecorticate, differentiated into prostrate axes attached to the substratum by unicellular rhizoids terminating in digitate or discoid holdfast pads, and simple or branched erect axes; paired opposite branches present in prostrate axes and either present or absent in erect axes; erect axes to 5 cm high, the branches alternate or secund, paired and opposite, or whorled with 3-4 laterals; indeterminate and determinate branches weakly differentiated in the erect system.

Growth by transverse division of apical cells, intercalary cells either naked or bearing single, paired or whorled laterals at their distal ends; unicellular rhizoids arising from proximal ends of axial cells in prostrate axes; vegetative cells cylindrical to barrel-shaped, multinucleate; plastids distinct, discoid to elongate.

Gametophytes monoecious or dioecious, with mixed phases present in some species. Spermatangial heads elongate to subglobular, sessile or pedicellate on upper branches, up to 5 cells long and with up to 4 periaxial cells per axial cell bearing branched filaments, each cell of which can function as a spermatangial mother cell bearing 2-3 spermatangia. Procarps subterminal in 3-celled fertile axes, borne terminally and commonly displaced laterally, overtopped by a lateral branch; fertile axial cell with 3 periaxial cells, the first abaxial, 1-celled, sterile, the second at a 90° angle to the first, fertile, consisting of a supporting cell bearing a terminal 1-celled sterile group and a horizontally curved, 4-celled carpogonial branch, the third opposite the second, 1-celled, potentially fertile; supporting cell and fertile periaxial cell each cutting off a broadly convex auxiliary cell terminally after fertilization; fertilized carpogonium cutting off 2 connecting cells on opposite sides that fuse with extensions of the auxiliary cells; cells of the carpogonial branch fusing and later degenerating; haploid nucleus migrating to the lower end of the auxiliary cell and usually cut off in an isolated cell (disposal cell) containing 1-2 haploid nuclei; auxiliary cell functioning directly as a gonimoblast initial and cutting off 2-5 gonimolobe initials that produce 2-3 orders of filaments, the terminal cells of which mature into clavate to pyriform carposporangia; inner gonimoblast cells and auxiliary cells fusing to axial cell around or alongside pit connections, usually forming a bilobed central fusion cell, also fusing to the cell below; gonimoblasts typically surrounded by 2-4 involucral filaments initiated from the cell subtending the fertile axis before or after fertilization, and sometimes also by filaments derived from cells below. Tetrasporangia initiated terminally on short lateral branches which rebranch in some species forming cymose clusters, subspherical, tetrahedrally divided, sometimes replaced by octosporangia.

References: Rosenvinge (1923-24), Kylin (1923, 1930), Feldmann-Mazoyer (1941), Gordon (1972).

Five species of *Spermothamnion* are listed for the British Isles in the most recent checklist (South & Tittley, 1986). Of these, *S. barbatum* is here transferred to *Ptilothamnion*, and placed in synonymy with *P. pluma*. We have found no evidence that *S. irregulare* (J. Agardh) Ardissone occurs in the British Isles, and British records are tentatively referred to *Ptilothamnion sphaericum*. We have not been able to confirm the validity of *Spermothamnion mesocarpum* (Carmichael ex Harvey in W. J. Hooker) Chemin, described from Ayr, because the original collections consist largely or entirely of *Ptilothamnion pluma*. *S. repens* is a highly variable species, but *S. strictum* lies outside its form range. The generic position of *S. strictum* is provisional as female reproduction is unknown (see Gordon, 1972, p. 119).

KEY TO SPECIES

Paired or whorled branches present on erect axes .... *S. repens*
Paired or whorled branching absent; branching irregular, alternate or secund ... *S. strictum*

Spermothamnion repens (Dillwyn) Rosenvinge (1923-24), p. 298.

Lectotype: BM-K. England, unlocalized.

Conferva repens Dillwyn (1802), pl. 18.
Ceramium turneri Mertens in Roth (1806), p. 127.
Conferva turneri (Mertens in Roth) Dillwyn (1809), pl. 100.
Callithamnion repens (Dillwyn) Lyngbye (1819), p. 128.
Callithamnion turneri (Mertens in Roth) C. Agardh (1828), p. 160.
?*Callithamnion roseolum* C. Agardh (1828), p. 181.
Spermothamnion turneri (Mertens in Roth) Areschoug (1847), p. 335.
?*S. turneri* β *roseolum* (C. Agardh) Areschoug (1850), p. 113.
S. turneri γ *repens* (Dillwyn) Areschoug (1850), p. 113.
?*Spermothamnion roseolum* (C. Agardh) Pringsheim (1862), p. 15.

Thalli growing as isolated tufts, or turfs up to 4 cm in extent, 0.3-2.5 cm high, spreading over the substratum by prostrate axes that bear numerous erect axes, rose-pink in colour, fairly delicate and flaccid.

Prostrate axes 30-40 μm wide, composed of cylindrical cells 2-5 diameters long, with walls 8 μm thick, giving rise laterally to further prostrate axes at irregular intervals, sometimes in a paired arrangement, also forming erect axes in an anterior position on cells of prostrate axis, either at irregular intervals or from every cell; *rhizoids* formed near the posterior end of cells of prostrate axes, unicellular, 14-28 μm wide, terminating in discoid to digitate holdfast pads c. 70 μm in diameter, consisting of several major lobes, each repeatedly bifurcate; *erect axes* enlarging slightly from apical cells 30-38 μm wide, up to 2 diameters long, with a blunt, rounded apex; subapical cell to 2 diameters long, to 34-42 μm in diameter; mature axes composed of cylindrical cells 1.5-6 diameters in length, with walls 8-10 μm thick, branching very irregularly, sometimes unbranched below for about 1/3 their length, or bearing a branch on every cell; *laterals* borne in a secund, irregularly alternate or paired pattern below, usually paired or in whorls of 3-4 towards apices, at an acute angle (30-45°), sometimes becoming equal in length to main axes or remaining much shorter, bearing 2-4 further orders of branching, ultimate branches either cylindrical and almost straight, or tapering markedly and curved inwards; hairs absent; *plastids* discoid near apices, sometimes becoming bacilloid or ribbon-like.

Gametophytes monoecious and usually also bearing tetrasporangia. *Spermatangial heads* terminal on short branches, the subtending cell sometimes branching laterally and giving rise to a further pedicellate spermatangial head, cylindrical or occasionally subglobular, (30-)42-68 μm long x 20-32 μm wide, composed of a central axis bearing whorls of short filaments terminating in spermatangial mother cells bearing 2-3 spherical spermatangia 3 μm in diameter. *Female axes* appearing terminal on first- to third-order laterals that bear a curved 4-celled carpogonial branch laterally; *cystocarps* subglobular, 150-240 μm in diameter, with a conspicuous central fusion cell and dense clusters of gonimoblast cells terminating in radiating carposporangia of different ages, not enclosed in a common membrane; carposporangia pyriform, 52-60 x 40-46 μm, surrounded by a whorl of 4 sparsely branched involucral branchlets borne on cell below female axis, with a second whorl of 2-3 branchlets sometimes developing from the cell below this. *Tetrasporangia* borne on branchlets formed by main and lateral axes, typically initially terminal on a short branch, later in a cymose cluster resulting from the pedicel giving rise

---

Fig. 46. *Spermothamnion repens*

(A) Habit on *Gracilaria* sp., showing paired branching in places (arrow) (Donegal, Dec.). (B) Prostrate axis bearing several erect axes with whorled branching (Clare, July). (C) Cells of prostrate axis forming erect axes in an anterior position and rhizoids in a posterior position (arrows) (Cork, Aug.). (D) Spermatangial heads (s) and procarps (p) on same thallus (D-E Cornwall, June). (E) Procarp, including carpogonial branch, and developing spermatangial head. (F) Cystocarp, with large T-shaped fusion cell (arrow) bearing few carposporangia (Cornwall, Aug.). (G) Tetrasporangia borne on whorled branches (as B).

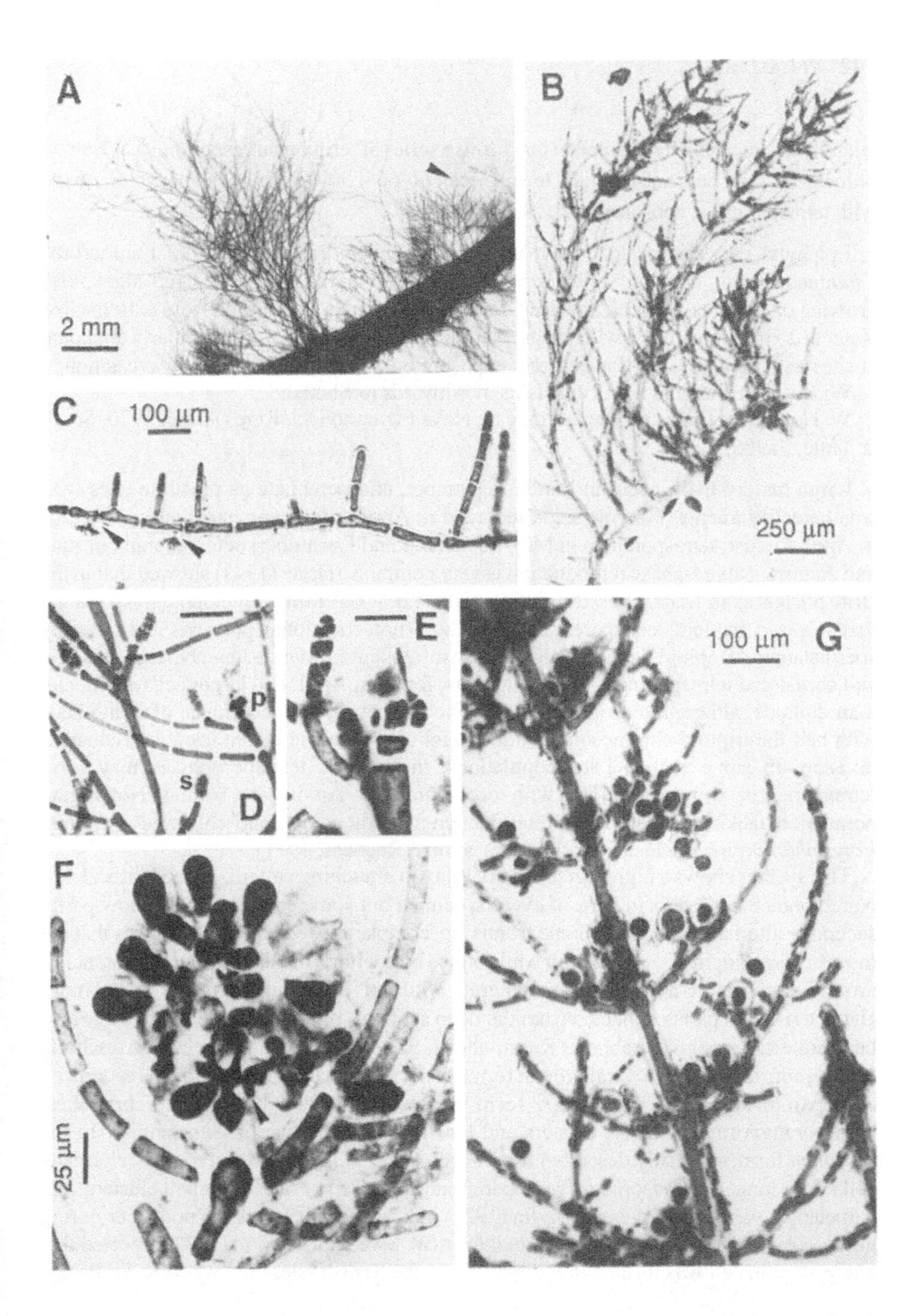
A
2 mm
B
250 µm
C
100 µm
p
s
D
E
100 µm
G
F
25 µm

laterally to a reproductive branchlet that forms a series of terminal tetrasporangia and lateral branchlets, 1 per cell, subspherical to spherical, 60-66 x 50-60 µm; *octosporangia* mixed with tetrasporangia, subspherical, c. 85 x 75 µm.

Epiphytic on various algae such as *Furcellaria lumbricalis* (Hudson) Lamouroux, *Cladophora* spp., stipes of *Laminaria hyperborea*, maerl and crustose red algae, also growing on pebbles, bedrock and dead shells, in lower-shore pools and from extreme low water to 25 m, found over a wide range of environmental conditions, particularly abundant at sites with moderate to strong current exposure but fairly sheltered from wave action.

Widely distributed in the British Isles, northwards to Shetland.

W. Norway to Canaries; Mediterranean; Nova Scotia to Delaware (Ardré, 1970; South & Tittley, 1986).

Large mature thalli occur in March-September, and perennate as prostrate axes with small erect filaments. Spermatangia recorded in April-August; procarps and cystocarps in April-August; tetrasporangia in May-September and December; octosporangia in June and August. Mixed-phase reproduction is very common. Drew (1943) showed that in the British Isles, as in Massachusetts, diploid plants ($2n = 60$) formed meiotic tetrasporangia that released haploid tetraspores, and also bore non-functional procarps and possibly spermatangia. Haploid thalli, which were most frequent in March-June, bore gametangia and occasional tetrasporangia. Triploid plants, found in April, had larger cells and nuclei than diploids; although tetrasporangia were not observed, the occurrence of individuals with half the triploid chromosome number might have resulted from irregular reduction division. In some British Isles populations, in contrast, tetrasporophytes may form tetrasporangia alone, or mixed with octosporangia. An isolate from a Norwegian population lacking mixed-phase reproduction was dioecious and followed a normal *Polysiphonia*-type life history without deviations (Rueness, 1971).

This species shows a high degree of variability in branching pattern. In the British Isles, paired branching occurs in parts of every specimen but some erect axes may show partly secund or alternate branching. Some fronds are complanate, with paired branches that are more or less straight and cylindrical, while others bear whorls of curved, attenuate branches around each main axial cell. In general, whorled branching appears to be more characteristic of plants collected from the deep subtidal, but complanate plants have also been collected in similar habitats. Rosenvinge (1923-24), in concurrence with Areschoug (1850), considered that several growth forms of *S. repens* had previously been recognized as separate species. The *turneri* form was characterized by opposite branches, tetrasporangia in corymbose clusters and the frequent presence of gametangia. In the *roseolum* form, originally described from W. Sweden, branching was generally alternate, cells were longer, tetrasporangia were borne singly, paired or rarely in small clusters, and gametangia were rarely present. Kylin (1923) indicated that fusions did not occur during gonimoblast development in *S. roseolum* from W. Sweden. Gordon (1972) reported that a large fusion cell was formed in *S. repens* from the British Isles, and we have confirmed this for a range of specimens, suggesting that *S. roseolum* may not be conspecific with *S. repens*. Material resembling *S. roseolum* has not been found in the British Isles.

Some forms of *S. repens* could be confused with *Ptilothamnion pluma* (q. v. for distinctions).

Spermothamnion strictum (C. Agardh) Ardissone (1883), p. 302.

Possible lectotype: BM-K, 'Algae Schousboeana', no. 219. Morocco (Tangiers), August 1817.

Callithamnion strictum C. Agardh (1828), p. 185.
Callithamnion crouanii Kützing (1849), p. 642 (see J. Agardh, 1851, p. 34).
Callithamnion semipennatum J. Agardh (1842), p. 72 (see J. Agardh, 1851, p. 34).

Thalli growing as isolated tufts, or turfs up to 1.5 cm in extent, 0.6-2.5 cm high, spreading over the substratum by prostrate axes that bear numerous erect axes, deep pinkish-red in colour, delicate to fairly stiff in texture.

Prostrate axes 38-100 µm wide, composed of cylindrical, barrel-shaped or subglobular cells 1-8 diameters long, with laminated walls 6-12 µm thick, giving rise laterally to further prostrate axes at irregular intervals, and forming erect axes at irregular intervals in an anterior position on cells, not normally on a cell bearing a rhizoid, little-pigmented, the cells densely packed with starch grains; *rhizoids* formed near the posterior end of cells of prostrate axes, unicellular, 14-44 µm wide, terminating in digitate to discoid holdfast pads 60-140 µm in diameter, consisting of several major lobes, each repeatedly bifurcate; *erect axes* unbranched below for at least half their length, growing from apical cells 35-48 µm wide and up to 3 diameters long, cylindrical or slightly conical, with a blunt rounded apex; subapical cell 1.5-3.5 diameters long, enlarging slightly to 40-72 µm in diameter; mature cells cylindrical to barrel-shaped, 2-13 diameters in length; *branches* borne in a regularly secund to very irregular arrangement, sometimes in a secund pattern for a few cells, then absent on the next few cells, or irregularly alternately in places, never paired, borne at an acute angle (<30°), up to 44 µm in diameter, with cells 2-4 diameters long, bearing branches in the same arrangement as main axes; hairs absent; *plastids* discoid near apices, becoming bacilloid.

Gametophytes unknown. *Tetrasporangia* formed on branchlets that develop secondarily in an adaxial position on main and lateral axes, sometimes in series on successive filament cells, typically terminal initially on a 1-celled branch that subsequently gives rise laterally to a reproductive branchlet, or in a cymose cluster of 6 or more tetrasporangia on branchlets 20-25 µm in diameter, one per cell, ellipsoid before division, subspherical to spherical when mature, 60-68 x 56-60 µm.

Epiphytic on algae, epilithic on pebbles, and epizoic on living gastropod shells, from extreme low water to 20 m depth at sites with slight to moderate wave exposure and moderate to strong currents; sometimes occurring in abundance in favourable habitats.

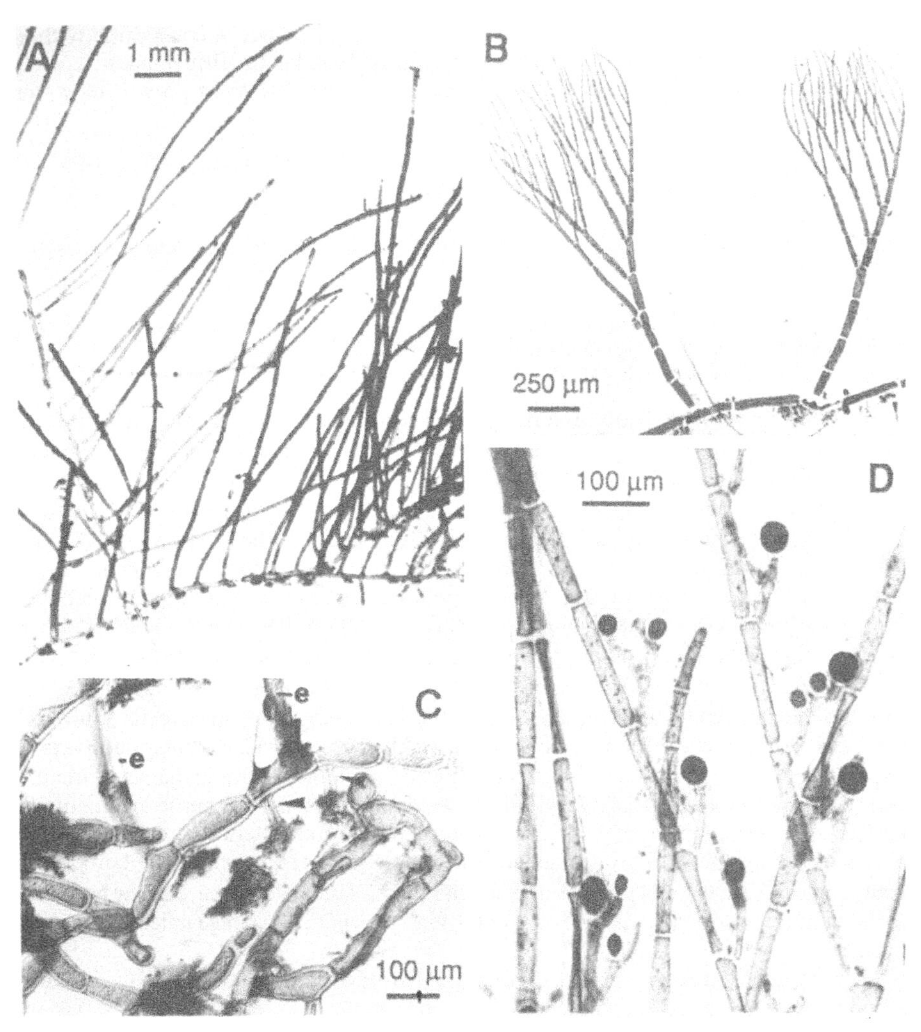

Fig. 47. *Spermothamnion strictum*

(A) Prostrate axis bearing numerous little-branched erect axes, collected at ELWST (Cornwall, March). (B) Small thallus with regularly flabellate, secund branching, collected at 15 m depth (Donegal, Aug.). (C) Inflated, starch-filled cells of prostrate axis forming erect axes (e) in an anterior position and rhizoids (arrows) in a posterior position (arrows) (C-D Dorset, July). (D) Clusters of tetrasporangia.

Dorset, N. Cornwall, Pembroke, Inverness, Down, Antrim.
British Isles, ?Netherlands, N. France, N. Spain, Morocco; W. Mediterranean (J. Agardh, 1851; South & Tittley, 1986).

Collected only in February and May-August, with tetrasporangia only in February and July; gametangia unknown.

Thallus size and the diameter of erect axes vary widely. A collection from Dorset consisted of plants less than 7 mm high, with erect axes only 30-40 μm in diameter, whereas specimens from Inverness and Down were 15-20 mm high and erect axes were 40-60 μm wide. The large and small thallus forms are tentatively considered conspecific, based on morphology of prostrate axes and branching pattern, but further studies, particularly of reproductive material, are required.

This species could be confused with small non-reproductive specimens of *Griffithsia devoniensis* (q. v.) or *Pleonosporium borreri* (q. v.). *G. devoniensis* has regularly pseudodichotomous branching and *P. borreri* branches regularly alternately whereas branching in *S. strictum* is irregularly alternate to secund. *S. strictum* differs from *Ptilothamnion sphaericum* (q. v.) in the position of rhizoids, which develop at the posterior end, rather than in the middle, of prostrate axial cells.

PTILOTHAMNION Thuret in Le Jolis

PTILOTHAMNION Thuret in Le Jolis (1863), p. 118.

Type species: *P. pluma* (Dillwyn) Thuret in Le Jolis (1863), p. 118.

Thalli uniseriate, ecorticate, differentiated into prostrate axes attached to the substratum by unicellular rhizoids terminated by digitate pads and simple or branched erect axes; paired opposite branches present in the prostrate system, and either present or absent in the erect system; erect axes to 2 cm high, alternately or sometimes oppositely branched, with the branches distichous, secund or irregularly arranged; indeterminate and determinate branches weakly differentiated in the erect system.

Growth by transverse division of apical cells, intercalary cells either naked or bearing single or paired laterals at their distal ends; unicellular rhizoids arising in proximal or median position on prostrate axes; vegetative cells cylindrical or barrel-shaped, multinucleate; plastids distinct, discoid to elongate or ribbon-shaped.

Gametophytes dioecious. Spermatangial heads subglobular, 2-3 cells long, terminal or lateral on the last 2 orders of branches, sessile or pedicellate, or clustered on small-celled lateral branches, each axial cell with up to 4 periaxial cells bearing branched filaments, each of which can function as a spermatangial mother cell bearing 2-3 spermatangia. Procarps subterminal on a potentially indeterminate axis in which the last 2 cells are smaller than the rest and densely filled with cytoplasm; fertile axial cell with 2-3 periaxial cells that develop in the same manner as in *Spermothamnion*; only one auxiliary cell produced after fertilization, cut off from the supporting cell; auxiliary cell functioning directly as a gonimoblast initial and cutting off 2-5 gonimolobe initials that branch 1-3 times, with all or only the terminal cells maturing into carposporangia; fusion cell weakly developed or absent; gonimoblasts surrounded by an involucre composed of 2-4 (-5) filaments initiated before or after fertilization from the cell subtending the fertile axial cell. Tetrasporangia sessile or pedicellate on upper branches, or clustered on small-celled lateral branches, pyriform to subspherical, tetrahedrally divided, or replaced by polysporangia of 8-16(-18) spores.

References: Kylin (1928), Gordon (1972), Gordon-Mills (1977), Wollaston (1984).

At present, only one species of *Ptilothamnion*, *P. pluma*, is recognized in the British Isles. A second species of *Ptilothamnion* is included here. Although gametangia and post-fertilization structures are unknown in *P. sphaericum*, we place this species in *Ptilothamnion* rather than *Spermothamnion* on the basis of the vegetative morphology (rhizoids and erect axes borne in a median position on prostrate axial cells) and the sessile polysporangia.

KEY TO SPECIES

Vegetative cells barrel-shaped, branching irregularly secund or alternate but not opposite; polysporangia lateral, sessile .... *P. sphaericum*

Vegetative cells cylindrical, branching mostly opposite, tetrasporangia and octosporangia terminal (rarely lateral) on branchlets .... *P. pluma*

Ptilothamnion sphaericum (P. Crouan & H. Crouan ex J. Agardh) Maggs & Hommersand, comb. nov.

Holotype: LD 35275. France (Anse du Minou, Brest), 25 viii 1836, *Crouan*.

Callithamnion sphaericum P. Crouan & H. Crouan ex J. Agardh (1851), p. 20.

Thalli growing as small tufts or extensive mats to 20 cm or more in extent, 1-2.5 cm thick, spreading over the substratum by prostrate axes that bear numerous erect axes; prostrate axes pinkish-red; erect axes dull brownish-red in colour, stiff and fairly rigid in texture.

Prostrate axes indeterminate, 25-70 μm wide, composed of barrel-shaped cells 2-4

Fig. 48. *Ptilothamnion sphaericum*

(A) Habit of tuft removed from turf (A-F Cornwall, March). (B) Detail of sparsely branched erect axes. (C) Inflated cells of irregularly branched erect axes. (D) Cell of prostrate axis bearing both an erect axis and a rhizoid in a median position. (E, F) Sessile polysporangia.

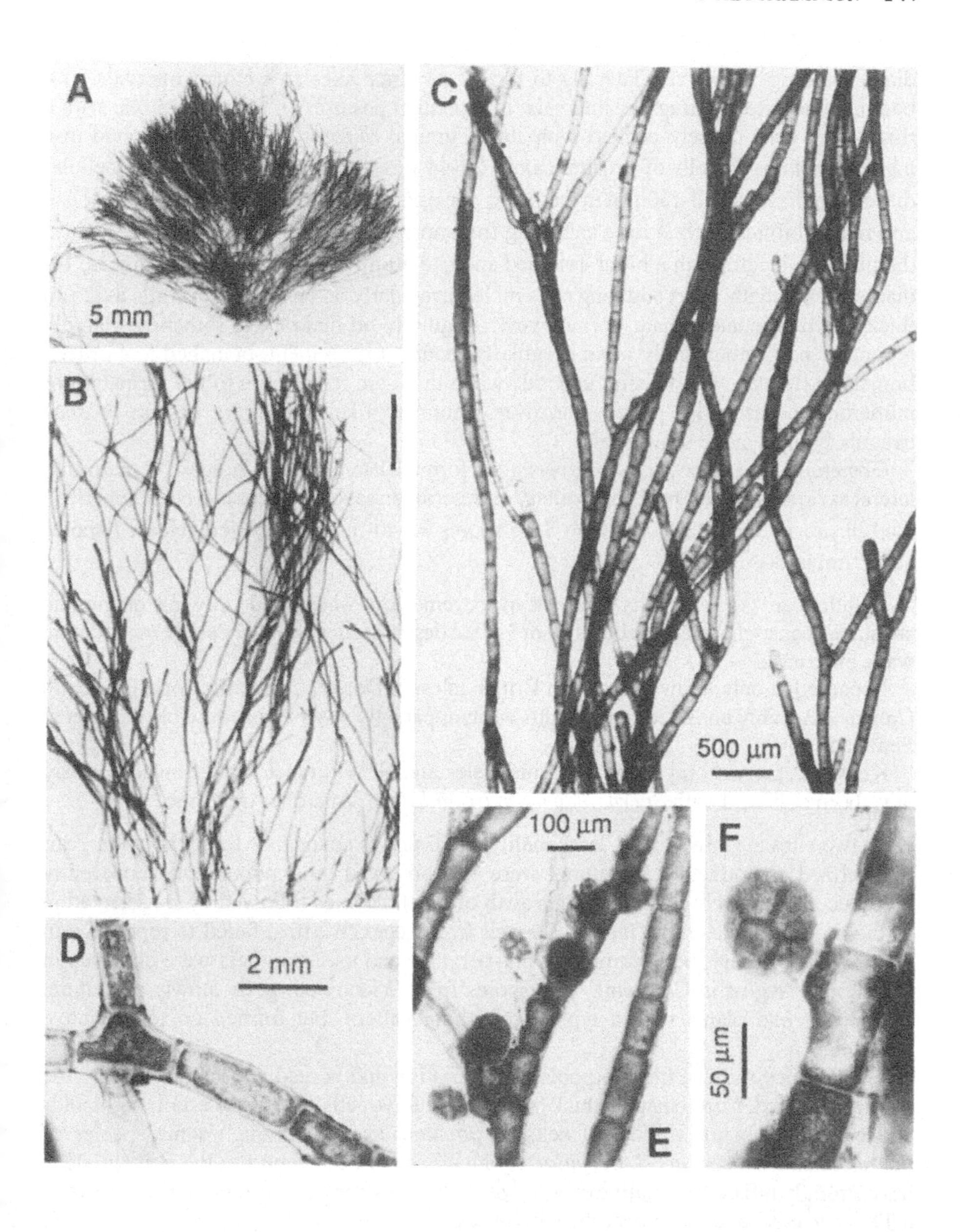
A
5 mm
B
C
500 µm
D
2 mm
100 µm
E
F
50 µm

diameters long, giving rise laterally to further prostrate axes at irregular intervals, also bearing erect axes at irregular intervals, in a median position on cells, often one with a rhizoid, all cells densely packed with starch grains; *rhizoids* unicellular, inserted in a median position on cells of prostrate axes, 28-44 μm wide, terminating in a unicellular discoid holdfast pad 70-130 μm in diameter, consisting of several major lobes, each lobed and finally bifurcate; *erect axes* enlarging from apical cells c. 60 μm wide and 0.7 to 1.5 diameters in length, with a blunt, rounded apex, to 40-85 μm, wider towards apices, 1-4 diameters in length, short and long cells mixed irregularly on an axis, with walls 8-12 μm thick; *laterals* indeterminate, formed very irregularly on main axes, sometimes largely absent, or borne abundantly in an irregularly secund arrangement, branched to 4 orders, branching frequency increasing upwards when thalli are fertile, due to the formation of numerous, occasionally paired, narrower secondary filaments; hairs absent; *plastids* irregularly elongate to polygonal.

Gametophytes unknown. *Polysporangia* formed laterally near apices of main and lateral axes, sessile, 1(-2) per cell, solitary or in series on several successive cells, spherical, 75-120 μm in diameter, containing 8-16 spores 32-40 μm in diameter; released spores 32-45 μm in diameter.

Epilithic on sandy or silty bedrock at extreme low water, and growing on mobile substrata such as maerl and pebbles from 3-10 m depth, at sites with very slight to moderate wave exposure.

Recorded at only a few sites in the British Isles, in Dorset, Cornwall, Cork, Clare and Galway. Possibly confined to the south-west; apparently absent from suitable habitats in Scotland.

Known at present only from the British Isles and N. W. France, but elsewhere it may have been confused with *Spermothamnion irregulare* (J. Agardh) Ardissone.

At two sites in Galway Bay, large thalli were found in abundance throughout the year. Specialized reproductive structures were not observed, and populations apparently reproduced effectively by extensive growth of prostrate axes followed by fragmentation of thalli. An isolate from Galway, derived from vegetative tips, failed to reproduce in culture (M. D. Guiry, pers. comm.). Polysporangia and tetrasporangia were observed in March and August in Cornwall. Polyspores from Widemouth grew slowly in culture, developing into plants with a typical branching pattern, but formed no reproductive structures.

This species shows little morphological variation and is readily recognizable by the strongly inflated, barrel-shaped, thick-walled vegetative cells, and the insertion of rhizoids and erect axes in the middle of cells of prostrate axes. Although some species of *Spermothamnion*, such as *S. irregulare* (which has not been found in the British Isles), have strongly inflated vegetative cells, rhizoids develop from the posterior end of prostrate cells while erect axes arise near the anterior end.

Ptilothamnion pluma (Dillwyn) Thuret in Le Jolis (1863), p. 118.

Lectotype: BM-K. Isotypes: BM-K, LD. Cork (Bantry Bay), *Hutchins* (see Dixon, 1962b).

Conferva pluma Dillwyn (1809), p. 72.
Callithamnion barbatum C. Agardh (1824), p. 181.
Callithamnion pluma (Dillwyn) C. Agardh (1828), p. 162.
Spermothamnion barbatum (C. Agardh) Nägeli (1862), p. 119.
Ptilothamnion lucifugum Cotton (1912), p. 139.

Thalli forming velvety patches up to 2 cm in extent, spreading over the substratum by prostrate axes that bear numerous complanate erect axes 0.2-0.8 cm high and 0.5-2 mm wide including the branches, dull purple in colour, erect fronds soft but not flaccid.

Prostrate axes indeterminate, 40-50 µm wide, composed of cylindrical or barrel-shaped cells 1-4 diameters long, giving rise to lateral indeterminate prostrate axes at irregular intervals, and forming erect axes from every second cell, in a median position, often opposite a rhizoid; *rhizoids* unicellular, inserted in a median position on cells of prostrate axes, 30-32 µm wide, terminating in extensive digitate ramifications; *erect axes* enlarging slightly from apical cells 22-32 µm wide, 1-1.3(-2) diameters long, with a blunt, rounded apex, to cells 40 µm diameter and 1-2 (-3) diameters in length; *branches* borne regularly or irregularly on main axes, singly or paired, distichous, in one plane, typically in series of pairs separated by a series of unbranched axial cells, borne at an acute angle, simple or with an abaxial branchlet from the first cell, and sometimes also the second and third cells, 5-20 cells long, tapering slightly to rounded apical cells, some branches indeterminate, resembling main axis and forming paired distichous branches; hairs absent; *plastids* discoid.

Spermatangial heads developing terminally on branches and branchlets, consisting of a short stalk cell and 2 axial cells bearing a dense cluster of spermatangial mother cells that cut off 1-2 spherical spermatangia 4-5 µm in diameter, enclosed in a mucilage sheath 5 µm thick. *Female axes* appearing terminal on branches, becoming partially enclosed by 2 curved involucral filaments 18 µm wide and up to 6 cells long formed by the subtending cell; involucral filaments sometimes giving rise to further female axes; *cystocarps* consisting of several small gonimolobes, initially globular, becoming irregularly rounded when mature, 100 µm wide x 80 µm long, each forming about 5 mature rounded carposporangia 28-42 µm wide. *Tetrasporangia* borne terminally and laterally on branches, sessile, 1(-2) per cell, spherical to ellipsoid before division, subspherical, ellipsoid or pyriform when mature, 60-74 x 52-70 µm; *octosporangia* formed on same thalli in equivalent positions, ellipsoid, 84-96 x 60-70 µm.

Epilithic on intertidal bedrock in pools and caves; epiphytic on *Laminaria hyperborea* stipes and occasionally on other algae including crustose corallines, sometimes forming a mixed turf with other small red algae such as *Spermothamnion* species, to 26 m depth, at sites with slight to strong wave exposure.

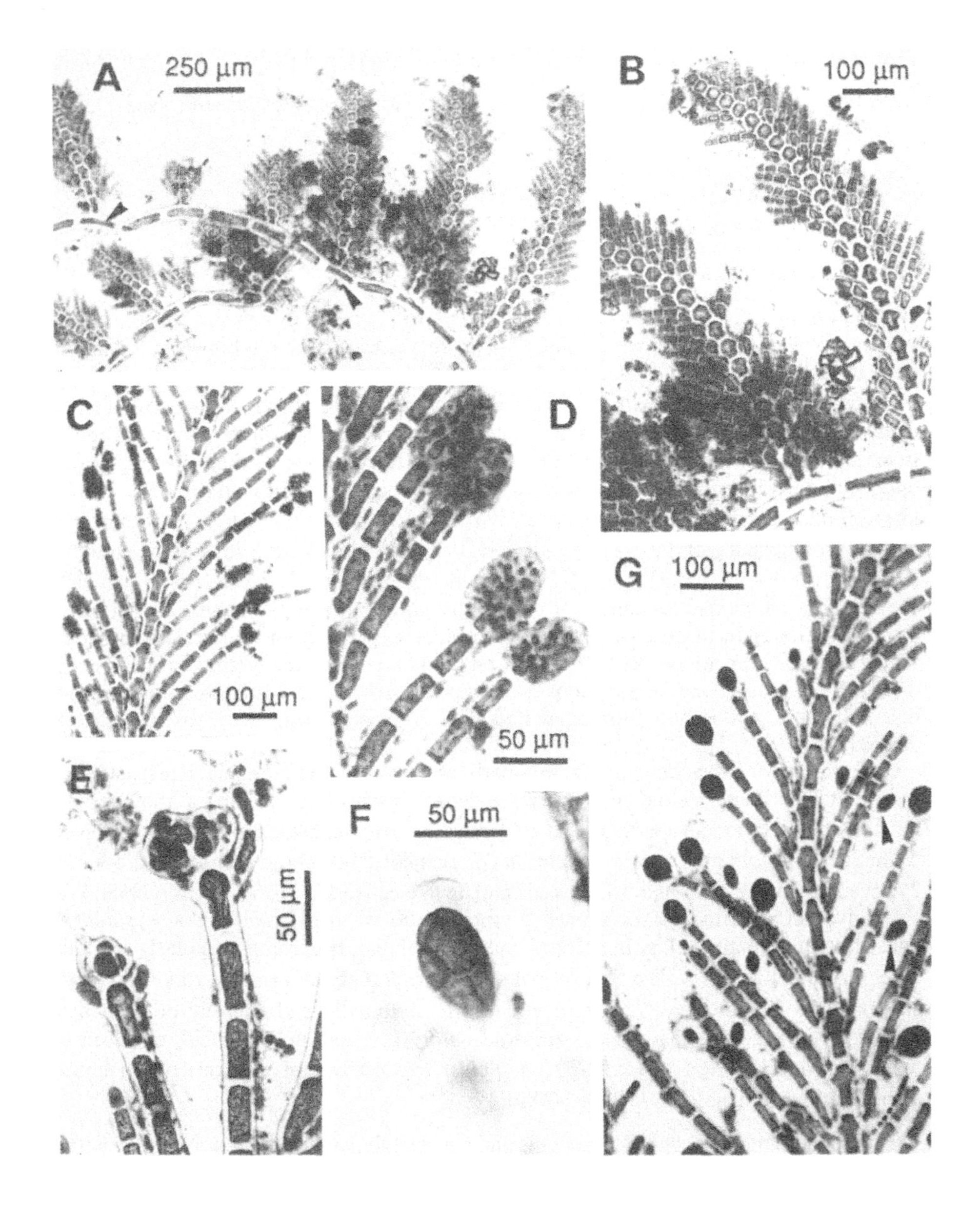
A
250 µm
B
100 µm
C
100 µm
D
50 µm
E
50 µm
F
50 µm
G
100 µm

Widely distributed in the British Isles, northwards to Shetland; although rarely recorded in northern Scotland and eastern England, this species was recently found to be abundant in W. Sutherland, and has probably been overlooked due to its small size.
British Isles to Portugal, Canaries; Mediterranean (Ardré, 1970).

Spermatangia recorded in April, June-August and October; procarps and cystocarps in April, July and October; tetrasporangia in April and July-October; and octosporangia in October.

P. pluma shows considerable variation in the numbers of paired branches formed, with some thalli consisting largely of naked axial filaments whereas others are clothed with branches. The branches themselves vary from simple to branched, usually bifid. *P. lucifugum*, characterized by bifid ramuli, was described on the basis of plants collected in a cave (Cotton, 1912), but Dixon (1962b) found no correlation between light levels and the formation of second-order laterals in *Ptilothamnion* thalli. Variation also occurs in the position of tetrasporangia. In general, tetrasporangia are predominantly terminal, and a few lateral sporangia may or may not be present. Plants with terminal tetrasporangia sometimes grow mixed with thalli bearing only lateral sporangia, which have been considered diagnostic of *Spermothamnion barbatum*.

The lectotype of *Callithamnion barbatum*, LD 17951, from Marseille, France (February 1820, *Schousboe*) bears opposite branches and is clearly attributable to *Ptilothamnion* because rhizoids and erect axes are formed in the middle of prostrate axial cells. Filaments of several other species, including very sparsely branched thalli of a species of *Spermothamnion*, are intermixed in this sample. Apparent syntypes in LD, C and BM consist largely or entirely of this *Spermothamnion* species. Although the lectotype is non-reproductive, no differences in vegetative morphology were apparent between it and British material of *P. pluma*, and we consider that *S. barbatum* is conspecific with *P. pluma*. British records of *S. barbatum* also appear to be based on specimens of *P. pluma*.

Some forms of *Spermothamnion repens* with paired branching may resemble *P. pluma*, but can be distinguished even when non-reproductive. In species of *Spermothamnion*, rhizoids are inserted near the posterior end, and erect axes near the anterior end, of prostrate axial cells, whereas in *Ptilothamnion*, both rhizoids and erect axes arise from near the middle of prostrate axial cells.

Fig. 49. *Ptilothamnion pluma*

(A) Cells of prostrate axes bearing both erect axes and rhizoids (arrows) in a median position (A-B Galway, Aug.). (B) Erect axes with paired branches, some of which are "bifid". (C) Paired branches terminating in spermatangial heads (C-G Sutherland, Oct.). (D) Ellipsoid spermatangial heads terminal on last-order branches. (E) Terminal procarps developing into cystocarps. (F) Ellipsoid octosporangium. (G) Tetrasporangia terminal and lateral (arrows) on branches.

SPHONDYLOTHAMNION Nägeli

SPHONDYLOTHAMNION Nägeli (1862), p. 380.

Type species: *Sphondylothamnion multifidum* (Hudson) Nägeli (1862), p. 380.

Thalli uniseriate, ecorticate above, attached to substratum by matted, filamentous holdfast and erect to 20 cm high, consisting of one to several irregularly branched main axes bearing 2-4(-6) determinate laterals (whorl-branches) at distal ends of axial cells; whorl-branches quadri-, tri- or dichotomously branched; basal rhizoidal filaments contributing to the holdfast.

Growth by transverse division of multinucleate apical cells, each axial cell normally producing an opposite pair of lateral initials 1-3 cells below the apex, with each successive pair at right angles in decussate arrangement, second pair arising later between first pair forming whorl, or the second pair absent, additional laterals sometimes interpolated between whorled laterals or produced below them; indeterminate branches arising from distal end of basal cells of whorl-branches or replacing them; rhizoidal filaments branched, interwoven, issuing from basal axial cells and sometimes also from apical cells of whorl-branches; cells multinucleate; plastids distinct, discoid to elongate.

Gametophytes dioecious. Spermatangial heads globose, sessile, adaxial at distal ends of cells of whorl-branches, 2-celled; spermatangial filaments whorled with 2-3 spermatangia per mother cell. Procarps subterminal at tips of potentially indeterminate axes 3-5 (-7) cells long; fertile axes solitary, in whorls, or compound, densely clustered on dwarfed lateral indeterminate branches; fertile axial cell with 3 periaxial cells, the first sterile, the second fertile bearing a terminal 1-celled sterile group and curved 4-celled carpogonial branch, the third potentially fertile; supporting cell and fertile periaxial cell each cutting off a broadly convex auxiliary cell terminally after fertilization; fertilized carpogonium cutting off 2 connecting cells on opposite sides that fuse with extensions of auxiliary cells; cells of carpogonial branch fusing incompletely; haploid nucleus migrating to far end of auxiliary cell and cut off as an isolated cell (disposal cell); gonimoblast initial cut off from opposite end; gonimoblast filaments monopodially branched, bearing terminal, clavate carposporangia; auxiliary cells and inner gonimoblast cells fusing to axial cell forming bilobed central fusion cell; cystocarp covered by branched sterile filaments formed from the apical cell, sterile periaxial cell, and sterile group of procarp after fertilization, and surrounded by whorled involucral filaments from axial cell below. Tetrasporangia sessile at distal ends of inner cells of whorl-branches, pyriform to spherical, tetrahedrally divided.

References: Feldmann-Mazoyer (1941), Gordon (1972).

One species in the British Isles.

Sphondylothamnion multifidum (Hudson) Nägeli (1862), p. 380.

Neotype: BM. Devon (Torquay), August 1902, *Batters* (see Gordon, 1972).

Conferva multifida Hudson (1778), p. 596.
Griffithsia multifida (Hudson) C. Agardh (1824), p. 143.
Griffithsia multifida var. *pilifera* C. Agardh (1828), p. 133.
Wrangelia multifida (Hudson) J. Agardh (1841), p. 38.
Callithamnion multifidum (Hudson) Kützing (1843), p. 373.
Corynospora multifida (Hudson) P. Crouan & H. Crouan (1867), p. 138.
Sphondylothamnion multifidum var. *pilifera* (C. Agardh) Batters (1902), p. 83.

Thalli erect, growing in dense tufts from dense, tangled filamentous holdfast 0.3-1 cm in diameter, 5-20 cm high, with single main axes below, branching irregularly above; all axes clothed with whorl-branches giving a total diameter of 1-2 mm, cylindrical or complanate depending on branching pattern, rose-pink in colour, crisp and rigid when fresh, decaying rapidly after collection to a greenish colour and flaccid texture.

Main axes growing from apical cells 20-25 µm in diameter; axial cells cylindrical, c. 600 µm in diameter and about 1 diameter long near holdfast, increasing upwards to 3-4(-6) diameters in length, each bearing a whorl of 2-4(-6) whorl-branches inserted distally, often in 2 opposite pairs of different sizes; *whorl-branches* c. 30 µm in diameter, branching in an irregularly whorled, dichotomous or trichotomous pattern 1-3 times; ultimate branchlets recurved, typically about 6 cells long, cylindrical, with bluntly rounded apices, occasionally bearing hairs near thallus apices; *indeterminate axes* developing from some whorl-branches in an irregularly alternate or paired arrangement; *rhizoidal filaments* arising in whorls from lowermost cells of main axis, and contributing to holdfast, branched, 160-200 µm in diameter; *plastids* discoid to bacilloid.

Spermatangial heads formed on last two orders of branching of whorl-branches, adaxially on basal 1-2 cells of branchlet, globular, about 65 µm in diameter, composed of a central filament of 2 cells each bearing a whorl of cells that cut off several spermatangial mother cells bearing 2-3 spherical spermatangia about 3 µm long (Gordon & Womersley, 1966). *Procarps* developing subterminally on whorl-branches or second-order laterals, usually several on each whorl-branch; *cystocarps* 140-600 µm in diameter when mature, composed of a large central fusion cell surrounded by a cluster of clavate carposporangia 70-100 µm long x 44-60 µm wide, enclosed in an involucre of branchlets formed by cells of female axis. *Tetrasporangia* formed adaxially on branchlets of last two orders, sessile on basal and suprabasal cells, 1-2 per cell, initially pyriform, rapidly becoming spherical, tetrahedrally divided when mature, to 85 µm in diameter.

Growing on bedrock in shaded lower-shore pools, subtidal on upward-facing bedrock to 30 m depth, frequently in kelp forest, also on mobile substrata such as maerl and pebbles, tolerant of some sand cover and sand-scour, on very sheltered to very wave-exposed coasts.

Common on south and south-west coasts of the British Isles, eastwards to Sussex, rare

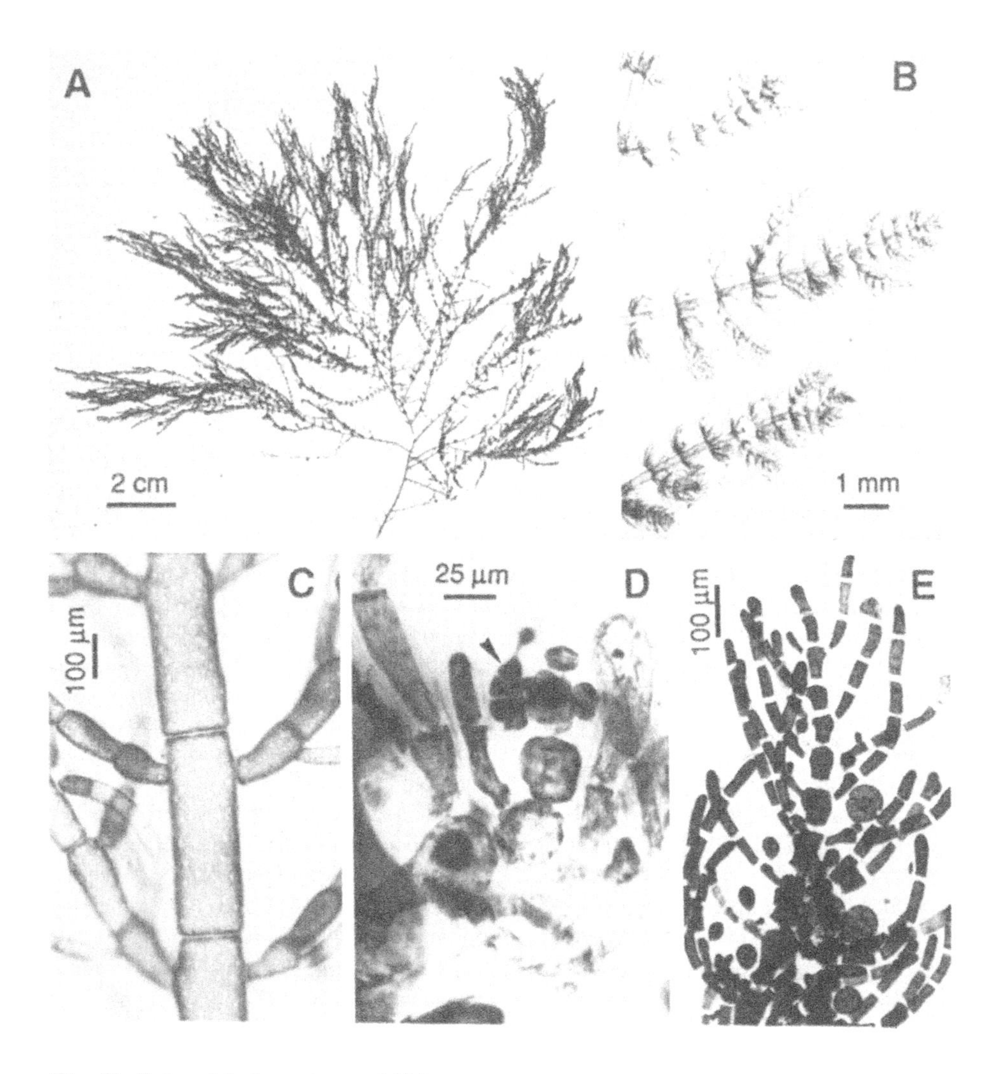

Fig. 50. *Sphondylothamnion multifidum*

(A) Habit of open-coast plant with whorled whorl-branches (Donegal, Aug.). (B) Detail of axes bearing whorled whorl-branches (Mayo, June). (C) Paired whorl-branches; numerous nuclei are visible in every cell of this live specimen (Donegal, Dec.). (D) Lateral carpogonial branch (arrow) on subterminal procarp (D-E Dorset, July). (E) Developing tetrasporangia.

in S.W. Scotland, with records from Bute, Ayr, Argyll, Inverness and Ross & Cromarty; widely distributed in Ireland but rare on east coasts; Channel Isles.

British Isles to Canaries; Mediterranean (Ardré, 1970).

Erect fronds probably annual. Spermatangia recorded in July; procarps and cystocarps in July and September; tetrasporangia in June-September. We have found gametangial plants only on the south coast of England.

The branching pattern shows a wide range of variation. Distichous thalli, with only two whorl-branches per axial cell, forma *disticha* Feldmann-Mazoyer (1941), differ strikingly from whorled plants but appear to be only a growth form (Dixon, 1963). Distichous branching appears to be associated with sheltered habitats, because open-coast plants are always whorled. A growth form, var. *pilifera*, in which determinate and indeterminate axes grow out into elongate, sparsely branched filaments, also appears to be a response to sheltered conditions.

S. multifidum could be confused with *Halurus equisetifolius* (q. v. for distinctions). The distichous form may resemble *Antithamnion* and *Pterothamnion* species, but lacks the gland cells characteristic of these genera, and differs in the form of the plastids in axial cells. In *S. multifidum*, these are discoid to bacilloid, whereas in *Antithamnion* and *Pterothamnion* they are ribbon-like to filiform.

BORNETIA Thuret

BORNETIA Thuret (1855), p. 159.

Type species: *B. secundiflora* (J. Agardh) Thuret (1855), p. 159.

Thalli uniseriate, ecorticate, consisting of prostrate indeterminate axes attached by a tangle of multicellular rhizoids and erect axes, or primarily erect to 10(-13) cm high, attached by rhizoids from basal cells; branching of vegetative axes unilateral to subdichotomous, not differentiated into indeterminate and determinate branches.

Growth by obliquely dividing apical cells; laterals initiated in rows from the high sides of axial cells, first to one side and then the other, the axis becoming sinusoidal as it flexes away from the laterals; vegetative cells large, 2-4 times longer than broad, cylindrical to somewhat inflated, multinucleate; plastids distinct, discoid to bacilloid.

Reproductive structures borne on modified branches composed of short cells. Gametophytes dioecious. Spermatangial heads ovoid to elongate, up to 6 cells long, sessile, borne on the inner sides of unilaterally branched or bifurcate condensed branches, and consisting of periaxial cells bearing branched filaments, each cell of which can function as a spermatangial mother cell bearing 2-3 spermatangia. Procarps 1-3 on modified lateral branches 6-8 cells long surrounded by an abaxial and 2 young lateral involucral branches; fertile axial cells with 2 periaxial cells, one sterile and the other fertile, consisting of a supporting cell, a 1-celled sterile group and a curved, 4-celled carpogonial branch;

supporting cell cutting off an auxiliary cell that is diploidized through fusion with a connecting cell cut off by the fertilized carpogonium; auxiliary cell evidently functioning directly as the gonimoblast initial, producing up to 6 gonimolobes which divide repeatedly to form elongate clusters of probably terminal carposporangia; involucral branches surrounding the gonimoblasts, 1 from the basal stalk cell and 2 more from the first and second cells of the fertile female branch; only one cystocarp maturing per fertile branch. Tetrasporangia sessile or pedicellate, borne on the inner sides of unilaterally branched or bifurcate condensed branches, some of which may be inflated forming an involucre, pyriform to subspherical, tetrahedrally divided.

References: Baldock & Womersley (1968); Stegenga (1985a).

One species in the British Isles.

Bornetia secundiflora (J. Agardh) Thuret (1855), p. 159.

Lectotype: LD 35341. Morocco (Tangiers), *Schousboe*.

Griffithsia secundiflora J. Agardh (1841), p. 39.

Thalli erect, 5-18 cm high, consisting of narrow to flabellate fastigiate tufts of filaments sparsely branched below, more densely branched towards apices, attached by a dense, tangled rhizoidal holdfast from which prostrate axes grow out and give rise to further erect axes, dark red in colour, firm and rigid when fresh.

Erect axes enlarging little from cylindrical apical cells with a blunt rounded apex, 260-360 µm in diameter, 2-4 diameters long, to a maximum diameter of 400-850 µm; mature cells 2-4 diameters long, cylindrical, with thin cell walls when alive, visible only at nodes, appearing very thick in preserved specimens due to shrinkage of cell contents; *branches* formed pseudodichotomously every 1-8 cells, in an irregularly alternate or secund pattern, at very narrow angles, almost parallel to parent axis; *rhizoidal filaments* formed by older axes, one or more per cell, typically at some distance from each end of the cell, pigmented, branched, forming secondary pit connections to nearby axial or rhizoidal cells, resulting in a rigid anastomosing lattice linking erect axes together, or growing out and forming erect axes, or reattaching to the substratum; *plastids* parietal, discoid to bacilloid, 1-8 x 1-3 µm.

Specialized reproductive structures unknown in the British Isles.

On boulders and bedrock, often under overhangs, from just below extreme low water to 3 m depth, tolerant of sand on rock, at moderately to very wave-exposed sites.

Confined to S. Devon, S. Cornwall (including Isles of Scilly) and Channel Isles. A report from Ireland (Cotton, 1912) was erroneous (Guiry, 1978).

England to Morocco; Mediterranean.

Large plants are found in June-October; some over-wintering plants observed in April. Most thalli apparently perennate as small basal fragments, and the small numbers of young plants are derived from vegetative fragments (Dixon, 1965). In the Mediterranean, *B. secundiflora* forms gametangia, cystocarps and tetrasporangia in September to December (Feldmann-Mazoyer, 1941), suggesting that there may be a short-day photoperiodic response. Reproduction could be inhibited in the British Isles by a requirement for high temperature in combination with short days.

There is little variation other than in overall size, and this species is one of the most easily recognizable members of the Ceramiaceae. The only possible confusion is with *Griffithsia corallinoides* (q. v.), from which it differs by the rigid rather than flaccid texture and the cylindrical rather than bead-like apical cells.

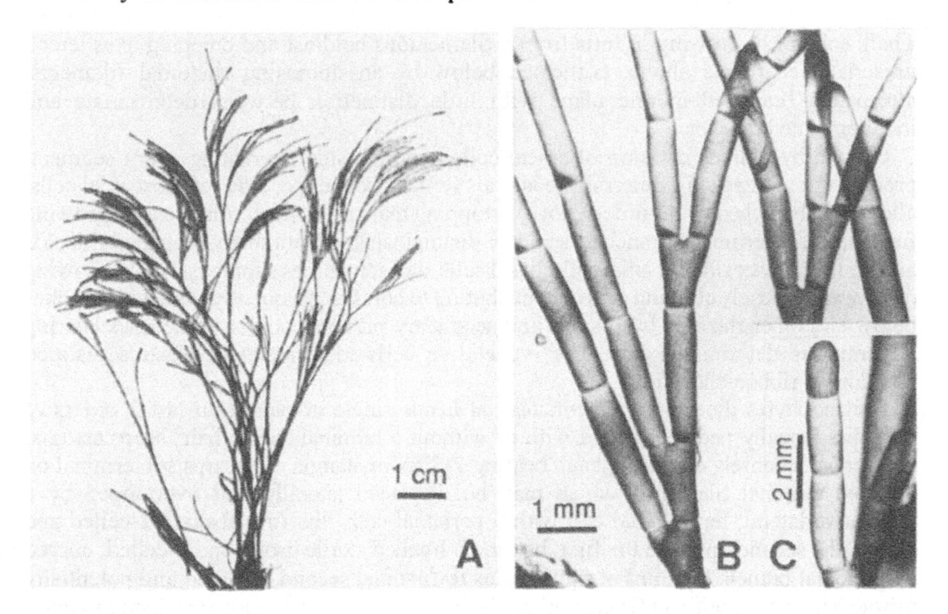

Fig. 51. *Bornetia secundiflora*

(A) Habit of herbarium specimen (Devon, undated, BM). (B) Detail of axes with secund branching pattern and very large cylindrical cells (B-C Channel Isles, Sep.). (C) Cylindrical apical and subapical cells.

Tribe COMPSOTHAMNIEAE Schmitz (1889), p. 450.

COMPSOTHAMNION Nägeli

COMPSOTHAMNION Nägeli (1862), pp. 326, 342.

Lectotype species: *C. thuyoides* (J. E. Smith) Nägeli (1862), p. 326 [see Schmitz (1889), p. 450].

Thalli solitary or growing in tufts from a filamentous holdfast and creeping axes, erect, uniseriate, ecorticate above, corticated below by anastomosing rhizoidal filaments, alternately branched in one plane with little distinction between determinate and indeterminate branches.

Growth by oblique division of apical cells, the high side alternating every segment, producing a zig-zag axis; determinate laterals formed from upper sides of most axial cells, alternately branched to 1-3 orders, not overtopping main axis; indeterminate axes growing from tips of determinate branches, scarcely distinguishable from them; branched rhizoids arising from lower sides of axial cells, basal cells, and sometimes suprabasal cells, growing downwards, loosely clothing axis or contributing to holdfast, frequently joined to branches below and other rhizoids by fusions and secondary pit connections, sometimes bearing adventitious determinate branches; vegetative cells multinucleate; plastids distinct, bacilloid to ribbon-shaped.

Gametophytes dioecious. Spermatangial heads single or paired on last 2 orders of branches, usually pedicellate and with or without a terminal sterile hair; spermatangial mother cells densely clustered, each bearing 2-3 spermatangia. Procarps subterminal on 3-celled terminal filaments, which may be displaced laterally and overtopped by a vegetative lateral; fertile axial cell with 3 periaxial cells, the first abaxial, 2-celled and sterile, the second opposite the first, bearing a l-celled sterile group and 4-celled, curved carpogonial branch, the third at right angles to first and second, 2-celled and potentially fertile; supporting cell and fertile periaxial cell each cutting off a dome-shaped auxiliary cell terminally after fertilization; carpogonial branch later degenerating; auxiliary cell functioning as the gonimoblast initial, cutting off two gonimolobe initials at opposite ends that branch repeatedly with most gonimoblast cells maturing into carposporangia; cystocarp mulberry-shaped, surrounded by sterile filaments derived by resumed growth of the apical cell and sterile cells of the procarp to produce filaments 5-7 cells long, with the lowermost cells bearing short or long rhizoids, some of which link to branchlet below by cell fusions or secondary pit connections. Tetrasporangia sessile or pedicellate on the last 2 orders of branches, or clustered in compound terminal or lateral branchlets, ovoid to subspherical, tetrahedrally divided or replaced by octosporangia.

References: Westbrook (1930a), Gordon-Mills & Wollaston (1990).

Taxonomic and nomenclatural confusion between *C. thuyoides* and *C. gracillimum*

(Dixon, 1960c) has led to identification problems. In the British Isles, most workers have regarded them as separate species, probably basing this on Harvey's (1846, pl. 5; 1850, pl. 269) treatment. Although Harvey acknowledged that these species could not be distinguished by 'any very definite character', he provided several differences in the descriptions and discussions. In *C. thuyoides* (as *Callithamnion*), branching was regular, tetrasporangia were globose and cystocarps were rare. In *C. gracillimum,* the fronds were larger, less regularly branched, and the cells of the main axes were shorter and more cylindrical; tetrasporangia were elliptical. Westbrook (1930c) stated that clustered tetrasporangia in *C. thuyoides* were 'said to constitute one of the main distinctions' from *C. gracillimum*, in which tetrasporangia were solitary. Newton (1931) used the greater regularity of branching in *C. thuyoides* as her main key character. More recently, Ardré (1970) treated these two entities as varieties of *C. thuyoides*, while Parke & Dixon (1976) and Price et al. (1981) suggested that they might be conspecific. Although distinctions between *C. thuyoides* and *C. gracillimum* are not always clear-cut, pending more detailed studies we concur with Harvey in recognizing them as separate species.

A third species is here attributed to *Compsothamnion*, based on *Callithamnion decompositum* J. Agardh (1851, p. 45). Maggs & L'Hardy-Halos (1993) showed that Agardh's species is multinucleate and shows several other features characteristic of *Compsothamnion*. Previous interpretations of the name *Callithamnion decompositum* excluded the holotype and appear to have been based on uninucleate plants from the rade de Brest, Brittany (Crouan & Crouan, 1867); part of this complex of species has been described as *Aglaothamnion priceanum* (q. v.).

KEY TO SPECIES

1 Tetrasporangia lateral and sessile .... *C. decompositum*
  Tetrasporangia terminal or pedicellate, never lateral 2
2 Some first-order branches lack branchlets on first 1-4 cells; second-order branches bear irregular arrangement of branchlets .... *C. gracillimum*
  Every cell of first and second-order branches bears branchlets in a perfectly regular alternate-distichous arrangement .... *C. thuyoides*

Compsothamnion decompositum (J. Agardh) Maggs & L'Hardy-Halos (1993), p. 528.

Holotype: LD 18964 (Maggs & L'Hardy-Halos, 1993, figs 1-7). Atlantic France, *Grateloup*.

Callithamnion decompositum J. Agardh (1851), p. 45.
Mesothamnion distichum Halos ex South & Tittley (1986), p. 52, nom. illeg.

Thalli erect, 1-2 cm in length, branched in one plane, with a single ecorticate main axis decreasing gradually from a maximum diameter of 0.1 mm at the base, and an irregularly rounded or triangular outline, rose-pink in colour, delicate and flaccid.

Main axes growing from cylindrical, usually binucleate, apical cells 8-10 μm wide and

1.5-2.5 diameters long; mature axial cells 3-4.5 diameters long, with median constrictions and thick walls; *first-order laterals* developing by oblique division of the third or fourth axial cell from the apex, one per cell, borne in a regularly alternate-distichous pattern, the basal cell remaining short, 1.5-2 diameters long, other cells elongating to 4-5 diameters in length; branching alternate-distichous from every cell, the basal cell bearing an adaxial branchlet; *second-order laterals* branched as those of first order, these third-order laterals bearing branchlets alternately or in a secund arrangement, with a further order of branching present on some thalli; *rhizoidal filaments* developing from basal cells of lateral branches and young axial cells, branched, composed of long cylindrical cells 16-24 µm wide, forming secondary pit connections and cell fusions with branchlet cells; *nuclei* increasing

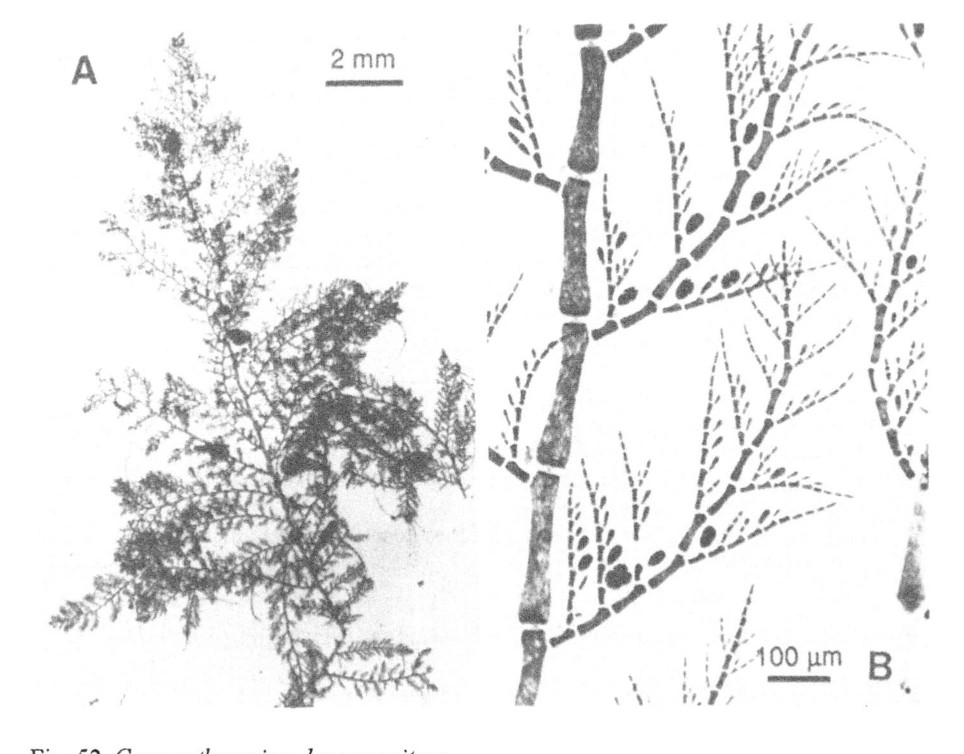

Fig. 52. *Compsothamnion decompositum*

(A) Habit (A-B Donegal, Aug.). (B) Main axes bearing regularly alternate laterals, with sessile ellipsoid tetrasporangia.

in number with cell size, up to 100-plus in axial cells; *plastids* ribbon-like or filiform, arranged longitudinally.

Spermatangial heads cylindrical, c. 60 μm long, pedicellate, borne terminally on lateral branchlets of the last two orders of branching, consisting of an elongate axis surrounded by spermatangial mother cells bearing spherical spermatangia. Female reproductive structures unknown in the British Isles. *Tetrasporangia* formed on last two orders of branching, generally adaxially, in series of up to 5 of different ages on successive branchlet cells, occasionally developing adaxially and abaxially on successive cells, sessile, ovoid or pyriform before division, subspherical to ovoid when mature, with a length: width ratio of 1.2-1.5, 36-44 μm long x 28-32 μm wide, tetrahedrally divided, containing four uninucleate spores; released spores c. 18 μm in diameter. *Octosporangia* occurring very rarely amongst tetrasporangia, ovoid, c. 50 x 30 μm.

Epilithic on vertical bedrock at about 15 m depth, at a moderately wave-exposed site, mixed with *Compsothamnion thuyoides*. In France, it has been collected in similar habitats on subtidal sponges, and also on rock near extreme low water (Maggs & L'Hardy-Halos, 1993).

Known in the British Isles only from one site in Donegal.

N.W. France (baie de Morlaix, Brittany).

Tetrasporangial and spermatangial thalli were collected in July, and tetrasporangia and a few octosporangia were observed in August. An isolate from Ireland formed tetrasporangia in culture at 15°C, 16: 8 h light: dark, but the offspring failed to reproduce.

We have insufficient material to comment on variability.

Compsothamnion decompositum could be confused with some distichous species of *Aglaothamnion*, but vegetative cells are multinucleate in contrast to the uninucleate cells of *Aglaothamnion*. In addition, none of the alternate-distichously branched *Aglaothamnion* species except *A. diaphanum* forms an adaxial branchlet on the basal cell of lateral branches, followed by an abaxial one on the second cell, as *C. decompositum* does. *A. diaphanum* can be distinguished from *C. decompositum* by the presence of bisporangia rather than tetrasporangia.

Compsothamnion gracillimum De Toni (1903), p. 1356.

Lectotype: LD 18907*. Syntype: BM. Devon (Torquay), February 1832, *Griffiths*.

non *Callithamnion gracillimum* C. Agardh (1828), p. 168.

---

* Selected because the valid use of *Compsothamnion gracillimum* by De Toni derives from the comparison by J. Agardh between C. Agardh's and Harvey's material of '*Callithamnion gracillimum*' (see Dixon, 1960c).

Thalli forming dense tufts of erect axes attached by loose filamentous holdfast and secondary creeping axes, 1-10 cm high, consisting of a single ecorticate main axis with a maximum diameter of 0.2 cm at the base, branched in one plane, but with lateral axes bunched together in mature plants, irregularly triangular to flabellate in outline, rose-pink to red in colour, delicate and flaccid.

Main axes growing from cylindrical apical cells 18 μm wide, composed of very thick-walled cells 0.5 diameter long basally, increasing to 3 diameters long, with median constrictions; *first-order laterals* developing from the fourth axial cell, one per cell, borne in a regularly alternate-distichous pattern, the basal cell remaining short, 1 diameter long, other cells elongating to 1.5 diameters in length, branched irregularly below, 1-4 of the first 4 cells often lacking branchlets, branching more regular distally, in an alternate-distichous pattern; *second-order laterals* branching similarly to those of first order; *rhizoidal filaments* developing from lateral branches and young axial cells, branched, growing downwards and forming secondary pit connections and cell fusions with branchlet cells or attaching to the substratum, composed of long cylindrical cells 18-40 μm wide; *plastids* discoid in young cells, ribbon-like or filiform in axial cells.

Spermatangial heads borne on last two orders of branching, terminal on branchlets 1-5 cells long, sometimes paired on single pedicels, cylindrical, 40-62 x 16-26 μm, lacking terminal sterile cells, consisting of a dense group of spermatangial mother cells each bearing 1-3 subspherical spermatangia 3.5 μm long x 2.5-3 μm wide. *Cystocarps* irregularly rounded, multilobed, about 500 μm in diameter, forming radiating rhizoidal filaments that connect to vegetative cells; carposporangia rounded, 16-22 μm in diameter. *Tetrasporangia* borne on last two orders of branching, pedicellate or terminal on branchlets up to 7 cells long, sometimes also formed terminally on short lateral branchlets that arise from the primary branchlet, and in clusters on pedicels, ovoid before division, subspherical to ovoid when mature, with a length: width ratio of 1.3-1.5, 36-46 μm long x 32-36 μm wide, tetrahedrally divided; *octosporangia* occasionally occurring amongst tetrasporangia, ovoid, 54-60 x 40 μm.

Intertidal on shaded vertical faces, and subtidal from extreme low water to 10 m depth, growing on rock, concrete and plastic structures, and on smaller perennial algae and epizoic on ascidians, at moderately to extremely wave-sheltered sites with moderate to strong exposure to tidal currents, tolerant of silty conditions and reduced salinity.

Fig. 53. *Compsothamnion gracillimum*

(A) Habit (Cornwall, Oct.). (B) Detail of thallus, with main axes bearing regularly alternate laterals, but having some irregularities in last-order branching, particularly near the base of branches (Donegal, Dec.). (C) Elongate spermatangial heads terminal on branchlets (C-D Pembroke, Oct.). (D) Large irregularly shaped cystocarp with rhizoidal filaments around it, some fused to vegetative cells (arrow). (E) Ellipsoid tetrasporangia (t) and larger octosporangia (o) terminal on last-order branches (Pembroke, undated).

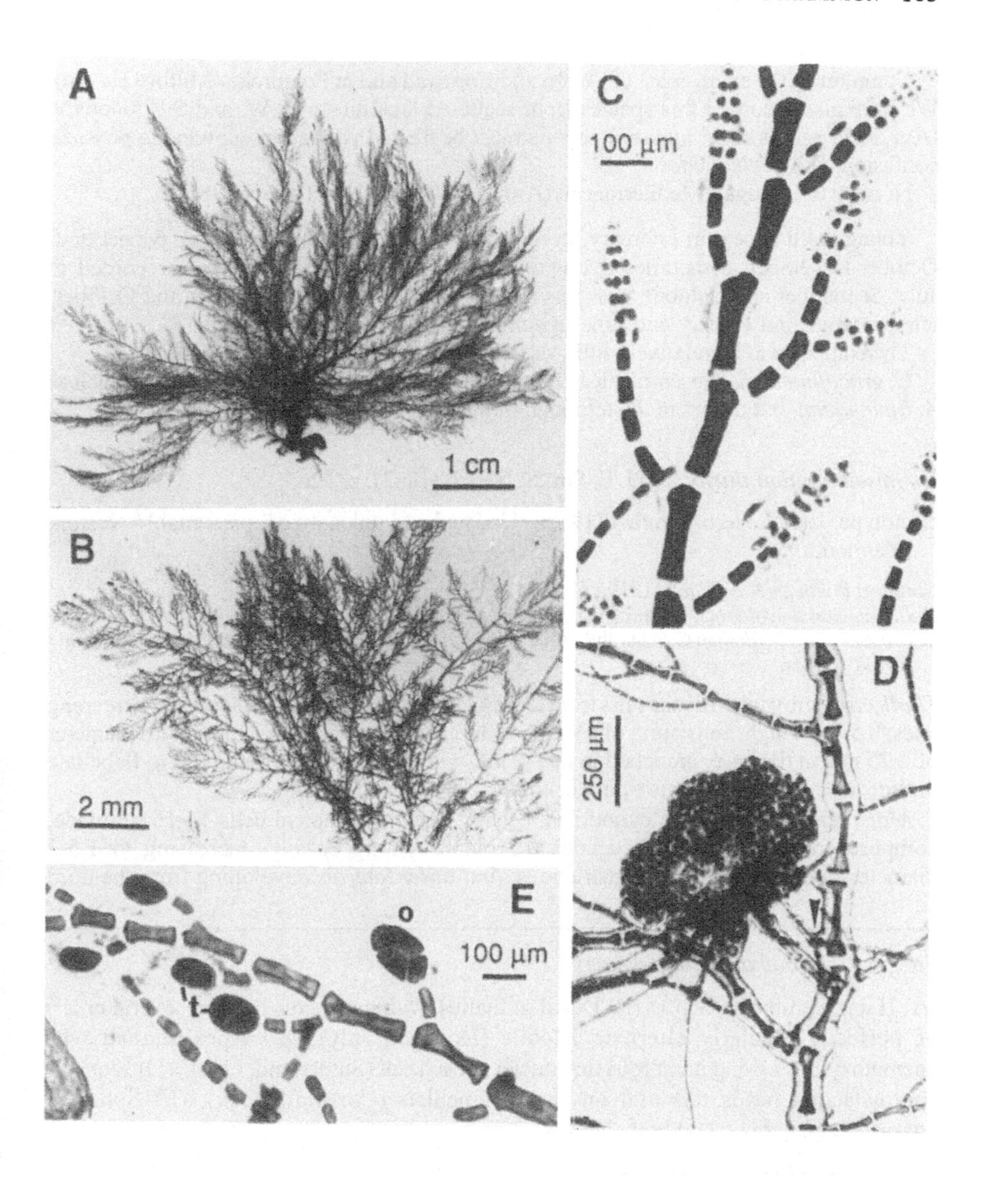
A
1 cm
B
2 mm
C
100 µm
D
250 µm
E
o
100 µm
t

Common on the south coast of Devon and Cornwall and in Pembroke (Milford Haven). We have also recorded this species from scattered localities in S.W. and S.E. Scotland (Ayr, Fife) and Orkney; literature reports must be treated with caution owing to possible confusion with *C. thuyoides*.

Norway to Portugal; Mediterranean (Ardré, 1970; South & Tittley, 1986).

Young thalli appear in February, become fertile from June onwards and persist until October-December, sometimes overwintering to February. Spermatangia recorded in July, September and October; procarps and cystocarps in July, September and October; tetrasporangia in February and June-December; and octosporangia in July.

There appears to be relatively little variation in form.

C. gracillimum may resemble distichously branched species of *Aglaothamnion*, such as *A. bipinnatum*, but differs in the terminal, rather than lateral, tetrasporangia.

Compsothamnion thuyoides (J. E. Smith) Nägeli (1862), p. 326.

Lectotype: LINN, Herb. Smith 1719.36. Undated and unlocalized, presumably Norfolk (Yarmouth).*

Conferva thuioides J. E. Smith (1810), pl. 2205.
Callithamnion thuyoides (J. E. Smith) C. Agardh (1828), p. 172.
Callithamnion gracillimum C. Agardh (1828), p. 168, non *Compsothamnion gracillimum* De Toni (1903), p. 1356.

Thalli erect, growing in dense tufts from loose filamentous holdfast and secondary creeping axes, 0.5-7 cm high, consisting of a single ecorticate main axis with a maximum diameter of 0.25 mm at the base, branched in one plane, with an irregularly triangular to flabellate outline, rose-pink to brownish-pink in colour, delicate and flaccid.

Main axes growing from cylindrical, usually binucleate, apical cells 14-16 μm wide, composed of very thick-walled cells 0.5 diameter long basally, increasing to 1.5-3 diameters long, with median constrictions; *first-order laterals* developing from the third

---

Fig. 54. *Compsothamnion thuyoides*

(A) Habit (Pembroke, Sep.). (B) Detail of thallus, with main axes bearing several orders of perfectly regularly alternate laterals (Donegal, July). (C) Apex stained with haematoxylin showing numerous tiny nuclei in each cell (Sutherland, Oct.). (D) Elongate spermatangial heads terminal on short branchlets (Cornwall, Oct.). (E) Spherical tetrasporangia terminal on branchlets (as A).

---

* Although this specimen has no collection data and is not that illustrated by Smith, no other authentic material has been located.

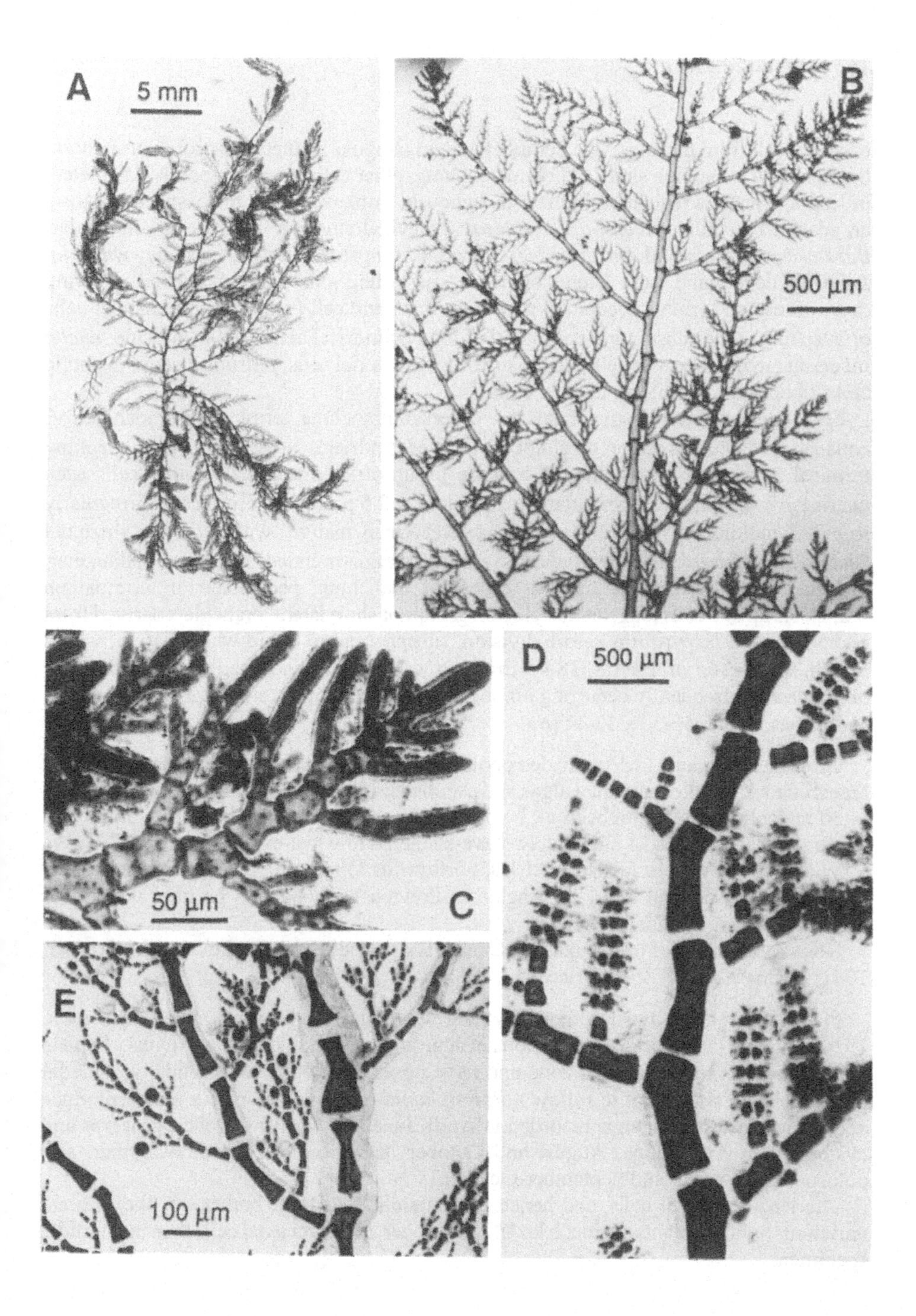
A
5 mm
B
500 µm
C
50 µm
D
500 µm
E
100 µm

or fourth cell from the apex, one per axial cell, in a regularly alternate-distichous pattern, the basal cell remaining short, 0.5 diameters long, other cells elongating to 1-2 diameters in length, branched regularly alternate-distichously from every cell, the basal cell bearing an adaxial branchlet; *second-order laterals* branched similarly to those of first order; *third-order laterals* and *fourth-order laterals* bearing branchlets alternately; *rhizoidal filaments* developing from lateral branches and young axial cells, branched, growing downwards and forming secondary pit connections and cell fusions with branchlet cells or attaching to the substratum, composed of long cylindrical cells 16-44 μm wide; *nuclei* increasing in number with cell size, up to 100-plus in axial cells; *plastids* discoid in young cells, ribbon-like or filiform in axial cells.

Spermatangial heads borne on last two orders of branching, terminal on branchlets 1-3 cells long, sometimes paired on single pedicels, cylindrical, 28-72 x 20-24 μm, lacking terminal sterile cells, consisting of a dense group of spermatangial mother cells each bearing 1-3 subspherical spermatangia 3 μm long x 2.5 μm wide. *Cystocarps* irregularly rounded, multilobed, 250-725 μm in diameter when mature, with radiating rhizoidal filaments that fuse with vegetative cells; carposporangia rounded, 16-24 μm in diameter. *Tetrasporangia* formed on last two orders of branching, pedicellate or terminal on branchlets up to 7 cells long, or clustered on groups of short lateral branchlets derived from pedicels, ovoid or pyriform before division, subspherical to ovoid when mature, with a length: width ratio of 1.1-1.3, 38-44 μm long x 32-36 μm wide, tetrahedrally divided; *octosporangia* frequently occurring amongst tetrasporangia, developing from binucleate sporocytes, ovoid, 46-50 x 37-44 μm.

Growing intertidally in deep shaded pools up to mid-shore and on damp, shaded vertical faces, on rock, smaller perennial algae and ascidians, and subtidal from extreme low water to 30 m depth, typically on bedrock below the lower limit of kelp, less often on mobile substrata such as maerl, at moderately wave-sheltered to wave-exposed sites.

Widely distributed in the British Isles, northwards to Orkney, although there are few records for E. Scotland and N.E. England. Previously thought to be rare, it is found frequently during subtidal surveys.

Norway, British Isles to Portugal and Canary Isles; Mediterranean (Feldmann-Mazoyer, 1941; South & Tittley, 1986; Price et al, 1986).

Plants apparently annual, although young individuals may over-winter (Westbrook, 1930); subtidal plants becoming obvious in June and persisting until late August. Female plants are rare in the intertidal zone and have never been observed subtidally; subtidal populations do not appear to follow a *Polysiphonia*-type life history but may reproduce apomictically. Spermatangia recorded in April, June and August-October; procarps and cystocarps in April, June, August and October; tetrasporangia in May-October; and octosporangia in July and September-October.

The length of axial cells, and hence the apparent density of branching, shows great variation. Subtidal plants are much laxer, with longer, narrower axial cells than in intertidal specimens.

C. thuyoides can resemble distichously branched species of *Aglaothamnion*, such as *A. bipinnatum*, but differs in the terminal, rather than lateral, tetrasporangia.

PLEONOSPORIUM Nägeli, nom. cons.

PLEONOSPORIUM Nägeli (1862), pp. 326, 339.

Type species: *Pleonosporium borreri* (J. E. Smith) Nägeli (1862), pp. 326, 339.

Thalli uniseriate, erect to 10(-20) cm high, ecorticate, corticated at base, or sometimes nearly to apex by descending rhizoids, attached to substratum by tangled rhizoids or rhizoids terminated by multicellular, digitate haptera; alternately branched in spiral, distichous or subdichotomous arrangement, one lateral per segment, often with little distinction between determinate and indeterminate branches.

Growth by slightly oblique division of apical cell, the high side alternating every segment, producing a zig-zag axis; determinate laterals alternately branched or secund, to 1-3(-4) orders of branching, sometimes strongly incurved at the apex; indeterminate branches growing from tips of determinate branches, scarcely distinguishable from them; rhizoidal filaments arising from basal cells of lateral branches and sometimes also from axial cells, growing downwards and forming secondary pit connections to axial cells, lateral branches, and other rhizoids below; vegetative cells multinucleate; plastids distinct, discoid to elongate.

Gametophytes dioecious. Spermatangial heads 5-8 cells long, terminal or lateral, often adaxial on the last 2 orders of branches; spermatangial mother cells on filaments clustered around axial cells, each bearing 2-3 spermatangia. Procarps subterminal on 2-celled terminal filaments, which may be displaced laterally and overtopped by a lateral branch; apical cell of fertile axis sometimes deflected and the procarp sometimes appearing to be terminal; fertile axial cell bearing 2-3 periaxial cells, with the first fertile, bearing a sterile group and 4-celled carpogonial branch, the second sterile, and the third sterile or absent; supporting cell cutting off a broad, dome-shaped auxiliary cell after fertilization, or the supporting cell enlarging and functioning directly as the auxiliary cell; fertilized carpogonium cutting off two connecting cells laterally, one of which fuses with an extension from the auxiliary cell; cells of the carpogonial branch fusing and later degenerating; auxiliary cell (or supporting cell) containing the haploid nucleus, which divides and the products either retained within the auxiliary cell or cut off inside 1-2 isolated cells (disposal cells) that degenerate, and cutting off a distal gonimoblast initial containing the diploid nucleus; primary gonimoblast cell forming 2-4 gonimolobe initials laterally that branch repeatedly, with most of the cells maturing into carposporangia; mature cystocarp consisting of 1 or more spherical masses of carposporangia, naked or surrounded by 1-several involucral filments produced from cells immediately below the fertile segment. Tetrasporangia or polysporangia solitary and often adaxial in secund series, sessile or pedicellate, or forming clusters on the last 1-2 orders of branches; sporocytes ellipsoid to pyriform, cleaving simultaneously into 4-32 spores.

References: Feldmann-Mazoyer (1941), Norris (1985), Stegenga (1986).

One species in the British Isles. Two forms of *Pleonosporium borreri* were previously recognized, but one of these, *P. borreri* f. *fasciculatum* (Harvey in W. J. Hooker) Holmes & Batters (1891, p. 97) [basionym: *Callithamnion fasciculatum* Harvey in W. J. Hooker (1833), p. 343], is based on a specimen of an *Aglaothamnion* species (Maggs & L'Hardy-Halos, 1993).

Pleonosporium borreri (J. E. Smith) Nägeli (1862), p. 326.

Lectotype: TCD. Norfolk (Yarmouth), 9 x 1806, *Borrer*.

Conferva borreri J. E. Smith (1807), pl. 1741.
Callithamnion seminudum C. Agardh (1828), p. 167 (see Harvey, 1848, pl. 159).
Callithamnion borreri (J. E. Smith) C. Agardh (1828), p. 170.

Thalli erect, forming loose to very dense tufts of few to many erect axes attached by tangled rhizoidal filaments, 1.5-5(-7.5) cm high and 1-5 cm broad, with a single principal axis below, complanate or bushy, with a pyramidal to irregularly flabellate outline, rose-pink in colour, fairly rigid when fresh.

Main axes enlarging from apical cells 20-26 μm wide to a maximum diameter of 125-150 μm, up to 500 μm in large thalli due to cover of loose intertwined downgrowing filaments that merge into the holdfast; lowermost cells 1-1.5 diameters long, cylindrical, increasing upwards to 2-4 diameters long, with median constrictions; *first-order laterals* developing from the third cell from the apex, the apices typically level with main apical cells, occasionally overtopping them, borne in a regularly alternate-distichous pattern except when damaged, unbranched below for a variable number of cells, then branched in a regular alternate-distichous pattern; *second-order laterals* and *third-order laterals* branched similarly to those of first order, the branches on each lateral lying parallel to each other, then curving inwards or outwards; hairs absent; *rhizoidal filaments* branched, multicellular, 20-46 μm in diameter, developing from basal cells of lateral branches, growing downwards and forming secondary pit connections to cells of other laterals and main axis; *plastids* discoid in young cells, bacilloid to ribbon-like in older cells.

Spermatangial heads formed adaxially on last two orders of branching, 1 per cell, on

Fig. 55. *Pleonosporium borreri*

(A) Habit with characteristic broken lower laterals (A-B Donegal, July). (B) Detail of thallus, with main axes bearing regularly alternate laterals that are naked below. (C) Conical spermatangial heads borne adaxially (C-D Dorset, July). (D) Procarp. (E) Cystocarp with several gonimolobes surrounded by involucral branches (Cornwall, July). (F) Sessile and pedicellate polysporangia (Pembroke, Sep.).

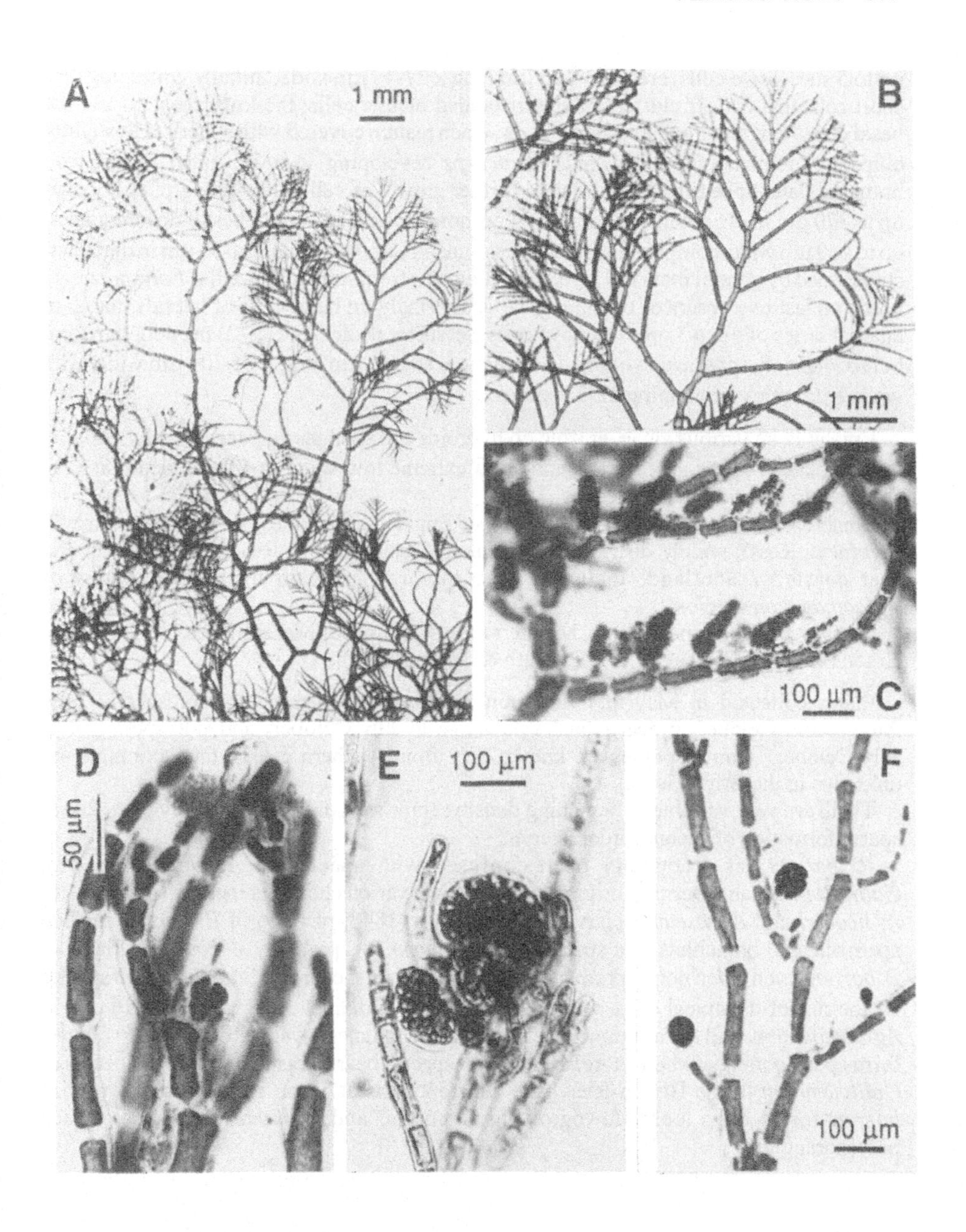
A
1 mm
B
1 mm
100 µm
C
D
50 µm
E
100 µm
F
100 µm

up to 5 successive cells, erect, 73-106 µm long x 27-46 µm wide, initially consisting of 3 short cells that cut off clusters of spermatangial mother cells, the lowermost part of the basal cell often remaining as a sterile stalk, when mature covered with spherical to slightly ellipsoid spermatangia 4 µm long. *Procarps* developing in a subapical position on branches, later appearing lateral due to further growth of cell below procarp; *cystocarps* up to 370 µm in diameter, consisting of 4 or more globular gonimolobes of different ages, up to 250 µm wide, composed of numerous rounded carposporangia 24-30 µm in diameter, surrounded by several incurved involucral filaments from subtending cells. *Polysporangia* borne on last two orders of branching, typically singly on basal cells of laterals, rarely in adaxial series of up to 3 on successive cells, sessile or pedicellate, 1(-2) per cell, pyriform before division, spherical to ovoid when mature, 80-106 µm long x 72-104 µm wide, with walls 8 µm thick, containing 12-32 spores.

Epiphytic on various algae, epizoic on hydroids and sponges and occasionally epilithic on bedrock, in lower-shore pools and from extreme low water to 30 µm depth at very sheltered to wave-exposed sites.

South-east, south and south-west coasts of England and Wales, W. Scotland (Argyll, Inverness, Ross); widely distributed in Ireland; Channel Isles. Records from north and east coasts of Scotland, including Orkney, are based on misidentifications of *Aglaothamnion* species.

Norway to Spain and Azores; Mediterranean; Morocco; Massachusetts to Delaware, Brazil (Ardré, 1970; Taylor, 1957; South & Tittley, 1986).

Plants collected in May and June non-reproductive; spermatangia present in July, August and October; procarps and cystocarps in July-October; and polysporangia in July-October. Gametophytes are known only from southern coasts; tetrasporangia are unknown in the British Isles.

Thalli are very variable in branching density; some subtidal plants are very open due to sparse formation of second-order laterals.

P. borreri has frequently been confused with species of *Aglaothamnion* and *Callithamnion*, and Borrer's original collections in various herbaria include specimens of *A. hookeri* and *A. roseum*. Harvey's illustrations (1848, pl. 159) of *P. borreri* include spermatangial branchlets of a species of *Aglaothamnion*, probably *A. roseum*. Plants of *P. borreri*, even when non-reproductive, can be distinguished from *Aglaothamnion* species by the size of the apical cells, which are 20-26 µm in diameter compared to ≤16 µm in *Aglaothamnion*, and the multinucleate rather than uninucleate vegetative cells. *P. borreri* forms polysporangia whereas *Aglaothamnion* species bear tetrasporangia. All species of *Callithamnion* in the British Isles have adherent cortication on main axes, whereas *P. borreri* forms only loose downgrowing filaments, and *Callithamnion* species lack polysporangia.

Tribe GRIFFITHSIEAE Schmitz (1889), p. 449.

HALURUS Kützing

HALURUS Kützing (1843), p. 374.

Type species: *H. equisetifolius* (Lightfoot) Kützing (1843), p. 374.

Thalli uniseriate, the tips setaceous with mucronate terminal cells, erect to 20 cm high, attached by a discoid rhizoidal holdfast or by tangled rhizoids that give rise secondarily to erect axes, ecorticate or corticated by ascending and descending, anastomosing rhizoids, consisting of numerous long and short axes, the axial cells bearing 5-8 whorled determinate branches, or the primary branching pseudodichotomous; whorled determinate branches 1-3 times di- to trichotomously branched; indeterminate branches developing directly from whorled determinate branches, or pseudodichotomies developing 5-9 cells behind the apex with second and third laterals later produced from the same axial cells; adventitious whorled branches, when present, issuing from the upper ends of axial cells or from corticating rhizoids.

Growth by extension from the apex and distal ends of subapical cells, followed by transverse or oblique divisions, the laterals developing into whorled determinate or indeterminate branches, or the primary branching pseudodichotomous; hyaline, hair-like branches (trichoblasts) absent on vegetative axes; rhizoidal filaments, when present, developing from basal cells of whorled determinate branches, corticating the axis and giving rise to adventitious whorled branches; vegetative cells multinucleate; plastids distinct, discoid to bacilloid.

Reproductive structures borne on short fertile axes produced secondarily from axial cells; gametophytes dioecious. Spermatangial heads borne adaxially on the upper ends of basal cells of whorled determinate branches, sessile, 3-4 times polychotomously branched; fertile segments forming whorls of tri- to dichotomously branched filaments terminating in spermatangial mother cells, each of which bears 2-3 spermatangia. Procarps subterminal in specialized 3-celled fertile axes clustered at the tips of determinate branches and surrounded by 6-7 whorled laterals bearing few to many deciduous hyaline hairs (trichoblasts); repeated branching from apical cells forming a compound structure containing several fertile female axes; fertile axial cell bearing 2 periaxial cells, the first sterile, the second consisting of a 1-celled sterile group and a curved, 4-celled, carpogonial branch; supporting cell evidently functioning directly as the auxiliary cell after fertilization, cutting off a gonimoblast initial distally that bears 1-4 subspherical gonimolobes in which most cells mature into carposporangia; gonimoblasts surrounded by an inner whorl of 1-celled involucral branches formed from the basal cell of the female axis after fertilization. Tetrasporangia sessile, developing adaxially from the distal ends of whorled determinate laterals, or in clusters on short adaxial filaments borne on basal cells of whorled determinate branches, spherical, tetrahedrally divided.

References: Baldock (1976), Millar (1986).

Halurus equisetifolius is at present the only species included in this genus. However, Baldock (1976) and Millar (1986) considered that there might be affinities between *Griffithsia flosculosa* and *Halurus*. Comparative reproductive studies of *H. equisetifolius* and *G. flosculosa* have shown remarkably close similarities, and *G. flosculosa* is here transferred to *Halurus*.

KEY TO SPECIES

Axes consisting of an axial filament bearing a whorl of whorl-branches on every cell .... *H. equisetifolius*
Axes consisting of dichotomously to trichotomously branched filaments lacking whorl-branches .... *H. flosculosus*

Halurus equisetifolius (Lightfoot) Kützing (1843), p. 374.

Lectotype: BM-K (see Dixon, 1983). Hampshire (Isle of Wight).

Conferva equisetifolia Lightfoot (1777), p. 984.
Conferva imbricata Hudson (1778), p. 603.
Ceramium simplicifilum De Candolle (1806), p. 8.
Griffithsia equisetifolia (Lightfoot) C. Agardh (1824), p. 143.
Griffithsia simplicifilum (De Candolle) C. Agardh (1828), p. 134.
Halurus simplicifilum (De Candolle) Kützing (1849), p. 663.
Halurus equisetifolius var. *simplicifilum* (De Candolle) J. Agardh (1851), p. 91.

Thalli consisting of erect axes growing singly or in groups of up to 7 from discoid holdfast 5-10 mm in diameter, 6-16(-22.5) cm high, cylindrical, branched irregularly, sparsely to very densely, to 4 orders of branching; main axes, including whorl-branches, tapering from 2-3 mm in diameter near the base to 1-2 mm in diameter at apices; lateral branches 1-3 mm wide, clothed with imbricate curved filaments except where denuded at the base of

---

Fig. 56. *Halurus equisetifolius*

(A) Habit of densely branched plant (Cornwall, Sep.). (B) Apex of main axis with whorls of whorl-branches (B-D Glamorgan, Mar.). (C) Axis with paler reproductive outgrowths (arrows). (D) Squashed spermatangial outgrowth, consisting of a central axis with whorls of branchlets bearing spermatangial heads (s). (E) One branchlet of spermatangial outgrowth with mucronate tips (arrow), bearing an adaxial spermatangial head (as A). (F) Tetrasporangial outgrowth consisting of central axis with densely whorled branchlets bearing tetrasporangia (F-G Cornwall, Mar.). (G) One branchlet of tetrasporangial outgrowth with tetrasporangia both sessile (arrows) and on pedicels (p).

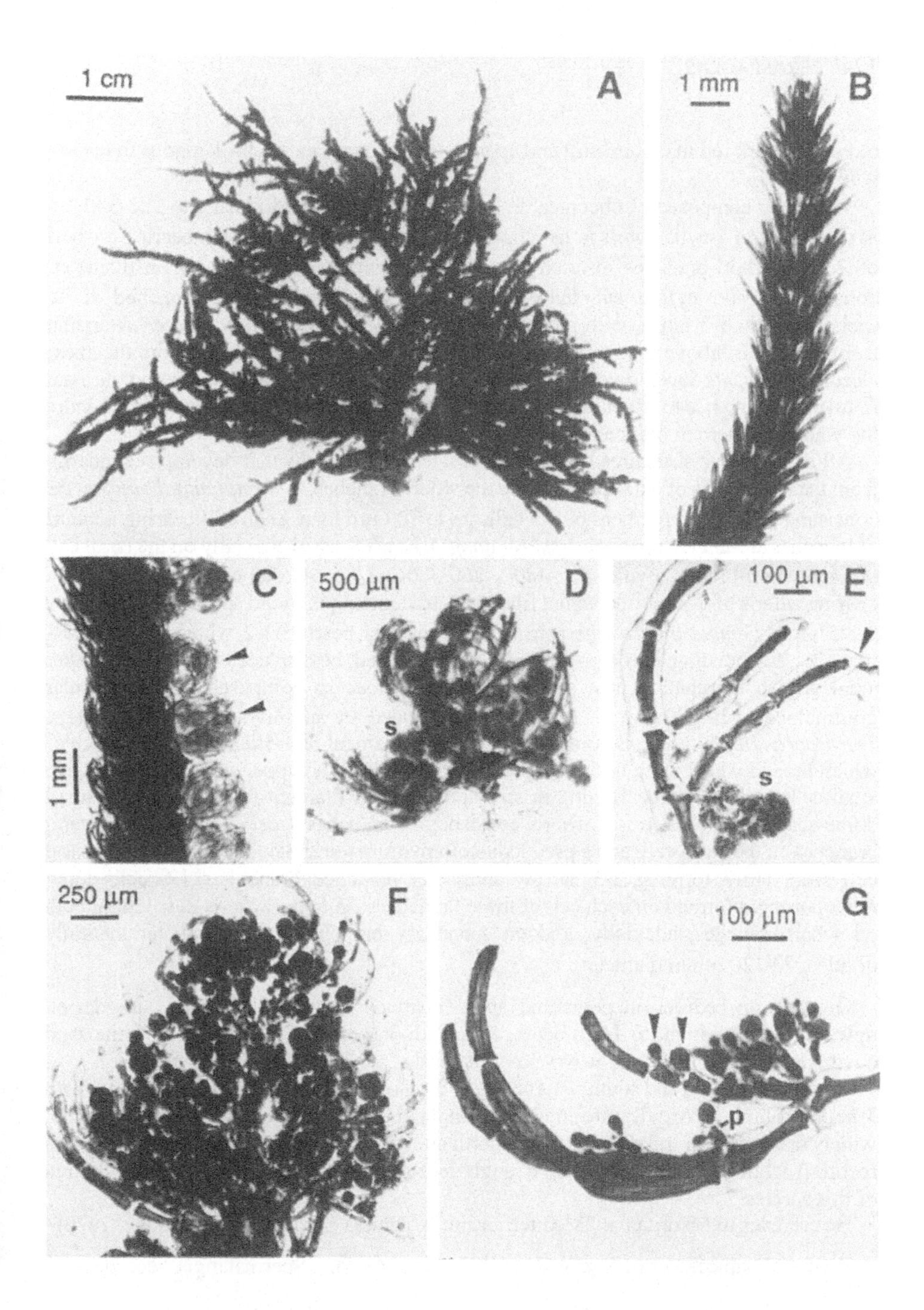
1 cm
A
1 mm
B
C
1 mm
500 µm
D
s
100 µm
E
s
250 µm
F
100 µm
G
p

old plants, dark red in colour, soft and spongy when young, more cartilaginous in texture when older.

Main axes composed of obconical to cylindrical axial cells increasing to 200-350 μm in diameter and 4-6 diameters in length, with walls to 70 μm thick, each bearing a whorl of 7 (5-8) whorl-branches inserted distally; *whorl-branches* 75-125 μm in diameter, consisting of 4-7 cylindrical, thick-walled cells 350-550 μm long, branched di- to trichotomously 1-3 times, with mucronate terminal cells; *indeterminate laterals* arising from axial cells, above the insertion of whorl-branches, at some distance from the apex; *rhizoidal filaments* developing from basal cells of whorl-branches and corticating the axial filaments, giving rise to adventitious branchlets that resemble whorl-branches and obscure the whorled pattern in older axes; *plastids* discoid to bacilloid.

All reproductive structures borne in short lateral outgrowths that develop secondarily from the upper ends of axial cells, above the whorl-branches. *Spermatangial outgrowths* consisting of an axial filament of 1-3 cells up to 700 μm long, each cell bearing a whorl of branches around the upper end; *spermatangial heads* formed adaxially on the basal cell of each whorl-branch, ovoid, 430-480 x 260-380 μm, composed of a central filament bearing whorls of 5-7 spermatangial filaments, terminating in ovoid spermatangia 6-6.5 x 3-3.5 μm. *Female outgrowths* consisting of an axis bearing 1-2 whorls of 7 whorl-branches, surrounding a procarp that may, if unfertilized, be displaced laterally by further axial growth terminating in a second procarp; *cystocarps* composed of 1-3 globular gonimolobes 430-720 μm in diameter, surrounded by an involucre of branchlets. *Tetrasporangial outgrowths* consisting of an axial filament of 4-5 turbinate cells, each of which bears a whorl of up to 6 incurved branches around the upper end, with successive axial cells and whorls decreasing in size and the axial filament terminating in a small dome-shaped cell; lowermost whorls consisting of branches 4 cells long, with 2 lateral branches on the basal cell, and a pseudodichotomy from the 2nd or 3rd cell, with the basal cell secondarily forming 2-3 narrow simple or branched branchlets 1-6 cells long; *tetrasporangia* formed on each cell of these branchlets, in upper whorls developing both on whorl-branches, adaxially; and on secondary branchlets, spherical, tetrahedrally divided, 70-120 μm in diameter.

Growing on bedrock in pools and open situations on the lower shore, subtidal on upward-facing bedrock to 14 m depth, frequently in kelp forest, tolerant of some sand cover, on moderately to very wave-exposed coasts.

Common on south and south-west coasts of the British Isles, rare in S.W. Scotland, with a northern limit in Argyll; rare in eastern England, occurring northwards to Yorkshire; widely distributed in Ireland but rare on north and east coasts. The original Firth of Forth record (Lightfoot, 1777) was either wrongly located or indicates a change in distribution of this species.

British Isles to Mauritania; ?Mediterranean; Argentina (Ardré, 1970; Baldock, 1976).

Perennial; apices resume growth in February-March. Spermatangia recorded for

February-June and September, procarps and cystocarps in March and July; tetrasporangia in February, March and June. Polysporangia were reported from Portugal (Ardré, 1970), but have not been found in the British Isles.

There is some variation in the degree of branching of whorl-branches; plants with simple whorl-branches correspond to var. *simplicifilum* (see Harvey, 1850, pl. 287 for further details).

H. equisetifolius can be distinguished from *Sphondylothamnion multifidum* (q. v.) by the whorl-branches, which are >75 µm in diameter, with a mucronate terminal cell, compared to c. 30 µm in diameter, with a bluntly rounded terminal cell, in *S. multifidum*. Two species that may superficially resemble *H. equisetifolius*, *Dasya hutchinsiae* (q. v.) and *Aglaothamnion sepositum* (q. v.), have spiral rather than whorled arrangements of branchlets.

Halurus flosculosus (Ellis) Maggs & Hommersand, comb. nov.

Lectotype: Ellis (1768), pl. 18, fig. E. Norfolk (Yarmouth).

Conferva flosculosa Ellis (1768), p. 425.
Conferva setacea Hudson (1778), p. 599.
Griffithsia setacea (Hudson) C. Agardh (1817), p. xxviii.
Griffithsia flosculosa (Ellis) Batters (1902), p. 84.

Thalli forming dense fastigiate obconical to hemispherical tufts up to 20 cm in length and 30 cm wide, consisting of numerous erect filaments attached by dense tangled rhizoidal filaments, from which further erect axes arise, dull purplish- to brownish-red in colour, with a rough, semi-rigid texture.

Erect axes growing from conical apical cells 70-100 µm in diameter and 0.5-1 diameter long; first few axial cells elongating little, increasing gradually to a maximum diameter of 360-420 µm; mature cells 3.5-6 diameters long, cylindrical or inflated distally, not constricted at cross-walls, with walls 30-40 µm thick when alive, appearing thicker if preserved; *branches* initially pseudodichotomous, developing 5-9 cells behind apex, later pseudodichotomous to quadrichotomous, every 1-4 cells, as second and third laterals develop later from same cells at acute angles, usually all lying in one plane, also developing adventitiously from lateral walls of older axial cells; *rhizoidal filaments* arising from lower ends of axial cells, one to several per cell, multicellular, 75-150 µm in diameter, branching and forming further rhizoidal filaments or erect filaments; *plastids* bacilloid.

All reproductive structures borne in short lateral outgrowths that develop secondarily from the upper ends of axial cells. *Spermatangial outgrowths* consisting of an axial filament of 2-3 cells terminating in a small dome-shaped or conical cell, the basal cell almost cylindrical, the second and third cells capitate, bearing distally a whorl of 8 incurved branches with mucronate tips, lower whorl larger than upper whorl, all whorl-branches 2-3 cells long; *spermatangial heads* ovoid, formed adaxially on basal cells of whorl-branches, 240-360 x 120-220 µm, composed of a central filament of 4-5 cells, each

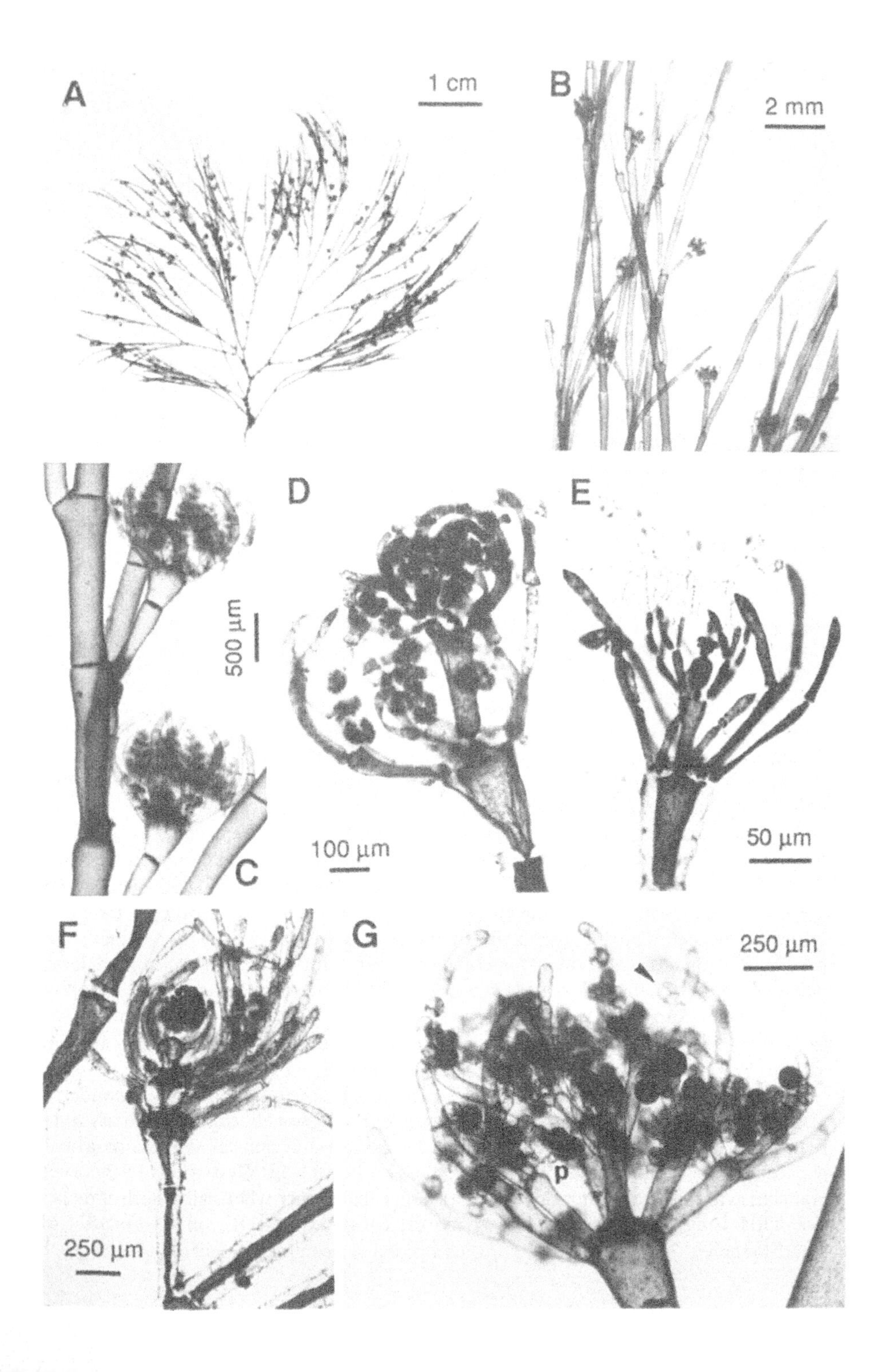
A
1 cm
B
2 mm
500 µm
C
D
100 µm
E
50 µm
F
250 µm
G
250 µm
p

with a whorl of 5 branchlets that consist of a rounded basal cell branching trichotomously and bearing dense clusters of ellipsoid spermatangia measuring 4 x 2 μm. *Female outgrowths* 6 cells long when mature, consisting of an elongate cylindrical basal cell, a second, turbinate, cell bearing a whorl of 8 simple 2-celled whorl-branches with mucronate tips, and another turbinate cell bearing a ring of 6 branched 3-celled whorl-branches with terminal trichoblasts that surround the central 3-celled female axis; further female axes developing from between whorl-branches if the first is unfertilized; *cystocarps* composed of 1-4 globular gonimolobes up to 250 μm in diameter containing numerous carposporangia 52-68 μm long, surrounded by an involucre of branchlets. *Tetrasporangial outgrowths* consisting of an axial filament of 3 cells terminating in a small dome-shaped cell, the basal cell almost cylindrical, the second and third cells turbinate, bearing distally a whorl of incurved branches with mucronate tips, lower whorl consisting of 7-8 branches, upper whorl of 5-6 branches, all branches 3 cells long, the basal cell forming 3 adaxial lateral branchlets of different lengths; *tetrasporangia* borne adaxially on all branchlet cells except the terminal ones, in addition to a cluster of 1-3 narrow branchlets, simple or branched, 1-6 cells long, and on each cell of these branchlets, spherical, tetrahedrally divided, 74-92 μm in diameter, with walls 12-15 μm thick.

On bedrock in pools and emergent near extreme low water, subtidal to 10 m, rarely to 23 m depth, on bedrock and pebbles, and epiphytic on *Laminaria hyperborea* stipes and other algae, occasionally on artificial substrata such as pier pilings, at very sheltered to very exposed sites; typically growing as scattered plants, occasionally abundant in sheltered habitats.

Generally distributed in the British Isles, although there are few records for eastern Scotland.

Norway and Faroes to S. Spain; Mediterranean; Morocco, ?Canaries, ?St Helena (Ardré, 1970; South & Tittley, 1986; Price et al., 1988).

Sporelings and developing and mature plants are found throughout the year. Spermatangia recorded for January-July and September; procarps and cystocarps in February, March, and May-July; tetrasporangia from January-July and October-

---

Fig. 57. *Halurus flosculosus*

(A) Habit (Galway, Feb.). (B) Detail of thallus showing cylindrical axial cells, tapered apices, and tetrasporangial outgrowths (Down, Jan.). (C) Spermatangial outgrowths when alive (as A). (D) Stained spermatangial outgrowth consisting of central axis with several whorls of branchlets bearing elongate spermatangial heads (D-F Dorset, July). (E) Female outgrowth with central procarp surrounded by whorls of branchlets, the inner ring bearing fine hairs. (F) Cystocarp surrounded by whorls of branchlets. (G) Tetrasporangial outgrowth consisting of central axis with densely whorled branchlets bearing sessile (arrow) and pedicellate (p) tetrasporangia (as B).

December. Male and tetrasporangial thalli are much more common than reproductive females.

There is little morphological variation.

Distinctions between *H. flosculosus* and *Griffithsia devoniensis* are given under *G. devoniensis* (q. v.).

ANOTRICHIUM Nägeli

ANOTRICHIUM Nägeli (1862), p. 397.

Type species: *A. barbatum* (C. Agardh) Nägeli (1862), p. 398.

Thalli uniseriate, composed of erect axes to 30 cm high, repeatedly pseudodichotomously branched, often tufted or corymbose at the tips, or the branching subsecund from lower ends of axial cells; ecorticate or corticated below by descending anastomosing rhizoids; erect axes segmented, the cells long, narrow, cylindrical to barrel-shaped; hyaline, tri-dichotomously branched vegetative hairs (trichoblasts) sometimes abundant, produced in whorls from the distal ends of cells near the apex.

Growth by extension from the apex and upper ends of apical and subapical cells followed by transverse or oblique divisions, continuing the main axis or forming a lateral branch; second and third laterals sometimes produced from the distal ends of axial cells; hyaline hairs (trichoblasts) initiated by protrusions from the distal ends of apical and subapical cells in vegetative and reproductive axes, usually formed in whorls; rhizoids, when present, issuing from upper and lower ends of axial cells, anastomosing or attaching to cell walls by terminal stellate haptera; cells multinucleate; plastids distinct, discoid to bacilloid.

Gametophytes dioecious or monoecious. Spermatangial heads borne adaxially on basal cells of whorled determinate laterals terminated by hyaline hairs (trichoblasts), pedicellate in whorls lacking trichoblasts, or solitary on pedicels and lacking trichoblasts; periaxial cells bearing polychotomously branched filaments terminated by spermatangial mother cells, each bearing 2-3 spermatangia. Procarps subterminal in 3-celled female axes formed singly at the ends of branches along with few to many trichoblasts, and usually displaced laterally, overtopped by a lateral vegetative branch; fertile axial cell bearing 2 periaxial cells, the first sterile, the second adaxial, at right angles to the first, consisting of a supporting cell, a terminal, 1-2-celled sterile group and a horizontally curved carpogonial branch; supporting cell functioning directly as the auxiliary cell after fertilization; the fertilized carpogonium cutting off 2 connecting cells from opposite sides, one of which fuses with an extension from the auxiliary cell; gonimoblast initial cut off distally from the auxiliary cell (= supporting cell) and producing few to many gonimoblast initials that mature successively into globose clusters of carposporangia; fusion cell incorporating the axial cell, supporting cell, primary gonimoblast cell, and inner gonimolobe cells; basal cell of fertile axis enlarging during procarp development and producing a whorl of 1-celled involucral branches that elongate around the gonimoblasts. Tetrasporangia pedicellate, subspherical, produced in whorls encircled by trichoblasts, or solitary, or borne in clusters, adaxially from the upper ends of cells near the thallus apex or in tufted lateral branches, subspherical, tetrahedrally divided.

References: Baldock (1976), Boudouresque & Coppejans (1982, as *Griffithsia*), Stegenga (1985b).

Only one species of *Anotrichium*, *A. barbatum*, was previously known in the British Isles (South & Tittley, 1986); *A. furcellatum* is reported for the first time here.

KEY TO SPECIES

Cells pyriform, swollen distally, apical cells c. 50 µm diameter; tetrasporangia borne with trichoblasts on whorls of elongate cells .... *A. barbatum*
Cells cylindrical, apical cells <25 µm diameter; tetrasporangia solitary on short pedicels lacking trichoblasts .... *A. furcellatum*

Anotrichium barbatum (C. Agardh) Nägeli (1862), p. 398.

Lectotype: BM-K. Syntypes: BM, LD. Sussex (Brighton), July 1807, *Borrer*.

Griffithsia [‘*Griffitsia*’] *barbata* C. Agardh (1828), p. 132.
Conferva barbata J. E. Smith (1808), pl. 1814, non *Conferva barbata* Zoega (1775), p. 202.

Thalli forming tufts 2-6 cm high; erect axes consisting of numerous fastigiate dichotomously branched filaments, with corymbose apices, rose-pink in colour, extremely flaccid and delicate.

Erect axes gradually increasing in width from pyriform apical cells c. 50 µm in diameter to a maximum diameter of 200 µm; mature cells swollen distally or almost cylindrical, 5-8 diameters long; *trichoblasts* borne in a distal whorl on younger vegetative cells, dichotomously branched, the basal cell 350-500 µm long; *branches* formed pseudo-dichotomously, numerous; angular protein crystals often present.

Gametophytes dioecious or monoecious. *Spermatangial heads* formed by whorls of elongate cells, borne singly, adaxially on an elongate cell that terminates in a branched trichoblast, ovoid, c. 50 µm long. *Cystocarps* consisting of 2-3 rounded gonimolobes surrounded by simple or branched incurved involucral filaments. *Tetrasporangia* formed by whorls of elongate cells, singly and adaxially on a swollen, elongate cell 250-300 µm long that terminates in a branched trichoblast, spherical, 50-60 µm diameter. [The above description is based largely on Harvey (1850, pl. 281), interpreted using Feldmann-Mazoyer (1941), and L'Hardy-Halos (1968a) as no recently collected material was available.]

Epiphytic on small algae in intertidal pools (Harvey, 1850, pl. 281). No further habitat data is available as most specimens appear to have been collected in the drift. There is

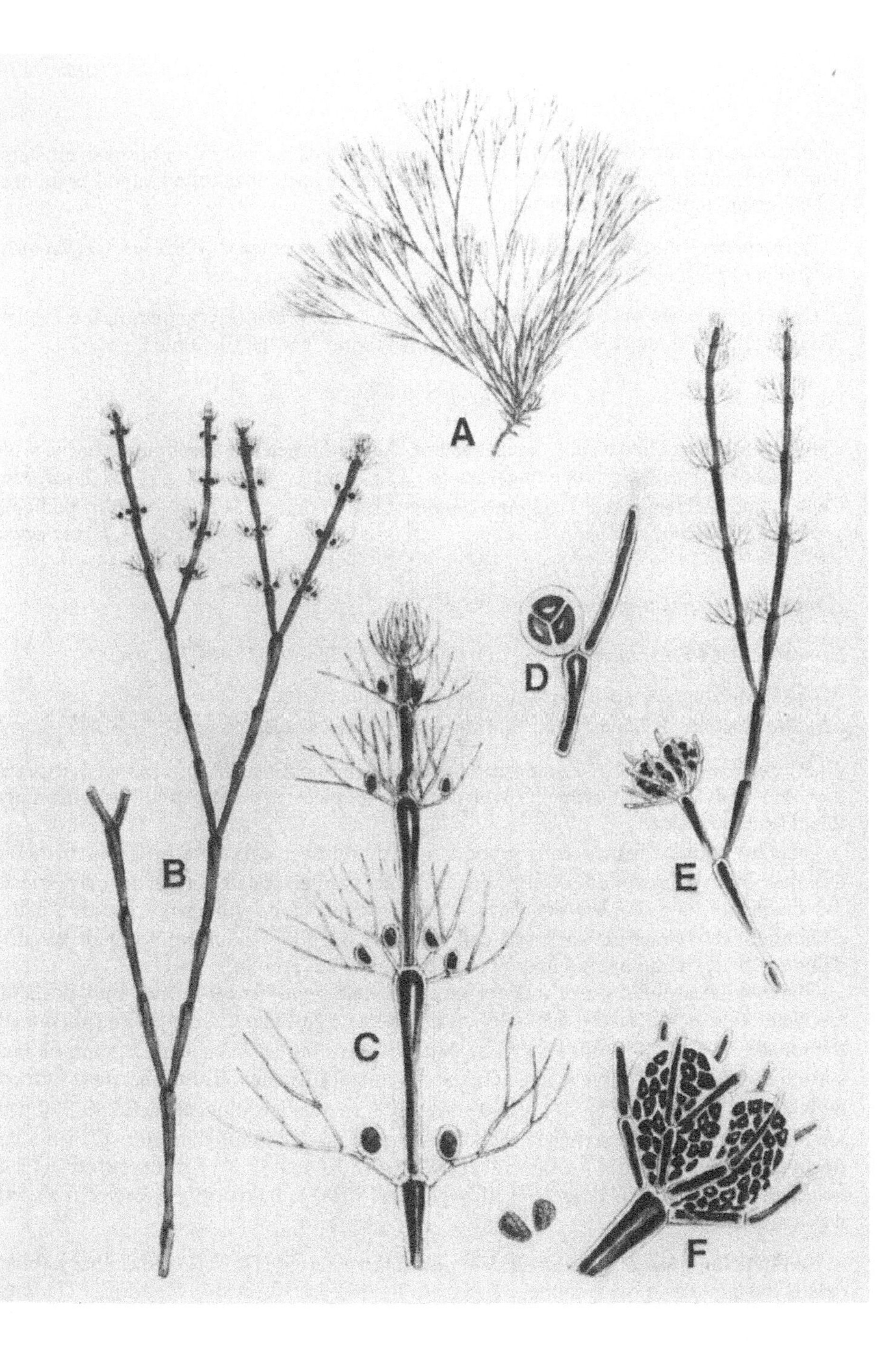
A
B
C
D
E
F

no evidence that this species has been found during this century; all recent reports represent misidentifications of *Anotrichium furcellatum*.

Found in the British Isles only on Channel coasts: Sussex, Hampshire, Dorset; Channel Isles.

British Isles to N. Spain; Mediterranean; Canaries, W. Africa (Børgesen, 1930; Feldmann-Mazoyer, 1941; South & Tittley, 1986; Price et al., 1986). Records from the western Atlantic (Virgin Is: Børgesen, 1916; North Carolina: Kapraun, 1980) are probably based on *Anotrichium tenue* (C. Agardh) Nägeli, which is characterized by whorls of terminal tetrasporangia.

Plants have been collected only from May-October. Spermatangia were observed in August (Buffham, 1891), cystocarps in May-October, and tetrasporangia in June-October.

This species can readily be distinguished from related algae by the occurrence of whorls of trichoblasts on vegetative cells.

Anotrichium furcellatum (J. Agardh) Baldock (1976), p. 560.

Lectotype: LD 19901; possible syntypes: BM, PC. Italy (Amalfi), 15 vi 1841, *J. Agardh*.

Griffithsia furcellata J. Agardh (1842), p. 75.
Neomonospora furcellata (J. Agardh) Feldmann-Mazoyer & Meslin (1939), p. 193.
?*Monospora tenuis* Okamura (1934), p. 24, non *Anotrichium tenue* (C. Agardh) Nägeli (1862), p. 398 (see Feldmann-Mazoyer & Meslin, 1939).

Thalli forming dense hemispherical cushions about 7 cm high, attached by loose rhizoidal filaments; erect axes consisting of numerous corymbose much-branched filaments, fanning out upwards, rose-pink in colour, extremely flaccid and delicate.

Erect filaments growing from cylindrical to conical apical cells with a blunt rounded apex, 14-20 μm in diameter, that increase in length after division from 2 up to 4 diameters; axes gradually increasing in width to a maximum diameter of 170-260(-450) μm; mature cells cylindrical or slightly constricted in the middle, with straight cross-walls, 5-12 diameters long, surrounded by walls 20-28 μm thick; *branches* formed pseudo-dichotomously 2-4(-7) cells behind apices, alternate; trichoblasts absent in vegetative and tetrasporangial thalli; *rhizoidal filaments* formed by axial cells, one or more per cell from near both ends, growing outwards, and either becoming hooked or fusing to a cell of an adjacent filament; *plastids* parietal, numerous, discoid, 3-10 μm in diameter.

Fig. 58. *Anotrichium barbatum*

Reproduced from Harvey (1850), pl. 281, based on specimens from the Channel Isles, undated (scales not available). (A) Habit (ca. 0.8 x natural size). (B) Detail of thallus showing pyriform axial cells and whorls of fertile trichoblasts towards apices. (C, D) Tetrasporangia borne on fertile trichoblasts. (E, F) Cystocarps surrounded by whorls of branchlets.

Gametophytes unknown in the British Isles. *Tetrasporangia* solitary, formed near apices, sometimes in addition to a vegetative branch, developing as oblique lateral projections that soon divide into a pedicel 20 µm wide, 1-2 diameters long, and a terminal pyriform reproductive cell, spherical when mature, 44-60 µm in diameter, with a wall 8 µm thick.

Epiphytic on other algae from extreme low water to 2 m depth, in very wave-sheltered, muddy habitats exposed to slight to moderate tidal currents, most frequently in non-estuarine inlets. Although rarely found, it can be abundant in suitable habitats.

Found in the British Isles at a few localities in Dorset, S. Devon and Pembroke; Channel Isles. The earliest known British collections were made in Dorset in 1976.

British Isles to Canaries; Mediterranean; North Pacific (Abbott & Hollenberg, 1976).

Plants have been collected only in September and October. All thalli were sterile, except for one plant obtained from the Channel Isles in October, which bore sparse developing and mature tetrasporangia. Tetrasporangia were reported to be rare in France (Feldmann-Mazoyer & Meslin, 1939) and the Netherlands (Stegenga & Mol, 1983). Female gametophytes have been observed in Brittany (L'Hardy-Halos, 1968a), but males and cystocarps are apparently unknown in Europe. European populations reproduce vigorously by secondary attachment and fragmentation (Feldmann-Mazoyer & Meslin,

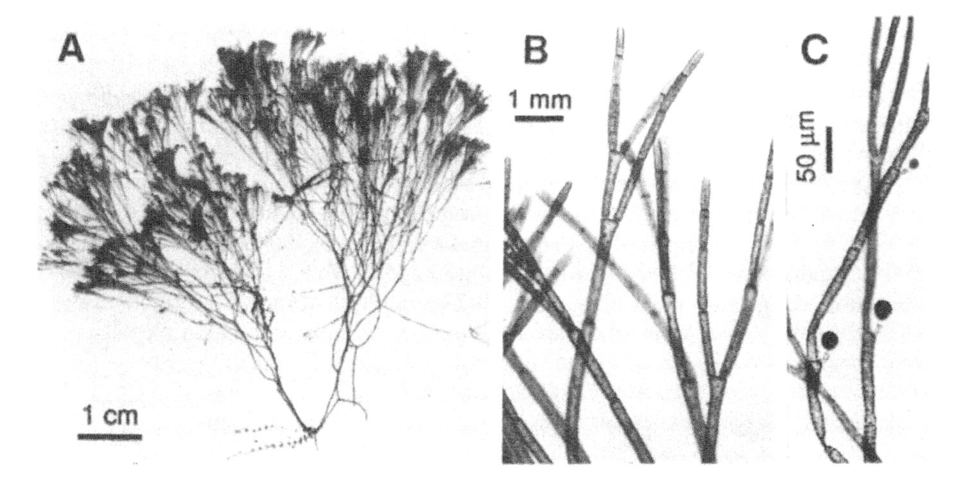

Fig. 59. *Anotrichium furcellatum*

(A) Habit of herbarium specimen, showing branching density increasing towards corymbose apices (Devon, Sep.). (B) Detail of thallus showing cylindrical axial cells and cylindrical to slightly conical apical cells (Pembroke, Sep.). (C) Developing and mature tetrasporangia borne on short pedicels (Channel Isles, Oct.).

1939). Gametangia and tetrasporangia, but not cystocarps, have been observed in California (Abbott & Hollenberg, 1976).

There is very little morphological variation in collections from the British Isles.

Feldmann-Mazoyer & Meslin (1939) reported that this species had been introduced from the Mediterranean to northern France prior to 1922. Its range in the Mediterranean seemed to be expanding rapidly and they queried whether the alga now present was identical to that described by J. Agardh. Since present populations resembled *Monospora tenuis* Okamura from the Pacific, they might have been derived from the introduction of a (possibly conspecific) Pacific form into the range of the original Mediterranean alga.

A. furcellatum has often been considered to be a synonym of *Griffithsia arachnoidea* C. Agardh (1828), p. 131 (e.g. by Børgesen, 1930; see Price et al., 1988). *G. arachnoidea* was described from French material sent by Duvau. Of the two Duvau specimens in Agardh's herbarium, LD 19905 is of *Halurus* (*Griffithsia*) *flosculosus*. The other, LD 19903, is difficult to identify, lacking all reproduction, but it is definitely not *A. furcellatum*, as it has sparse branching and much wider apical cells than *A. furcellatum*. The name *G. arachnoidea* can thus be excluded from considerations of the nomenclature of *A. furcellatum*.

Sterile thalli of *A. furcellatum* may resemble some forms of *Callithamnion corymbosum* (q. v.). In *C. corymbosum*, however, lateral branching shows a spiral divergence, and numerous terminal hairs are usually present, whereas branching of *A. furcellatum* is pseudodichotomous in one plane, and hairs are absent. Tetrasporangia are sessile in *C. corymbosum* and pedicellate in *A. furcellatum*. Cell diameter near apices of *Griffithsia devoniensis* (q. v.) is 65-100 µm in contrast to <25µm in *A. furcellatum*, and hooked rhizoids are formed only in *A. furcellatum*.

GRIFFITHSIA C. Agardh

GRIFFITHSIA C. Agardh (1817), p. xxviii.

Lectotype species: *G. corallina* (Lightfoot) C. Agardh (1817), p. xxviii [= *G. corallinoides* (Linnaeus) Batters (1902), p. 84].

Thalli uniseriate, ecorticate, differentiated into erect axes to 20 cm high and stolon-like creeping axes attached by multicellular rhizoids, pseudodichotomously (rarely pseudotrichotomously) or unilaterally branched; erect indeterminate axes segmented, often moniliform, the cells large, cylindrical to barrel-shaped or clavate to subglobose; whorled pigmented vegetative determinate branches absent; hyaline poly-, tri-, or dichotomous, vegetative hair-like branches (trichoblasts) formed in whorls from the upper ends of cells, or absent.

Growth by extension from the apex and upper sides of apical cells, followed by transverse or oblique divisions, continuing the main axis or forming a lateral branch; hyaline hair-like branches (trichoblasts) initiated by protrusion from the upper ends of apical and subapical cells, often in whorls; rhizoids issuing from the proximal ends of axial cells, sometimes in whorls; vegetative cells highly vacuolate, often inflated, multinucleate, the number of nuclei increasing by synchronous mitoses; plastids distinct, discoid to bacilloid.

Gametophytes dioecious, occasionally monoecious. Spermatangial heads clustered in whorls at the distal ends of apical and subapical cells, compound, consisting of a basal cell and polychotomously branched filaments bearing spermatangial mother cells, each with 1-3(-4) spermatangia, either naked or surrounded by whorled involucral branches. Procarps subterminal in specialized 3-celled fertile axes; fertile axes 1-2, produced terminally along with few to many deciduous hyaline hairs (trichoblasts) from an apical cell, terminal or displaced laterally, overtopped by a lateral vegetative branch; fertile axial cell bearing 2 (rarely 3) periaxial cells, the first abaxial, sterile, the second at right angles to the first, fertile, consisting of a supporting cell, a terminal, 1-celled sterile group and 1 or 2 horizontally curved 4-celled carpogonial branches; the third, when present, potentially fertile, opposite the second; supporting cell functioning directly as the auxiliary cell after fertilization; the fertilized carpogonium cutting off 2 connecting cells from opposite sides, one of which fuses with an extension from the auxiliary cell; gonimoblast initial cut off distally from the auxiliary cell (= supporting cell) and producing few to many gonimoblast initials that mature successively into globose clusters of carposporangia, surrounded by a whorl of involucral branches borne on the basal cell of the fertile branch, each consisting of a small ovoid cell and an inflated apical cell. Tetrasporangia clustered in whorls at the distal ends of apical or subapical cells, each cluster consisting of a stalk cell and few to many sessile tetrasporangia, or compound, resembling the spermatangial clusters, naked or surrounded by whorled fertile or sterile involucral branches, pyriform, tetrahedrally divided.

References: Lewis (1909), Kylin (1916), Baldock (1976).

Three species of *Griffithsia* are listed for the British Isles by South & Tittley (1986). Of these, *G. flosculosa* is here transferred to *Halurus*, while the positions of *G. devoniensis* and *G. corallinoides* remain unchanged.

KEY TO SPECIES

Subapical cells >125 µm in diameter, mature cells pyriform; rhizoidal filaments sparse except near holdfast .... *G. corallinoides*

Subapical cells <75 µm in diameter, mature cells cylindrical; rhizoidal filaments common, formed at upper ends of axial cells .... *G. devoniensis*

Griffithsia corallinoides (Linnaeus) Batters (1902), p. 84.

Holotype: OXF, Herb. Dillenius (L. Irvine, pers. comm.). Undated, unlocalized specimen labelled 'Conferva marina gelatinosa corallinae instar geniculata, crassior'.

Conferva corallinoides Linnaeus (1753), p. 1166.
Conferva geniculata Ellis (1768), p. 425.
Conferva corallina Lightfoot (1777), p. 988.
Griffithsia corallina (Lightfoot) C. Agardh (1817), p. xxviii.

Thalli forming dense tufts 2.5-20 cm long, consisting of numerous much-branched erect filaments, attached by tangled rhizoidal filaments, deep red to orange-red in colour, soft and rather slippery, often with a strong chemical smell when fresh.

Erect axes growing from spherical apical cells 50-100 µm in diameter, increasing gradually in diameter to a maximum of 500 µm; cells 3-4 diameters long when mature, initially inflated at both ends, becoming markedly obpyriform, constricted near cross-walls, with thick cell walls; *branches* developing 1-3 cells behind apex, pseudodichotomous (rarely pseudotrichotomous) every 1-3 cells, all lying in the same or a different plane; *trichoblasts* short, present near apices of female plants only; *rhizoidal filaments* developing from lateral walls of some older cells, branched, 50-100 µm in diameter, formed in whorls from bases of cells near holdfast and contributing to the holdfast; *plastids* parietal, discoid to bacilloid.

Gametophytes dioecious. *Spermatangial heads* developing near apices in a distal position on young cells and enlarging as axes mature, forming dense bands up to 300 µm wide around nodes, each branchlet consisting of an obconical basal cell that gives rise to a filament up to 100 µm long, bearing on every cell a whorl of densely but irregularly branched filaments terminating in spermatangial mother cells that form spherical to ovoid spermatangia 6 µm long (Grubb, 1926). *Cystocarps* 1000-1300 µm in diameter, composed of 3-4 rounded gonimolobes of different ages, up to 700 µm in diameter, containing numerous rounded carposporangia, 40-75 µm in diameter, surrounded by an involucre of incurved 1-2 celled filaments to 720 µm long x 360 µm wide, with lobed or bifid apices. *Tetrasporangia* borne in bands around upper ends of axial cells, developing near apices and enlarging as axis matures, borne in clusters on narrow 1-3-celled filaments and on basal cells of 2-celled incurved involucral filaments c. 200 µm long, clavate before and after division, with long extensions connected to bearing cell, 50-70 µm in diameter.

Epiphytic on *Zostera* leaves and smaller algae, less often on kelp blades and stipes, epizoic on shells, ascidians and polychaete worm tubes, and occasionally growing on pebbles and boulders, in wet channels in the intertidal, and subtidal to 20 m depth at moderately to extremely wave-sheltered sites, sometimes with exposure to tidal currents.

Generally distributed in the British Isles.

Norway to S. Spain; ?Morocco, ?Canaries (Ardré, 1970; Price et al., 1986; South & Tittley, 1986).

Mature plants can be found in abundance throughout the year, although the majority of thalli are probably short-lived; sporelings were noted in June and August-November. Spermatangia recorded for February, April and June-October, cystocarps in May-October, and tetrasporangia throughout the year.

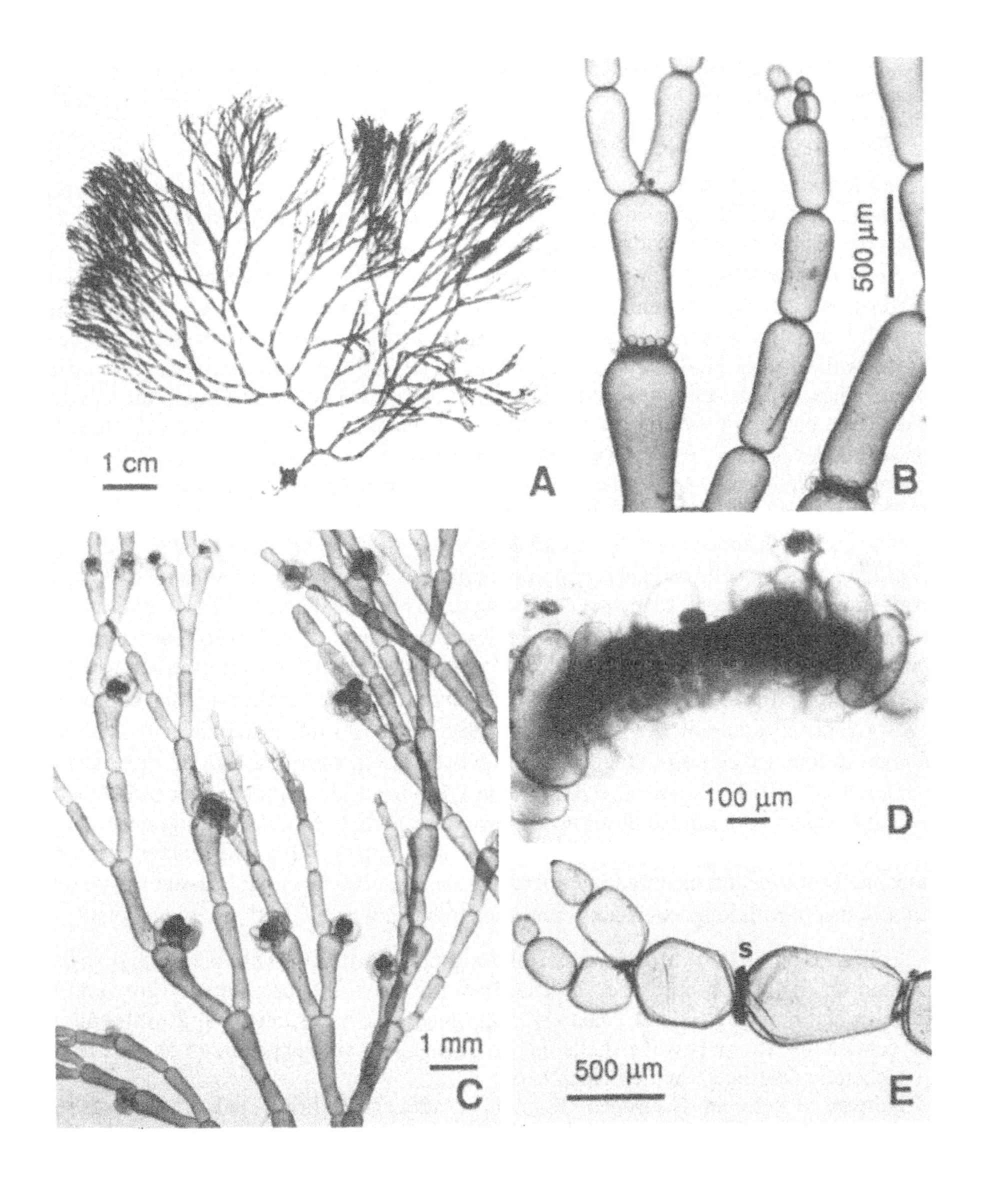
1 cm
A
500 µm
B
1 mm
C
100 µm
D
s
500 µm
E

There is little variation other than in overall size.

Chromosome number was reported as *n* = 20 (Kylin, 1916), but this requires confirmation. Some specimens of *G. corallinoides* may resemble *Bornetia secundiflora* (q. v. for distinctions).

Griffithsia devoniensis Harvey (1846), pl. 16.

Lectotype: BM-K. Syntypes: TCD. Devon (Salcombe), September 1840, *Wyatt*.

Thalli forming dense hemispherical clumps up to 12 cm in diameter when luxuriant, consisting of numerous erect filaments 3-5.5(-10) cm long; sparser thalli consisting of only a few erect filaments, attached by tangled rhizoidal filaments, deep pink to orange-pink in colour, extremely soft and flaccid.

Erect axes growing from apical cells that elongate after division from hemispherical to 1.5 diameters long, with a blunt rounded apex, gradually increasing in diameter from 65-100 μm near apices to a maximum of 150-325 μm; mature cells 5-14 diameters long, cylindrical or slightly inflated distally or at both ends, constricted near cross-walls, with very thin cell walls visible only at nodes; *branches* developing by oblique division of third cell behind apex, pseudodichotomous every 1-5 cells when mature; trichoblasts absent; *rhizoidal filaments* abundant, formed at distal ends of older cells, pigmented, branching and reattaching to the substratum; *plastids* parietal, in long bead-like chains.

Gametophytes dioecious. *Spermatangial heads* borne near apices, developing distally on young cells, forming dense whorls up to 150 μm long around nodes, consisting of pseudodichotomously branched spermatangial branchlets, each spermatangial mother cell bearing a group of spermatangia. *Cystocarps* 280-380 μm long x 360-530 μm wide, composed usually of 3 gonimolobes of different ages, up to 330 μm in diameter, bearing carposporangia 26-40 (-50) μm in diameter, surrounded by a few unicellular incurved involucral cells 360-450 x 130-150 μm. *Tetrasporangia* borne in whorls around nodes of mature axes, developing in clusters on small cells formed at distal ends of axial cells, slightly pyriform, 44-60 x 38-52 excluding the loose mucilaginous walls, intermixed with little-pigmented, incurved involucral cells 72-180 x 40-44 μm.

On stones, dead shells and loose-lying on mud, epizoic on ascidians and polychaete worm tubes on silty bedrock and boulders, from near extreme low water to 14 m depth in

---

Fig. 60. *Griffithsia corallinoides*

(A) Habit (Cork, Nov.). (B) Detail of thallus showing large swollen axial cells and spherical apical cells; developing tetrasporangia present on some nodes (B-D Devon, Nov.). (C) Detail of female thallus with mature pyriform or centrally constricted axial cells, bearing cystocarps partially enclosed by involucral branches. (D) Tetrasporangia borne in a ring around node, enclosed by incurved involucral branches. (E) Spermatangial branchlets (s) borne around nodes (Galway, Nov.).

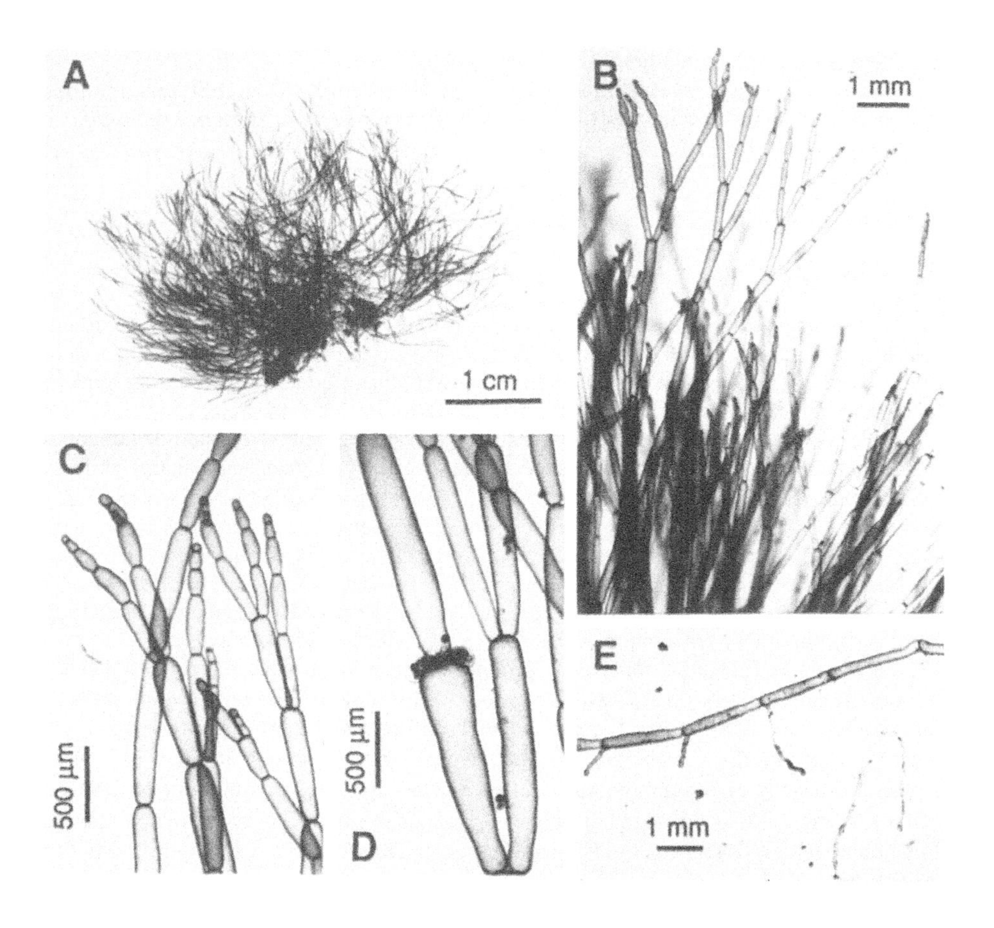

Fig. 61. *Griffithsia devoniensis*

(A) Habit (Devon, Oct.). (B) Detail of thallus showing frequent pseudodichotomous branching and cylindrical axial cells (B-E Dorset, May). (C) Apices showing cylindrical axial cells and globular to cylindrical apical cells. (D) Tetrasporangia borne in a ring around node, enclosed by incurved involucral branches. (E) Lower part of axis with rhizoids decreasing in age towards the left-hand side, borne in an anterior position on cells.

moderately to extremely wave-sheltered bays and inlets, sometimes with exposure to fairly strong tidal currents; occasionally found to 10 m at sites exposed to moderate wave action.
South and south-west coasts of the British Isles: Essex, Sussex, Hampshire, Dorset, S. Devon, Cornwall; Pembroke; Cork, Dublin; Channel Isles.
British Isles, Denmark (as drift), Netherlands, France.

Mature plants found from April to October; only small fragments have been collected in winter. Gametangial thalli are rare in populations on the south coast of England and have not been observed in Wales or Ireland. Spermatangia recorded for June, cystocarps in June and October, and tetrasporangia in April, June, July and September.
There is little variation other than in overall size and density of thalli.
G. devoniensis can readily be distinguished from *Halurus* (*Griffithsia*) *flosculosus* (q. v.), even when sterile, by its flaccid texture, thin cell walls, obtuse apical cells and the regular formation of narrow rhizoidal filaments. *H. flosculosus* is rigid, with thick cell walls and pointed apical cells. *Anotrichium barbatum* bears whorls of trichoblasts around younger vegetative cells, whereas trichoblasts are lacking in *G. devoniensis*. Some forms of *Spermothamnion* species can resemble *G. devoniensis*, when sterile, but are distinguishable by the long apical cells, approximately equal in length to subapical cells, whereas apical cells in *G. devoniensis* are never more than 2 diameters long.

MONOSPORUS Solier in Castagne

MONOSPORUS Solier in Castagne (1845), p. 242.

Lectotype species: *M. pedicellatus* (J. E. Smith) Solier in Castagne (1845), p. 242 [see Schmitz (1889), p. 450].

Thalli uniseriate, erect to 20 cm high, attached to substratum by a tangled rhizoidal holdfast; ecorticate, except at the base; main axes pseudodichotomous, lateral branches alternate-distichous to spirally branched, consisting of one to several branches from the distal ends of axial cells; apical cells mucronate, axial cells cylindrical, elongate.
Growth by extension from the apex and upper ends of apical and subapical cells followed by oblique divisions; branching thought to be sympodial with the lateral initial forming the main axis and the apical initial producing a lateral branch; hyaline hairs (trichoblasts) absent; rhizoids issuing from the lower ends of basal axial cells; vegetative cells multinucleate; plastids distinct, bacilloid to elongate.
Sexual reproduction and tetrasporangia unknown. Propagules developing as outgrowths from the upper ends of axial cells; pedicellate, often appearing axillary due to the formation of an additional lateral branch; shed by dehiscence of the pedicel from its bearing cell; germination direct, without release of a spore.

References: L'Hardy-Halos (1967), Baldock (1976), Huisman & Kraft (1982), Kim & Lee (1990).

Monosporus has been treated as a form-genus containing species that produce pedicellate propagules and lack sexual reproduction. Their numbers have decreased as sexual stages have been demonstrated in more and more species. At present two morphological types are attributed to *Monosporus*: those that are wholly erect as in the type species, and those that consist of prostrate and erect axes that resemble *Spermothamnion*. The description given above applies only to the erect type.

One species in the British Isles.

Monosporus pedicellatus (J. E. Smith) Solier in Castagne (1845), p. 242.

Possible type: BM-K. Sussex (Brighton), undated, *Borrer*.

Conferva pedicellata J. E. Smith (1808), pl. 1817.
Corynospora pedicellata (J. E. Smith) J. Agardh (1851), p. 69.
Monospora pedicellata (J. E. Smith) J. Agardh (1876), p. 608.
Neomonospora pedicellata (J. E. Smith) Feldmann-Mazoyer & Meslin (1939), p. 195.

Thalli erect, consisting of densely tufted groups of several fastigiate to divaricate branched erect axes 1.5-20 cm high, attached by tangled rhizoidal holdfast, bright pink to brownish-red in colour, crisp and rigid when fresh, rapidly becoming flaccid, often with a sweet chemical smell.

Main axes growing from clavate, mucronate apical cells, increasing in diameter from 30-39 μm to 180-300 μm and in length from 2.5 to 10 diameters, cylindrical or slightly inflated at both ends, the walls 25 μm thick (appearing thicker in preserved material); *branches* formed distally by each axial cell, one or more per cell, borne in an irregularly alternate to pseudodichotomous arrangement, forming a further series of irregularly alternate-distichously to spirally arranged branches; *rhizoidal cells* formed by basal ends of main axial cells and basal cells of laterals, one to several per cell, non-corticating, tapering from 100 to 30 μm in diameter.

Gametangial and tetrasporangial structures unknown. *Propagules* developing as outgrowths from upper end of laterals, often appearing axillary to a branchlet, 1(-2) per cell, unicellular when mature, ovoid to pyriform, 180-250 μm long x 60-140 μm wide, with deeply pigmented opaque contents, borne on a non-pigmented obconical pedicel 50-60 μm x 50-65 μm, shed by dehiscence of pedicel from its bearing cell.

On bedrock and epiphytic on maerl, *Laminaria* blades and other algae, in deep lower-shore pools and subtidal to 16 m depth, tolerant of silt cover on rock, at sheltered to exposed sites with or without current exposure; usually occurring as scattered thalli, rarely in abundance.

Widely distributed on west coasts, northwards to Orkney, apparently rare on east coasts, recorded sparsely from E. Scotland and N. E. England.

S. Norway to S. Spain; Mediterranean (South & Tittley, 1986; Feldmann-Mazoyer, 1941).

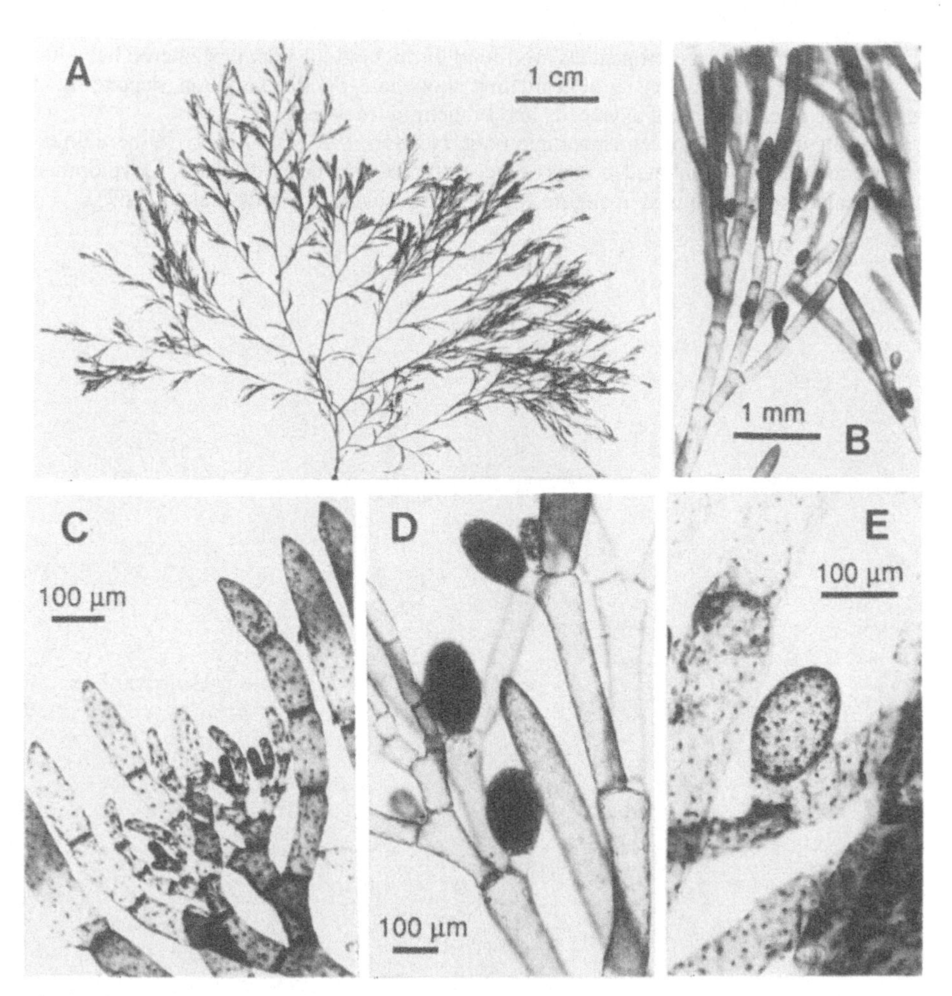

Fig. 62. *Monosporus pedicellatus*

(A) Habit (Down, June). (B) Detail of thallus showing frequent pseudosympodial branching and clavate apical cells (Cork, Nov.). (C) Apex stained with haematoxylin showing highly multinucleate cells (C-E Donegal, Dec.) (D) Propagules consisting of dense ovoid body on little-pigmented stalk cell; some nuclei are visible in this live specimen. (E) Propagule stained with haematoxylin.

Mature plants bearing propagules are found throughout the year in sheltered habitats.

The overall size and density of branching show considerable variation, depending on the degree of elongation of axial cells and branching frequency.

Despite illustrations of tetrasporangia (e. g. Newton, 1931, fig. 224D, E), there do not appear to be any substantiated records of these structures in the British Isles. Development of propagules was followed in culture for a French isolate (L'Hardy-Halos, 1967).

DELESSERIACEAE

DELESSERIACEAE Bory (1828), p. 181 [as Delesseriae].

Germination bipolar with the formation of a narrow primary rhizoid and a broad upper cell that divides transversely and longitudinally in one plane forming a flattened juvenile thallus surmounted by a transversely dividing apical cell; apical initial uninucleate with a single cutting face initiating a primary cell row, or replaced by initials with two cutting faces and a marginal meristem. Adult thallus attached by a primary disc and often secondarily by rhizoids, and bearing linear to broadly ovate or orbicular membranous monostromatic or polystromatic erect blades; blades with or without a central midrib and macroscopic or microscopic veins, organized in a solid sheet, less often perforated or forming a lattice-work, undivided to variously lacerated, commonly branching from the margin or proliferating from the midrib or surface of the blade.

Growth fundamentally by concavo-convex division of a dome-shaped apical cell initiating a primary cell row and central axis; periaxial cells 4 with the first and second cut off laterally and the third and fourth transversely, rarely the first abaxial, the second and third lateral and the fourth adaxial (rhodomelacean sequence); lateral periaxial cells dividing obliquely forming second-, third- and sometimes higher-order cell rows and producing a membrane in which the cells unite laterally by means of secondary pit connections; transverse periaxial cells either remaining undivided or dividing transversely and anticlinally forming the cortex and central midrib; descending rhizoidal filaments present or absent, if present either growing between the periaxial and cortical cells or overlying them; second-, third- and higher-order cell rows branching primarily abaxially, primarily adaxially or both, and either reaching the thallus margin or ending inside the blade. Lateral branches arising from the margins by transformation of terminal cells of second- or third-order cell rows into new apical initials, or adventitiously from marginal cells; surface proliferations issuing from periaxial cells or cortical cells, or arising endogenously from axial cells; in a few instances the branching exogenous by oblique division of the apical cell. Cells of the first-, second-, third-, and occasionally higher-order cell rows undergoing transverse or both transverse and longitudinal intercalary divisions. Blades lacking transversely dividing apical cells growing by oblique divisions of marginal cells with two cutting faces, and/or by intercalary divisions of marginal and internal cells. Microscopic veins, where present, formed by elongation of primary (central) cells, either naked or covered by transversely dividing surface cells, sometimes transformed into macroscopic veins through the production of a cortex.

Spermatangia formed superficially in patches or sori, usually inside the margin on both sides of the blade or blade-like lateral or superficial proliferations; spermatangial mother cells formed by periclinal and anticlinal divisions of primary (central) cells in monostromatic portions of a blade, and from surface cells in polystromatic portions of a blade, each forming 2-3 protuberances that are cut off as spermatangia; spermatia released individually or produced continuously and accumulated under the outer, cuticular layer. Procarps formed near the thallus apex or growing margin, either on primary cell rows distal to the midrib (Delesserioideae), or scattered randomly over the thallus surface (Nitophylloideae), and consisting of a supporting cell (= transverse periaxial cell or surface cell), a lateral sterile group containing 1-7 cells, a 4-celled carpogonial branch in which the second cell is often larger than the rest, and a basal sterile group containing 1-4 cells. The fertilized carpogonium either cutting off two connecting cells distally, or a distal capping cell and a posterio-lateral connecting cell which fuse with the auxiliary cell or a process extending from it; sterile cells either remaining undivided after fertilization or each dividing one or more times and containing enlarged nuclei and amplified levels of DNA; auxiliary cell dividing transversely into a residual cell containing both a diploid and 1-2 haploid nuclei, and a single distal (rarely lateral) gonimoblast initial, which may divide transversely forming a row of primary gonimoblast cells; gonimoblasts branched monopodially and bearing terminal carposporangia or carposporangia in simple or branched terminal chains, rarely shifting to sympodial growth after initiation of the first carposporangia; pericarp formed by resumed growth of ordinary cortical filaments or, less often, by the de novo production of laterally united modified axes; cystocarp cavity formed schizogenously by dissolution of pit connections and rupture between central cells and cortical cells in most species; a single ostiole typically present, usually situated directly above cells of the first (lateral) sterile group; fusion cell absent or present, small, involving only the residual auxiliary cell, supporting cell and adjoining central cell, or large, incorporating inner gonimoblast cells and often cells in the floor of the cystocarp; more than one crop of gonimoblasts bearing carposporangia produced in some species.

Tetrasporangial sori mostly originating on both sides of the blade or blade-like marginal or superficial proliferations, formed inside the margin from surface cells in polystromatic portions, or by dedifferentiation and periclinal and anticlinal divisions of primary (central) cells in monostromatic portions; tetrasporangia originating from primary cells (= central cells), surface cells, or inner cortical cells and covered by one or more layers of presporangial or postsporangial cortical cells or filaments; tetrahedrally divided.

References: Kylin (1923, 1924, 1956), Wagner (1954), Wynne (1983).

HYPOGLOSSUM GROUP

HYPOGLOSSUM Kützing

HYPOGLOSSUM Kützing (1843), p. 444.

Type species: *H. woodwardii* Kützing (1843), p. 444 [= *H. hypoglossoides* (Stackhouse) F. Collins & Hervey (1919), p. 116].

Thalli attached by a discoid holdfast and prostrate axes bearing erect or decumbent blades to 40 cm high, or prostrate attached by peg-like haptera from the midrib or by terminal or marginal rhizoids; blades monostromatic with percurrent midribs, mostly linear-lanceolate, tapering at the tips and bases, simple to repeatedly branched on one or both sides from the midrib with up to 5(6) orders of branches; midrib initially three-layered and uncorticated, later corticated by internal descending rhizoids over most of its length, or the cortication slight, restricted to basal portions; main blades commonly eroded in large thalli leaving a heavily corticated, stipe-like midrib which may be perennial.

Growth by concavo-convex division of a prominent dome-shaped apical cell; primary cell row forming central axis, 2-3 orders of lateral cell rows and paired transverse periaxial cells; apical initials of first-, second-, and third-order cell rows reaching thallus margin, with all or only some second-order cell rows bearing third-order cell rows; branches originating endogenously from distal end of axial cell, often in pairs, sometimes adventitious from periaxial or cortical cells; midribs composed of radially branched cortical filaments with rhizoids absent or formed internally, intermingled amongst larger cells in centre of axis; microscopic veins absent; apical and marginal cells uninucleate, intercalary cells oblong to polygonal at maturity, multinucleate, the nuclei distributed in two bands along the cell margins or beneath the surface, proportional to number of secondary pit connections; cell walls conspicuous, often appearing reticulate; intercalary cell divisions rare or absent; plastids parietal, dissected into small platelets.

Plants dioecious. Spermatangial sori formed in islands, chevrons, or continuous bands alongside midrib bordered by sterile margin of monostromatic blade. Procarps formed along midrib, consisting of two 1-celled sterile groups and a 4-celled carpogonial branch with the second cell especially larger than the rest; fertilized carpogonium cutting off two connecting cells distally, one of which fuses with a process from the auxiliary cell; gonimoblasts filamentous, subdichotomously branched, with inner gonimoblast cells either remaining distinct or incorporated into a central fusion cell; carposporangia formed sequentially in branched chains, or solitary and terminal; cystocarp broad, emergent, solitary and subterminal, or several and dispersed along midrib, the ostiole with or without a beak. Tetrasporangial sori interrupted or continuous, situated alongside midrib or extending over it, bordered by a monostromatic sterile margin, 3-7 cell layers thick; tetrasporangia cut off from primary cells in plane of blade, sometimes including lateral or transverse periaxial cells, or from cortical cells only or both primary cells and cortical cells.

References: Kylin (1923, 1924); Womersley & Shepley (1982); Wynne (1988).

One species in the British Isles.

Hypoglossum hypoglossoides (Stackhouse) F. Collins & Hervey (1919), p. 116.

Lectotype: CN. Cornwall (Polridmouth Cove) (see Wynne, 1984a).

Fucus hypoglossum Woodward (1794), p. 30.
Fucus hypoglossoides Stackhouse (1801), p. 76.

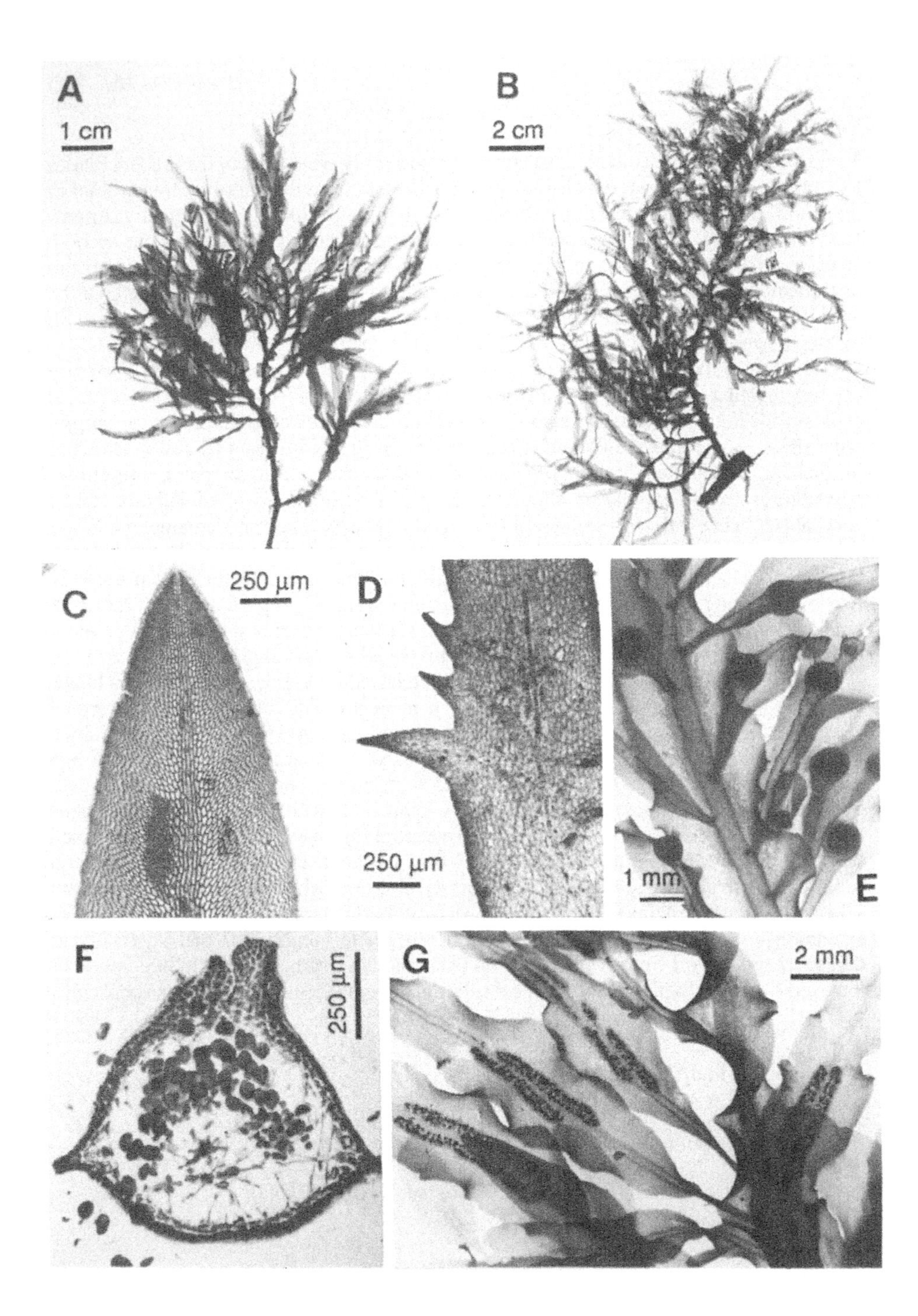
A
1 cm
B
2 cm
C
250 µm
D
250 µm
1 mm
E
F
250 µm
G
2 mm

Delesseria hypoglossum (Woodward) Lamouroux (1813), p. 124.
Hypoglossum woodwardii Kützing (1843), p. 444.

Thalli consisting of blades 2-30 cm high, formed singly or in dense tufts, attached by holdfast consisting of small solid disc and extensive, branched, cylindrical or winged prostrate axes that give rise to new blades; *primary blades* 2-18 (-28) cm in length, 0.1-0.8 cm wide, with a short cylindrical stipe and conspicuous unbranched midrib 0.2-0.8 mm in width that bears numerous single or paired lateral blades on both sides of blades, eroding towards base of plant to terete stipe-like midribs up to 1 mm in diameter with tattered blade remnants; *lateral blades* bearing on midribs a further 1-4 orders of successively shorter and narrower blades, stipitate, lanceolate to ovate with acute to obtuse apices and entire or scalloped, flat or ruffled margins, lacking lateral veins but occasionally forming dentations or pinnate bladelets directly from the margin; colour rose-pink to brownish-red; texture of young blades membranous, of old midribs flexible and tough.

Blades growing from apical cell 14-18 µm in diameter; *midrib* composed in TS of large pseudoparenchymatous cells 80-120 µm in diameter, interspersed with smaller cells and rhizoidal filaments, covered by an outer layer of cortical cells 20-30 µm in diameter; *blade laminae* entirely monostromatic, 30-50 µm thick when young, composed of angled second-order cell rows terminated by prominent 3-sided apical cells, 10-14 µm x 6-10 µm, bearing a third-order cell row on every cell; cells of blade initially diamond-shaped, later rectangular, increasing to 60-200 µm long x 20-100 µm wide, with sinuous walls and conspicuous secondary pit connections; thickness of blade increasing on either side of midrib to about 200 µm by development of rhizoidal filaments from blade cells; *plastids* discoid in young lamina cells, becoming elongate or beaded in midrib and reproductive tissues.

Spermatangial sori formed on either side of midrib on both sides of blades, up to 0.7 mm wide, continuous over length of blade or divided by sterile cell rows, with obliquely dentate margins, about 70 µm thick, consisting of a layer of rounded spermatangial mother cells 6-8 µm wide bearing cylindrical spermatangia 6 µm long x 3 µm wide. *Procarps* borne on midrib near apex of female blades; *cystocarps* typically developing singly on blades, hemispherical when mature, 850-1100 µm in diameter, with projecting ostiole 250-350 µm long, flared out terminally, the ostiolar pore 70-150 µm in diameter; inner

Fig. 63. *Hypoglossum hypoglossoides*

(A) Habit of small plant (Cork, Nov.). (B) Habit of large, much-branched thallus (Down, Jan.). (C) Apex of blade showing midrib and large-celled lamina lacking veins (Down, Nov.). (D) Unusual growth form with conspicuous marginal teeth (Inverness, June). (E) Cystocarps on midribs with prolonged tube-like ostioles (as B). (F) T.S. of cystocarp showing small central fusion cell, chains of carposporangia, and elongate inner pericarp cells (Cornwall, Sep.). (G) Detail of tetrasporophyte with linear tetrasporangial sori on either side of midribs (as C).

gonimoblast cells incorporated into large, branched fusion cell; gonimoblast filaments terminating in branched chains of 1-3 mature spherical to ellipsoid carposporangia, 40-56 µm x 40 µm. *Tetrasporangial sori* borne on either side of midrib, continuous along both sides of blade, 1.5-30 mm long x 0.2-0.5 mm wide and 140 µm thick, multilayered when mature, covered by sterile cortical cells; *tetrasporangia* spherical, primary ones 60-90 µm in diameter, cut off from cortical cells in two layers, secondary ones smaller, arising from central cells and some cortical cells.

Growing on bedrock and pebbles, crustose corallines, maerl, larger algae and *Laminaria hyperborea* stipes, in shaded lower-shore pools and subtidal to at least 30 m depth, most abundant in *L. hyperborea* forest and on subtidal cliffs, tolerant of extremely sand-scoured conditions, at moderately to extremely wave-exposed sites and at extremely sheltered sites with strong current exposure.

Generally distributed in the British Isles, northwards to Shetland (Norton, 1985).

British Isles to Spain, Azores and Canaries; Mediterranean; W. Atlantic from N. Carolina to Brazil (Schneider & Searles, 1991).

On sheltered shores, mature plants can be found throughout the year; in exposed-coast kelp forest, most thalli appear to be annual, young plants becoming conspicuous in March-April. Spermatangia recorded in February, April, May and July-October; cystocarps in January, February and May-September, and tetrasporangia in February-November. Chromosome number reported as $n = 20$ (Kylin, 1923), but requires confirmation.

There is considerable variation in blade width, narrower blades apparently occurring at more sheltered sites. Marginal branching is common in some populations but absent in others; its significance is unknown but branched thalli resemble members of the genus *Branchioglossum* Kylin (see Wynne, 1988).

H. hypoglossoides is frequently confused with *Apoglossum ruscifolium* (q. v.), because the shape of blades and apices can be misleading. However, these species can readily be distinguished. *H. hypoglossoides* lacks lateral veins, whereas lateral microveins occur in *A. ruscifolium*, and the blade cells of the former are >60 µm long compared to <20 µm long in *A. ruscifolium*. Large plants of *H. hypoglossoides* may resemble small blades of *Delesseria sanguinea* (q. v.) but lack the large macroscopic lateral veins of *D. sanguinea*.

MEMBRANOPTERA GROUP

MEMBRANOPTERA Stackhouse

MEMBRANOPTERA Stackhouse (1809), pp. 57, 85.

Lectotype species: *M. alata* (Hudson) Stackhouse [See Kylin (1924), p. 15].

Thalli attached by a discoid holdfast and creeping basal blades bearing few to many erect blades up to 20 cm high; blades monostromatic to broadly polystromatic, with percurrent multilayered midribs and with or without oblique microscopic or macroscopic lateral veins; branching alternate to subdichotomous from the margin, sometimes with proliferous bladelets along margins, in branch axils, or from midribs or macroscopic veins; tips of blades depressed, blunt, or pointed; midribs coarse or weak to inconspicuous, with or without internal rhizoids; older blades eroding towards base forming stipe-like midribs in large species.

Growth initiated by concavo-convex division of a dome-shaped apical cell; primary cell row forming central axis bearing 3(-4) orders of lateral cell rows and paired transverse periaxial cells, second-order cell rows reaching margin, third-order cell rows mostly abaxial with a few also adaxial, not all reaching margin, fourth- and higher-order cell rows mostly abaxial, ending internally within blade; intercalary cell divisions absent; periaxial cells elongating parallel to axial cell, not transversely divided; microscopic and macroscopic veins, where present, formed from second- and sometimes third-order cell rows; midribs and macroscopic veins composed of radially branched cortical filaments, with rhizoids absent or formed internally, intermingled amongst larger cells in centre of axis; lateral branches originating by transformation of apical cells of second-order cell rows into dome-shaped apical initials of primary cell rows; adventitious bladelets, where present, originating from periaxial or cortical cells; apical cells, marginal cells and axial cells uninucleate, intercalary cells becoming multinucleate with nuclei proportional to number of secondary pit connections; plastids parietal, dissected into minute platelets in young blades, reticulate or catenate in polystromatic tissues.

Reproductive organs borne on ordinary branches or adventitious bladelets; plants dioecious, rarely with mixed tetrasporangial and sexual phases. Spermatangial sori on both sides of blade, scattered alongside midrib or localized at tips and continuous over midrib, bordered by a sterile margin. Procarps formed on both sides of blade near apex, consisting of a lateral 1-2(-3) celled sterile group, a 4-celled carpogonial branch, and a basal 1-celled sterile group; sterile groups enlarging but not dividing after fertilization; cystocarps mostly one per branch tip or bladelet; gonimoblast filaments subdichotomously branched; carposporangia in short terminal chains, maturing sequentially. Tetrasporangial sori scattered on both sides of blade alongside midrib, sometimes between lateral veins, or localized at tips and continuous over midrib, bordered by a sterile margin; fertile areas polystromatic, initially 3-5 layers thick, tetrasporangia originating from surface cells, arranged in two rows at maturity on either side of blade covered by an external cortex.

References: Kylin (1923, 1924).

One species in the British Isles. The seaweed recorded for the British Isles as *Pantoneura angustissima* (Griffiths ex Harvey) Kylin (1924, p. 18), is a species of *Membranoptera*. It appears to be a growth form of *M. alata* but further studies, preferably of freshly collected material, are required.

Membranoptera alata (Hudson) Stackhouse (1809), p. 85.

Lectotype: BM-K. Undated, unlocalized.

Fucus alatus Hudson (1762), p. 472.
Delesseria alata (Hudson) Lamouroux (1813), p. 124.

Thalli consisting of very dense tufts of much-branched blades, 5-20 cm high, attached by a solid discoid holdfast that forms lobed margins and gives rise to new blades; *blades* 0.6-2.4 mm wide, with a conspicuous midrib 0.15-0.5 mm in width, branching in one plane irregularly alternately to dichotomously to 5 or more orders, also forming axillary tufts of simple or branched blades from midribs; laminae with translucent microscopic lateral veins and entire, ruffled margins, eroding towards base of plant to terete stipe-like midribs up to 2.5 mm in diameter bearing tattered blade remnants; *apices* obtuse to pointed, incurved, forcipate when branched, with asymmetric laminae well-developed on abaxial side of midrib but leaving midrib naked adaxially; colour bright red to dark brownish-red, texture of young blades membranous but fairly tough, of old midribs flexible and cartilaginous.

Blades growing from apical cells 11-12 µm in diameter; *midribs* rapidly increasing in width and composed largely of rhizoidal filaments 6-20 µm in diameter, surrounded by 1-2 layers of cortical cells; *blade laminae* monostromatic, c. 25 µm thick, composed of rectangular to polygonal cells, 6-18 µm x 6-9 µm, with numerous secondary pit connections; *microscopic veins* oblique, branched abaxially when older, 3 cells thick; *plastids* discoid in young blade cells, reticulate or catenate in midrib and reproductive tissues.

Plants dioecious; all reproductive structures formed near apices and on smaller reproductive bladelets that develop in axillary clusters. *Spermatangial sori* oval, up to 1 x 0.5 mm, entirely covering axillary bladelets except for narrow sterile margins and apices, 60-80 µm thick, consisting of a layer of cuboid spermatangial mother cells 5-7 µm wide, bearing cylindrical spermatangia 6-7 µm long x 2-3 µm wide. *Cystocarps* usually developing singly on bladelets or branches, subspherical, 400-560 µm in diameter when mature, with smooth pericarp wall and non-projecting ostiole, the ostiolar pore 25 µm in diameter; gonimoblast filaments terminating in chains of 1-3 mature subspherical carposporangia, 38-55 µm x 34-38 µm. *Tetrasporangial sori* formed centrally on blades

---

Fig. 64. *Membranoptera alata*

(A) Habit (Donegal, May). (B) Detail of thallus showing branching, with asymmetric blade development resulting in keyhole-shaped gaps between axes at tips (Devon, Nov.). (C) Apex of blade showing midrib and small lateral veins (C-G Antrim, Jan.). (D) Spermatangial bladelets with narrow sterile borders. (E) Cystocarp on midrib. (F, G) T.S. of young and mature cystocarps showing central fusion cell, elongate inner pericarp cells, and non-protuberant ostiole. (H) Detail of tetrasporophyte with axillary tetrasporangial bladelets (arrows) (Pembroke, Dec.). (I) Tetrasporangial sori at apex of blade (as A).

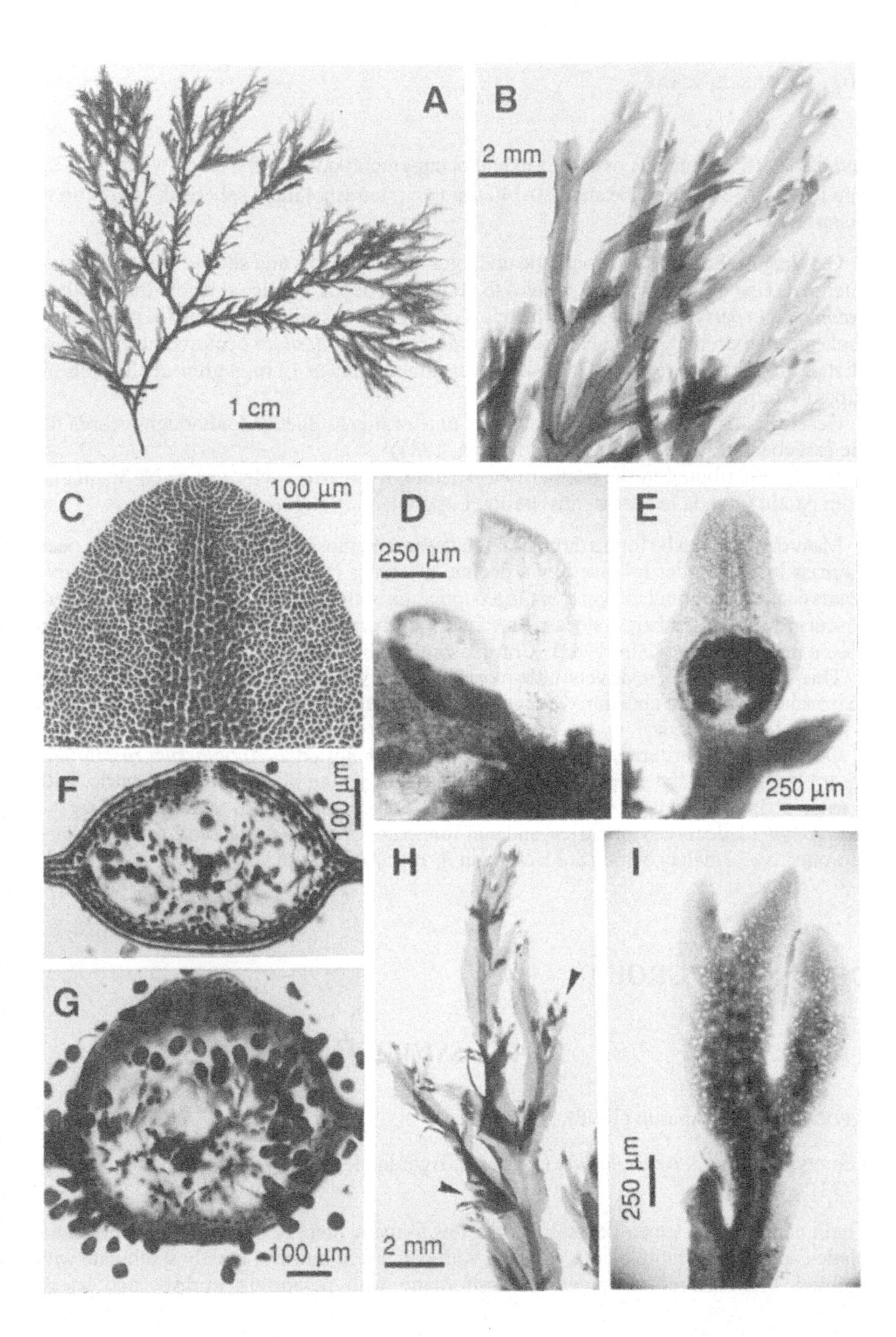
A
1 cm
B
2 mm
C
100 µm
D
250 µm
E
250 µm
F
100 µm
G
100 µm
H
2 mm
I
250 µm

and bladelets, continuous over midrib and occupying about half the blade width, 0.3-2.5 mm long, 100-350 µm wide and 120-140 µm thick; tetrasporangia spherical, 45-75 µm in diameter.

Growing on bedrock and epiphytic on larger algae in pools and shaded places in lower intertidal, also on bedrock and pebbles to about 3 m depth, restricted below this depth to *Laminaria hyperborea* stipes, on which it is one of the most abundant algae at moderately sheltered to extremely wave-exposed sites. In Berwick, *M. alata* occurred near the base of stipes in shallow water, but near the lower limit of kelp at 12 m, it grew at all levels on stipes (Whittick, 1983).

Generally distributed in the British Isles, northwards to Shetland, although records for the east coast of England are sparse (Norton, 1985).

Widely distributed in the northern N. Atlantic, from Arctic Norway to N. Spain and from Arctic Canada to Massachussets and Long Island (South & Tittley, 1986).

Mature plants can be found throughout the year, and those epiphytic on kelp reach peak biomass in September followed by a decline in winter (Whittick, 1983). Reproductive plants occur throughout the year, but in a population at the Isle of Man spermatangia were absent in July-September, cystocarps absent in October and November, and tetrasporangia absent in September (Kain, 1982). Chromosome number is $n = 32$ (Austin, 1956).

This species shows relatively little morphological variation except that resulting from seasonal changes and abrasion damage. Thalli on kelp stipes become heavily encrusted with epiphytic bryozoa.

Occasional herbarium plants of *M. alata* are misidentified as *Apoglossum ruscifolium* (q. v.) but differ in the shape of apices. In *M. alata,* young blades are asymmetric, with greater abaxial blade development resulting in forcipate apices, whereas young *A. ruscifolium* blades are symmetrical and non-forcipate. In addition, *M. alata* forms axillary reproductive bladelets, which are lacking in *A. ruscifolium*.

DELESSERIA GROUP

APOGLOSSUM J. Agardh

APOGLOSSUM J. Agardh (1898), p. 190.

Lectotype species: *A. ruscifolium* (Turner) J. Agardh (1898), p. 190 [See Kylin (1924), p. 23].

Thalli consisting of prostrate axes bearing few to many erect or decumbent membranous blades up to 10 cm high; blades stipitate, with percurrent midribs, mostly spathulate with pointed, blunt or retuse apices, monostromatic with percurrent midribs and lateral microscopic veins, simple or branched from the midrib on either side of blade with up to four orders of branches; midribs uncorticated, or lightly to heavily corticated by external descending rhizoids; lower blades commonly eroded in older thalli forming tough stipes corticated by external rhizoids.

Growth initiated by concavo-convex division of a dome-shaped apical cell often situated in an apical depression; primary cell row forming central axis, 4 (-5) orders of lateral cells rows and paired transverse periaxial cells, second-order cell rows all reaching margin, third-order cell rows formed abaxially, not all reaching margin, fourth-order cell rows mostly adaxial, ending internally within blade, additional cell rows sometimes present; intercalary divisions absent in central axis, frequent in second-, third- and sometimes higher-order cell rows, with new intercalary cells cut off distally; periaxial cells elongating parallel to axial cell, not transversely divided, linked to cells below by secondary pit connections; microscopic veins derived by cell elongation from second- and some third-order cell rows, extending laterally in an ascending arc and sometimes forming networks; branches originating endogenously from distal end of axial cell, often in pairs, or adventitiously from periaxial or cortical cells; rhizoids issuing from surface cells forming cylinder around core of larger central cells; apical cells, marginal cells and axial cells uninucleate, intercalary cells multinucleate with nuclei in median plane, proportional to number of secondary pit connections; plastids parietal, dissected into platelets.

Plants dioecious. Spermatangial sori in radially arranged striae on both sides of blade, bordered internally by the periaxial cells, externally by a broad sterile margin, and separated into patches by sterile microscopic veins. Procarps formed on both sides of midrib near apex, consisting of two 1-celled sterile groups and a 4-celled carpogonial branch; sterile groups dividing immediately after fertilization, each forming branched filaments with up to 6-10 cells; fertilized carpogonium cutting off two connecting cells distally, one of which fuses with a process from the auxiliary cell; gonimoblast filaments repeatedly subdichotomously branched, bearing round to pyriform carposporangia in branched terminal chains, maturing sequentially; fusion cell small, incorporating the central cell, supporting cell, and innermost gonimoblast cells; pericarp 4-5 cell layers thick with a central ostiole and projecting beak. Tetrasporangial sori 5-9 cell layers thick, continuous or discontinuous in paired arrangement alongside midrib, or circular to oval at distal end of blade; tetrasporangia initiated from inner cortical layer on both sides of blade, arranged in two rows or multilayered at maturity, covered by an external cortex.

References: Kylin (1923), Wynne (1984b).

One species in the British Isles.

Apoglossum ruscifolium (Turner) J. Agardh (1898), p. 190.

Lectotype: BM-K. Norfolk (Yarmouth), undated.

Fucus ruscifolius Turner (1802), p. 27.
Delesseria ruscifolia (Turner) Lamouroux (1813), p. 124.

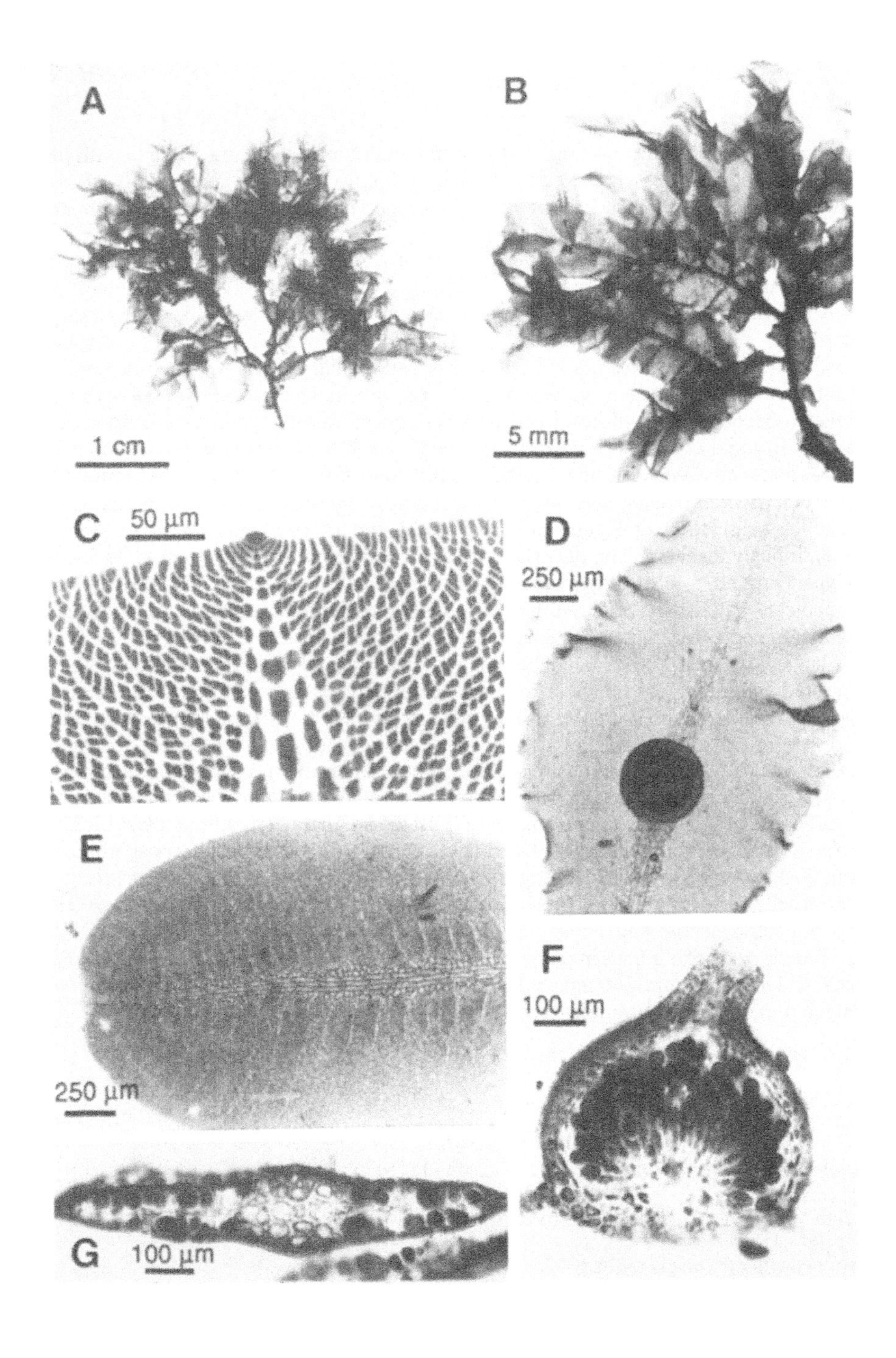
A
B
1 cm
5 mm
C
50 µm
D
250 µm
E
250 µm
F
100 µm
G
100 µm

Thalli densely tufted, attached by holdfast consisting of branched prostrate axes that give rise to new erect axes and may fuse into a solid disc up to 0.5 cm in diameter; *primary blades* 2-10 cm in length, up to 0.8 cm wide, with a conspicuous midrib 250-400 μm in width, normally simple but occasionally dichotomous or alternately branched, eroding towards base of plant to produce terete stipe-like midribs up to 650 μm in diameter bearing tattered blade remnants; *lateral blades* arising from the midrib at irregular intervals on both sides of blades, sometimes paired, bearing a further 3-4 orders of branches, stipitate, ovate to lanceolate with obtuse to pointed apices, strongly ruffled in male thalli, 1-2.5 cm long and 0.2-0.8 cm wide, with translucent microscopic lateral veins and entire, ruffled margins, rose-pink to deep red in colour, the texture of young blades membranous, and old midribs tough and cartilaginous.

Blades growing from apical cell 10-14 μm in diameter; *midrib* composed in TS of a row of 4-6 cells c. 40 μm in diameter, with 1-2 layers of smaller cells above and below this, covered by several layers of rounded cells 6-15 μm in diameter; *blade laminae* monostromatic, 24-35 μm thick, composed of polygonal cells 10-18 μm in diameter; *microscopic veins* angled forward, branching and forming a network, increasing in width from one to several cells, each 10-25 μm in diameter and 2-12 diameters long, with numerous secondary pit connections to lamina cells; *plastids* discoid in young blade cells, reticulate or beaded in midrib and reproductive tissues.

Spermatangial sori formed on either side of midrib on both sides of blades, covering about half the blade width, divided by sterile striae over the microveins, 35-55 μm thick, consisting of a layer of cuboid spermatangial mother cells 4-6 μm wide, bearing clavate to ellipsoid spermatangia 6-9 μm long x 3-3.5 μm wide. *Cystocarps* formed singly or several per blade, sometimes becoming fused together, hemispherical when mature, 540-720 μm in diameter x 500 μm high, with projecting ostiole 250 μm long x 250 μm wide and pore 50-100 μm in diameter; gonimoblast filaments much-branched, 8-15 μm in diameter, with cells 2-3 diameters long, terminating in chains of 1-3 mature spherical to pyriform carposporangia, 35-55 μm x 28-35 μm. *Tetrasporangial sori* borne on either side of midrib, continuous along both sides of blade or discontinuous, 1-6 mm long, 250-360 μm wide and 130-185 μm thick; tetrasporangia spherical to ellipsoid, 45-63 μm long x 40-52 μm wide.

Growing on bedrock and epiphytic on *Laminaria hyperborea* stipes and smaller algae,

---

Fig. 65. *Apoglossum ruscifolium*

(A) Habit (A-B Cork, Nov.). (B) Detail of thallus showing branching of blades from midrib, and linear tetrasporangial sori on either side of midribs. (C) Apex of blade showing midrib and development of cell rows of lamina (Ross, Aug.). (D) Cystocarp on midrib (D-E Devon, Nov.). (E) Spermatangial sorus divided by narrow sterile striae over veins. (F) T.S. of cystocarp showing carposporangia in chains and tube-like ostiole (Cornwall, Oct.). (G) T.S. of tetrasporangial sori on both sides of blade (Cork, Dec.).

in pools and shaded places in lower intertidal, subtidal to at least 17 m depth; occurring as a characteristic component of the underflora in *L. hyperborea* forest but most abundant in the shallow subtidal where kelp forest is absent, at extremely sheltered to extremely wave-exposed sites with little to strong current exposure.

Generally distributed in the British Isles, northwards to Shetland, although records for the east coast of England are sparse (Norton, 1985).

Norway to Portugal; Mediterranean; widely distributed in W. and S. Atlantic and Indian Oceans (Schneider & Searles, 1991).

Mature plants can be found throughout the year, apparently perennial at sheltered sites, old axes becoming heavily encrusted with epiphytic invertebrates and algae; spermatangia recorded in February, June-August and October-December, cystocarps in February and June-December, and tetrasporangia in February-December. Occasional plants are found bearing both tetrasporangia and cystocarps. Chromosome number reported as n = c. 20 (Kylin, 1923), but requires confirmation.

This species shows relatively little morphological variation.

A. ruscifolium thalli occasionally bear plants of the parasite *Apoglossocolax pusilla* (q. v.).

Hypoglossum hypoglossoides (q. v.) can readily be distinguished from *A. ruscifolium* by the absence of lateral microveins, and the much larger blade cells, which are >60 µm long compared to <20 µm in *A. ruscifolium*. Large plants of *A. ruscifolium* resemble small blades of *Delesseria sanguinea* (q. v.) but lack the large macroscopic lateral veins of *D. sanguinea*. Occasionally, plants of *Membranoptera alata* (q. v. for differences) are misidentified as *A. ruscifolium*.

APOGLOSSOCOLAX Maggs & Hommersand, gen. nov.

APOGLOSSOCOLAX Maggs & Hommersand, gen. nov.

Thalli minute, hemiparasitic on *Apoglossum ruscifolium*, consisting of endophytic filaments and emergent blades; blades formed singly or in stellate groups, ovate with indistinct midribs; differing from *Phitycolax* Wynne & Scott (1989), the only parasite known in the subfamily Delesserioideae, in that third-order cell rows do not all reach the margin, and carposporangia are borne in chains, not singly; differing from the host in habit and in the much smaller sizes of most vegetative and reproductive structures.

Thalli minuti, hemiparasitici in Apoglosso ruscifolio, *e filamentis endophyticis et laminis emergentibus constantes. Laminae singulatim aut in catervis stellatis formatae, ovatae costis indistinctis. Differt a* Phitycolax *Wynne & Scott (1989), solus parasiticus in subfamilia Delesserioideis, seriebus cellularum tertii ordinis ad marginem non attingentibus, et carposporangiis in catenis sed non singulatim portatis. Differt a hospite habito et magnitudine multo minore plurimarum structurarum vegetativarum reproductivarum.*

Type species: *A. pusilla* Maggs & Hommersand, sp. nov.

Thalli less than 2 mm in diameter, consisting of basal tissue endophytic in host, and emergent blades formed singly or in stellate groups; blades less than 1 mm in length, shortly stipitate, flattened, ovate or branched, with indistinct midribs and entire or dentate margins.

Endophytic filament cells uninucleate or multinucleate after formation of secondary pit connections between them, also forming secondary pit connections to host cells; blades growing from single prominent apical cell, cutting off crescent-shaped cell that gives rise to primary axial row and basal cells of second-order rows, axial cells remaining uninucleate, linked by broad primary pit connections, lacking intercalary divisions, cells of second-order rows becoming multinucleate due to formation of secondary pit connections (no intercalary divisions seen in second-order rows either); third-order filaments formed sparsely and not reaching margins; periclinal divisions resulting in increases in blade thickness; plastids reticulate.

All blades becoming reproductive. Spermatangial sori covering both sides of male blades except for meristematic regions and stipe. Procarps formed along midrib, consisting of supporting cell borne on axial row, 1-celled sterile group-1, curved 4-celled carpogonial branch, and 1-celled sterile group-2; cystocarps lacking large fusion cell, consisting of much-branched gonimoblast filaments terminating in short chains of mature carposporangia; pericarp with a central raised ostiole and beak-like structure formed by apex of female blade. Tetrasporangial sori covering blades except for meristems; tetrasporangia developing from cortical cells on either side of blade, exposed when mature, ellipsoid, tetrahedrally divided.

Wynne & Scott (1989) described *Phitycolax inconspicua*, the first known parasite in the subfamily Delesserioideae, from the southern Indian Ocean, where it was collected on *Phitymophora amansioides* (Sonder) Womersley. It is strikingly similar to *Apoglossocolax pusilla* in general appearance, but differs in vegetative and reproductive morphology. *P. inconspicua* was assigned to the Hypoglossum group of the Delesserioideae, as is its host, on the basis of features such as all third-order cell rows reaching the margin. In *A. pusilla*, in contrast, third-order cell rows are formed sparsely and do not reach the margins; we assign it to the Delesseria group in which its host *Apoglossum ruscifolium* is also placed. A further significant difference between *Phitycolax* and *Apoglossocolax* lies in the formation of carposporangia singly in *Phitycolax* but in short chains in *Apoglossocolax*.

One species in the British Isles.

Apoglossocolax pusilla Maggs & Hommersand, sp. nov.

Thalli 0.2-1.8 mm in diameter, consisting of ovate blades with indistinct midribs, formed

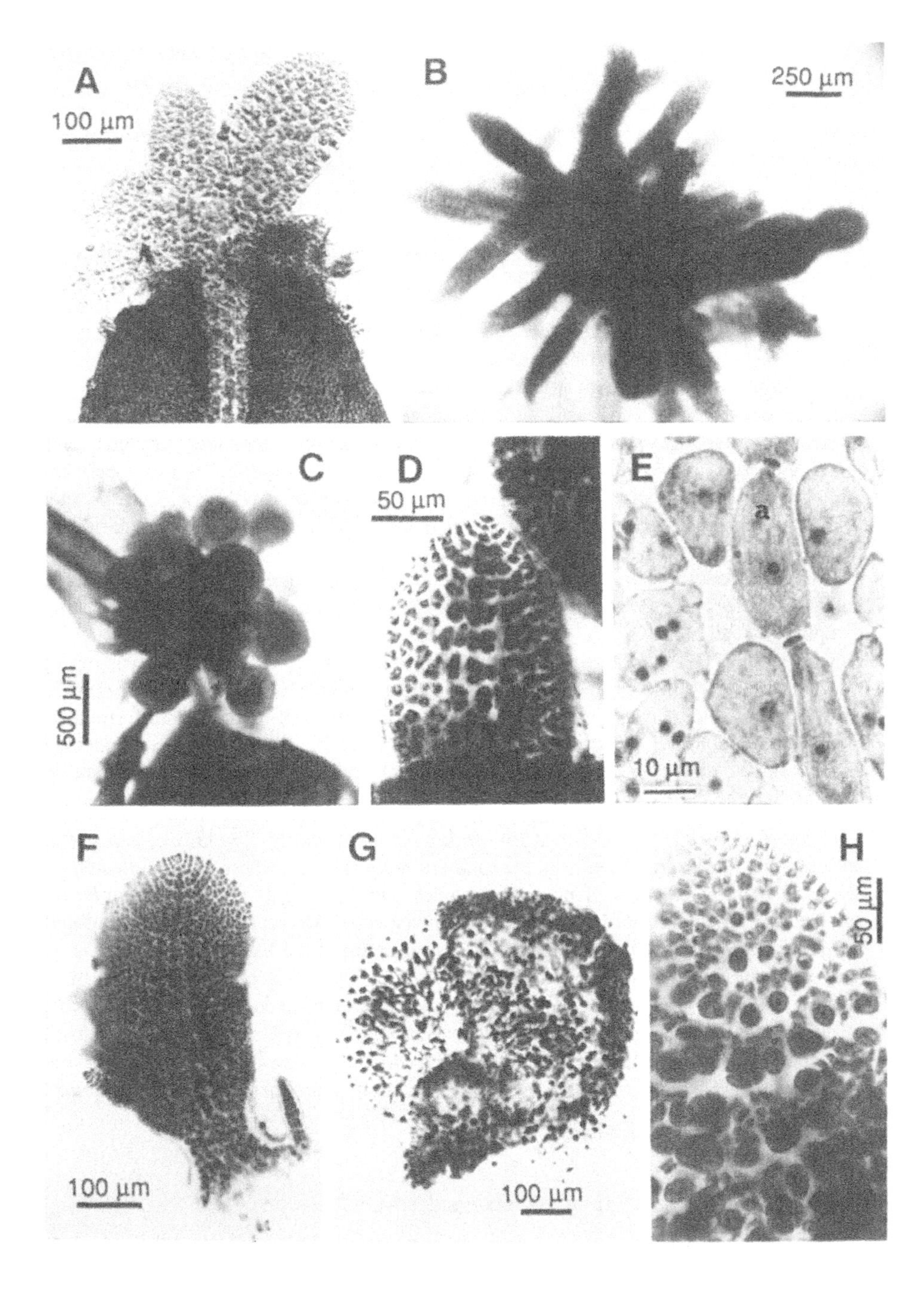
A
100 µm
B
250 µm
C
500 µm
D
50 µm
E
a
10 µm
F
100 µm
G
100 µm
H
50 µm

singly or in stellate groups, hemiparasitic on *Apoglossum ruscifolium*, differing from the host in habit and in the much smaller sizes of most vegetative and reproductive structures.

Thalli 0.2-1.8 diametro, e laminis ovatis habentibus costas indistinctas constantes, singulatim aut in catervis stellatis formati, hemiparasitici in Apoglosso ruscifolio *e quo differt habito et magnitudine multo minore plurimarum structurarum vegetativarum reproductivarumque.*

Holotype: BM. Isotypes: BM. Dorset (Old Harry Rocks, Studland), 18 vii 1990, *Maggs*.
Paratypes: GALW, PC. Galway (Flannery Bridge), 13 ix 1988, *Maggs*.

Thalli comprising basal tissue endophytic in host, and emergent thallus 0.2-1.8 mm in diameter, consisting either of a conical or irregular outgrowth bearing a stellate group of 5-30 or more blades, or a single blade attached by a short cylindrical stipe; *blades* flattened, 0.2-0.6 mm long x 0.1-0.4 mm wide, ovate or bearing lateral branches, with a midrib visible in male and tetrasporangial blades, margins smooth in females, smooth or dentate in males and tetrasporophytes, frequently with teeth developing into irregularly arranged lateral branches; red in colour, paler than host, texture soft and becoming almost gelatinous when preserved.

Endophytic filaments branched, 3-6 μm in diameter, with cells 2-8 diameters long, penetrating between host cells and forming secondary pit connections to host and other parasite cells, densely packed with starch grains; *emergent tissue* composed of medulla of rounded cells to 35 μm in diameter, and filamentous cortical tissue with outer layer of rounded cells 6-12 μm in diameter, from which erect blades arise; *blades* growing from apical cell 8-9 μm in diameter that cuts off axial cells, these enlarging to 24 μm long x 12 μm wide; basal cells of second-order cell rows diamond-shaped, c. 25 μm x 25 μm; blades increasing to 3-9 cells in thickness, 40-50 μm thick in males, up to 200 μm thick in females; *plastids* reticulate.

All blades becoming reproductive. *Spermatangial sori* covering entire surface of male blades, except for stipe and main and lateral apices, consisting of a layer of conical to cuboid spermatangial mother cells 5-6.5 μm long x 2.5-3.5 μm wide, bearing clavate to

Fig. 66. *Apoglossocolax pusilla*

(A) Habit of tetrasporophyte on midrib of *Apoglossum* blade (A-C, E & G paratype material, Galway, Sep.). (B) Female thallus with numerous narrow blades prior to cystocarp development. (C) Female thallus with densely crowded cystocarps. (D) Apex of female blade showing development of midrib and lateral cell rows from single dome-shaped apical cell (D, F & H holotype and isotype material, Dorset, July). (E) Axial cell row (a) and first- and second-order lateral cell rows. (F) Spermatangial blade with marginal dentations (holotype slide, in BM). (G) Partly squashed cystocarp showing carposporangia in chains. (H) Apex of tetrasporangial blade with rows of developing and mature tetrasporangia (holotype slide, in BM).

ovoid spermatangia 4.5 μm long x 2-3 μm wide. *Procarps* borne near apex on midrib, on both sides of blade, in series of up to at least 3 on successive segments; *cystocarps* typically developing singly on blades, less often paired on either side of blade, spherical, 360-500 μm in diameter, with a smooth pericarp wall, a compressed spine from the persistent blade apex, and a projecting ostiole 50 μm long x 100 μm wide, the pore 50 μm in diameter; gonimoblast filaments much-branched, 6.5-8 μm in diameter, cells 1.5-3 diameters long, terminating in chains of 1-2 mature spherical carposporangia, 11-20 μm in diameter. *Tetrasporangial sori* covering blades except for apex; tetrasporangia formed in chevrons near apices, ellipsoid, 22-36 μm long x 22-28 μm wide.

Hemiparasitic on the stipes and midribs, rarely blades, of *Apoglossum ruscifolium* and on other plants of *Apoglossocolax pusilla*, in depths of 1-5 m at moderately to extremely wave-sheltered sites, with slight to strong current exposure.

Known only from Dorset and Galway, *A. pusilla* has been found at only two localities. It was apparently quite common at one of these sites but searches of the host plant from several other localities were unsuccessful.

Developing and mature thalli with spermatangia, cystocarps and tetrasporangia were collected in July and September.

Male and tetrasporangial plants vary greatly in size because growth continues after reproduction has started; small thalli are easily overlooked. Female plants, which sometimes occur in extensive patches, are relatively conspicuous when cystocarpic due to the comparatively large size and opacity of the cystocarps. *A. pusilla* is the only known parasite of *Apoglossum ruscifolium*.

DELESSERIA Lamouroux, nom. cons.

DELESSERIA Lamouroux (1813), p. 122.

Lectotype species: *D. sanguinea* (Hudson) Lamouroux (1813), p. 122 [see Kützing (1843), p. 445]

Thalli attached to solid substratum by a discoid holdfast and prostrate axes bearing one to several erect primary blades up to 30 cm high; blades stipitate, lanceolate, tapering at base and apex, monostromatic with percurrent midribs and macroscopic lateral veins; branching from midrib on either side of blade, up to 4(5) orders of branches; midribs heavily corticated by descending internal rhizoids; older blades often eroded from margin leaving tough, stipe-like midribs.

Growth initiated by concavo-convex division of a dome-shaped apical cell; primary cell row forming central axis, 4 or more orders of lateral cell rows and paired transverse periaxial cells; cells of second-order cell rows reaching margin, third-order cell rows mostly abaxial with a few also adaxial, not all reaching margin, fourth- and higher-order cell rows mostly abaxial, usually ending internally within blade; intercalary cell divisions absent in central axis, present in cell rows of second, third and sometimes higher orders, new intercalary cells cut off either distally or proximally; periaxial cells elongating parallel to axial cell and dividing transversely, linked to cells below by secondary pit connections; lateral macroscopic veins produced from third-order cell rows; midribs and macroscopic veins multilayered, composed of radially branched cortical filaments lacking intercalary cell divisions; rhizoids formed internally from periaxial and cortical cells, intermingled amongst larger cells in centre of axis and linked to them by secondary pit connections; branches originating adventitiously from periaxial and cortical cells on either side of midrib; apical and marginal cells uninucleate, intercalary cells and axial cells becoming multinucleate with nuclei usually proportional to number of secondary pit connections; plastids parietal, dissected into platelets in young blades, reticulate in polystromatic tissues.

Reproductive structures in adventitious bladelets, formed mostly on denuded midribs; plants dioecious, rarely with mixed tetrasporangial and sexual phases. Spermatangial sori multilayered, initiated on either side of midrib, later extending over midrib and bordered by a sterile margin. Procarps formed on both sides of midrib near apex, consisting of two sterile groups and a 4-celled carpogonial branch; sterile groups containing 2-6 cells, approximately doubling in number after fertilization; gonimoblast filaments repeatedly subdichotomously branched; carposporangia in branched terminal chains, smaller than the gonimoblast cells and maturing simultaneously; fusion cell conspicuous, incorporating the central cell, supporting cell and inner gonimoblast cells; pericarp 7 or more cell layers thick with a central ostiole and projecting beak. Tetrasporangial sori irregularly shaped, polystromatic, extending over midrib and bordered by sterile margin; tetrasporangia initiated from surface cortical cells on both sides of blade, arranged in two rows at maturity covered by an external cortex; bisporangia and irregularly divided tetrasporangia reported in forms having mixed phases.

Delesseria is monotypic in the above description, containing only *D. sanguinea*; as usually circumscribed at present the genus is polyphyletic.

Reference: Kylin (1923).

One species in the British Isles.

Delesseria sanguinea (Hudson) Lamouroux (1813), p. 122.

Lectotype: Morison (1699), p. 645, pl. 8, fig. 6 [based on specimen at OXF (L. M. Irvine, pers. comm.)].

Fucus sanguineus Hudson (1762), p. 475.

Thalli attached by holdfast consisting of a solid disc up to 1 cm in diameter and branched stolon-like axes that give rise to new blades; *primary blades* formed singly or in small groups, simple, flat and lanceolate when young, lanceolate to ovate when mature, with acute to blunt apices and entire, deeply ruffled margins, 8-25 cm long x 3-8 cm wide, with a conspicuous midrib up to 3.5 mm wide, paired lateral veins up to 1.5 mm wide, and a

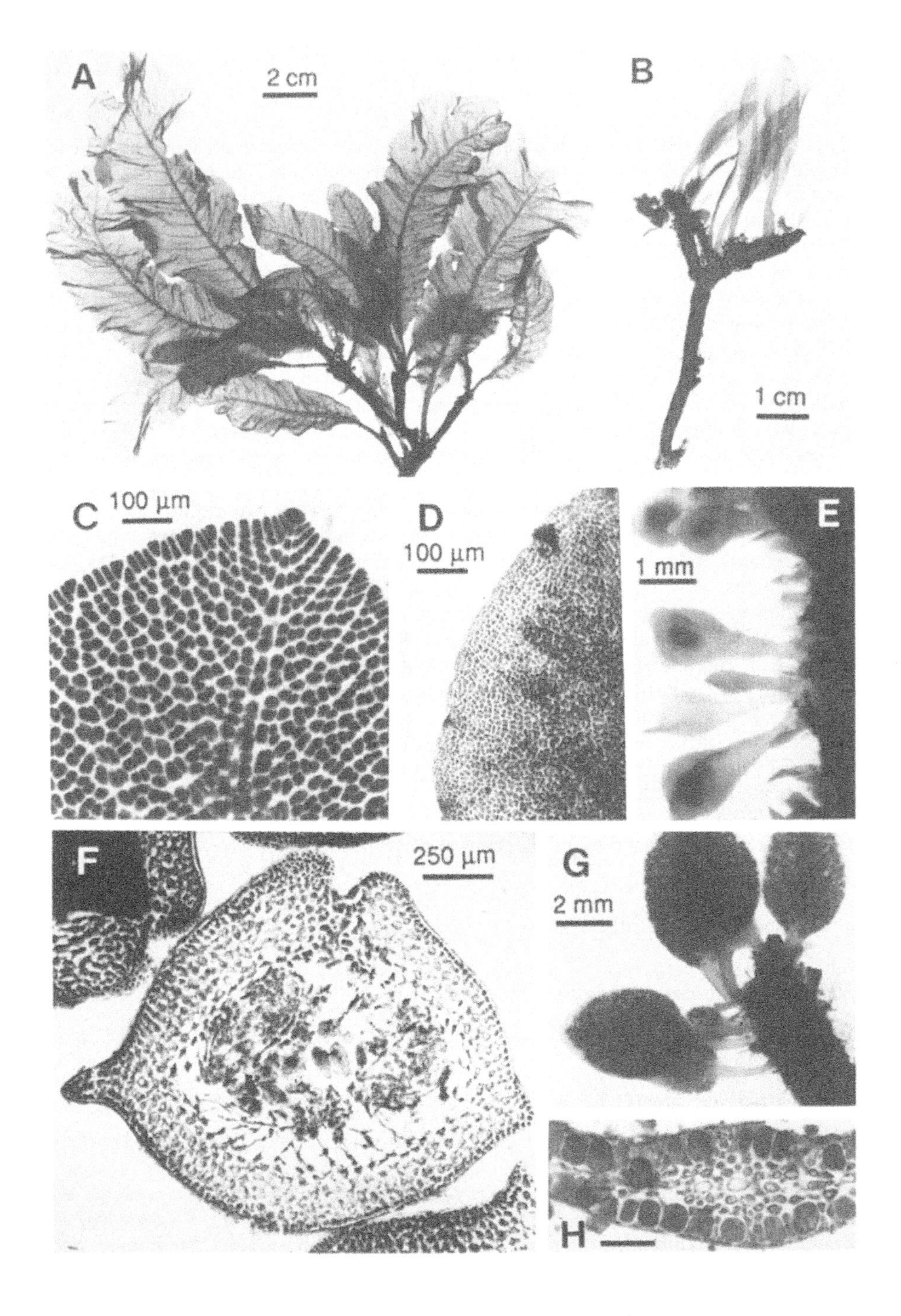

A
2 cm
B
1 cm
C
100 µm
D
100 µm
E
1 mm
F
250 µm
G
2 mm
H

further series of macroscopic veins running abaxially from these towards margins; *secondary blades* formed in winter on stipe-like eroded midribs up to 15 cm in length and 0.5 cm in diameter, the process being repeated in successive years so that several orders of branching may occur; bright red in colour, blades membranous in texture, old midribs and stipes tough and cartilaginous.

Blades growing from apical cells 12-14 µm in diameter; *midrib* composed in TS of scattered large rounded cells 80-120 µm in diameter embedded among cells 16-40 µm wide, with smaller cells above and below them; *blade laminae* monostromatic, 32-40 µm thick, composed of polygonal cells 6-28 µm in diameter, including occasional hyaline cells with refractive wall thickenings; *plastids* discoid in young blade cells, filiform and reticulate in midrib.

All reproductive bladelets formed on midribs, male bladelets formed on upper and lower surfaces while blade lamina still present, female and tetrasporangial bladelets borne on all sides of denuded midribs. *Male bladelets* stipitate, ovate, with midrib and strongly ruffled margins, 1.5-6 mm long x 0.7-2 mm wide; *spermatangial sori* developing initially on either side of midrib on both sides of blades, later extending over midrib, with apices and margins remaining sterile, about 55 µm thick, bearing spermatangial mother cells 4-6 µm wide that form cylindrical to ellipsoid spermatangia 7-10 µm long x 4-5 µm wide. *Female bladelets* lanceolate, with midrib and narrow lamina, each forming a single cystocarp on midrib near apex, with the apex and margins of bladelet remaining as terminal and lateral frills on pericarp; *cystocarps* subglobular, 1140-1400 µm in diameter, with projecting ostiole 50-100 µm long x 250 µm wide and a pore 60-70 µm in diameter; gonimoblast filaments c. 80 µm in diameter, sparsely branched, consisting of a few isodiametric cells terminating in branched chains of mature spherical carposporangia 24-35 µm in diameter. *Tetrasporangial bladelets* stipitate, ovate, 3-8 x 1.5-4 mm, 160-265 µm thick; tetrasporangia formed in a continuous sorus over both surfaces except at apex and margins, spherical to ellipsoid, 80-95 µm in diameter.

Growing on bedrock and boulders, occasionally epiphytic on *Laminaria hyperborea* stipes, in deep lower-shore pools and subtidal to at least 30 m depth, forming a

---

Fig. 67. *Delesseria sanguinea*

(A) Habit of mature plant (Donegal, May). (B) Old midribs bearing tetrasporangial bladelets and young vegetative blades (B-C & E-G Down, Dec.). (C) Apex of blade showing development of midrib and lateral cell rows from single dome-shaped apical cell. (D) Part of spermatangial bladelet with central sorus and sterile border (Sutherland, Oct.). (E) Female bladelets with developing and mature cystocarps. (F) T.S. of mature cystocarp showing carposporangia and elongate inner pericarp cells. (G) Tetrasporangial bladelets, covered with tetrasporangia except for narrow sterile margins. (H) T.S. of tetrasporangial bladelets with tetrasporangia on both faces (Isle of Man, Dec.).

characteristic component of the underflora in *L. hyperborea* forest, at moderately sheltered to extremely wave-exposed sites.

Generally distributed in the British Isles, northwards to Shetland, although records for the east coast of England are sparse (Norton, 1985), presumably due to lack of suitable substratum.

Arctic Norway and Iceland to Spain (South & Tittley, 1986).

Plants perennial, forming young blades and reproductive bladelets in winter. Blades reach their maximum size in May-June, then become increasingly battered and torn, and are reduced to midribs by December. Old stipes and midribs are heavily encrusted with algae and epiphytic invertebrates such as bryozoa, sponges and ascidians. Formation of new blades is stimulated by low temperatures (7-10°C), few being formed at 14°C, and the southern geographical limit of *D. sanguinea* may be determined by winter temperatures (Kain, 1987). Stipes and old midribs contain storage materials and new blades can be produced in darkness (Lüning, 1990). Spermatangia have been recorded in September-December, cystocarps in November-May and tetrasporangia in December-May. At the Isle of Man, male bladelets were present in September-December, cystocarps in December-March, and tetrasporangia in December-February (Kain, 1982). Short-day photoperiodic responses are involved in the formation of gametangia and tetrasporangia (Kain, 1987; 1991). In culture, male bladelets were stimulated by 11-12 h daylengths, and spermatangia developed within 4 weeks, the response being inhibited by a light-break during the dark period. Under field conditions, male bladelets were formed at longer daylengths, apparently as a result of low underwater light levels at dawn and dusk.

Chromosome number is $n = 31$ (Magne, 1964).

This species shows relatively little morphological variation.

Mature plants of *D. sanguinea* are readily identifiable. Young plants could be mistaken for large blades of *Apoglossum ruscifolium* (q. v.) or *Hypoglossum hypoglossoides* (q. v.), but these species lack the large macroscopic lateral veins of *D. sanguinea*. Very battered plants of *D. sanguinea* could resemble *Phycodrys rubens* (q. v.) but *P. rubens* blades have lobed or toothed, rather than entire, margins, and bear reproductive structures on mature blades rather than on old denuded midribs as in *D. sanguinea*.

PHYCODRYS GROUP

PHYCODRYS Kützing

PHYCODRYS Kützing (1843), p. 444.

Type species: *Phycodrys sinuosa* (Goodenough & Woodward) Kützing (1843), p. 444 [= *Phycodrys rubens* (Linnaeus) Batters (1902), p. 76.]

Thalli attached by a discoid holdfast and prostrate, flattened axes bearing rhizoids and giving rise to 1-several erect, stipitate, leaf-like, monostromatic blades up to 20 cm high with a percurrent central midrib and subopposite lateral nerves; blades unbranched, or suboppositely branched from margin at tips of nerves; microscopic veins absent.

Growth initiated by concavo-convex division of apical cell forming a primary cell row and central midrib; second-order cell rows reaching thallus margin, sometimes transforming into primary cell rows terminated by apical cells capable of producing side branches; third- and fourth-order cell rows opposite or abaxial, occasionally adaxial; transverse and longitudinal intercalary cell divisions frequent in primary cell row and cell rows of higher orders, not obscuring the basic growth pattern; all meristematic cells initially uninucleate, becoming multinucleate upon formation of secondary pit connections; adjoining cells linked by 1-2 median primary or secondary pit connections, later by superficial pit connections; plastids parietal, dissected into platelets.

Gametophytes dioecious. Spermatangial sori continuous in a band inside margin of blade or on lateral proliferations, initially 3-layered, formed on both sides of blade by periclinal division of mature vegetative cells, followed by anticlinal divisions of surface cells; intercalary cell divisions and secondary pit connections absent; surface cells functioning as spermatangial mother cells, each bearing 2-3 spermatangia. Procarps scattered in opposite pairs in monostromatic areas near tips of blades or on lateral proliferations; consisting of a supporting cell, a 4-6 (-9) celled sterile group-1, a curved, 4-celled carpogonial branch, and a 2-3 (-4) celled sterile group-2; sterile groups remaining undivided after fertilization, enlarging beneath ostiole; fertilized carpogonium cutting off a distal and a latero-posterior connecting cell, the latter fusing with a process from the auxiliary cell; auxiliary cell large, dome-shaped, dividing transversely, forming a primary cell row bearing repeatedly branched gonimoblasts and carposporangia in terminal, branched chains; fusion cell initially small, composed of central cell, supporting cell, foot cell and innermost gonimoblast cells, later expanding to include sterile groups and basal branches of gonimoblast filaments; pericarp 6-8 cell layers thick with a central ostiole. Tetrasporangial sori formed in monostromatic portion of blade between veins, or on lateral or superficial proliferations, initially 3-layered, later 5-layered; tetrasporangia produced laterally from inner cortical cells, rarely from central cells, arranged in two rows covered by an outer layer of cortical cells at maturity.

Reference: Kylin (1923).

One species in the British Isles.

Phycodrys rubens (Linnaeus) Batters (1902), p. 76.

Holotype: L 910.128.1044 (Dixon, 1964, fig. 1). Unlocalized, undated.

Fucus rubens Linnaeus (1753), p. 1162.
Fucus sinuosus Goodenough & Woodward (1797), p. 111.
Delesseria sinuosa (Goodenough & Woodward) Lamouroux (1813), p. 124.

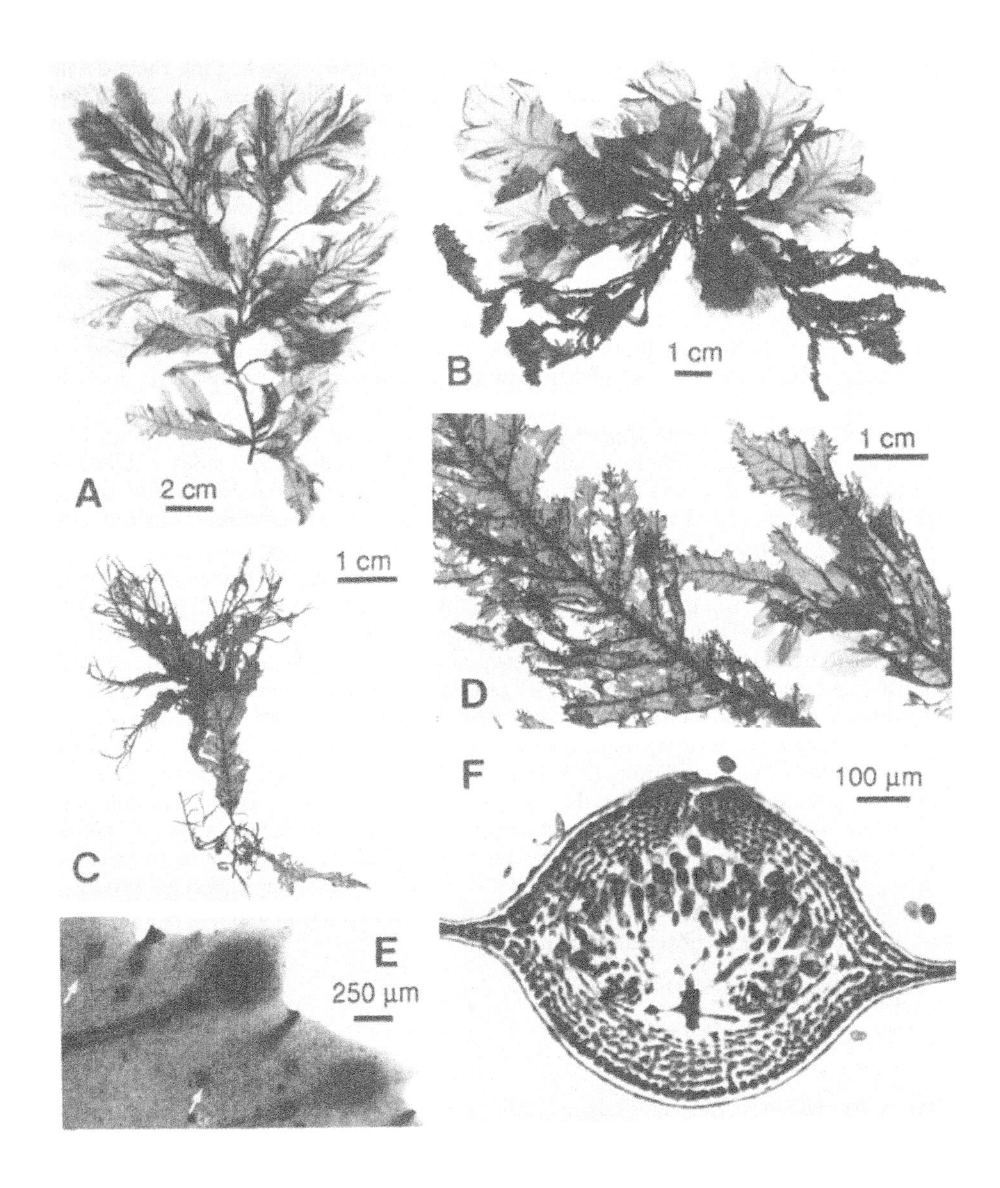
A
2 cm
B
1 cm
1 cm
D
1 cm
C
F
100 µm
E
250 µm

Thalli 5-20 cm high, consisting of small groups of blades attached by a holdfast composed of a small solid disc and branched stolon-like axes formed basally by the stipe; *blades* initially simple, stipitate, 1-3 cm wide, with prominent midrib up to 1.2 mm in diameter and paired oblique lateral veins up to 0.4 mm wide that may bear a further series of opposite veins, ruffled and lobed when young due to the formation of lateral apices from lateral veins, with dentate margins, later becoming cleft into numerous leaf-like blades by erosion of the lamina between lateral veins, finally consisting of branched veins with tattered blade remnants, branching from midribs, veins and margins, the lobes and dentations sometimes extending out as narrow branched fringes up to 3 cm long that may reattach to substratum; dull brownish-red in colour, blades membranous in texture, tearing easily, old midribs and stipes strong and flexible.

Apical cell of blades 12-14 µm in diameter, *midrib* in TS consisting of a row of large rounded cells 50-65 µm in diameter surrounded by pseudoparenchymatous cells decreasing in size to cortical cells 15-25 µm in diameter; *blade laminae* monostromatic, c. 25 µm thick, composed of polygonal cells 14-37 x 12-20 µm, with hyaline cells occurring singly or in small groups among them, lacking microscopic veins; *plastids* discoid.

Reproductive structures borne near apices and margins of blades and on small bladelets formed by both surfaces of midribs, on lateral veins, and along margins and torn edges of blades. *Male bladelets* stipitate, round to ovate, strongly ruffled, 0.5-1.2 mm long x 0.3-0.9 mm wide, almost covered by spermatangial sorus, with only apices and margins remaining sterile; *spermatangial sori* 60-85 µm thick, with conical spermatangial mother cells, 5-6 µm wide, bearing cylindrical spermatangia 10-12 µm long x 3.5 µm wide. *Procarps* formed abundantly near margins and on lanceolate female bladelets; *cystocarps* scattered or, typically, formed singly on bladelets, subglobular, 600-900 µm in diameter, with smooth pericarp wall, non-projecting ostiole and pore 30-40 µm in diameter; gonimoblast filaments 12-12 µm in diameter, terminating in branched chains of 2-4 mature spherical to ellipsoid carposporangia 40-70 x 32-60 µm. *Tetrasporangial sori* formed near apex of larger bladelets or covering small (<2 mm long) lanceolate, simple to pinnately branched bladelets, c. 220 µm thick; tetrasporangia spherical, 55-75 µm in diameter.

Growing in Scotland on subtidal bedrock, boulders, pebbles, shells and larger algae down to 27 m depth, frequently the most abundant species, apparently resistant to *Echinus*

Fig. 68. *Phycodrys rubens*

(A) Habit of plant collected from *Laminaria hyperborea* stipe (A-B Donegal, May). (B) Habit of plant collected from bedrock. (C) Habit of herbarium specimen with long fringed marginal extensions (Argyll, July). (D) Detail of blades showing midribs, lateral veins and fringes of tetrasporangial bladelets (D-E Down, Dec.). (E) Margin of female blade with numerous procarps (arrows) and developing cystocarps. (F) T.S. of mature cystocarp showing fusion cell with gonimoblast filaments terminating in chains of carposporangia (Sutherland, Aug.).

grazing; found in the southern British Isles on bedrock in lower-shore pools and subtidally to about 3 m depth, less often down to 30 m; abundant on *Laminaria hyperborea* stipes thoughout the British Isles, at moderately sheltered to extremely wave-exposed sites. In Berwick, *P. rubens* occurred near the base of stipes in shallow water, increased in biomass below 6 m, and grew at all levels on stipes near the lower limit of kelp at 12 m (Whittick, 1983).

Generally distributed in the British Isles, northwards to Shetland, although records for the east coast of England are sparse, presumably due to lack of suitable substrata (Norton, 1985).

Widely distributed in the northern N. Atlantic, from Spitzbergen to N. Portugal and from Arctic Canada (Ellesmere Is.) to New Jersey (Kain, 1982).

Mature plants can be found throughout the year, and those epiphytic on kelp reach peak biomass in September, most of which is lost by December at Berwick (Whittick, 1983). Old plants are heavily epiphytized by crustose bryozoans. At the Isle of Man, spermatangial plants were observed in January-May, July, and September-December; cystocarps and tetrasporangia were found throughout the year (Kain, 1982).

Plants growing in Scotland develop long, branched fringes from the margins; elsewhere in the British Isles margins remain lobed and finely dentate. Chromosome number reported as n = c. 20 (Kylin, 1923), but requires confirmation.

The only possible confusion is with battered *Delesseria sanguinea* (q. v.), but *D. sanguinea* blades have entire margins and tough, semi-rigid midribs, whereas *P. rubens* blades have lobed, toothed or fringed margins and narrow, flexible midribs.

ERYTHROGLOSSUM J. Agardh

ERYTHROGLOSSUM J. Agardh (1898), p. 174.

Lectotype species: *E. schousboei* (J. Agardh) J. Agardh (1898), p. 174 [? = *E. sandrianum* (Kützing) Kylin (1924), p. 31; see Athanasiadis (1985b), p. 461].

Thalli erect or decumbent, 1-10(-20) cm tall, consisting of few to many membranous blades arising from simple discoid or compacted, creeping holdfasts; blades stipitate, often branched from stipe, simple or branched from margin, narrowly lanceolate to flabellate; principal blades and side branches of juvenile thalli initially provided with percurrent, microscopic midveins, later with basal or percurrent polystromatic midribs, or subdichotomously branched nerves below and anastomosing microscopic veins above; margins dentate, the teeth situated at ends of microscopic veins, or lateral veins absent.

Growth initiated by concavo-convex division of apical cells generating primary cell rows; transverse and longitudinal divisions frequent in primary cell rows and cell rows of higher orders, not obscuring basic growth pattern; marginal meristems sometimes replacing apical meristems in older thalli; meristematic cells initially uninucleate, becoming multinucleate upon establishment of secondary pit connections; microscopic veins at first differentiating acropetally along primary cell rows, later formed randomly and anastomosing, or anastomosing veins absent, tristromatic, the central cells elongate, tubular with broadened proximal and distal pit connections; macroscopic veins and midribs derived by lateral extension of tristromatic tissue bordering veins, later polystromatic with cells arranged in horizontal tiers and radially branched rows; plastids parietal, at first dissected into platelets, later sausage-shaped.

Gametophytes dioecious. Spermatangial sori in bands inside thallus margin, at first elliptical, later confluent, linear, developing as in *Phycodrys*. Procarps scattered in opposite pairs in monostromatic areas of blade, consisting of a supporting cell, a 4-7(-9) celled sterile group-1, and two opposite, curved, 4-celled carpogonial branches; sterile group-2 absent; cells of sterile group-1 remaining undivided after fertilization, enlarging beneath ostiole; auxiliary cell dividing transversely into primary cell row bearing subdichotomously branched gonimoblasts and carposporangia in terminal branched chains; fusion cell large, incorporating the supporting cell, cells in cystocarp floor, sterile group-1, and inner gonimoblast cells; pericarp 5-7 layers thick with a central ostiole. Tetrasporangial sori in bands inside thallus margin, initially elliptical, later confluent and linear, developing as in *Phycodrys*; tetrasporangia formed laterally from inner cortical cells, rarely from central cells, arranged in two rows at maturity covered by an outer layer of cortical cells.

Reference: Kylin (1924).

One species of *Erythroglossum*, *E. sandrianum* (Kützing) Kylin, is currently included with some doubt in checklists of the British Isles flora (Parke & Dixon, 1976; South & Tittley, 1986), based mainly on collections from the south coast of England. All specimens we have examined from the British Isles and Atlantic Europe are indistinguishable from *Polyneura laciniata* (Lightfoot) P. Dixon. The holotype of *E. sandrianum* from Dalmatia (L 941.183.396) [which takes precedence over the neotype designated by Athanasiadis (1985b)] may also be conspecific with *P. laciniata*, but further studies of *E. sandrianum* from the Mediterranean are required. On the basis of its reproductive morphology, *P. laciniata* is here transferred to *Erythroglossum*, of which it is the only representative in the British Isles.

Erythroglossum laciniatum (Lightfoot) Maggs & Hommersand, comb. nov.

Holotype: BM-K (Fig. 66A). Argyll (Iona), undated (see Dixon, 1983).

Ulva laciniata Lightfoot (1777), p. 974.
Delesseria gmelinii Lamouroux (1813), p. 124.
Nitophyllum gmelinii (Lamouroux) Greville (1830), p. 82.
Polyneura gmelinii (Lamouroux) Kylin (1924), p. 40.
Polyneura laciniata (Lightfoot) P. Dixon (1983), p. 13.

Thalli erect, consisting of one or more blades, either arising from a cylindrical stipe 0.8-1.6

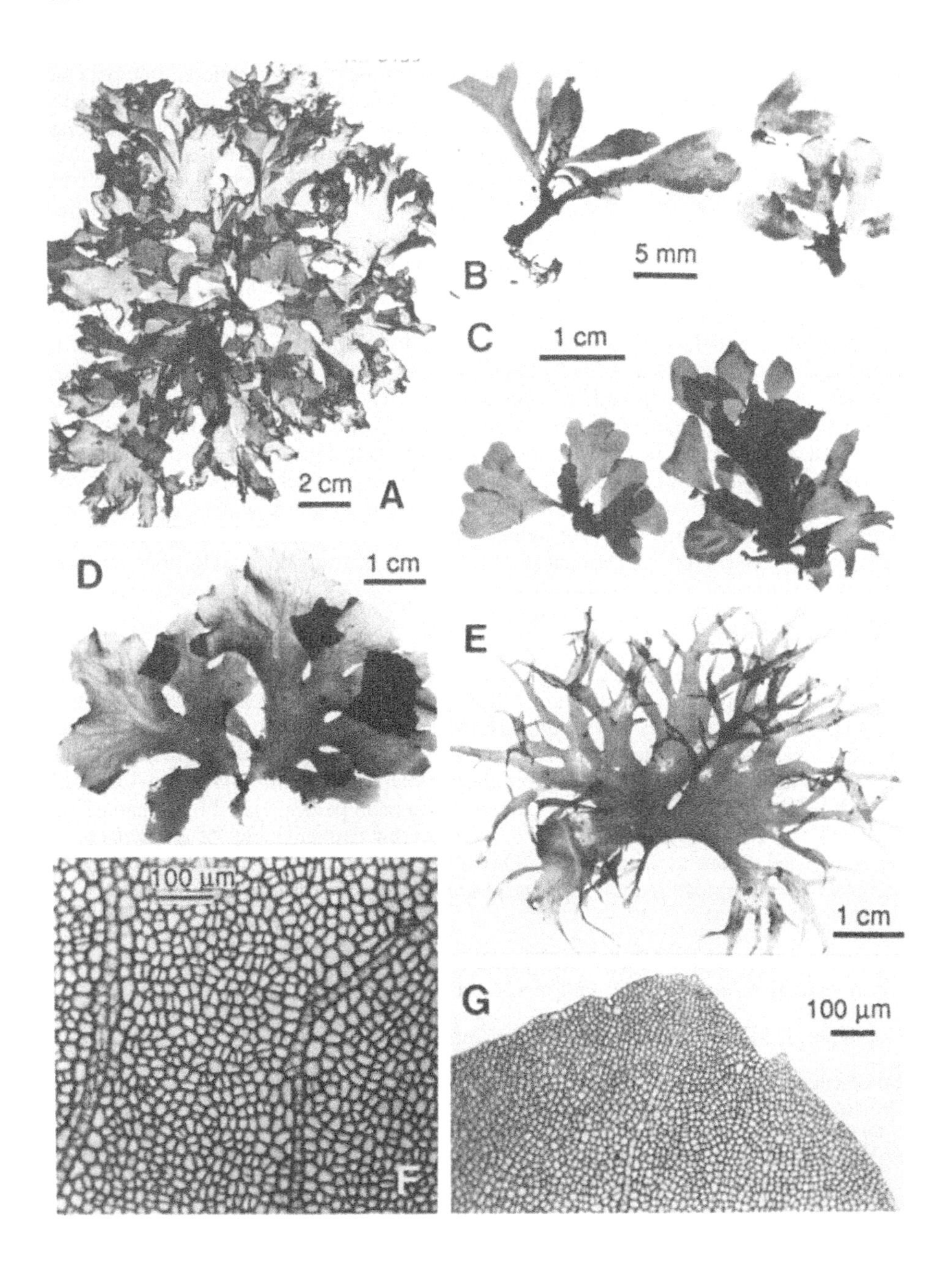
2 cm
A
5 mm
B
C
1 cm
D
1 cm
E
1 cm
100 µm
F
G
100 µm

mm in diameter and up to 5 mm long, or attached directly to solid discoid holdfast up to 2 mm in diameter, from which cylindrical rhizome-like extensions grow out and sometimes give rise to further blades; *blades* 1.5-16 cm high and 3-15 (-20) cm broad, very variable in shape, fan-shaped and partially divided into broad lobes, or deeply dissected into few to numerous lanceolate to overlapping flabellate segments, which may be divided again, with rounded to acute apices and the margins entire, dentate, or fringed with a pinnate arrangement of lanceolate proliferations 1-5 mm wide and up to 3 cm long, some of which curve downwards, becoming prostrate and attaching by marginal haptera and bearing pinnate branches; *veins* macroscropic, usually conspicuous near base of blade, up to 0.5 mm wide, appearing as a single midrib in narrower blades, blade segments, and prostrate axes, branching and fanning out in wider blades, decreasing in width towards apices, where a network of fine veins is usually visible; thalli brownish-red in colour, sometimes with a conspicuous blue iridescence underwater; texture membranous, crisp, with blades not lying flat out of water, and becoming rather brittle when older.

Blade laminae with a variable number of triangular apical cells along apices and margins, monostromatic when young, 50-100 μm thick, increasing to 2-8 or more cells thick, 100-500 μm in thickness, composed of polygonal cells 20-80 μm long x 20-55 μm wide, *veins* rapidly becoming multicellular, branching and anastomosing at irregular intervals; *plastids* discoid in young cells, becoming sausage-shaped to ribbon-like.

Spermatangial sori formed along margins, just behind apices of young blades or on either side of midrib, oval, triangular, elongate or V-shaped, 0.5-3.5 mm x 0.3-2.5 mm, consisting of a layer of small, rounded spermatangial mother cells 5-7 μm in diameter on both sides of blade, bearing elongate spermatangia 8-10 μm long x 3-4 μm wide. *Cystocarps* scattered on both sides of blades near apices, 900-1200 μm in diameter when mature, with a smooth outer pericarp and little-protruding ostiole with a pore 70 μm in diameter, containing a large fusion cell and branched gonimoblast filaments 12-20 μm in diameter terminating in rows of 4-6 mature, ovoid to pyriform carposporangia, 88-130 μm long x 56-80 μm wide. *Tetrasporangial sori* developing in discontinuous lines inside margins of large blades, paired on either side of midrib in narrow marginal outgrowths, or between veins of small blades and near apices, occasionally covering entire surface of

---

Fig. 69. *Erythroglossum laciniatum*

(A) Holotype, with numerous small marginal sori (Iona, Argyll, undated in BM-K). (B) Thalli regenerating from old stipes, forming sori in small blades (Pembroke, June). (C) Plants collected from intertidal bedrock forming sori over most of the surface of young blades (Glamorgan, Mar.). (D) Typical young subtidal thallus, showing branched veins and denticulate margins (Donegal, May). (E) Typical mature subtidal specimen, from same site as D, with long fringed marginal extensions (Donegal, July). (F) Blade lamina with veins (as D). (G) Margin of blade showing veins developing towards dentations (as D).

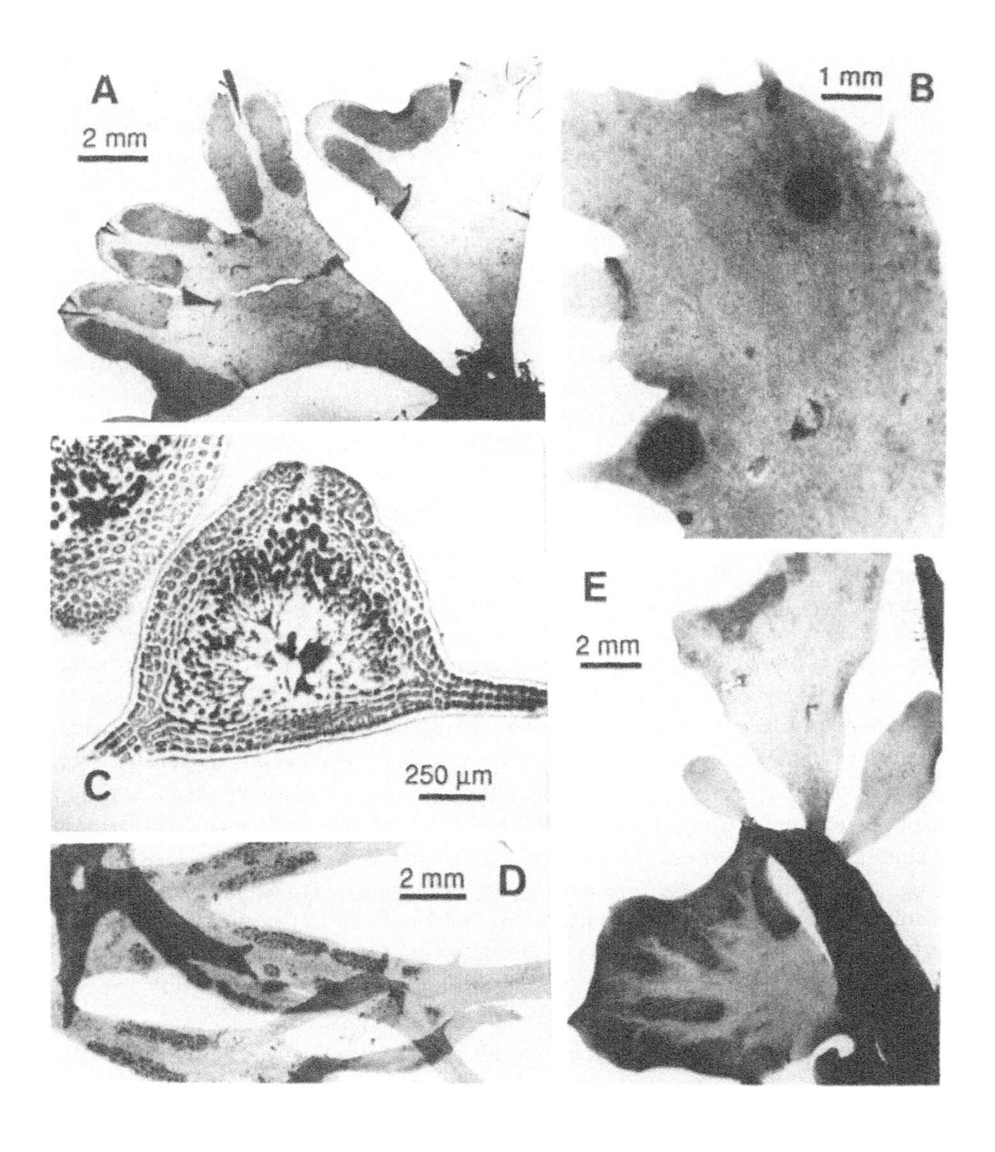
A
2 mm
1 mm
B
C
250 µm
E
2 mm
2 mm
D

young blades, circular, linear or triangular, 2-8.5 mm long x 0.8-3.5 mm wide, 140-170 µm in thickness; tetrasporangia spherical to ellipsoid, 44-95 x 38-55 µm.

Growing on bedrock in pools near extreme low water and subtidal to at least 30 m depth, abundant on cliffs and under overhangs in the shallower part of this depth range, where it is a characteristic understorey component in *Laminaria hyperborea* forest, more common on upward-facing surfaces near its lower depth limits, at sheltered to extremely wave-exposed sites, with or without tidal currents.

South coast of England eastwards to Kent, west coasts northwards to Ross & Cromarty; widely distributed in Ireland; Channel Isles. Records from N. Scotland and Shetland require confirmation as herbarium specimens from these areas are often misidentifications of *Haraldiophyllum bonnemaisonii*.

British Isles to Portugal; W. Mediterranean (Athanasiadis, 1987, p. 463); probably more widespread.

Mature plants can be found throughout the year at sheltered sites. At wave-exposed sites, some plants perennate as thickened stipes with blade remnants which form new blades in March-June. Blades enlarge until July-August, and by late September most are damaged and degenerating. There appear to be two reproductive periods, as some blades become fertile when very small and others enlarge before reproducing. Spermatangia recorded in April-June and August, cystocarps in March, April and June-November, and tetrasporangia in March, April and June-December.

Erythroglossum laciniatum is an extremely variable species. Variation is correlated, in part at least, with environment, seasonal changes, and age of the individual plant. Under sheltered conditions, very broad, lobed blades reach 15-20 cm in width and margins are entire; there is little change in form throughout the season. On exposed coasts, young plants collected in May to early June are up to 2 cm in width and consist of entire or broadly lobed blades, sometimes with small marginal teeth. By July-August, all thalli have developed fringed apices and narrow marginal outgrowths; they are rarely more than 5-10 cm in total width. Plants in their first year of growth lack cylindrical stipes; after overwintering, thalli consist of thickened stipes bearing new blades. Perennating plants often reproduce on very small, thickened blades that grow out from the stipes, and apparently undergo a further period of reproduction later when blades have enlarged, whereas in young thalli reproductive structures develop only on large blades. Overall habit

Fig. 70. *Erythroglossum laciniatum*

(A) Detail of male thallus with paired elongate spermatangial sori (Kerry, Aug.). (B) Detail of female thallus with two cystocarps (Donegal, July). (C) T.S. of mature cystocarp showing large fusion cell and candelabra-like gonimoblast filaments terminating in chains of carposporangia (Sussex, Dec.). (D) Paired elongate tetrasporangial sori in narrow marginal extensions (as B). (E) Tetrasporangial sori covering young blades except for sterile striae over veins (Glamorgan, Mar.).

determines the pattern of venation, to some extent: extremely narrow blades have midribs only, whereas wider blades may lack midribs, forming instead a fan-like pattern of ribs, with anastomosing veins between and around these.

Erythroglossum laciniatum thalli occasionally support plants of the parasite *Asterocolax erythroglossi* (q. v.).

Distinctions betwen *E. laciniatum* and fan-shaped plants of *Cryptopleura ramosa* are listed under *C. ramosa* (q. v.). Thalli of *Polyneura bonnemaisonii* (q. v.) differ from *E. laciniatum* in the formation of small scattered rather than marginal tetrasporangial and spermatangial sori and by the spiny ornamentation on cystocarps, in contrast to the smooth pericarps of *E. laciniatum*. Young plants of *Acrosorium venulosum* (q. v.) may resemble *E. laciniatum,* but can be distinguished by the lack of multicellular macroscopic veins. *Haraldiophyllum bonnemaisonii* (q. v.) differs from *E. laciniatum* by the absence of macroscopic or microscopic veins, although a few radiating thickened nerves may be present at the base of the plants.

ASTEROCOLAX J. Feldmann & G. Feldmann

ASTEROCOLAX J. Feldmann & G. Feldmann (1951), p. 1139.

Type species: *A. erythroglossi* J. Feldmann & G. Feldmann (1951), p. 1139.

Thalli parasitic on members of the Phycodrys group (Delesseriaceae), consisting of stellate clusters of minute blades arising from a basal cushion and endophytic filaments that penetrate the host tissue and link to host cells by means of secondary pit connections; blades initially lanceolate to ovate, with smooth or dentate margins; unbranched or suboppositely branched from the margin; thickened secondarily and needle-like or club-shaped in some species; central midrib present, often inconspicuous; microscopic veins absent.

Spore germination bipolar, the primary rhizoid penetrating between host cells, branching endophytically and producing a cushion-like pad externally, bearing blades. Growth of blade initiated by concavo-convex division of apical cell forming a primary cell row, or replaced by adventitious tips originating from surface cortical cells; second-order cell rows reaching thallus margin, sometimes converting into primary cell rows that may form lateral branches; higher-order cell rows developing as in *Phycodrys*; transverse and longitudinal intercalary cell divisions frequent in primary cell rows and cell rows of higher orders, not obscuring the basic pattern of growth; blades thickened secondarily by cortical growth above and alongside the central midrib in some species; plastids discoid, variably pink, yellow or colourless with few to many traversing thylakoids in different species or in the same species on different hosts.

Gametophytes dioecious; all blades reproductive, with the phases or sexes sometimes intermingled due to coalescence of genetically different endophytic systems.

Spermatangia produced from surface cells covering male blades. Procarps appearing to be scattered over surface, or more probably developing apically on short lateral branches, consisting, where known, of a 4-5 celled sterile group-1, a 4-celled carpogonial branch, and a sterile group-2 initial; gonimoblasts branched, bearing carposporangia terminally in branched chains. Tetrasporangial sori covering the fertile blades; origin and development of the tetrasporangia unknown.

References: J. Feldmann & G. Feldmann (1951, 1958), Wagner (1954, as *Polycoryne*), Goff (1982).

One species in the British Isles.

Asterocolax erythroglossi J. Feldmann & G. Feldmann (1951), p. 1139.

Neotype* : PC, Herb. J. Feldmann 9204. France (Astan, baie de Roscoff), August 1955, *Magne*.

Thalli 0.3-4.2 mm in diameter, consisting of endophytic basal tissue and emergent cushion-like pad 0.1-3 mm in diameter, bearing a stellate group of 5-40 complanate, lanceolate to ovate blades 0.2-1.8 mm long x 0.1-0.5 mm wide, with a smooth or dentate margin; pale brownish-pink in colour, with a solid texture.

Endophytic filaments branched, 15-20 µm in diameter, penetrating between host cells and forming numerous secondary pit connections to them; *emergent tissue* composed in TS of a dense pseudoparenchymatous medulla of irregularly rounded cells 20-36 µm in diameter, surrounded by a cortical region of small rounded cells that give rise to erect blades; *blades* growing from a conspicuous apical cell 9-12 µm in diameter that divides by a curved wall, increasing in thickness rapidly to c. 150 µm, pseudoparenchymatous, consisting in TS of a row of axial cells surrounded by rounded to elongate medullary cells, 40-75 x 40-50 µm, and a cortical zone of cells decreasing to 8-15 µm in diameter; uniaxial lateral proliferations frequently formed along margins and developing into lateral branches in a pinnate arrangement, sometimes forming a further order of laterals; starch grains abundant in all cells; *plastids* discoid to bacilloid.

All blades and marginal outgrowths becoming reproductive. *Spermatangial sori* covering entire surface of male blades, except for growing apex, consisting of a layer of round to ellipsoid spermatangial mother cells 4-6 µm long x 5-8 µm wide, bearing cylindrical to clavate spermatangia 8-12 µm long x 2.5-3.5 µm wide. *Procarps* borne in lateral branches which frequently form a further order of branches that remain as short spines radiating from the developing pericarp wall; *cystocarps* developing singly or in

* The type collection was from rade de Brest, France, 21 ix 1951, designated as *J. Feldmann* 8008, in PC, but no material of this collection is present in PC (F. Ardré, pers. comm.).

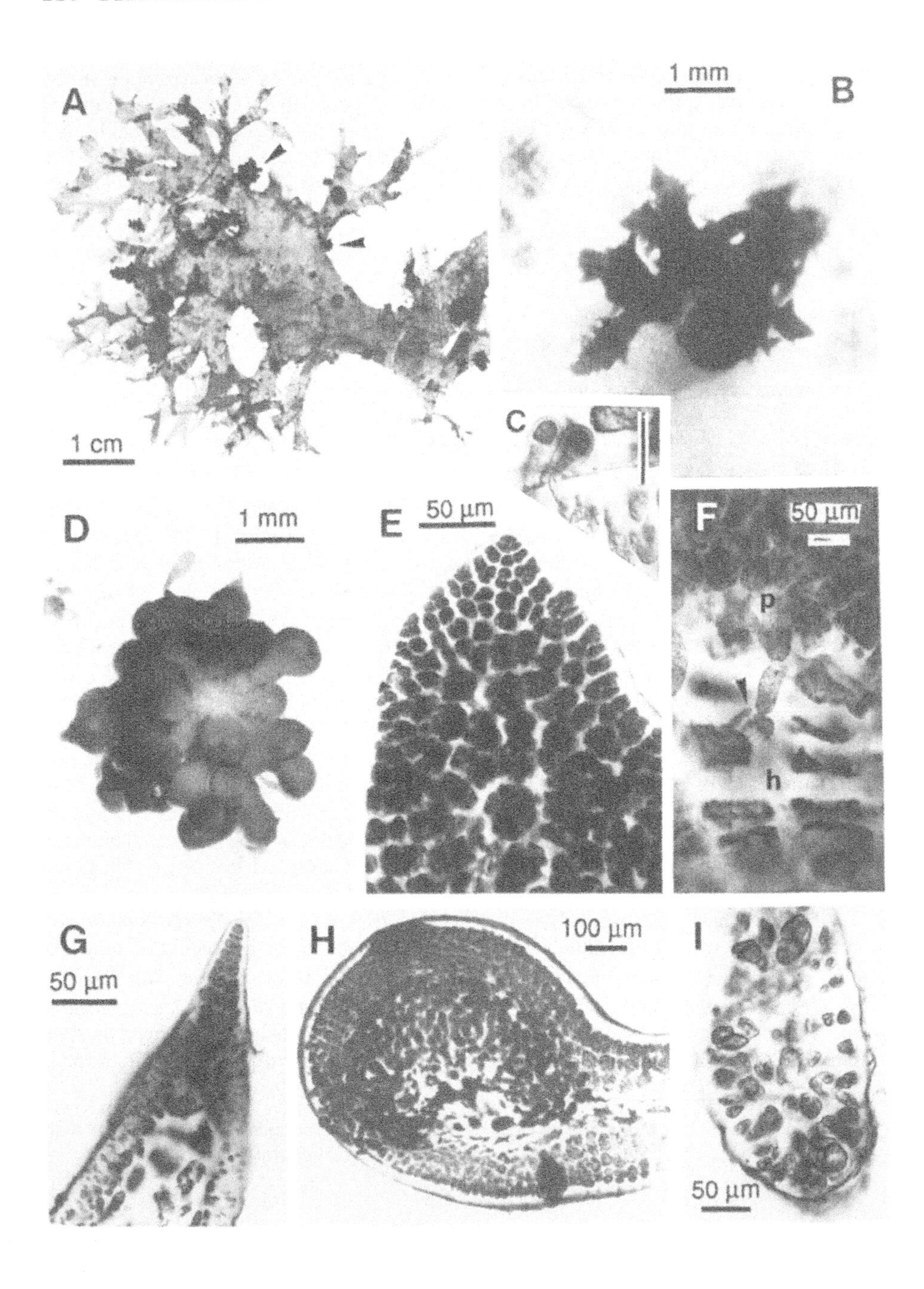
A
1 cm
1 mm
B
C
D
1 mm
E
50 µm
F
50 µm
p
h
G
50 µm
H
100 µm
I
50 µm

catenate series, spherical when mature, 650-840 μm in diameter, with a non-protruding ostiole 100-125 μm in diameter; gonimoblast filaments terminating in chains of 6-8 spherical to ellipsoid carposporangia, 20-36 μm long x 18-20 μm wide. *Tetrasporangia* formed in sori that cover blades except for apex, ellipsoid, 34-48 μm long x 28-38 μm wide.

Hemiparasitic on stipes and blades of *Erythroglossum laciniatum* and on other plants of *A. erythroglossi* at depths of 1-15 m on moderately exposed to moderately wave-sheltered coasts. Distribution is patchy: thalli are present on almost every plant in some *E. laciniatum* populations and absent from nearby populations.

Known in the British Isles only from Kent, Somerset, Pembroke, Kerry, Mayo and Donegal.

British Isles to N.W. France.

Young thalli become noticeable in April; mature plants with spermatangia, cystocarps and tetrasporangia occur in June-August. Released tetraspores apparently germinate on the surface of sporophytic blades, form penetrating filaments, and develop into gametophytes. Dwarf males reproduce rapidly in this situation, when less than 0.5 mm in length, while female plants probably do not survive long enough to reproduce.

Mature plants vary greatly in size because growth continues after reproduction has started. Female plants with cystocarps are particularly conspicuous due to the relatively large size and opacity of the cystocarps.

J. Feldmann & G. Feldmann (1951, 1958) reported the host species as *Erythroglossum sandrianum* (Zanardini) Kylin, but material from Brittany attributed to this species appears to be a growth form of *Erythroglossum laciniatum*.

Fig. 71. *Asterocolax erythroglossi*

(A) *Erythroglossum laciniatum* blade with numerous parasitic thalli of *A. erythroglossi* (arrows) (Donegal, Aug.). (B) Male thallus with marginal dentations on blades (Kent, Aug.). (C) Sporelings showing penetration of filaments into host (scale bar = 50 μm) (C, E-F, H & I Mayo, June). (D) Female thallus with densely crowded cystocarps (as A). (E) Apex of blade showing development from single apical cell. (F) Interface between parasitic thallus (p) and host tissue (h), showing secondary pit connections between them (arrow). (G) T.S. of spermatangial blade with sterile margin (as B). (H) T.S of cystocarp showing central fusion cell and tightly packed carposporangia. (I) T.S. of tetrasporangial blade.

POLYNEURA (J. Agardh) Kylin, nom. cons.

POLYNEURA (J. Agardh) Kylin (1924), p. 33.

Type species: *P. hilliae* (Greville) Kylin (1924), p. 33 [= *P. bonnemaisonii* (C. Agardh) Maggs & Hommersand, comb. nov.]

Thalli consisting of one or more erect blades up to 30(-45) cm tall attached by a discoid rhizoidal holdfast or by secondary rhizoids from flattened, basal bladelets, stipitate, ovate to orbicular when young, later fan-shaped and variously lobed to deeply cleft; margin entire or fringed with multicellular spines; midrib absent; anastomosing macroscopic nerves present, restricted to basal parts or extending to upper parts of blade; anastomosing microscopic veins present, anterior to macroscopic veins and giving rise to them.

Growth of juvenile thallus initiated by concavo-convex division of apical cell; intercalary longitudinal and transverse divisions abundant, beginning 1-2 segments below apical cell, obscuring primary cell row; older blades growing by marginal and intercalary meristems, with the meristematic cells mostly multinucleate, linked by median pit connections; microscopic veins differentiating acropetally, not reaching margin, tristromatic, the central cells elongate with broadened proximal and distal pit connections; macroscopic veins derived by lateral extension of tristromatic areas bordering veins, later polystromatic, the cells arranged in horizontal tiers and radially branched rows; plastids parietal, dissected into platelets.

Gametophytes dioecious. Spermatangial sori formed in monostromatic portions of blade between veins, initially small, circular, later expanding, becoming confluent, developing as in *Phycodrys*. Procarps scattered in opposite pairs in monostromatic parts of blade, consisting of a supporting cell, a 4-6 celled sterile group-1, and two opposite, curved, 4-celled carpogonial branches; sterile group-2 absent; cells of sterile group-1 remaining undivided after fertilization, enlarging beneath ostiole; gonimoblasts at first compact, repeatedly branched, bearing carposporangia in short, terminal chains, later enlarging with the production of secondary gonimoblasts; fusion cell initially small, consisting of the central cell, supporting cell, foot cell, and inner gonimoblast cells, later branched, candelabra-like, incorporating additional gonimoblast filaments through broadening and dissolution of primary pit connections; pericarp 6-8 layers thick with a central ostiole, smooth or covered with spines. Tetrasporangial sori small, circular, scattered over blade between microscopic or macroscopic veins, developing as in *Phycodrys*; tetrasporangia formed primarily from central cells, rarely from inner cortical cells, arranged in two rows alongside central layer and covered by outer layer of cortical cells at maturity.

Reference: Kylin (1924).

In the most recent checklists for the British Isles, three species of *Polyneura* are listed. Of these, British specimens of *P. litterata* (J. Agardh) Kylin are misidentifications of other species, and *P. laciniata* is transferred to *Erythroglossum*. An older epithet for *P. hilliae* is based on *Delesseria bonnemaisonii* C. Agardh (1822). As noted by J. Agardh (1852), the type of *D. bonnemaisonii* is a specimen of the alga known then as *Delesseria hilliae* Greville and currently as *Polyneura hilliae* (Greville) Kylin.

Polyneura bonnemaisonii (C. Agardh) Maggs & Hommersand, comb. nov.

Holotype: LD 30519. France (Finistère), undated, *Bonnemaison*.

Delesseria bonnemaisonii C. Agardh (1822), p. 186.
Delesseria hilliae Greville (1827-8), pl. 351.
Nitophyllum hilliae Greville (1830), p. 80.
Nitophyllum ulvoideum W. J. Hooker (1833), p. 287 (see Harvey, 1848, pl. 169).

Thalli 6-20 cm high and 7-40 cm broad, composed of one or more blades attached by small solid discoid holdfast up to 3.5 mm in diameter with prostrate rhizoidal outgrowths that give rise to new blades; *blades* terminal on a cylindrical stipe 3-10 mm long and 1-1.5 mm thick, which may continue as a midrib, also formed in series along its length, circular, fan-shaped, or linear, sometimes divided into diverging or overlapping lobes, or deeply cleft nearly to base into wedge-shaped segments, with rounded apices and undulating to strongly ruffled, entire, fringed or lobed margins, when older usually bearing fan-shaped proliferous lobes and often perforated with numerous holes; *macroscropic veins* conspicuous near base of blade, branching in a dendroid pattern and anastomosing, decreasing from 1 mm in width towards margins; colour deep pink when young, becoming dark brownish-red when older or bleaching to yellowish-brown, sometimes with purple iridescence underwater; texture crisp and rigid, blades not lying flat out of water, rather cartilaginous near holdfast.

Blade laminae monostromatic when young, 35-72 μm thick, later increasing to 200 μm in thickness and consisting of 3-4 layers of polygonal cells 35-60 μm long x 25-40 μm wide; *veins* 3 cells thick initially, some increasing to up to 700 μm in thickness; *plastids* plate-like.

All reproductive structures formed in radiating rows between veins on distal two-thirds of blades. *Spermatangial sori* initially round to elliptical, 0.4-5 x 0.4-2 mm, later coalescing between veins, 80-95 μm thick, consisting of a layer on either side of blade of cuboid spermatangial mother cells 7-10 μm wide, bearing ovoid-clavate spermatangia 8-12 μm long x 5 μm wide. *Cystocarps* formed on both sides of blade, strongly protuberant, flat-topped with spiny ornamentation on upper and sometimes lower pericarp surfaces, 900-1300 μm in diameter when mature, with non-protruding ostiole and pore 45-60 μm in diameter; carposporangia borne in branched chains of 2-4, becoming rounded as they mature, 45-60 μm long x 35-48 μm wide. *Tetrasporangial sori* round, occasionally coalescing, 0.2-0.6 mm in diameter, protruding on either side of blades, 200 μm thick; tetrasporangia spherical, 48-65 μm in diameter.

Epilithic on bedrock and boulders in shaded lower-shore pools, subtidal on bedrock and

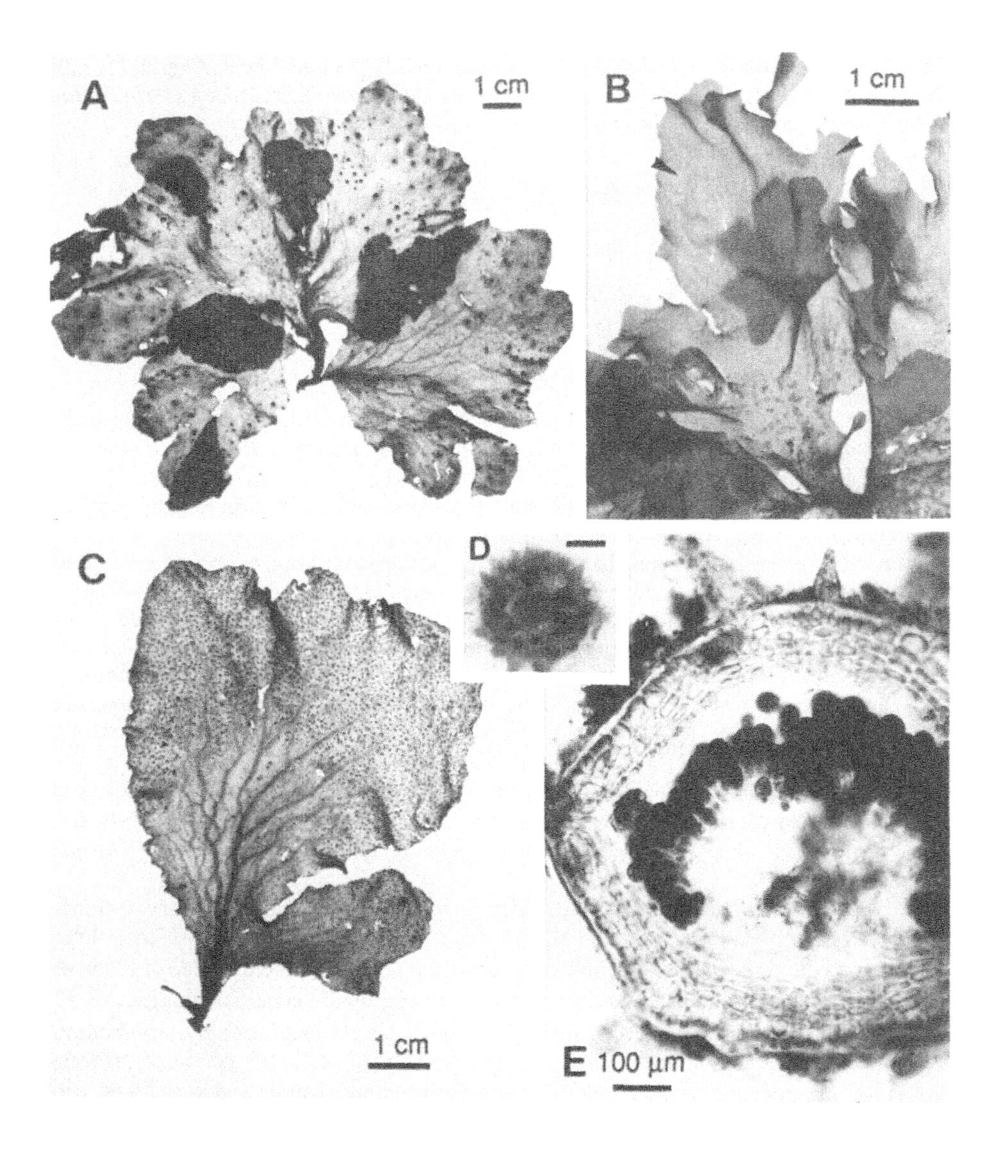
A
1 cm
B
1 cm
C
D
1 cm
E 100 μm

boulders and epiphytic on perennial algae and *Laminaria hyperborea* stipes from extreme low water to 19 m depth, sometimes abundant, at sheltered to moderately wave-exposed sites, tolerant of strong current exposure and muddy situations.

Widely distributed in the British Isles, northwards to Shetland; much more common on southern and south-western coasts than in Scotland.

Shetland to Portugal (South & Tittley, 1986).

Holdfasts and blade bases perennial, new blades developing in December-January, becoming fertile from February onwards and covered with epiphytic bryozoa from June onwards. Spermatangia recorded in February-April, June-October and December; cystocarps in April and June-September; tetrasporangia in April and June-October.

There is little variation in overall morphology.

For distinctions between *P. bonnemaisonii* and two species with which it could be confused, *Erythroglossum laciniatum* and *Haraldiophyllum bonnemaisonii*, see entries for these latter species.

SCHIZOSERIS GROUP

DRACHIELLA Ernst & J. Feldmann

DRACHIELLA Ernst & J. Feldmann (1957), p. 458.

Type species: *D. spectabilis* Ernst & J. Feldmann (1957), p. 458.

Thalli annual or perennial, if perennial consisting of one to many erect annual blades up to 7 cm long arising from a perennial, irregularly branched, rhizomatous basal system, briefly stipitate or with a prominent cylindrical stipe; blades initially erect, obovate to lanceolate, becoming fan-shaped and dichotomously, subdichotomously or irregularly lobed or divided, free ends decumbent or repent, sometimes secondarily attached by haptera or rhizoids; margins entire, minutely toothed, or forming clusters of rhizoids or polystromatic tendrils or haptera; blade initially monostromatic, becoming distromatic or tristromatic away from the margin in some species, polystromatic towards stipe or base, with cells arranged in horizontal tiers and vertical rows; microscopic veins absent.

Fig. 72. *Polyneura bonnemaisonii*

(A) Habit of female plant with numerous cystocarps (A & C-E Kerry, Aug.). (B) Detail of male blades with small, inconspicuous sori (arrows) (Galway, Feb.). (C) Habit of tetrasporangial plant with numerous small sori between large branched veins. (D) Cystocarp with spiny wall ornamentation. (E) T.S. of mature cystocarp showing central fusion cell, carposporangia, and spiny pericarp wall ornamentation.

Growth diffuse by marginal and intercalary cells, marginal cells triangular or rectangular, mostly uninucleate, inner meristematic cells irregularly rectangular, uninucleate or multinucleate; nuclei and pit connections initially restricted to median plane of blade, nuclei later redistributed to surface area just beneath the chloroplast, with numerous minute secondary pit connections forming between adjacent cells parallel to surface; plastids discoid in young cells, fusing to form a single dissected parietal plate or several convoluted, ribbon-like plastids; cells of rhizomes, stipes and blade thickenings filled with distinctive starch granules.

Sexual reproduction infrequent, unknown in some species. Gametophytes dioecious. Spermatangial sori minute, irregularly distributed on monostromatic and polystromatic parts of blade. Procarps formed only on upper side of blade, circular in outline, consisting of a supporting cell bearing a 1-celled sterile group, a strongly curved, 4-celled carpogonial branch and trichogyne emerging distal to the lateral sterile group, and a small, 1-celled basal sterile group; cystocarps hemispherical, with a large central fusion cell and carposporangia in branched chains; pericarp 4 cell layers thick, with distal protruding ostiole. Tetrasporangial sori solitary or aggregated, formed on ordinary blades or specialized reproductive bladelets; tetrasporangia originating from central and cortical cells, arranged in 2 layers or irregularly dispersed, tetrasporangia tetrahedrally divided. Asexual reproduction by specialized callus-like marginal thickenings in some species.

References: Kylin (1924), Magne (1957).

The genus *Drachiella* is at present monotypic, containing only the type species, *D. spectabilis*. However, Ernst & Feldmann (1957) drew attention to similarities between *D. spectabilis* and the alga known at that time as *Nitophyllum versicolor* Harvey, which they showed was conspecific with the type of *Halymenia heterocarpa* Chauvin ex Duby. They were unable to place *H. heterocarpa* generically owing to lack of reproductive material. Comparison between reproductive specimens of these species shows many close similarities, and we therefore propose the transfer of *H. heterocarpa* to *Drachiella*. A third species, previously known as *Myriogramme minuta* Kylin, is also transferred to *Drachiella* on the basis of its vegetative and tetrasporangial morphology, in particular the shape of the plastids.

KEY TO SPECIES

1 Mature blades 1 cell thick, forming opaque, callus-like nodules along margins and apices .... *D. heterocarpa*
Mature blades 3 cells thick, lacking callus-like nodules 2

2 Plants <5 cm wide; one plastid per mature cell, with deep sinuous indentations; tetrasporangial sori formed near apices of mature blades .... *D. minuta*
Plants up to 14 cm wide; several ribbon-like, convoluted plastids per mature cell; tetrasporangial sori confined to small specialized bladelets formed in winter .... *D. spectabilis*

Drachiella heterocarpa (Chauvin ex Duby) Maggs & Hommersand, comb. nov.*

Holotype: CN. France (Calvados), undated (see Ernst & Feldmann, 1957).

Halymenia heterocarpa Chauvin ex Duby (1830), p. 942.
Nitophyllum versicolor Harvey (1841), p. 59.
Myriogramme versicolor (Harvey) Kylin (1924), p. 58.
Nitophyllum heterocarpum (Chauvin ex Duby) L'Hardy-Halos (1973), p. 33.
Haraldiophyllum versicolor (Chauvin ex Duby) Zinova (1981), p. 14.
Haraldiophyllum heterocarpum (Chauvin ex Duby) Wynne (1983), p. 444.

Thalli 5-7 cm high and up to 9 cm wide, attached by branched holdfast up to 10 mm in diameter consisting of discoid primary holdfast and rhizome-like prostrate axes 3 mm in diameter that form rhizoidal outgrowths and give rise to new blades; *blades* formed singly or in small groups, each initially terminal on a cylindrical stipe 5-10 mm long and 1 mm in diameter, later formed on old branched stipes up to 2.5 mm in diameter and in dense tufts on remnants of old blades, more-or-less erect when young, fan-shaped, dichotomously to irregularly lobed or deeply divided into wedge-shaped segments with entire or toothed margins, later decumbent and developing opaque thickenings in a triangular basal area, which extends out into the blade as rays, and forming oblong to oval callus-like growths at blade apices and margins, 1-3 mm wide x 2-3 mm long, that give rise to branched rhizoids when in contact with the substratum; rose-pink to bright red in colour, translucent, with cells visible to the naked eye as a granular pattern; texture of blades membranous, fairly rigid but tearing and creasing easily, stipes cartilaginous.

Blade laminae growing from rectangular marginal cells 10-20 µm long x 9-22 µm wide, mostly monostromatic, about 55 µm thick, lacking veins, composed when mature of elongate-polygonal cells, 80-100 x 32-60 µm, linked by numerous secondary pit connections, up to 1100 µm thick basally, consisting in TS of rows about 20 cells long of little-pigmented quadrate cells 40-70 µm in diameter, densely packed with starch grains; *plastids* varying with the season, appearing during winter-spring as a single dissected plastid in each cell and in summer as a composite structure composed of fused, sausage-shaped units.

Gametophytes unknown. *Tetrasporangial bladelets* specialized, developing on old stipes and thickened blade remnants, stipitate, lanceolate, ovate or reniform, often with pointed tips, or lobed, 0.7-6 mm long x 0.5-2.7 mm wide, deformed into a bowl shape when fertile; *tetrasporangial sori* formed singly in a central position on specialized bladelets or paired laterally, 0.7-3.3 mm long x 0.6-0.8 mm wide, occasionally also formed near base of young blades, often linear, paired on either side and sometimes fusing into arc across blade; tetrasporangia borne on axial and cortical cells, each occupying a

* Although this species is currently known as *Myriogramme heterocarpum* (Chauvin ex Duby) Ernst & Feldmann, this combination was not made by Ernst & Feldmann (1957).

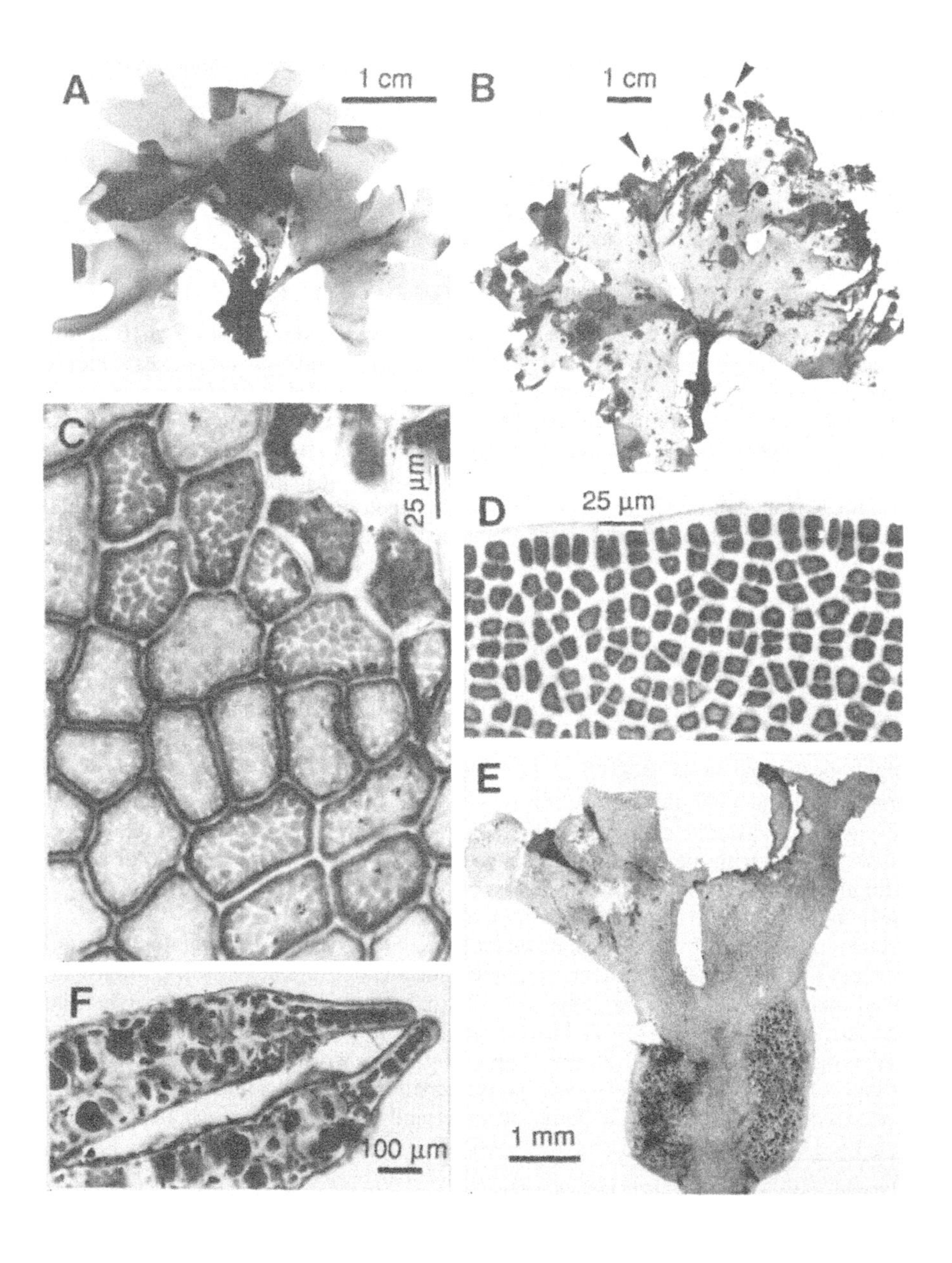
A
1 cm
B
1 cm
C
25 µm
D
25 µm
E
1 mm
F
100 µm

near base of young blades, often linear, paired on either side and sometimes fusing into arc across blade; tetrasporangia borne on axial and cortical cells, each occupying a rectangular locule framed by elongated inner cortical cells, spherical, tetrahedrally divided, 46-52 μm in diameter.

Epilithic on bedrock, rarely in shaded lower-shore pools, typically on vertical faces and under overhangs from 2-30 m depth, also on horizontal, sometimes silty, rock in the deeper part of this range, below the lower limit of kelp, where it is often one of the deepest-growing algae, at moderately to extremely wave-exposed sites, rarely on sheltered subtidal cliffs.

Widely distributed on south-western coasts (Hiscock & Maggs, 1984), eastwards to Dorset and Hampshire (Isle of Wight), northwards to Caernavon; Cork, Kerry, Clare, Donegal and Antrim (Rathlin Island); Channel Isles.

British Isles to N. Spain (South & Tittley, 1986).

Perennial; new blades arise from old stipes and thickened blade bases in April-May, expanding during June-July to about 3 cm in length and forming marginal lobes. Thalli reach their maximum size in July-September, and develop callus-like thickenings. The thickened blades and stipes apparently act as storage tissues and the apical callosities as vegetative propagules. Vegetative reproduction results in the development of dense monospecific stands of *D. heterocarpa* in deep water and on cliffs. Large sterile blades occur until October, when they are heavily epiphytized by bryozoans, then are eroded back to the thickened bases. Tetrasporangial sori are present in early April to early June.

This species shows little form variation other than that associated with seasonal changes.

D. heterocarpa is readily identifiable when the unique apical callosities are present. Younger plants could be confused with large plants of *Rhodophyllis divaricata* (Stackhouse) Papenfuss, but blades are only 1 cell thick in contrast to 2 cells in *R. divaricata*. *D. heterocarpa* has been mistaken for *Haraldiophyllum bonnemaisonii*, but *H. bonnemaisonii* thalli nearly always bear scattered reproductive structures, whereas *D. heterocarpa* bears tetrasporangia only at the bases of young blades and on specialized bladelets. These species can also be distinguished by the plate-like plastids in *H. bonnemaisonii* compared to their convoluted form in *D. heterocarpa*.

Fig. 73. *Drachiella heterocarpa*

(A) Habit of plant forming young blades on old stipe (Donegal, May). (B) Mature blades bearing numerous epiphytes, with basal rays and marginal callus-like thickenings (arrows) (Cornwall, Sep.). (C) Blade cells with convoluted plastids (Donegal, Aug.). (D) Blade margin showing secondary pit connections (D-E Pembroke, Dec.). (E) Tetrasporangial bladelet with paired sori. (F) T.S. of tetrasporangial bladelet showing monostromatic margins (Pembroke, May).

Drachiella minuta (Kylin) Maggs & Hommersand, comb. nov.

Lectotype: Kylin (1924), fig. 44a, based on material from Naples, May 1924.*

Myriogramme minuta Kylin (1924), p. 56.

Thalli 0.5-5 cm wide and 0.7-5 cm high, erect to mainly prostrate, attached by prostrate rhizome-like cylindrical axes 200-250 µm in diameter that form rhizoidal haptera; *blades* developing from rhizomes and remnants of old blades, attached by compressed stipe 1-2 mm long and 0.3 mm wide, initially erect, lanceolate to ovate, becoming decumbent, fan-shaped, sometimes deeply divided into several segments that expand towards irregularly dichotomously lobed apices, with entire or fringed margins giving rise to branching rhizoidal extensions that reattach to the substratum; rose-pink to wine-red in colour, occasionally with a strong blue iridescence, membranous in texture, delicate, tearing and creasing easily; stipes and stolons cartilaginous.

Blade laminae growing from square or rectangular marginal cells 6-14 µm long x 4-8 µm wide, lacking veins, tristromatic when mature, 50-60 µm in thickness, increasing in thickness near stipe to 130 µm and 4-5 cells thick, composed of cells that are elongate-polygonal in face view, 14-44 µm long x 8-24 µm wide, linked by numerous conspicuous secondary pit connections; plastids occuring singly in surface cells, parietal, plate-like with lobed margins and sinuous indentations, several present in internal cells.

Gametophytes unknown in British Isles. *Tetrasporangial sori* subapical, just inside the marginal lobes, round to elongate, 120-360 µm long, 75-250 µm wide; tetrasporangia spherical, tetrahedrally divided, 28-52 µm in diameter.

Epilithic on bedrock, epiphytic on small perennial algae and epizoic on sessile invertebrates in deep shaded lower-shore pools and subtidal to 15 m at very wave-exposed sites.

Known in the British Isles from two localities in N. Devon (Lundy and a mainland site)

Fig. 74. *Drachiella minuta*

(A) Plant with large fan-shaped blade (Clare, July). (B) Plant with much-dissected blades, strongly iridescent (Cornwall, Sep.). (C) Small blades collected intertidally (C-G Channel Isles, Sep.). (D) Rhizoidal proliferations from margin. (E) T.S. of tristromatic blade. (F) Blade margin showing secondary pit connections. (G) Developing tetrasporangial sori. (H) Cells with single strongly indented plastids (as B). (I) Mature tetrasporangial sorus (as A).

---

* The only material in Kylin's herbarium at LD was collected in 1929; we have not designated this as neotype because all specimens are non-reproductive.

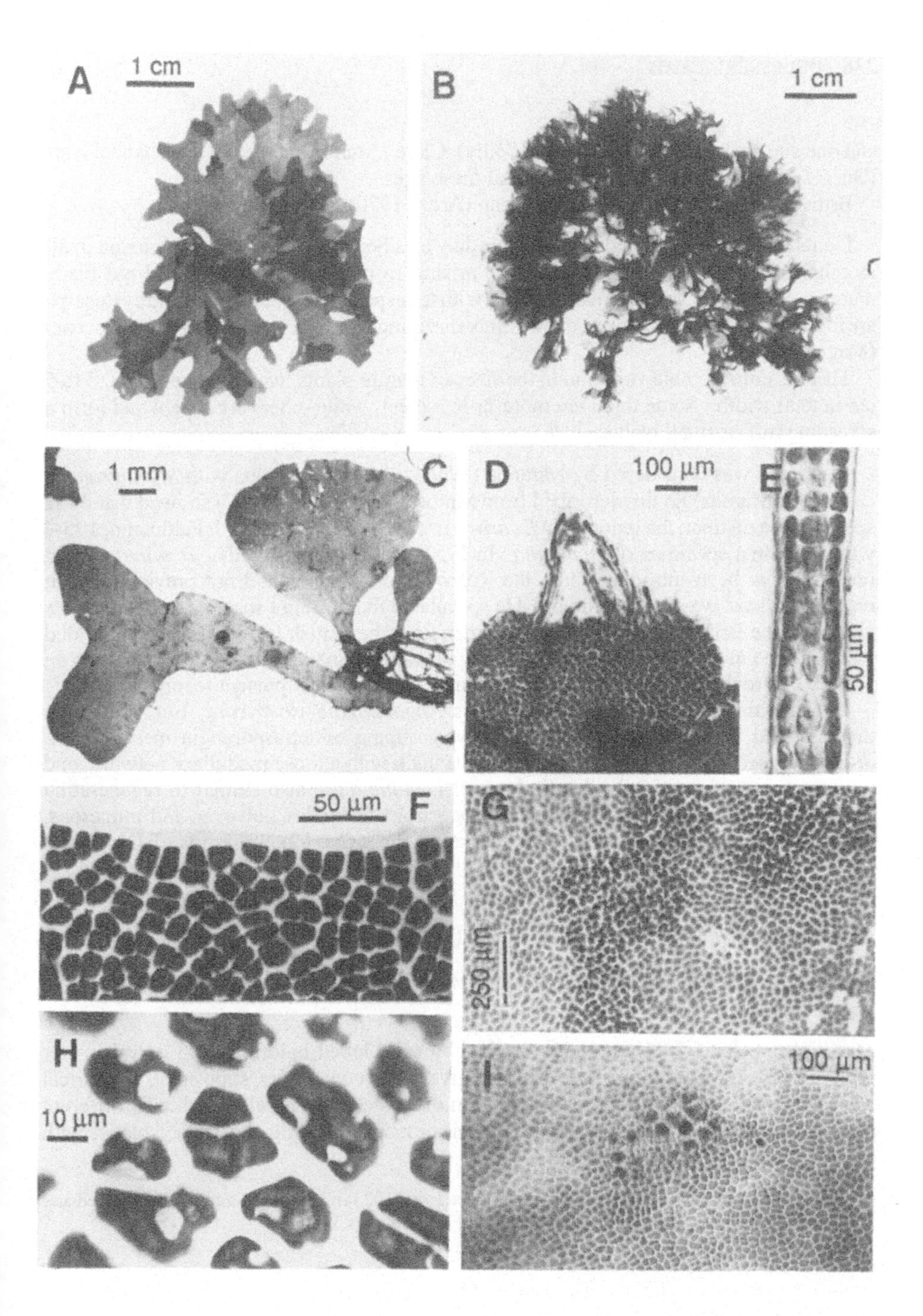
A
1 cm
B
1 cm
1 mm
C
D
100 µm
E
50 µm
50 µm
F
G
250 µm
H
10 µm
I
100 µm

and one site each in Cornwall (Isles of Scilly), Clare (Aran Islands), and the Channel Isles (Guernsey); it appeared to be rare at all of these sites.
British Isles to Morocco; Mediterranean (Ardré 1970).

Plants have been collected only in June, July and September, with tetrasporangia in all months. In September, new blades were arising from prostrate stolons and old blade margins. In Brittany, France, mature plants with tetrasporangia, gametangia and cystocarps are found from August-October, a few plants surviving to January when new growth occurs (Magne, 1957).
There is considerable variation in the sizes of mature plants, which range from 0.5 to 5 cm in total width. Some thalli are more-or-less erect, while others are repent and form a spreading turf of small blades.

D. *minuta* was considered by Miranda (1936) to be synonymous with *Myriogramme carnea* (Rodriguez) Kylin, described from Minorca, but Magne (1957) showed that these species were distinct; the record of *M. carnea* from Roscoff, Brittany (J. Feldmann, 1954) was based on a specimen of *D. minuta*. In 1957, Magne suggested that *D. minuta* might recently have been introduced into the Roscoff area, since it had not previously been reported and yet was quite common. He speculated that it might subsequently extend its range into the British Isles. The first specimen from the British Isles that we have located was collected in 1978, but *D. minuta* is easily overlooked in the field and it may have been present before this; likewise its distribution may be wider than present records indicate.
In general habit, *D. minuta* resembles *Rhodophyllis divaricata*, but it can be distinguished by the tristromatic blades and grouping of tetrasporangia in sori; in *R. divaricata* (see Vol. 1, 1), blades are two cells thick with a loose medullary network, and tetrasporangia are scattered. Small plants of *D. minuta* are also similar to regenerating *Nitophyllum punctatum* (q. v.), which has monostromatic blades, however, and numerous, small and discoid to bacilloid plastids instead of the large, lobed plastids of *Drachiella* species.

Drachiella spectabilis Ernst & J. Feldmann (1957), p. 446.

Lectotype: PC, J. Feldmann 9060. Isotype: BM, Ernst 1023. France (Les Cochons Noirs, Roscoff), 28 xii 1954, *Ernst.**

Thalli semi-peltate, 2-14 cm wide, attached by solid lobed holdfast 2-5.5 mm diameter with prostrate rhizome-like outgrowths that give rise to new blades; stipe erect, cylindrical to compressed, 5-10 mm long and 1-2.5 mm thick, bearing lateral branches of different ages, main and lateral stipes terminating in one or more spreading, fan-shaped blades with

---

* Ernst & Feldmann (1957) designated as holotype Ernst 1023 in PC, but this cannot be located and a lectotype has been chosen from the isotypes (F. Ardré, pers. comm.).

entire margins, deeply dissected into several wedge-shaped segments that expand outwards and divide irregularly into undulating ribbon-like segments with pointed tips, these curving downwards and reattaching to the substratum by branching rhizoidal extensions; blades reddish-purple, with brilliant blue and purple iridescence, translucent except for an opaque patch near the stipe that extends outwards as faint rays; membranous in texture, fairly rigid but tearing and creasing easily; stipe tough and cartilaginous.

Blade laminae growing by marginal meristems and by intercalary divisions, lacking veins, increasing in thickness from 65 μm near apices to 80 μm when mature, young blades monostromatic, older blades three cells thick, surface cells elongate-polygonal in face view, with abundant secondary pit connections, 24-84 μm long x 24-40 μm wide, blade thickness increasing near stipe to 450 μm, 8-10 cells thick, internal cells starch-filled, little-pigmented, cuboid to rounded in TS, 36-62 μm wide and high; blade cells multinucleate; *plastids* parietal, reticulate, ribbon-like to irregularly forked and lobed, several per cell.

Male and female plants unknown. *Tetrasporangial bladelets* specialized, developing on blade remnants 2-4 cm in length, along margins, on upper surface, and occasionally on lower surface, stipitate, initially oval, occasionally divided, 1.2-1.7 mm long, with a single central sorus, later forming a constriction above the sorus and renewing growth, increasing to 5.5 mm in length; *tetrasporangial sori* initially oval, sometimes extending upwards into the constriction, 0.7-1.6 mm wide, up to 3 mm long, and 200 μm thick, consisting of the axial layer and about 6 layers of cortical cells; tetrasporangia formed in two layers by cortical cells, ovoid or nearly spherical, 60-70 μm long x 44-54 μm wide.

Forming large patches on bedrock and boulders, particularly on horizontal to gently sloping rock, less commonly on cliffs, at 2-30 m depth, in *Laminaria hyperborea* forest and below the lower limit of kelp, at moderately to extremely wave-exposed sites.

South-western and western coasts, most commonly around offshore islands and reefs: Cornwall, S. Devon, N. Devon (Lundy), Pembroke, Caernavon, Argyll (Islay); Cork, Clare, Mayo, Donegal, Antrim; Channel Isles.

British Isles to N. France.

Perennial; new blades arise from old stipes in April-May, reach a maximum size in July-September and are usually reduced to small remnants by November. Young blades are erect, lobed and strongly iridescent; older blades bend downwards and develop pointed apices. Iridescence has disappeared by September and blades become heavily epiphytized by bryozoans. Tetrasporangial bladelets are present on old blade remnants in April and May.

There is little variation in form, other than that associated with seasonal changes.

Gonimocolax roscoffensis J. Feldmann & G. Feldmann (1961, p. 18) was described as a parasite of *D. spectabilis* plants collected at Roscoff, Brittany. It has not been observed in the British Isles, but small galls are common along blade margins of *D. spectabilis*.

The brilliant purple and blue iridescence of young *D. spectabilis* blades is easily recognizable underwater. Older fronds, and preserved specimens, are often similar in

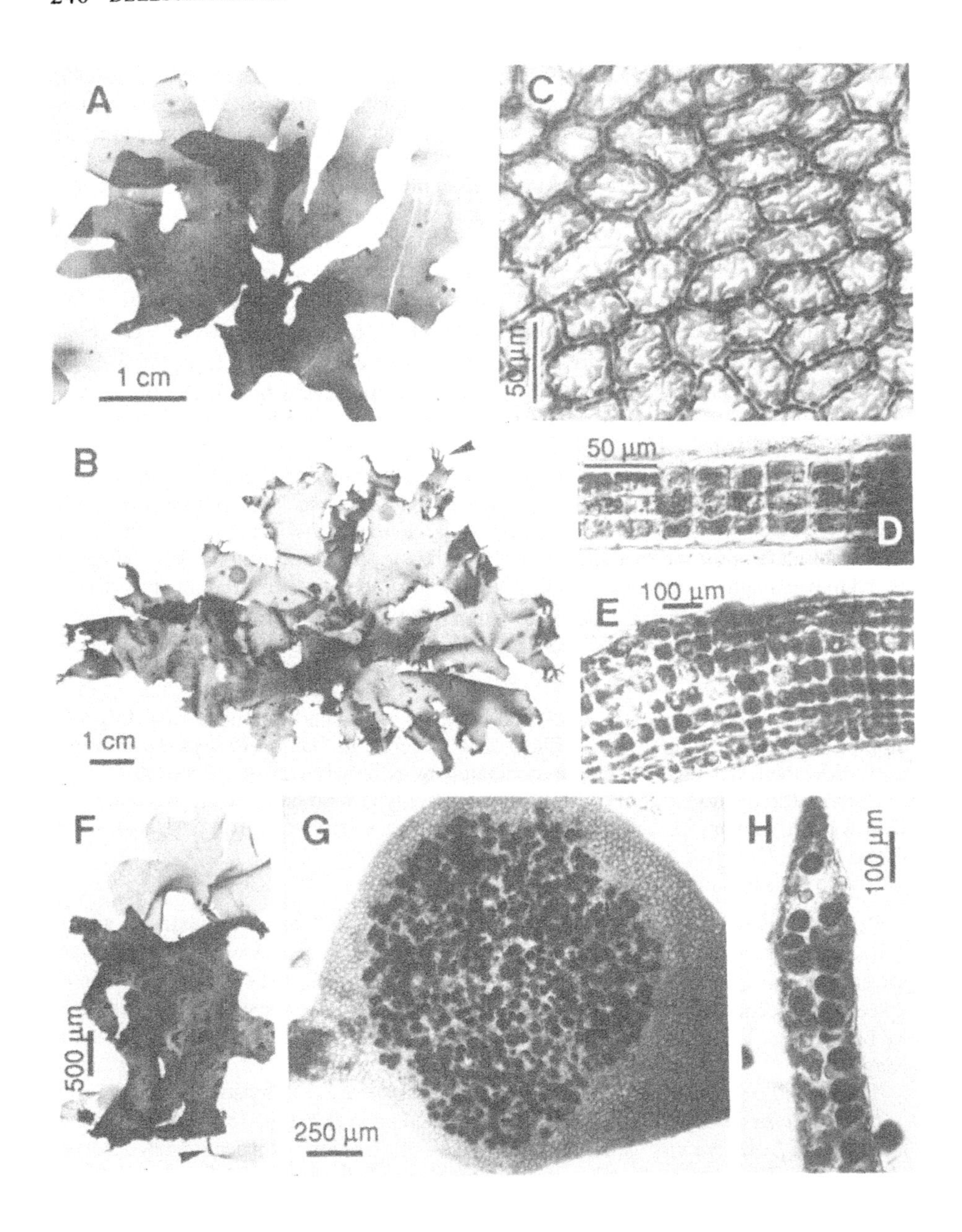
A
1 cm
C
50 µm
B
1 cm
50 µm
D
100 µm
E
F
500 µm
G
250 µm
H
100 µm

shape and texture to large plants of *Rhodophyllis divaricata*. They can be distinguished from *R. divaricata* by the tristromatic fronds, convoluted, ribbon-like plastids, and the formation of tetrasporangia only on special bladelets. *R. divaricata* is two cells thick, with sparse medullary filaments, has discoid plastids, and forms tetrasporangia and conspicuous spherical cystocarps on mature blades.

MYRIOGRAMME GROUP

HARALDIOPHYLLUM Zinova

HARALDIOPHYLLUM Zinova (1981), p. 13.

Type species: *H. bonnemaisonii* (Kylin) A. Zinova (1981), p. 13.

Thalli erect, to 15 cm high, consisting of one to several fan-shaped, dichotomously divided blades with rounded tips, arising from a discoid primary holdfast and short cylindrical stipes; blades thin, membranous, monostromatic above, polystromatic towards the base and in fertile ares, with cells arranged in horizontal tiers and vertical rows; microscopic veins absent, laminae eroding basally in older plants exposing a tough stipe-like midrib, sometimes bearing secondary proliferations.

Growth diffuse, by means of marginal and intercalary meristems, the divisions often perpendicular to one another, producing clusters of rectangular cells that elongate or become polygonal after enlarging; meristematic cells uninucleate or multinucleate, the nuclei initially median, later arranged in two rings on either side of median plane along the side walls, ultimately lying in a broad band along side walls between the chloroplasts; primary and secondary pit connections initially median, later formed near cell surfaces, especially abundant in polystromatic areas; plastids parietal, initially simple, later dissected into platelets.

Gametophytes dioecious. Spermatangial sori circular to elliptical or coalescent, inside sterile margin, formed on both sides of blade by anticlinal and periclinal divisions of mature, multinucleate vegetative cells. Procarps formed on both sides of blade near margin, directly opposite each other, the central cell first cutting off sterile vegetative cells

---

Fig. 75. *Drachiella spectabilis*

(A) Thallus with well-developed blades (Mayo, June). (B) Plant with old blades bearing marginal rhizoidal proliferations (arrow) (Cornwall, Sep.). (C) Cells with strongly convoluted plastids (C-E Donegal, Aug.). (D) T.S. of tristromatic blade. (E) T.S. of thick blade near base. (F) Old blade remnant bearing young blades, one with a basal tetrasporangial sorus (arrow) (F-H Donegal, May.) (G) Detail of tetrasporangial bladelet with single sorus. (H) T.S. of tetrasporangial sorus.

(cover cells) apically, followed by the fertile periaxial cells (supporting cells); mature procarp consisting of the supporting cell, a 1-2-celled lateral sterile group on one side, a 4-celled carpogonial branch on the other, and a 1-celled basal sterile group; fertilized carpogonium cutting off a distal and a posteriolateral connecting cell, the latter fusing with a process from the auxiliary cell; auxiliary cell dividing into an anterior gonimoblast initial containing a diploid nucleus and a posterior residual cell (foot cell) containing a diploid and 1 or 2 haploid nuclei; sterile groups enlarging but not dividing after fertilization; gonimoblast filaments initially branching pseudodichotomously in plane of blade, encircling supporting cell, then radiating outwards and bearing solitary terminal, clavate to pyriform carposporangia, thereafter the gonimoblasts branching sympodially from subterminal cells below the maturing carposporangia; fusion cell large, incorporating supporting cell, central cell, foot cell and gonimoblast initial and extending radially to include inner gonimoblast filaments and cells in floor of cystocarp; pericarp 4-5 cell layers thick with a non-projecting ostiole overlying the first sterile group. Tetrasporangial sori circular to elliptical, sometimes coalescent inside sterile margin; formed through dedifferentiation and anticlinal and periclinal division of mature vegetative cells, up to 5 cell layers thick; tetrasporangia originating primarily from central cells, arranged in a zig-zag on both sides of the central layer.

Reference: Kylin (1924).

In the most recent checklist for the British Isles (South & Tittley, 1986), two species of *Myriogramme* were listed. *M. heterocarpum* has now been transferred to *Drachiella*. The other species, *M. bonnemaisonii* is here transferred to *Haraldiophyllum*. *Myriogramme bonnemaisonii* Kylin should be regarded as a new species validated by Greville's description, and is not based on *Delesseria bonnemaisonii* C. Agardh (P. Silva, pers. comm.).

Haraldiophyllum bonnemaisonii (Kylin) A. Zinova (1981), p. 13.

Lectotype: E. Syntype: LD 30318. Orkney, 1826, *Clouston*.

Myriogramme bonnemaisonii Kylin (1924), p. 258.
Delesseria bonnemaisonii sensu Greville (1827), p. 322, non *Delesseria bonnemaisonii* C. Agardh (1822), p. 186*.

Thalli 2-15 cm high and 3-20 cm broad, consisting of one or more blades attached by small solid holdfast with prostrate rhizoidal outgrowths; *blades* fan-shaped, terminal on short cylindrical stipe 1-2 mm long and 1-1.2 mm thick, more-or-less regularly dichotomously divided into overlapping lobes with rounded or obtuse apices and entire or dentate

* See under *Polyneura bonnemaisonii*.

margins, frequently proliferous when older, sometimes cleft nearly to base into long ribbon-like segments, with one or more midrib-like or nerve-like thickenings developing basally and extending upwards, eroding to simple or branched stipe-like thickened nerves up to 10 mm long with attached blade remnants; rose-pink to brownish-red in colour, becoming darker with age, and drying purplish-red; texture of blades membranous and fairly rigid, stipes cartilaginous.

Blade laminae lacking veins, monostromatic when young, increasing in thickness from 40 μm near apices to 150 μm basally, up to 4 or more cells thick, up to 250 μm thick in the ray-like thickenings, composed of cells elongate-polygonal in surface view, 30-90 μm long x 30-45 μm wide, multinucleate, linked by secondary pit connections that increase in abundance in polystromatic regions; *plastids* plate-like to bacilloid.

Gametophytes dioecious. *Spermatangial sori* formed on younger parts of blades, initially round to oval, 0.25-1.8 x 0.25-0.6 mm, each sorus later expanding up to 5 x 5 mm, covering much of the apical third of blades, consisting on each side of blade of a layer of cuboid spermatangial mother cells 7-10.5 μm wide, bearing elongate-conical spermatangia 8-9 μm long x 4-5 μm wide. *Procarps* numerous, formed on both sides of younger blades; *cystocarps* scattered, hemispherical, 430-600 μm in diameter when mature, with a smooth outer pericarp and non-protruding ostiole about 25 μm in diameter, containing a fusion cell c. 130 μm in diameter and little-branched gonimoblast filaments, 12-14 μm in diameter, terminating in single mature clavate to pyriform carposporangia, 56-75 μm long x 25-40 μm wide. *Tetrasporangial sori* numerous, scattered over upper two-thirds of blades and on proliferations from the margins of old blades, round to oval, occasionally coalescing, 0.24-1.2 mm long x 0.2-0.6 mm wide, protruding on either side of blades, c. 150 μm thick; tetrasporangia spherical, 50-70 μm in diameter.

Epilithic on bedrock and boulders in strongly scoured or current-exposed habitats, more commonly epiphytic on *Laminaria hyperborea* stipes and holdfasts, from extreme low water to 21 m depth, at moderately to extremely wave-exposed sites.

South-west and west coasts, northwards to Shetland, widely distributed in Ireland; rarely reported from E. Scotland, possibly absent from E. England.

Shetland to N. Spain (South & Tittley, 1986); tentative records from S. and W. Africa (Wynne, 1986) probably represent another entity.

Basal parts of thalli are probably perennial, forming new blades in early spring. Spermatangia recorded in April, August and September; cystocarps in June, July, October and November; tetrasporangia in April and June-October.

Variation in overall morphology appears to be related in part to habitat. Thalli growing on kelp stipes are divided into numerous narrow, wedge-shaped segments, often badly shredded, whereas epilithic plants form broad fan-shaped blades.

Specimens of *Drachiella* species may resemble *H. bonnemaisonii*, but they have large ribbon-like to convoluted plastids rather than small discoid plastids. *D. heterocarpa* and *D. spectabilis* do not form reproductive structures on mature blades but only on small

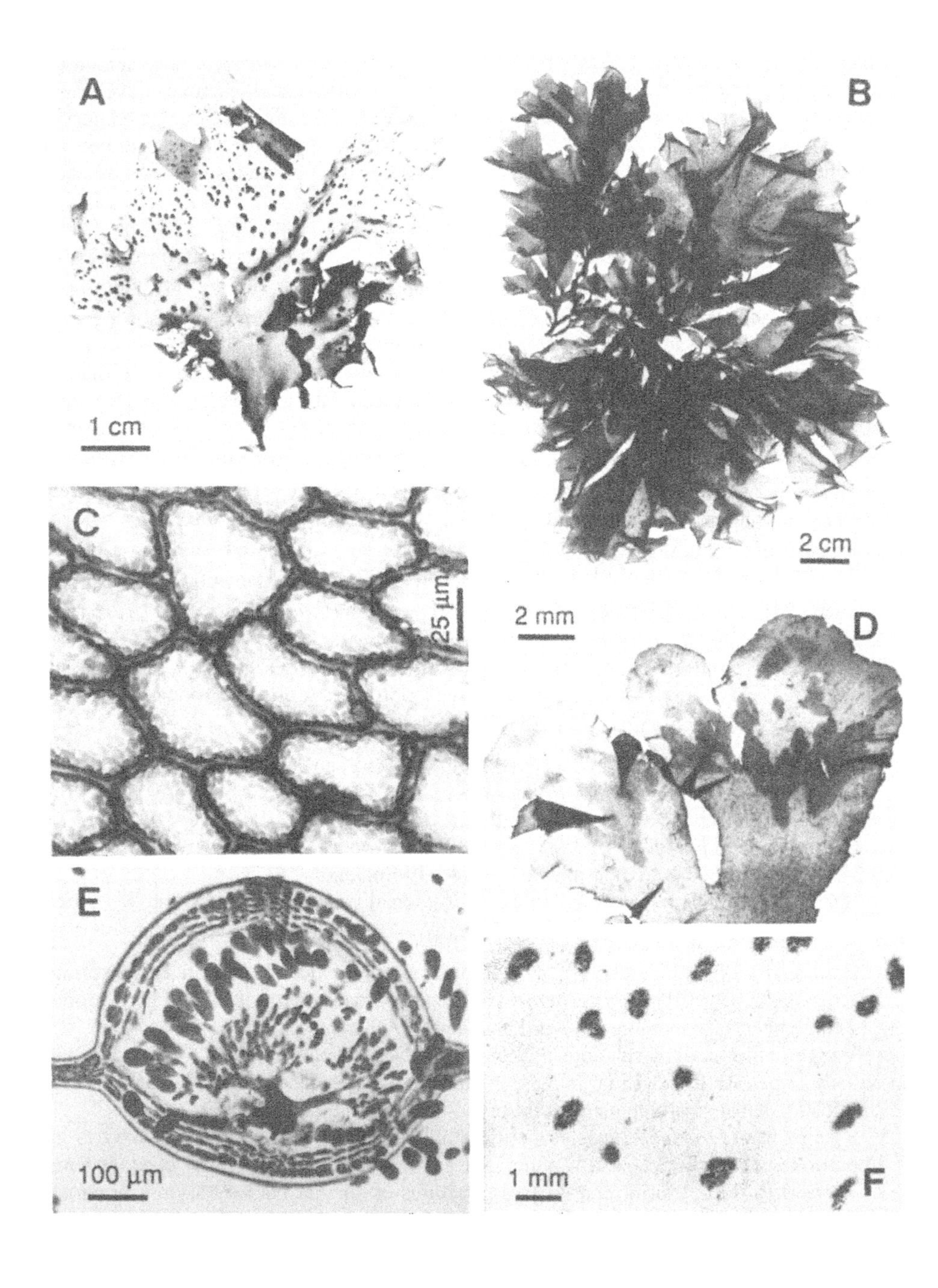
A
B
1 cm
2 cm
C
25 µm
2 mm
D
E
100 µm
1 mm
F

bladelets in winter; *D. minuta* thalli are less than 5 cm wide. Specimens of *H. bonnemaisonii* are occasionally misidentified as *Erythroglossum laciniatum* (q. v.), but *E. laciniatum* blades form a network of veins in contrast to the lack of veins in *H. bonnemaisonii*. Broad fan-shaped plants of *Nitophyllum punctatum* (q. v.) may closely resemble *H. bonnemaisonii*, but reproductive structures are larger in *N. punctatum*: cystocarps are ≥650 µm diameter, with a projecting ostiole, in contrast to ≤600 µm, lacking a projecting ostiole, in *H. bonnemaisonii*, and tetrasporangia are 110-175 µm diameter as opposed to 50-70 µm in *H. bonnemaisonii*.

CRYPTOPLEURA GROUP

CRYPTOPLEURA Kützing, nom. cons.

CRYPTOPLEURA Kützing (1843), p. 444.

Type species: *C. lacerata* (S. G. Gmelin) Kützing [= *C. ramosa* (Hudson) Kylin ex Newton (1931), p. 69].

Thalli consisting of prostrate and erect membranous blades, or erect blades absent; prostrate blades attached from lower sides by short to long, peg-like haptera composed of bundles of unicellular or multicellular rhizoids; erect blades up to 30 cm high, strap-shaped to flabellate, subdichotomously branched or lacerated with rounded apices, and smooth or ruffled, entire or serrulate margins, with or without lateral proliferations, monostromatic above, at least near apex, polystromatic below; microscopic and macroscopic veins present, either conspicuous or obscure, subdichotomously branched with free or anastomosing ends; lamina eroding basally in larger species, exposing a tough, stipe-like midrib.

Growth initiated by triangular marginal cells with two cutting faces and continued by marginal and intercalary meristems, meristematic cells uninucleate or multinucleate, the nuclei initially median as seen in optical section, later spreading uniformly beneath the chloroplasts; microscopic veins differentiating in acropetal direction, flabellately

---

Fig. 76. *Haraldiophyllum bonnemaisonii*

(A) Tetrasporophyte with numerous oval sori (Inverness, July). (B) Large, much-branched thallus (B-C & F Antrim, June). (C) Cells with numerous small plastids. (D) Irregularly shaped spermatangial sori (Antrim, Apr.). (E) T.S. of mature cystocarp showing conspicuous, branched central fusion cell, a single layer of mature carposporangia, and smooth pericarp with non-protuberant ostiole (Sutherland, Aug.). (F) Tetrasporangial sori on lamina lacking veins.

branched, the ends free or anastomosing; initially monostromatic, becoming tristromatic, with the central cells elongate, tubular, linked by broad, secondarily thickened primary pit connections, ultimately transforming into macroscopic nerves by lateral extension of tristromatic tissues bordering veins which become polystromatic; macroscopic veins 5-12 cell layers thick with the cells arranged in horizontal tiers and vertical rows; plastids parietal, dissected into minute, discoid platelets.

Reproductive structures on ordinary blades or marginal proliferations. Gametophytes dioecious. Spermatangial sori lunate, elliptical or linear, formed behind apex or inside smooth or ruffled margins. Procarps formed on both sides of blade near apex, consisting of a supporting cell, a 1-2(-3) celled sterile group-1, a curved, 4-celled carpogonal branch, and a 1-2 -celled sterile group-2; sterile groups usually dividing once after fertilization, remaining distinct and enlarging beneath ostiole; auxiliary cell large, dividing horizontally to form an initial cell row; gonimoblasts repeatedly subdichotomously branched; carposporangia terminal, spherical to ellipsoidal, maturing singly in basipetal succession; fusion cell moderately large, incorporating several central cells, the supporting cell and innermost gonimoblast cells; pericarp 5-6 cell layers thick with a non-projecting ostiole. Tetrasporangial sori lunate, ovate or linear, formed behind apex or inside smooth or ruffled margins, polystromatic, up to 7 layers thick; tetrasporangia originating from central cells and inner cortical cells, arranged in two layers at maturity; tetrasporangia tetrahedrally to irregularly divided.

Reference: Kylin (1924).

One species in the British Isles.

Cryptopleura ramosa (Hudson) Kylin ex Newton (1931), p. 332*.

Neotype: BM-K, ex Herb. Hudson. Yorkshire (Scarborough), undated, *Frankland.*

Ulva ramosa Hudson (1762), p. 476.
Fucus laceratus S.G. Gmelin (1768), p. 179.

Fig. 77. *Cryptopleura ramosa*

(A) Young plant from bedrock (Donegal, May). (B) Large thallus with marginal hooks (arrows) (Clare, June). (C) Thallus collected from kelp stipe (C-D as A). (D) Microscopic veins in blade lamina. (E, F) T.S. of blades through monostromatic lamina and polystromatic ribs (Sutherland, July). (G) Elongate marginal spermatangial sori (s) (Pembroke, Feb.). (H) T.S. of spermatangial sorus with mature spermatangia (Cornwall, Sep.).

* Newton attributed the new combination to Kylin, without explanation.

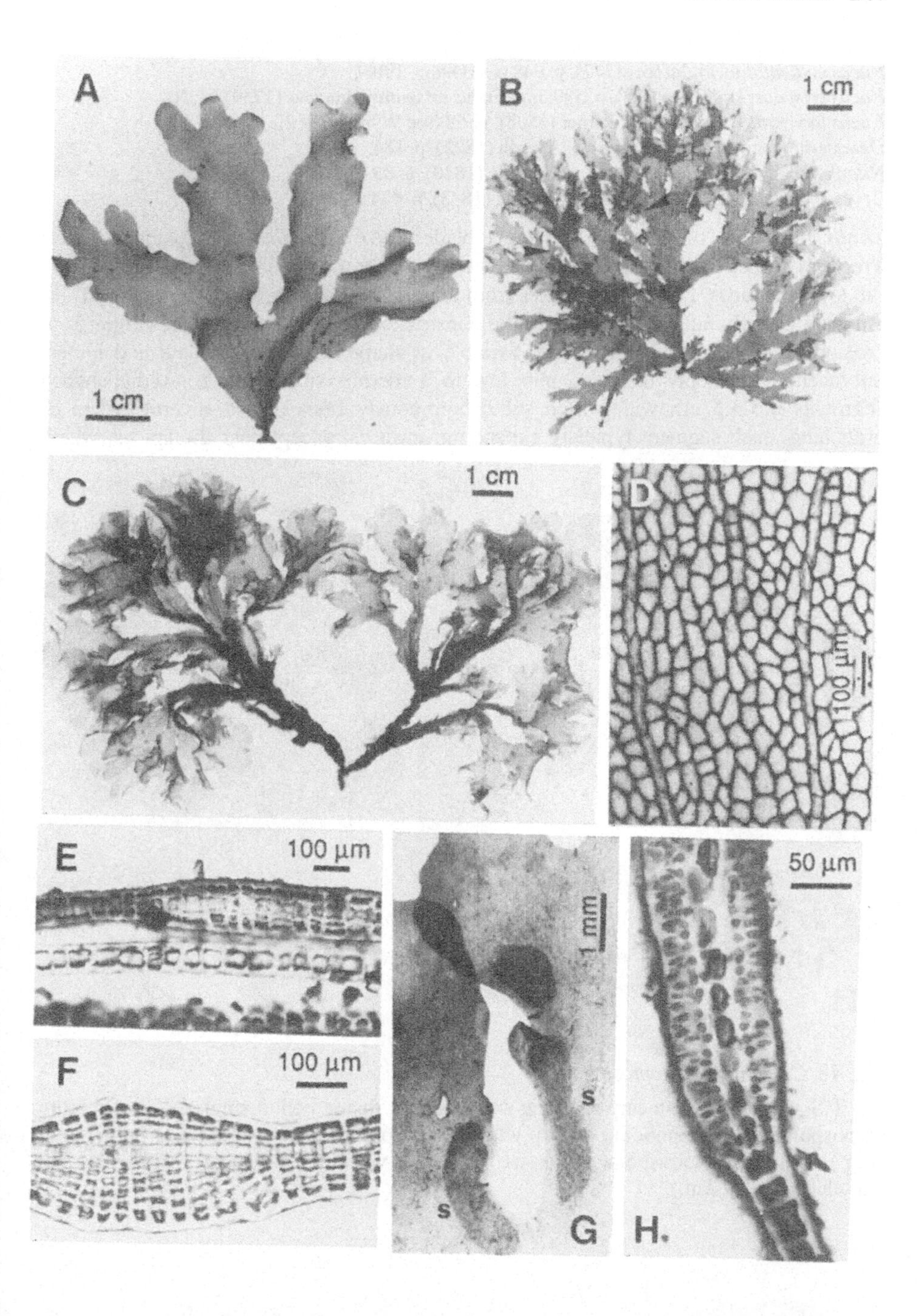
A
1 cm
B
1 cm
C
1 cm
D
100 µm
E
100 µm
F
100 µm
G
1 mm
s
s
H.
50 µm

Fucus endiviifolius Lightfoot (1777), p. 948 (see Dixon, 1983).
Fucus crispatus Hudson (1778), p. 580, non *Fucus crispatus* Linnaeus (1759), p. 718.
Fucus laceratus var. *uncinatus* Turner (1808), p. 68 (see Wynne, 1989).
Delesseria lacerata (S.G. Gmelin) C. Agardh (1822), p. 184.
Nitophyllum laceratum (S.G. Gmelin) Greville (1830), p. 83.
Cryptopleura lacerata (S.G. Gmelin) Kützing (1843), p. 444.

Thalli consisting of prostrate and erect blades; *prostrate blades* strap-shaped, with irregularly lobed margins and apices, attached by numerous peg-like haptera formed from the lower surfaces, forming erect axes from margins and apices; *erect blades* 3-15 (-20) cm high, usually attached by a stipe-like constriction about 1 mm thick resulting from erosion of the lamina basally, very variable in shape, broadly flabellate and almost unbranched to deeply dissected into few to numerous ribbon-like to wedge-shaped segments 0.5-3.5 cm wide, often subdichotomously branched to several orders of branching, each segment typically expanding upwards, except near the apices, where

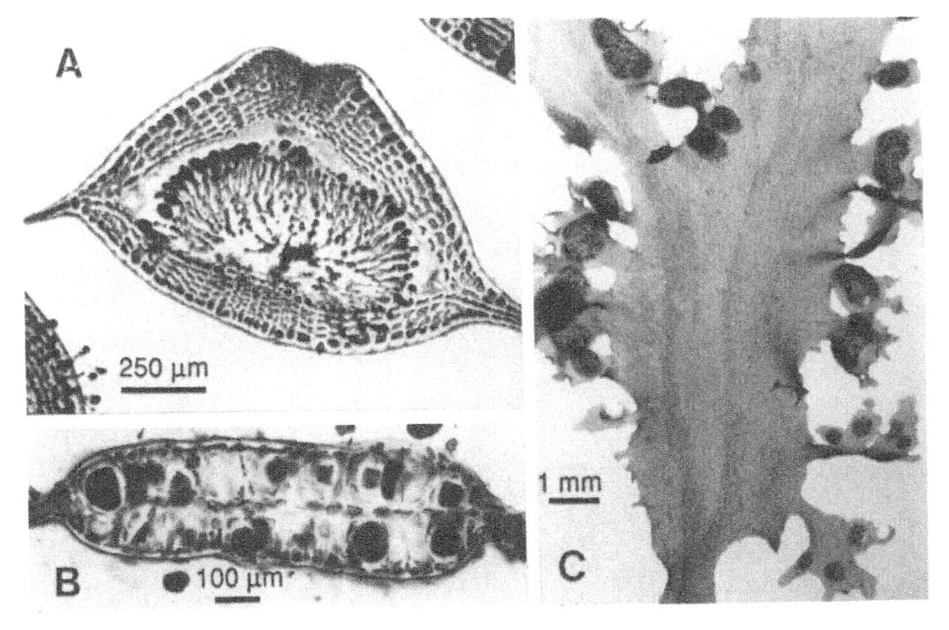

Fig. 78. *Cryptopleura ramosa*

(A) T.S. of mature cystocarp showing small central fusion cell, a single layer of mature carposporangia, and smooth pericarp with slightly protuberant ostiole (A-B Sutherland, Aug.). (B) T.S. of tetrasporangial sorus. (C) Blade with tetrasporangial sori in marginal bladelets (Down, Jan.).

segments taper gradually to rounded tips; *blade margins* smooth, ruffled, lobed, or bearing hooks with a thickened inner margin; *nerves* usually conspicuous near base of blade, up to 0.8 mm wide, one or more, simple or branched, disappearing towards apices; *veins* fine, usually visible against the light as a more-or-less parallel array; colour brownish-red, less often pink, or bleached yellowish-pink under high-light conditions, frequently with a blue iridescence underwater; texture of young blades membranous, older blades flexible and fairly tough.

Blade laminae monostromatic, about 40 µm thick, composed of rectangular to polygonal cells 30-75 µm long x 20-45 µm wide, with marginal cells triangular to rectangular or wedge-shaped, increasing in thickness with age to 8 or more cells thick, 200 µm in thickness, with the nerves up to 12 or more cells thick; *microscopic veins* branching and anastomosing at irregular intervals, 3 cells thick, composed of elongate cells 40-90 µm long x 12-20 µm wide; *plastids* discoid.

Spermatangial sori formed along margins, just behind apices, or in small marginal outgrowths, pale pink in colour, round, elongate or semi-circular, 0.5-10 mm x 0.5-3 mm, when mature twisting or deforming blades into deep ruffles, developing on both sides of blade, consisting of a layer of small, rounded spermatangial mother cells bearing elongate spermatangia 8-12 µm long x 4.5 µm wide. *Cystocarps* scattered on blades, often near apices, 900-1300 µm in diameter when mature, with a smooth outer pericarp and non-protruding ostiole about 100 µm in diameter; mature carposporangia ellipsoid, 45-68 µm long x 30-65 µm wide. *Tetrasporangial sori* developing in discontinuous lines inside margins, just behind apices, or singly or catenate in small rounded to elongate marginal outgrowths, round to linear, 0.3-6 mm long x 0.2-2 mm wide, protruding equally on both sides of blade, 160-225 µm in thickness; tetrasporangia 60-95 µm diameter.

Growing on a wide variety of substrata, including large and small algae, pebbles, shells and maerl, in pools and damp places on the lower shore and subtidal to at least 30 m depth, particularly abundant in *Laminaria hyperborea* forest, both on stipes and beneath kelp canopy on bedrock and crustose corallines, occurring at sites with a wide range of environmental conditions, from extremely sheltered to extremely wave-exposed, with or without tidal currents.

Generally distributed in the British Isles, although records for the east coast of England are sparse (Norton, 1985).

S. Norway and Faroes to Spain (South & Tittley, 1986); Mediterranean (Ardré, 1970); ?W. Africa (Price et al., 1986); Brazil (Wynne, 1986).

Mature plants can be found throughout the year, although the majority of blades in kelp forest populations develop in May-June and are badly damaged and encrusted with bryozoa by October. Spermatangia have been recorded in February-November, cystocarps in January and March-December, and tetrasporangia throughout the year. A population on subtidal bedrock at the Isle of Man had its highest proportion of tetrasporangial thalli in July-December; the smaller numbers of female plants with cystocarps were greatest in

October-January (Kain, 1982). Less than 10% of plants were reproductive in March-May. Spermatangial plants are usually less frequent than those with female or tetrasporangial structures (Grubb, 1926; Kain, 1982). Chromosome number is $n = 32$ (Austin, 1956).

Cryptopleura ramosa shows a remarkable degree of variation in overall shape, marginal features, and the development of venation. Blades range from broad and almost undivided to narrow and subdichotomously divided or much-dissected. Erosion of laminae, particularly marked in thalli epiphytic on kelp stipes, results in long branched or unbranched stipe-like structures, bearing regenerating remnants of old blades. Margins are smooth, undulating or occasionally denticulate; marginal hooks are characteristic of the form 'var. *uncinata*' (see Wynne, 1989). It is not known whether hook formation is genetically or environmentally determined. Macroscopic veins can form greatly thickened ribs extending towards apices, or develop only near the base of thalli, or be entirely lacking.

C. ramosa thalli frequently support plants of the parasite *Gonimophyllum buffhamii* (q. v.).

Distinctions between *Cryptopleura ramosa* and *Acrosorium venulosum* are listed under *A. venulosum* (q. v.). *Erythroglossum laciniatum* (q. v.) may closely resemble fan-shaped plants of *C. ramosa*, but can be distinguished by several features, including angular marginal denticulations in young plants. Whereas *E. laciniatum* forms narrow prostrate blades with acute tips and marginal teeth, prostrate *C. ramosa* blades have lobed or smooth margins. In addition, the development of venation in *E. laciniatum* differs from that of *C. ramosa*: multicellular macroscopic veins occur just behind apices whereas in *C. ramosa* these veins are only one cell wide. Thalli of *Polyneura bonnemaisonii* (q. v.) differ from *C. ramosa* in the formation of scattered rather than marginal tetrasporangial and male sori and by the spiny ornamentation on cystocarps, in contrast to the smooth outer wall of *C. ramosa* cystocarps. Prostrate blades are formed only sparsely in *P. bonnemaisonii*, in contrast to *C. ramosa,* and they are attached by marginal haptera rather than the peg-like haptera formed over the lower surfaces of prostrate blades of *C. ramosa*.

GONIMOPHYLLUM Batters

GONIMOPHYLLUM Batters (1892), p. 66.

Type species: *G. buffhamii* Batters (1892), p. 67.

Thalli hemiparasitic on Delesseriaceae belonging to the Cryptopleura group, consisting of small globose clusters of compressed to flattened or membranous blades arising on both surfaces of host from a basal cushion composed of hypertrophied host cells and monosiphonous endophytic filaments that penetrate the host tissue, linking to host cells by means of secondary pit connections; blades monostromatic in sterile regions at the base and margins, polystromatic in fertile areas, up to 6(-10) cell layers thick; microscopic veins present in monostromatic regions towards base of blade; macroscopic veins absent.

Growth initiated by triangular marginal cells with 2 cutting faces, rarely by transversely dividing apical cells, and continued by marginal and intercalary meristems; meristematic cells uninucleate or multinucleate, mature cells multinucleate; microscopic basal veins mostly monostromatic, subdichotomously branched, disappearing into the polystromatic fertile areas above; plastids parietal, dissected into minute platelets.

Blades from individual cushions usually all male, female or tetrasporangial. Spermatangial sori covering both sides of blade except the sterile margin and base. Procarps numerous, scattered over both surfaces of blade, consisting of a 1-2 -celled sterile group-1, a curved carpogonial branch, and a 1(-2) -celled sterile group-2, or sterile group-2 absent; sterile groups usually dividing once after fertilization; gonimoblast filaments subdichotomously branched; carposporangia terminal, pyriform to ellipsoidal, maturing singly in basipetal succession; fusion cell conspicuous, incorporating the central cell, supporting cell and inner gonimoblast cells; pericarp 5-6 cell layers thick, with non-protruding ostiole. Tetrasporangial sori in broad marginal bands or covering blade on both sides except the sterile margin and base, 5-7 or more cell layers thick; tetrasporangia originating from central cells and inner cortical cells or mostly from cortical cells, typically arranged in 2 layers at maturity; polysporangia with 20-50 spores present in one species.

References: Kylin (1924), Wagner (1954), Mendoza (1970).

One species in the British Isles.

Gonimophyllum buffhamii Batters (1892), p. 67.

Lectotype: BM, microscope slide 8746. Syntypes: BM. Kent (Deal), October 1891, *Neeve*.

Thalli consisting of basal tissue endophytic in host, and emergent thallus 1-4.6 mm in diameter, composed of cushion-like pad 1-1.7 mm in diameter bearing a rosette of few to numerous blades; blades rounded, lobed or reniform, typically broader than long, 0.4-3.6 mm long x 0.5-3 mm wide, 120-210 µm in thickness, with a smooth or ruffled margin, pale brown in colour, with a solid texture.

Endophytic filaments consisting of round to ovoid cells, 14-20 µm in diameter and 1-2 diameters long, that penetrate between host cells and form numerous secondary pit connections to them; *emergent tissue* composed of a callus-like development of host cells, with a dense mass of parasitic filaments growing among them, giving rise externally to blades with margins 1 cell thick, rapidly increasing in thickness to 6-10 cells thick in females and tetrasporophytes.

Plants dioecious, all blades becoming reproductive. *Spermatangial sori* formed singly on male blades, surrounded by growing margin 2-6 cells wide, covering both sides of

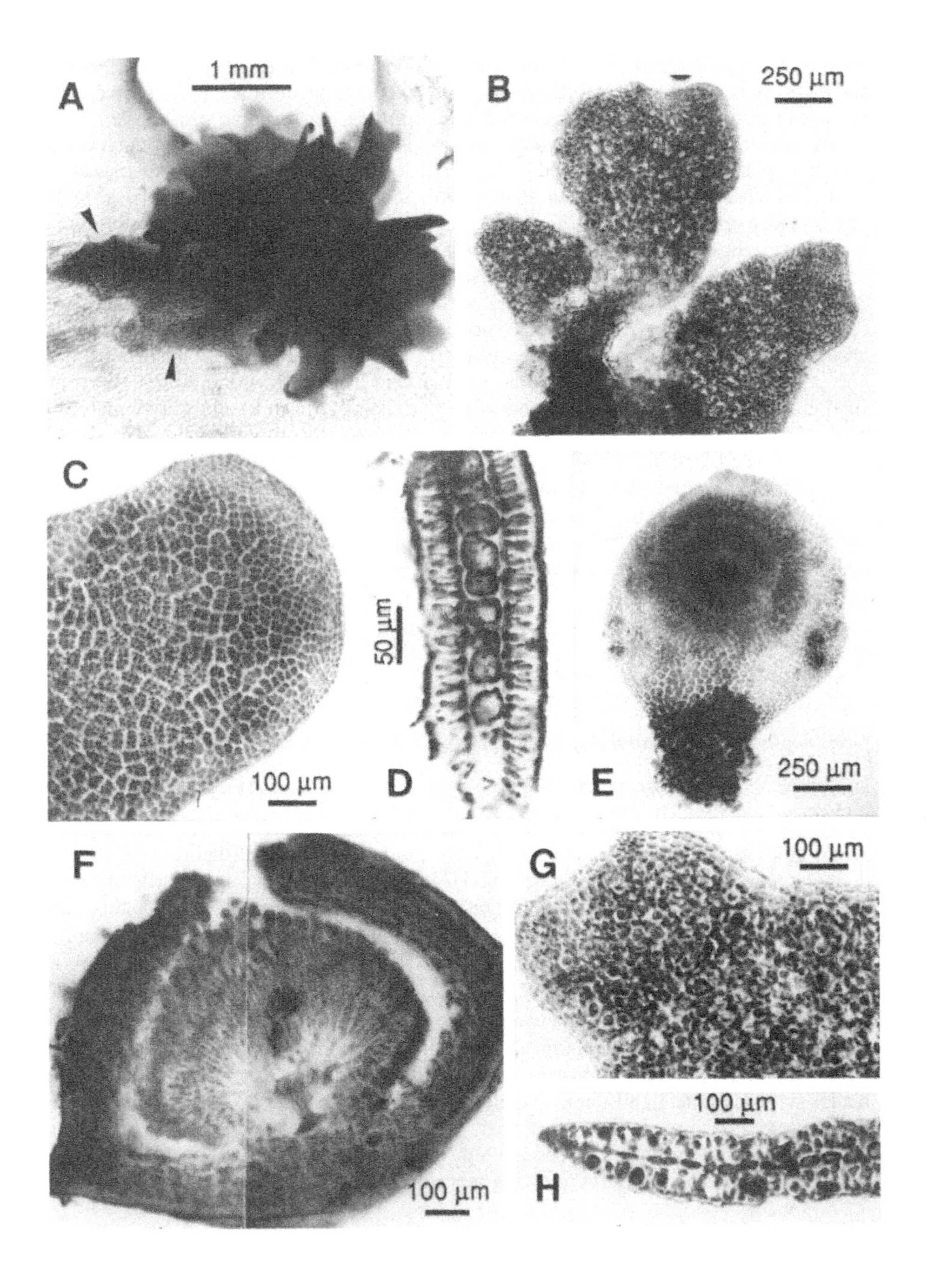
A
1 mm
B
250 µm
C
100 µm
D
50 µm
E
250 µm
F
100 µm
G
100 µm
H
100 µm

Plants dioecious, all blades becoming reproductive. *Spermatangial sori* formed singly on male blades, surrounded by growing margin 2-6 cells wide, covering both sides of blade, consisting of a layer of rounded spermatangial mother cells 6-8 µm in diameter, bearing cylindrical to clavate spermatangia 12-16 µm long x 3-4 µm wide. *Procarps* numerous; *cystocarps* developing 1-4 on each female blade, all orientated downwards to lower surface of blade, 1000 µm in diameter, with non-protruding ostiole about 75 µm in diameter; mature carposporangia pyriform to ellipsoid, 24-36 µm long x 24-30 µm wide. *Tetrasporangial sori* covering both sides of blade except for marginal 3-6 cells; tetrasporangia 60-95 µm in diameter.

Commonly hemiparasitic on all parts of *Cryptopleura ramosa* plants, including margins and laminae of erect fronds, on their reproductive structures, and on prostrate blades, also on other *G. buffhamii* plants, rarely on *Acrosorium venulosum*, on hosts growing on bedrock and mobile substrata at depths of 2-10 m, most frequently at moderately wave-sheltered sites.

South and east coasts of England (Cornwall, Devon, Dorset, Kent), west Wales (Pembroke, Caernavon), apparently rare in Scotland (Inverness); widely distributed in Ireland; Channel Isles.

British Isles to Spain (South & Tittley, 1986).

Mature plants can be found throughout the year, but are less frequent in winter as *C. ramosa* declines in abundance. Spermatangia recorded for February-December, cystocarps in January, March, April and June-December, and tetrasporangia in February, April and June-October.

There is considerable difference in habit between male and female thalli. Male blades are twisted and folded, whereas female blades are usually flat plates, with a glossy surface, that continue to increase in size when procarps remain unfertilized.

Grubb (1926) noted that the anatomy of *G. buffhamii* was extremely similar to that of *C. ramosa*, and suggested that the 'parasites' were abnormal outgrowths from the blades of *C. ramosa*. Although there have been no life history studies on *G. buffhamii*, it is now recognized as an adelphoparasite (J. Feldmann & G. Feldmann, 1958).

Fig. 79. *Gonimophyllum buffhamii*

(A) Male plant on *Cryptopleura ramosa*, with emergent blades and thick endophytic tissue (arrows) (Kerry, July). (B) Tetrasporangial plant with fertile blades (B-C, E & G-H Clare, Oct.). (C) Spermatangial blade. (D) T.S. of spermatangial sorus with mature spermatangia (as A). (E) Female blade with central cystocarp and several procarps near margin. (F) T.S. of cystocarp showing central fusion cell, carposporangia and pericarp with non-protuberant ostiole (as A). (G) Detail of tetrasporangial blade with narrow sterile margins. (H) T.S. of tetrasporangial sorus.

ACROSORIUM Zanardini in Kützing

ACROSORIUM Zanardini in Kützing (1869), p. 4.

Type species: *A. aglaophylloides* Zanardini in Kützing (1869), p. 4. [= *A. venulosum* (Zanardini) Kylin (1924), p. 77].

Thalli consisting of prostrate and erect or decumbent membranous blades; prostrate blades strap-shaped, attached from lower surface by short to long peg-like haptera composed of bundles of multicellular rhizoids; erect blades arising from tips and marginal lobes of prostrate blades, up to 15 (-20) cm high, broadly flabellate and subdichotomously branched to strap-shaped and irregularly alternately branched with rounded, blunt or recurved tips and entire or irregularly toothed margins; blades monostromatic except at the base; microscopic veins present, prominent, with free or anastomosing ends; macroscopic veins absent.

Growth initiated by triangular marginal cells with two cutting faces and continued by marginal and intercalary meristems; meristematic cells uninucleate or multinucleate, the nuclei initially median, later spreading beneath the surface; microscopic veins differentiating acropetally, subdichotomously branched, the ends free or anastomosing, initially monostromatic, becoming tristromatic with elongate, tubular central cells; polystromatic towards base, 5-7 cell layers thick with the cells arranged in horizontal tiers and vertical rows; plastids parietal, dissected into platelets.

Gametophytes dioecious. Spermatangial sori lunate to elliptical, formed inside margin behind apex at tips or lobes of branches. Procarps formed on both surfaces of blade along margin near apex, consisting of a supporting cell, a 1-3 celled sterile group-1, a curved, 4-celled carpogonial branch, and a 1-2 celled sterile group-2; sterile groups usually dividing once after fertilization; auxiliary cell large, dividing horizontally to form an initial cell row; gonimoblast filaments subdichotomously branched; carposporangia terminal, ellipsoidal, maturing singly in basipetal succession; fusion cell large, incorporating several central cells, the supporting cell, foot cell and many inner gonimoblast cells; pericarp 5-6 cell layers thick with a central non-projecting ostiole. Tetrasporangial sori developing near apex, round to elliptical, 7 or more cell layers thick, bordered by a sterile margin; tetrasporangia originating from central cells and inner cortical cells, arranged in two layers at maturity.

Reference: Papenfuss (1939).

In algal checklists for the British Isles, Parke & Dixon (1976) and South & Tittley (1986) included two species of *Acrosorium*, *A. uncinatum* (Turner) Kylin and *A. reptans* (P. Crouan & H. Crouan) Kylin. Wynne (1989) showed that the name *A. uncinatum* was based on a hook-forming variety of *Cryptopleura ramosa*, and that the species previously known in the British Isles as *A. uncinatum* appeared to be conspecific with *A. venulosum* (Zanardini) Kylin from the Mediterranean. *A. reptans*, based on *Nitophyllum reptans* P. Crouan & H. Crouan (1851, p. 365) was regarded by Wynne (1989) as a prostrate growth form of *Cryptopleura ramosa*. *A. venulosum* is thus the only representative of this genus in the British Isles.

Acrosorium venulosum (Zanardini) Kylin (1924), p. 77.

Types: unknown*. Type locality: Zara & Dalmatia, Adriatic.

Nitophyllum venulosum Zanardini (1866), p. 33.
Acrosorium uncinatum sensu Kylin (1924), p. 77, non *Fucus laceratus* var. *uncinatus* Turner (1808), p. 68.

Thalli consisting of prostrate and erect blades forming tangled clumps 3-15 (-20) cm high; *prostrate blades* irregularly lobed to strap-shaped, attached by numerous peg-like haptera formed from the lower surface of the blades, giving rise to groups of erect blades from margins; *erect blades* broadly flabellate, to 15 cm wide, expanding rapidly above a narrow constriction into a fan-shaped lamina deeply dissected into few to numerous ribbon-like to wedge-shaped segments 3-10 mm wide, each bearing an irregular, pinnate or alternate arrangement of ribbon-like axes that can give rise to 1-2 further orders of branching, tapering gradually towards apices, which are initially pointed but may be transformed into hooks with a thickened inner margin; *blade margins* bearing irregular series of rounded lobes, small dentations or large pointed (acuminate) teeth, some of which continue growth and give rise to small tightly incurved hooks or long narrow projections terminating in open hooks; *veins* fine, usually visible against the light as a more-or-less parallel array, although large macroscropic veins are absent; colour clear rose-pink in young thalli, sometimes becoming brownish-red in older plants, or bleaching yellowish-pink under high-light conditions; texture crisp and membranous, tearing and creasing readily.

Blade laminae monostromatic, 52-65 µm in thickness, composed of rectangular to polygonal cells, 40-130 long x 25-70 µm wide, with triangular to rectangular or wedge-shaped marginal cells; *microscopic veins* branching and anastomosing at irregular intervals, 3 cells thick, composed of elongate cells 80-225 µm long x 17-35 µm wide; *plastids* discoid to bacilloid.

Gametophytes unknown. *Tetrasporangial sori* developing near apices and in rounded or pointed outgrowths from blade margins, sometimes paired on either side of blade, round to elliptical, 0.3-2.5 mm long x 0.3-0.8 mm wide and about 200 µm in thickness; tetrasporangia spherical, 40-65 µm in diameter.

Growing on flat and vertical bedrock and pebbles, usually attached to crustose Corallinaceae rather than the rock itself, epiphytic on small algae and kelp stipes and hooked around *Zostera* plants, and epizoic on invertebrates, e.g. hydroids, in deep

* Although Zanardini's types should be in the Museo Civico, Venice, De Toni & Levi (1888) did not list any relevant specimen in this collection. Other possibilities are W, HBG and L.

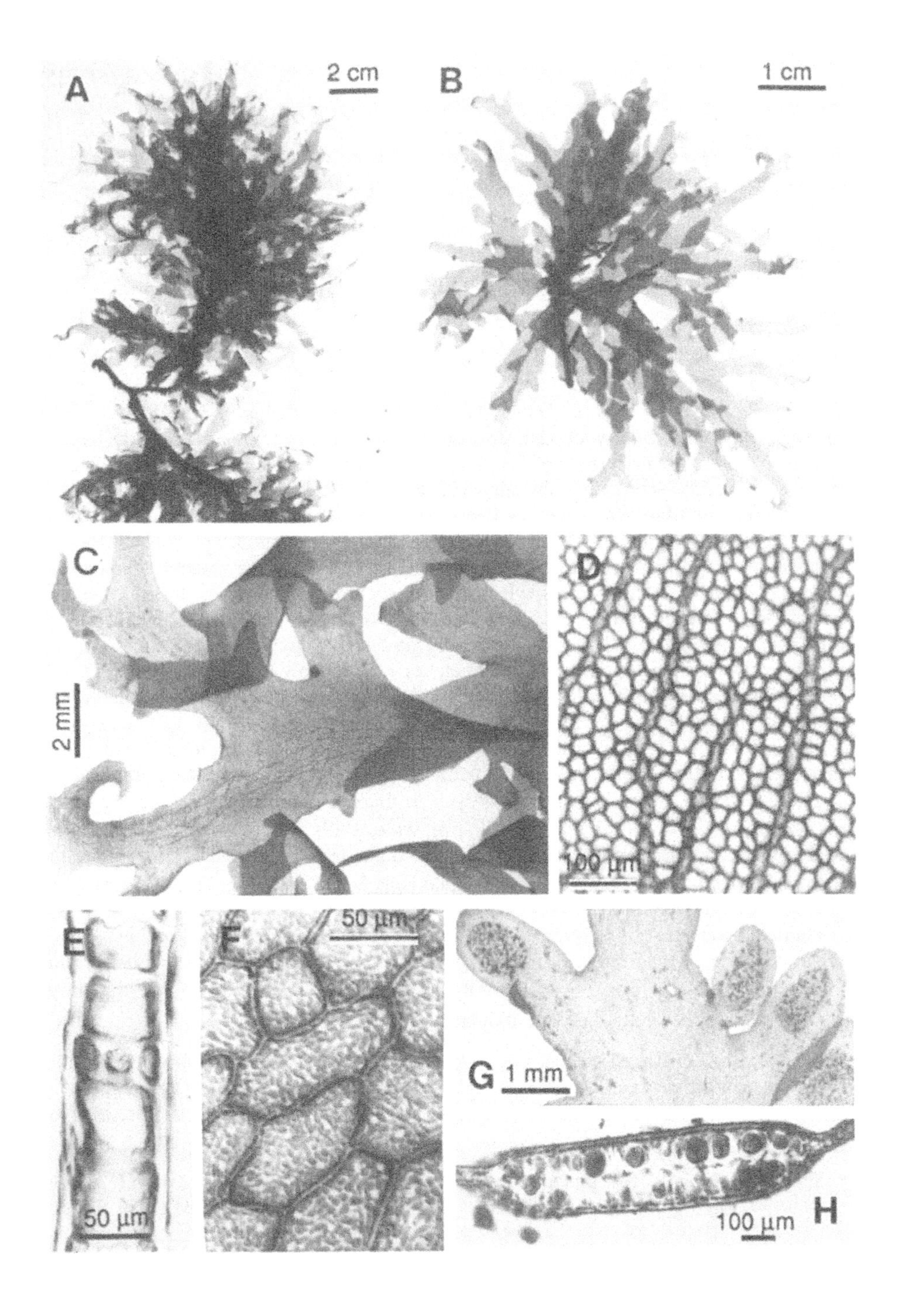
A
2 cm
B
1 cm
C
2 mm
D
100 µm
E
50 µm
F
50 µm
G
1 mm
H
100 µm

lower-shore pools and subtidal to 30 m depth, growing in greatest abundance on bedrock under *Laminaria hyperborea* canopy, frequently on sandy rock, at moderately to extremely wave-exposed sites, occasionally with moderate tidal currents.

South coast of England eastwards to Hampshire, west coasts of England, Wales and Scotland northwards to W. Sutherland and Orkney; widely distributed in Ireland except for central east coast; Channel Isles.

British Isles to Sénégal; Mediterranean; North Carolina to Brazil (Price *et al.*, 1986; Schneider & Searles, 1991); 'worldwide distribution' (Wynne, 1989).

Individual blades are annual, becoming obvious in May-June and forming tetrasporangia in July-October; old blades eroded and encrusted with bryozoa by September and October. Gametangia have never been observed.

The number of hooks varies both among the blades of a single plant and between plants growing together. Whereas some individuals lack hooks, the margins instead being lobed or bearing acuminate teeth, others form numerous conspicuous terminal and marginal hooks.

A. venulosum thalli occasionally bear plants of the parasite *Gonimophyllum buffhamii* (q. v.).

Cryptopleura ramosa 'var. *uncinata*' (q. v.) may closely resemble *A. venulosum*. *C. ramosa* can be distinguished by the positions of the hooks, which are marginal only rather than both terminal and marginal as in *A. venulosum*. When non-reproductive, *C. ramosa* lacks the angular marginal projections characteristic of *A. venulosum*. Cotton (1912) made a detailed comparison between these species, but some of his points of distinction are not always valid. These include the colour, which is generally a brighter pink in *A. venulosum* compared to brownish in *C. ramosa*, and the frequent occurrence of large macroscopic veins in *C. ramosa* but their absence in *A. venulosum*. Prostrate blades of the two species, often known as '*Acrosorium reptans*', cannot easily be separated morphologically. *Radicilingua thysanorhizans* (q. v.) can be distinguished from delicate plants of *A. venulosum* by its attachment structures, which are small marginal projections formed at regular intervals. Prostrate *A. venulosum* blades, in contrast, are attached by peg-like holdfasts formed over the entire lower surface of prostrate blades. In addition, the plastids of *R. thysanorhizans*, although initially discoid to plate-like, rapidly become ribbon-like and convoluted, whereas they remain discoid to bacilloid in *A. venulosum*.

Fig. 80. *Acrosorium venulosum*

(A) Dense clusters of epiphytic plants (Antrim, June). (B) Single thallus with marginal and apical hooks (Clare, June). (C) Detail of blades with apical hooks and net-like venation (as A). (D) Microscopic veins in blade lamina (Pembroke, May). (E) T.S. of blade through monostromatic lamina and tristromatic vein (Cornwall, Sep.). (F) Cells with numerous small bacilloid plastids, with nuclei visible as clear spots (Donegal, Aug.). (G) Tetrasporangial sori in marginal bladelets (G-H Sutherland, July). (H) T.S. of tetrasporangial sorus.

NITOPHYLLUM GROUP

NITOPHYLLUM Greville, nom. cons.

NITOPHYLLUM Greville (1830), pp. xlvii, 77.

Lectotype species: *N. punctatum* (Stackhouse) Greville (1830), p. 79, pl. 12 [see Schmitz (1889), p. 445; Kylin (1924), p. 69].

Thalli sessile or shortly stipitate, erect to 50 cm high or decumbent, cuneate-flabelliform to deeply dissected, typically with rounded, dichotomizing tips, membranaceous, monostromatic above, polystromatic at base and in fertile areas; microscopic and macroscopic veins or midribs absent.

Growth initiated by an apical cell in very young thalli, otherwise by marginal and intercalary meristems, meristematic cells mostly multinucleate, the nuclei initially parietal, alongside lateral walls, later uniformly distributed beneath cell surface, medial and superficial pit connections abundant between adjoining cells; plastids parietal, dissected into minute, discoid platelets linked by fine strands, and either distributed uniformly beneath cell surface or forming bead-like, branched chains.

Reproductive structures scattered throughout thallus, except the base. Gametophytes dioecious. Spermatangial sori elliptical, separate or coalescent, formed on both sides of central layer by dedifferentiation and anticlinal and periclinal divisions of mature, multinucleate central cells; spermatia accumulating beneath outer membrane before release. Procarps typically formed in diagonally opposite pairs on both sides of thallus, the central cell first cutting off sterile vegetative cells (cover cells), followed by the fertile pericentral cells (supporting cells); mature procarp consisting of the supporting cell, a 1-celled lateral sterile group, and a straight, 4-celled carpogonial branch flanked laterally by the cover cells and sterile group, second sterile group absent; cover cells and sterile group dividing or forming cortex after fertilization; auxiliary cell cut off apically from supporting cell, diploidized by a connecting cell before dividing into a foot cell and gonimoblast initial; gonimoblasts irregularly subdichotomously branched, bearing carposporangia sequentially, either terminally or in short, branched chains; fusion cell small, consisting of the supporting cell, foot cell, and basal gonimoblast cells; pericarp 3-4 cell layers thick with a non-protruding ostiole. Tetrasporangial sori distinct, round to elliptical; tetrasporangia formed by dedifferentiation of multinucleate central cells linked to adjoining cells by secondary pit connections, all but one nucleus degenerating, the remaining one undergoing meiosis; mature tetrasporangia tetrahedrally to irregularly divided, covered by a thin outer cortex.

Reference: Kylin (1924).

One species in the British Isles.

Nitophyllum punctatum (Stackhouse) Greville (1830), p. 79.

Type: ?CN. Dorset (Weymouth), September 1792, ?*Stackhouse*.

Ulva punctata Stackhouse (1797), p. 236.
Delesseria punctata (Stackhouse) C. Agardh (1822), p. 184.

Thalli consisting of tufts of blades attached by solid discoid holdfast up to 4 mm in diameter that spreads over the substratum and gives rise to new blades at its margins; *blades* 3-50 cm high and up to 20 cm broad, occasionally becoming decumbent, reattaching by secondary holdfasts and giving rise to new fronds, very variable in shape, sometimes with a short compressed stipe up to 2 mm long, broadly flabellate or linear, almost undivided with branching at apices only, partly cleft into wedge-shaped segments, or deeply dissected into numerous ribbon-like segments, the most-divided blades consisting of fairly regularly dichotomously branched segments 0.2-0.4 cm wide with widely divergent branches, always with dichotomously branched apices rounded at the tips; macroscopic and microscopic veins absent; blade margins entire and undulating, forming dichotomous proliferations, or developing thickened ruffles when older; rose-pink in colour when young, becoming brownish-red, or bleached yellowish-brown under high-light conditions, occasionally with a blue iridescence underwater; texture of young blades delicate and membranous, tearing easily, older blades becoming slightly tougher.

Blade laminae mostly monostromatic, 50-100 μm thick, composed of elongate polygonal cells 30-110 μm long x 20-60 μm wide, with triangular to rectangular or wedge-shaped marginal cells; older blades and opaque spots on female blades increasing to 3-4 cells thick, up to 200 μm in thickness; *plastids* discoid to bacilloid.

All reproductive structures scattered on blades except near stipe. *Spermatangial sori* arranged longitudinally or in fan-shaped arrays, oval, elliptical, or coalescing into convoluted shapes, 0.2-3 mm x 0.2-0.6 mm, c. 50 μm thick, consisting on either side of blade of a layer of spermatangial mother cells 6 μm in diameter, bearing cylindrical spermatangia 8 μm long x 3-4 μm wide. *Cystocarps* developing on both sides of blades, 650-960 μm in diameter when mature, hemispherical, with smooth outer pericarp and central projecting ostiole with pore 50-100 μm in diameter; carposporangia produced sequentially, usually terminal, sometimes in short branched chains, pyriform, 45-65 μm long x 30-45 μm wide. *Tetrasporangial sori* arranged longitudinally or in fan-shaped arrays, round to elliptical, 0.3-3.5 mm long x 0.2-1 mm wide, protruding equally on both sides of blade, 250-300 μm in thickness; tetrasporangia spherical, 110-175 μm diameter.

Growing on a wide variety of substrata, including large and small algae, blades and stipes of *Laminaria hyperborea*, pebbles, shells, maerl and artificial material, in pools and damp places on lower shore and subtidal to at least 24 m depth, particularly abundant in tidal rapids and beneath kelp canopy on bedrock and crustose corallines, at sites with exposure to moderate to strong wave action or tidal currents.

Generally distributed in the British Isles, northwards to Shetland, although records for the east coast of England are sparse.

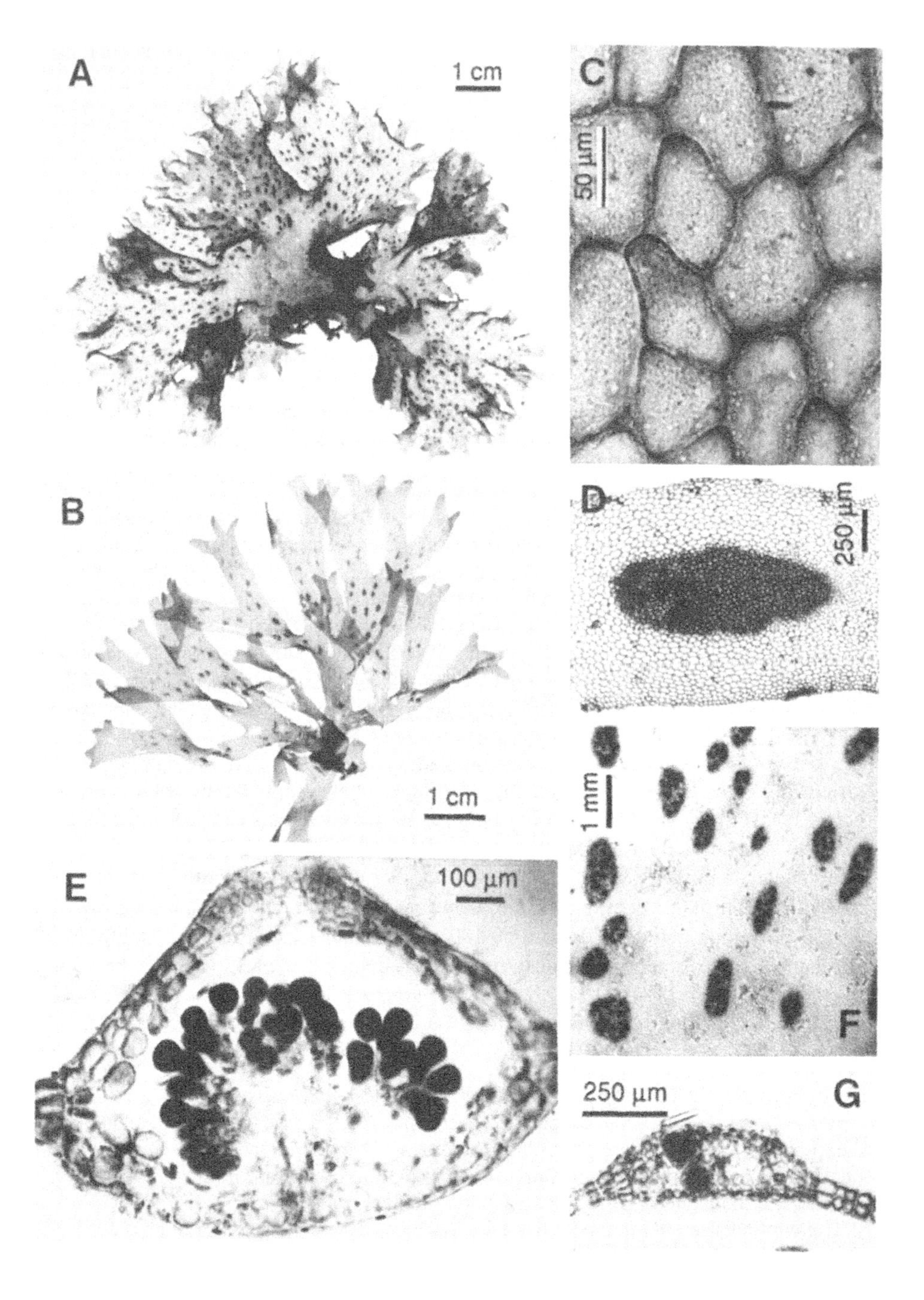
A
1 cm
B
1 cm
C
50 µm
D
250 µm
E
100 µm
1 mm
F
250 µm
G

Norway, British Isles to Morocco; Mediterranean and Black Sea (Ardré, 1970).

Mature plants can be found throughout the year in sheltered habitats. Kelp-forest populations are seasonal, young thalli becoming conspicuous in April-May and persisting until November. Spermatangia recorded in February and May-September and December; cystocarps in February, April-October and December; and tetrasporangia in January, February, April-September and November-December. Chromosome number was reported as $n = 20$ (see Magne, 1964), but requires confirmation.

There is a remarkable range of variation in overall shape, size and texture. Harvey (1847, pls 102-103) illustrated a range of morphologies that had previously been regarded as distinct varieties, characterized by their branching pattern and degree of marginal proliferations. Some of the variation appears to be influenced by environment: plants from kelp forests are generally less divided than those growing in more wave-sheltered habitats.

N. punctatum plants may closely resemble *Haraldiophyllum bonnemaisonii* (q. v. for distinctions), and non-reproductive thalli of *Drachiella minuta* (q. v.) and *Rhodophyllis divaricata*. *D. minuta* and *N. punctatum* can be distinguished by the shape of plastids, which are single dissected or convoluted plates in the former and numerous, discoid to bacilloid, in the latter. *R. divaricata* fronds are more than one cell thick even at apices, in contrast to the monostromatic young blades of *N. punctatum*. The texture of *N. punctatum* resembles that of *Porphyra* spp., but *Porphyra* thalli have very small quadrate cells and form no macroscopic reproductive structures.

RADICILINGUA Papenfuss

RADICILINGUA Papenfuss (1956), p. 160.

Type species: *R. thysanorhizans* (Holmes) Papenfuss (1956), p. 160.

Thalli arising from discoid holdfasts, sessile or briefly stipitate, essentially prostrate, consisting of alternately to subdichotomously branched overlapping membranous blades

---

Fig. 81. *Nitophyllum punctatum*

(A) Broad fan-shaped plant (A & E-F Donegal, May). (B) Dissected thallus with narrow ribbon-like blades and dichotomous apices (Devon, Nov.). (C) Cells with numerous small bacilloid plastids, with nuclei visible as clear spots (as A). (D) Elongate spermatangial sorus in narrow blade (Down, Dec.). (E) T.S. of cystocarp showing small central fusion cell, single layer of large carposporangia and pericarp with slightly protuberant ostiole. (F) Elliptical to elongate tetrasporangial sori. (G) T.S. of tetrasporangial sorus with large mature tetrasporangia (Down, Nov.).

with rounded, upturned tips, sometimes secondarily attached by marginal unicellular rhizoids, monostromatic except at the base and traversed by a rectangular network of microscopic veins; midribs and macroscopic veins absent.

Growth initiated by concavo-convex division of a depressed-obovate apical cell, subterminal cells cutting off a pair of lateral periaxial cells with subsequent segmental cells dividing transversely and vertically to form an indistinct primary cell row; cell rows of the second order bearing third-order cell rows primarily adaxially; fourth-order cell rows lateral, parallel to second-order cell rows; fifth-order cell rows adaxial, parallel to third-order cell rows, etc., up to 6-7 orders; vertical and transverse intercalary divisions occurring in cell rows of all orders beginning 2-3 segments behind the apex; cells initially rectangular, arranged in curved, horizontal tiers and vertical rows, polygonal after thallus expansion; microscopic veins differentiating acropetally, the ends free or anastomosing, initially monostromatic, later tristromatic with tubular central cells having broad primary pit connections; new apical initials differentiating alternately on either side of apex from apical cells of second order cell rows, spreading as the thallus undergoes intercalary cell divisions and either remaining dormant or forming lateral branches; adventitious proliferations sometimes formed from marginal cells; meristematic cells initially uninucleate, becoming multinucleate prior to and after the formation of the secondary pit connections; plastids parietal, dissected into platelets which may unite to form ribbons.

Gametophytes dioecious. Spermatangial sori widespread in tips of ultimate branches, forming small, separate or confluent patches on both sides of blade separated by microscopic veins and monostromatic vegetative cells; individual clusters initiated at apex and also secondarily by anticlinal and periclinal divisions of mature central cells forming two spermatangia-bearing layers replacing the central layer. Procarps formed in diagonally opposite pairs near apices on both sides of thallus, often developing normally on only one side; central cells first cutting off sterile vegetative cells (cover cells) followed by the fertile pericentral cells (supporting cells); mature procarp consisting of the supporting cell, a 1-celled lateral sterile group, and a straight, 4-celled carpogonial branch flanked laterally by the cover cells and sterile group, second sterile group absent; lateral sterile group situated beneath ostiole, remaining undivided after fertilization; gonimoblasts irregularly subdichotomously branched, bearing carposporangia in branched chains; carposporangia maturing simultaneously followed by production of a second cluster derived from an inner gonimoblast cell; fusion cell absent, central cells and inner gonimoblast cells with broadened pit connections; pericarp thin, mostly 2-layered with non-protruding ostiole. Tetrasporangial sori formed near apex, lunate to triangular, bordered by a sterile margin; tetrasporangia superficial, originating first from central cells and later from cortical cells, arranged in two layers at maturity, flanked by cortical cells and exposed at the surface, tetrahedrally to irregularly divided.

Radicilingua has traditionally been placed in the Phycodrys group; however, procarp development takes place in the same way as in *Nitophyllum* and the genus is properly placed in the Nitophyllum group. Its nearest relative is *Calonitophyllum* from the western North Atlantic Ocean.

References: Kylin (1924), Huvé & Riouall (1970), L'Hardy-Halos (1973).

Radicilingua thysanorhizans (Holmes) Papenfuss (1956), p. 160.

Lectotype: BM. Cornwall (Torpoint), undated, *Holmes* (see Holmes, 1873, pl. 12, fig. 1).

Nitophyllum thysanorhizans Holmes (1873), p. 2.
Rhizoglossum thysanorhizans (Holmes) Kylin (1924), p. 28.

Thalli largely prostrate, forming patches up to 15 cm in extent, consisting of groups of partly overlapping blades spreading over the substratum, becoming erect at the apices, each 2-5 cm long and broad, composed of axes 1-4 mm wide, with rounded or pointed apices, branching irregularly dichotomously to alternately, to 5 or more orders, the branches diverging widely, with entire, undulating margins bearing at regular intervals pointed inrolled projections 0.5 mm long, which curve downwards and attach terminally to the substratum; colour clear rose-pink; texture extremely delicate and filmy, tearing and creasing readily.

Blades growing from several apical cells 8 µm in diameter, with product cells arranged in semi-circular arrays around apical cells, monostromatic, 40 µm in thickness, with square to rectangular marginal cells that become elongate at marginal projections, composed of rectangular to polygonal cells 40-120 long x 28-55 µm wide, with conspicuous secondary pit connections; *microscopic veins* mostly parallel, longitudinal, connected by a trellis-like pattern of transverse veins, 3 cells thick, composed of elongate cells 30-90 µm long x 12-25 µm wide; *plastids* initially discoid, becoming ribbon-like or convoluted and catenate.

Spermatangial sori formed over large areas of young blades, causing ruffling when mature, differentiating at apices by development of quadrate groups of small cells separated by veins and large sterile blade cells, 30 µm in thickness, consisting of 2 layers of quadrate groups of spermatangial mother cells 2-3 µm in diameter, each bearing 3 ovoid spermatangia 2.5-5 x 1-2 µm. *Cystocarps* scattered, developing on both surfaces of blades, 475-530 µm in diameter when mature, almost spherical, with protruding ostiole 75 µm long and pore 50-75 µm in diameter; carposporangia ovoid to spherical, 28-55 x 24-40 µm. *Tetrasporangial sori* developing near apices, singly or paired on either side of apex, triangular, round or semi-circular, 0.3-1 mm long x 0.3-3 mm wide and 75 µm in thickness; tetrasporangia spherical to ellipsoid, 24-36 x 20-28 µm.

Growing on unstable substrata, including stones, gravel, shells, maerl and wreck plates, forming large patches on sand- or wave-scoured bedrock, from extreme low water to 30 m depth, at moderately to extremely wave-exposed sites, often where exposed to strong tidal currents. It is a relatively common seaweed in these habitats (Hiscock & Maggs, 1984).

South-west coasts, eastwards to Dorset, northwards to Pembroke and Caernavon; widely distributed in Ireland except for central east coast.

British Isles to Portugal; W. Mediterranean (Ardré, 1973).

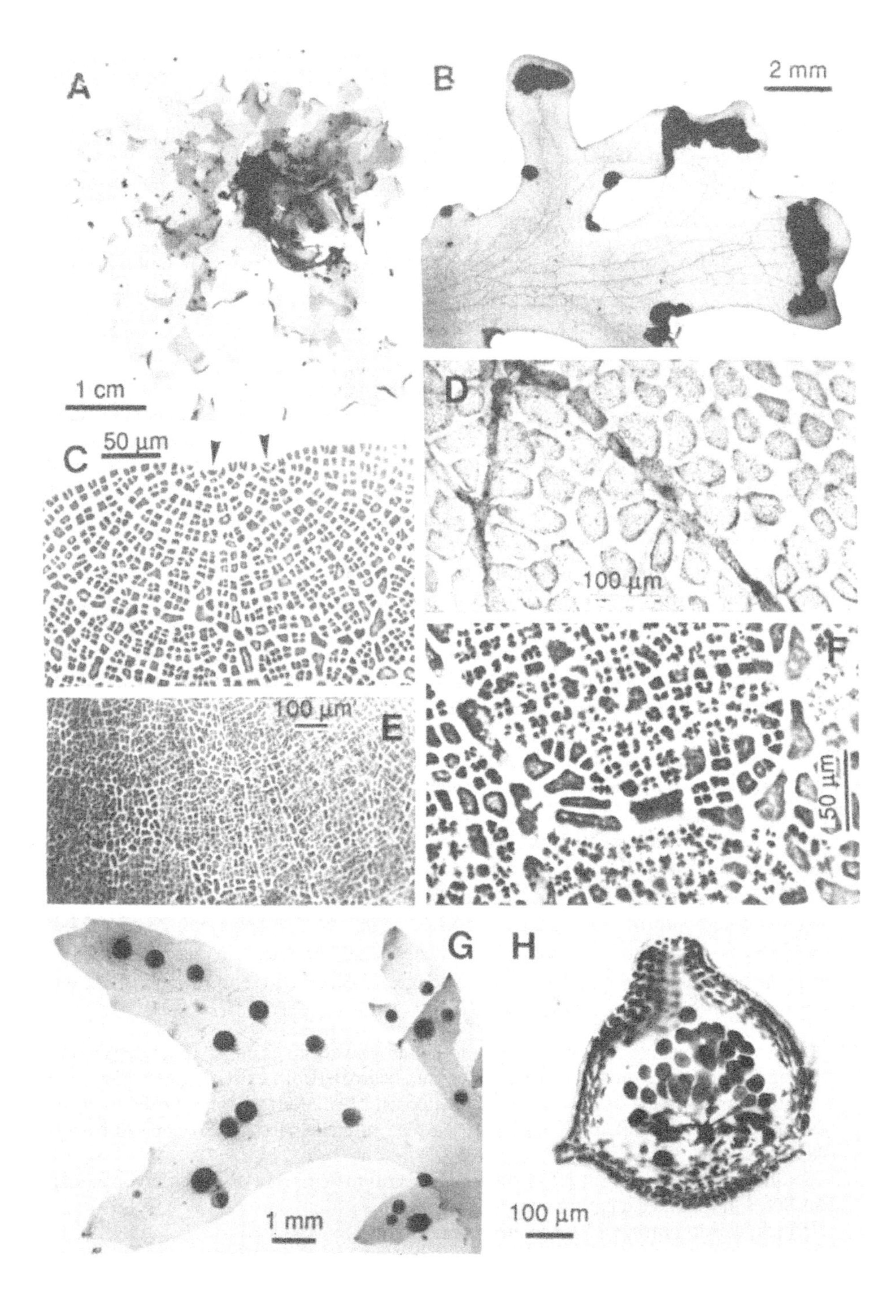
A
1 cm
B
2 mm
C
50 μm
D
100 μm
E
100 μm
F
50 μm
G
1 mm
H
100 μm

Individual fronds are probably short-lived. On open coasts, small plants and regenerating fragments become apparent in April and May, mature thalli occur in June-August, and only fragments remain by September. Spermatangia have been recorded in June; cystocarps in June-July; tetrasporangia in January and May-August.

Little morphological variation occurs in material from the British Isles.

R. thysanorhizans and *Acrosorium venulosum* (q. v.) were thought to be conspecific by Holmes & Batters (1891) and Batters (1902) (see Wynne, 1989). They can be distinguished by several features, however, even when non-reproductive (see under *A. venulosum* for details).

Fig. 82. *Radicilingua thysanorhizans*

(A) Habit of tetrasporophyte (A-C & E-H Mayo, June). (B) Detail of tetrasporophyte with net-like venation and sori just behind apices. (C) Apex with 2 apical cells (arrows) surrounded by semi-circular arrays of cells. (D) Microscopic veins among mature blade cells linked by secondary pit connections and containing reticulate ribbon-like plastids (Cornwall, July). (E) Young spermatangial sorus. (F) Mature spermatangial sorus divided into rectangular areas by sterile striae. (G) Female blade with cystocarps. (H) T.S. of cystocarp showing small central fusion cell, chains of carposporangia and pericarp with protuberant ostiole.

DASYACEAE Kützing

DASYACEAE Kützing (1843), pp. 413, 414 [as Dasyeae].

Germination bipolar, the sporeling initially filamentous, monosiphonous and monopodially branched; erect axes shifting to sympodial branching early in development with the apical cell diverted laterally producing a pseudolateral and the subterminal cell cutting off an initial that continues the main axis. Growth by oblique or transverse division of a uninucleate or multinucleate apical cell; axes radially, bilaterally, or dorsiventrally organized, polysiphonous with 4 to many periaxial cells cut off either in a circle with the last periaxial cell next to the first, or in an alternating sequence with the last periaxial opposite the first; periaxial cells elongating to approximately the same length as the axial cell or dividing transversely with the upper cell retaining primary pit connection; axes ecorticate or corticated by descending rhizoids originating from basal cells of pseudolaterals and/or lower ends of periaxial cells, forming a thick cortex in some species; pseudolaterals pigmented, monosiphonous throughout or partly to largely polysiphonous except the tips, up to 7 times pseudodichotomously branched in secund, adaxial series, and either free or united by small lateral cells into a spiral network; adventitious monosiphonous filaments originating from periaxial and cortical cells in some species; lateral sympodial axes arising from the basal cell or the first or second adaxial branch of a pseudolateral, or adventitiously on cortical cells or monosiphonous filament.

Spermatangia borne in regenerated axes on ultimate branches of pseudolaterals or adventitious filaments; periaxial cells 4-5 per segment, cut off in alternating sequence, each dividing transversely and anticlinally to produce a single layer of spermatangial mother cells, each bearing 3-4 spermatangia. Procarps formed on cells of pseudolaterals or sympodially developed axes containing five periaxial cells cut off in either a circular or an alternating sequence, with the third, fourth or fifth periaxial cell fertile and consisting of a supporting cell, a lateral sterile group containing 1-7 cells, a 4-celled carpogonial branch, and a basal sterile group containing 1-4 cells; pericarp initials either present or absent before fertilization. Fertilized carpogonium cutting off 2 connecting cells, one of which fuses with the auxiliary cell; sterile groups each usually dividing once after fertilization and their nuclei enlarging; gonimoblasts monopodially branched and bearing carposporangia in terminal branched chains or converting to sympodial branching with the first carposporangia terminal followed by branching from subterminal cells; fusion cell small, containing the auxiliary cell, supporting cell and central cell or large, incorporating additional inner gonimoblast cells; pericarp composed of up to 14 modified axes bearing periaxial cells and 1-6 layers of cortical filaments united laterally by secondary pit connections; cystocarps ostiolate, often with a prominent beak.

Tetrasporangia formed in whorls on regenerated axes (stichidia) on the ultimate branches of pseudolaterals or adventitious monosiphonous filaments, or on modified polysiphonous branches; each fertile segment containing 5-7 (-10) periaxial cells cut off in alternating sequence and 4-6 tetrasporangia, each surrounded by 2-4 postsporangial cover cells which may divide once transversely, or by 2-3 presporangial cover cells; tetrasporangia tetrahedrally divided.

References: Rosenberg (1933), Kylin (1956), Parsons (1975).

DASYA C. Agardh, nom. cons.

DASYA C. Agardh (1824), p. 211.

Type species: *D. elegans* (Martens) C. Agardh [=*D. baillouviana* (S. G. Gmelin) Montagne (1841), p. 164.

Thalli radially organized, terete, polysiphonous, consisting of 1-several erect, spirally branched main axes arising from a loosely matted, filamentous prostrate basal system; pseudolaterals one per segment, typically arranged in a 2/5 left-handed spiral, monosiphonous throughout, or sometimes polysiphonous at the base, pigmented, persistent, 3-7 times subdichotomously branched; lateral axes arising from the basal or suprabasal cell of a pseudolateral, or from the first or second segment of an adventitious monosiphonous filament; periaxial cells (4-)5, longitudinally connected by secondary pit connections, uncorticated or corticated by descending rhizoidal filaments issuing from basal cells of pseudolaterals and lower ends of periaxial cells; adventitious monosiphonous filaments absent or present, formed on basal cells of pseudolaterals, or on periaxial cells and/or cortical cells.

Growth sympodial, initiated by oblique division of the apical cell with the apical initial diverted laterally forming a pseudolateral, and the subapical cell continuing the main axis; pseudolaterals at first branched adaxially, subdichotomous at maturity; vegetative periaxial cells cut off in a left-handed circle, with the first to the right of the pseudolateral, the second to the left, and so on, so that the fifth lies next to the first.

Gametophytes dioecious. Spermatangial axes formed on monosiphonous pseudolaterals or adventitious monosiphonous filaments; periaxial cells 4 per segment, cut off in alternating sequence and dividing transversely and anticlinally to produce a single layer of spermatangial parent cells, each bearing 3-4 spermatangia. Procarps formed from the third periaxial cell in segments of sympodial axes containing 5 periaxial cells arranged in a circle, and consisting of a supporting cell, a 1-celled lateral sterile group, a 4-celled carpogonial branch, and a 1-celled basal sterile group; pericarp initials absent before fertilization; fertilized carpogonium cutting off 2 connecting cells, one of which fuses with the auxiliary cell; sterile groups each dividing once after fertilization and their nuclei enlarging; gonimoblasts monopodially branched and bearing carposporangia in terminal branched chains; fusion cell moderately large, incorporating the supporting cell, auxiliary cell, innermost gonimoblast cells, axial cell, and sometimes the adjacent periaxial cells; pericarp composed of modified axial filaments originating from periaxial cells of the fertile segment, with each segment forming two periaxial cells and additional cortical cells; cystocarp typically 4 or more cell layers thick, ostiolate, usually with a prominent beak. Tetrasporangia borne in whorls, 4-6 per segment, in modified axes (stichidia) formed on pseudolaterals or adventitious monosiphonous filaments, periaxial cells 5-7 per segment, cut off in alternating sequence; fertile periaxial cell dividing transversely into a stalk cell and a tetrasporangium, the stalk cell then cutting off 2-4 (usually 3) postsporangial cover cells which remain undivided or divide again, but do not completely cover the tetrasporangium.

References: Rosenberg (1933), Parsons (1975).

There has been a great deal of confusion between the British species, and determinations of herbarium material frequently appear to be erroneous. We have found that the most useful taxonomic character is the detailed branching pattern of the pigmented, hair-like pseudolaterals. The basal cell, which is often immersed in the axis due to the development of a layer of corticating filaments around the periaxial cells, is either branched or unbranched. If the basal cell is branched, the pseudolaterals appear to emerge in pairs from the axis, while if it is unbranched, the suprabasal cell emerges singly, and bears a pair of filaments; branching of the next few cells is also important in delimiting species. Another feature that discriminates between some species is the number of tetrasporangia formed in each segment of the stichidia.

We have confirmed previous reports that there are four species of *Dasya* in the British Isles. *D. hutchinsiae* is the only common and widespread species, and is especially variable in different habitats. We have not found any evidence in the British Isles of *Dasya baillouviana*, described from the Mediterranean and also widespread on Atlantic coasts of North America, which has been introduced into Holland and Sweden (Dixon & Irvine, 1970). It is readily distinguishable from British *Dasya* species by the wider main axes (up to 2 mm in diameter), and its complete cover of long silky pigmented pseudolaterals and numerous adventitious monosiphonous filaments.

KEY TO SPECIES

1 Tetrasporangia 4 per stichidial segment; pseudolaterals branched from suprabasal and all of next few cells .... *D. corymbifera*
Tetrasporangia 5 per stichidial segment; pseudolaterals branched from basal or suprabasal cell but if from suprabasal cell then not from **all** of next few cells 2

2 Pseudolaterals branched from immersed basal cell, appearing to arise in pairs (see Fig. 84F), and also branched from **all** of next few cells .... *D. hutchinsiae*
Pseudolaterals branched either from basal cell or suprabasal cell, but not from **all** of next few cells 3

3 Pseudolaterals branched from suprabasal cell and then from alternate cells; main axes simple or with a few simple lateral branches .... *D. ocellata*
Pseudolaterals branched either from basal cell or suprabasal cell, then unequally at pseudodichotomies, from first cell on one side, from 2nd (-3rd) cell on other side; main axes branched to 4-5 orders .... *D. punicea*

Dasya corymbifera J. Agardh (1841), p. 31.

Lectotype: LD 44053a. Morocco (Tangier), July 1827, *Schousboe* (see Schlech & Abbott, 1989).

Dasya venusta Harvey (1849a), pl. 225.

Thalli pyramidal in outline, consisting of erect axes 4.5-10 cm tall, attached by a discoid rhizoidal holdfast, tapering gradually from 0.8 mm in diameter near base, bearing up to 5 orders of branches, the first two of which are alternate and devoid of pseudolaterals, the last two orders being pinnate, fully clothed with hair-like pseudolaterals and ecorticate for up to 2 cm behind the apex; deep red in colour, with a delicate, flaccid texture.

Main axes composed of 5 periaxial cells, linked basally in files by 1-3 (usually 2) secondary pit connections, partly corticated by narrow rhizoidal filaments 4-12 µm in diameter that extend along the edges of the periaxial cells, bearing one pseudolateral per segment in a 2/5 spiral; *pseudolaterals* consisting of fine-textured filaments of nearly uniform diameter, c. 25 µm throughout, the basal cell unbranched, initially inflated and emergent, gradually stretching longitudinally in plane of periaxial cells and finally immersed within cortex, suprabasal cell and next 3-4 segments branching pseudodichotomously, forking at 45° angles, initially Y-shaped, becoming U-shaped, ultimately deciduous, breaking down progressively from tips, forming adventitious polysiphonous branches from basal cells only; adventitious monosiphonous branches absent.

Spermatangial branches cylindrical, developing from the last order of branching of pseudolaterals (Harvey, 1849a, pl. 225). *Cystocarps* sessile, urceolate with a long beak-like ostiole (Harvey, 1849a, pl. 225) [details lacking as gametangial material is not available]. *Tetrasporangial stichidia* borne on pseudolaterals on 1-2 -celled, often very elongate pedicels, variable in shape and length, initially conical, later cylindrical with long pointed apex, 200-525 µm long and 60-85 µm in diameter, consisting of up to 30 tetrasporangia-bearing segments; initially 5 periaxial cells per segment, one aborting, the remaining 4 bearing tetrasporangia; tetrasporangia with 2-3 undivided cover cells covering 1/3 of sporangium, spherical, 25-30 µm in diameter.

No information on habitat is available, as all specimens collected in the British Isles appear to have been drift. There are no verified records from the twentieth century.

All correctly identified specimens are from the Channel Isles (Jersey and Guernsey); records from the south coast of England cannot be confirmed at present. Reports from Ireland (with the possible exception of a specimen from Kinsale in BM, slide 11485) are

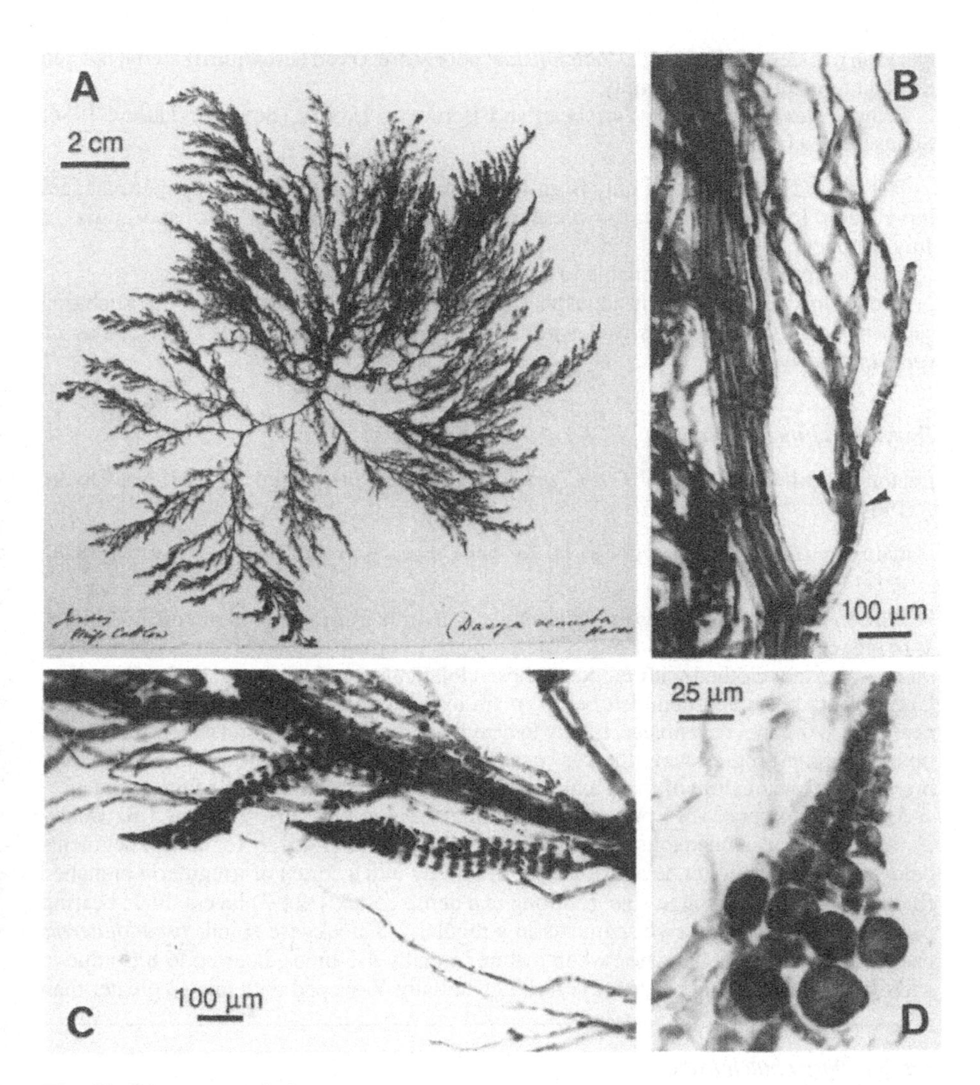

Fig. 83. *Dasya corymbifera*

(A) Habit of herbarium specimen (A-D Channel Isles, undated, in BM). (B) Axis bearing pseudolateral branched above basal cell (arrows). (C) Elongate tetrasporangial stichidia with fine pointed apices. (D) Apex of tetrasporangial stichidium squashed slightly to show 4 tetrasporangia per segment.

based on misidentifications of *D. hutchinsiae*; poorly preserved herbarium material is often difficult to identify with certainty.

British Isles to Morocco; Caribbean and Bermuda; Hawaii (South & Tittley, 1986; Schlech & Abbott, 1989).

Plants have been collected only from July-October. No dated records of spermatangia have been located; cystocarps were recorded in August, and tetrasporangia in July-October.

Insufficient material is available to comment on variability.

The lectotype specimen is tetrasporangial, and has 4 tetrasporangia per stichidial segment; it thus corresponds well with material from the British Isles described as *D. venusta* Harvey.

Dasya hutchinsiae Harvey in W. J. Hooker (1833), p. 335.

Lectotype: Dillwyn (1809), pl. G, "*C. arbuscula*", based on material in BM-K (see Dixon, 1964).

Dasya arbuscula sensu Harvey (1849a), pl. 224, non *Dasya arbuscula* (Dillwyn) C. Agardh (1828), p. 121 (see Dixon, 1960c).

Thalli consisting of erect axes growing singly or in tufts from a discoid rhizoidal holdfast, 2-10 (15) cm high, forming 4-5 orders of irregularly alternate to subdichotomous branches, all axes densely clothed with pigmented pseudolaterals, producing a cylindrical outline; larger plants becoming denuded below; main axes tapering gradually from 0.5-0.6 mm near base to 0.2 mm near apices, lightly to heavily corticated throughout except at extreme tips, becoming progressively thicker basally; colour red to reddish-brown, occasionally brown or purple; texture of main axes tough and flexible, pseudolaterals fairly rigid.

Main axes composed of 5 periaxial cells connected longitudinally in files by 1-3 (usually 2) secondary pit connections, forming rhizoidal cortication 15-20 (or more) segments below apex, which either develops slowly or rapidly into a cortex of irregularly branched filaments of cells varying in size, resulting in a dense cortex 1-2 (-3) layers thick, bearing pseudolaterals on successive segments in a regular 2/5 clockwise spiral; *pseudolaterals* pseudodichotomously branched when mature, initially 4-5 times, later up to 8 (9) due to additional branching, divaricate, the forkings initially Y-shaped with angles greater than

---

Fig. 84. *Dasya hutchinsiae*

(A) Thallus from sheltered site (Donegal, Dec.). (B) Axes bearing spiral arrangements of hair-like pseudolaterals (Galway, Feb.). (C) Spermatangial stichidia (Cornwall, July). (D) Tetrasporangial stichidia (Cornwall, Sep.). (E) Apex of tetrasporangial stichidium squashed slightly to show 5 tetrasporangia per segment (arrow) (Mayo, Aug.). (F) T.S. of axis with pseudolateral branched from embedded basal cell (thus resembling a pair of pseudolaterals) and from each of next few cells above (as B).

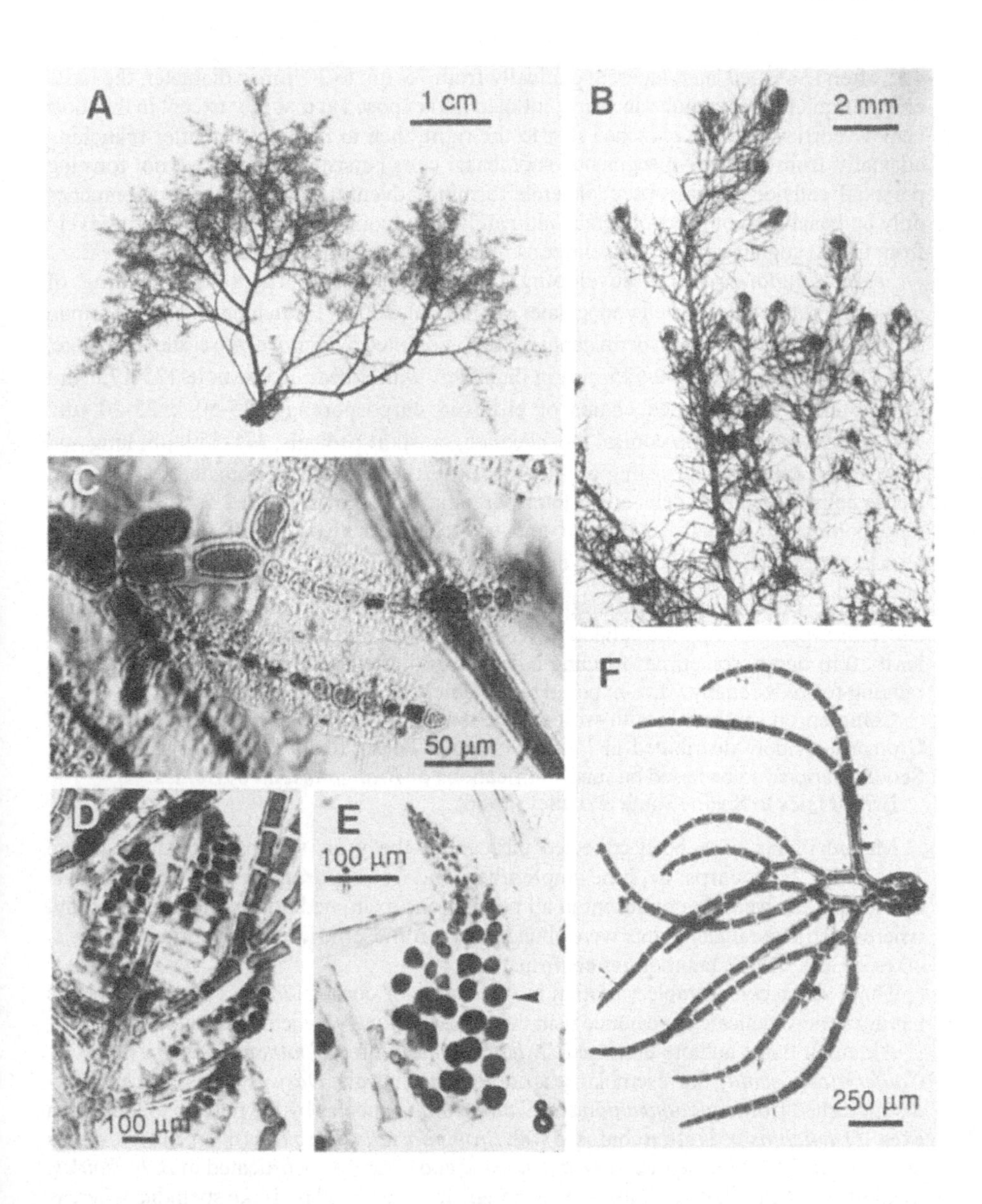
A
1 cm
B
2 mm
C
50 µm
D
100 µm
E
100 µm
F
250 µm

45°, often U-shaped later, tapering gradually from 75 μm to 30 μm in diameter, the basal cell triangular to rectangular in shape, inflated and exposed at maturity except in the most heavily corticated axes, branched first to the right, then to the left, thereafter branching adaxially from successive segments; suprabasal cells persistent and naked, not forming periaxial cells or rhizoids; pseudolaterals forming adventitious polysiphonous branches only on basal cells of pseudolaterals, ultimately deciduous, breaking down progressively from tips to suprabasal cells; adventitious monosiphonous filaments absent.

Spermatangial branches developing from the last 1-2 orders of branching of pseudolaterals, conical when young, later cylindrical, 200-375 μm long and 30-40 μm in diameter, bearing ellipsoid spermatangia 2 μm in diameter. *Cystocarps* sessile, urceolate, 525-800 μm long and 450-525 μm in diameter, with a beak-like ostiole 125-175 μm long, containing branched chains of ellipsoid carposporangia 35-50 x 25-30 μm. *Tetrasporangial stichidia* conical to lanceolate, on short pedicels, 375-550 μm long and 110 μm in diameter, consisting of 12-17 (-20) fertile segments terminated by 1-several sterile cells; 5 periaxial cells and 5 tetrasporangia per segment, with 2-3 cover cells per tetrasporangium, usually undivided and covering less than 1/3 of tetrasporangium; tetrasporangia spherical, 35-50 μm in diameter.

Epilithic on bedrock or pebbles, epiphytic on smaller perennial algae, particularly *Corallina* spp., on damp rock and in lower-shore pools, and from extreme low water to at least 10 m depth, sometimes forming a dense band at extreme low water level, at sites ranging from extremely wave-exposed to extremely sheltered.

Common on south and south-west coasts, rare in W. Scotland, northwards to Ross & Cromarty; widely distributed in Ireland; Channel Isles. Records from Orkney and N. Scotland appear to be based on misidentifications of *Brongniartella byssoides*.

British Isles to Spain (South & Tittley, 1986).

Mature plants have been collected throughout the year; spermatangia recorded in June-July, cystocarps in June-September and tetrasporangia in May-October. Gametophytes were rare or absent in all populations examined during the present study, whereas tetrasporangial plants were abundant. Chromosome number reported as $2n$ = c. 40 (see Cole, 1990), but requires confirmation.

Thalli show considerable variation in the degree of cortication, less corticated plants having a more delicate appearance than the robust, heavily-corticated plants.

Although early authors confused *D. hutchinsiae* with *Aglaothamnion sepositum* (as *Conferva arbuscula*), the resemblance is only superficial, and *Dasya* species can be readily distinguished from *Aglaothamnion* and *Callithamnion* species by the polysiphonous main axes. *D. hutchinsiae* is often confused with *Brongniartella byssoides* (q. v.), but main axes are corticated to some degree in *D. hutchinsiae* and entirely uncorticated in *B. byssoides*. In addition, *D. hutchinsiae* forms tetrasporangia in specialized pod-like stichidia, whereas in *B. byssoides* tetrasporangia are borne in lateral branches.

Dasya ocellata (Grateloup) Harvey in W. J. Hooker (1833), p. 335.

Lectotype: LD 43847. Unlocalized, undated Grateloup specimen.

Ceramium ocellatum Grateloup (1807), p. 34.
Hutchinsia ocellata (Grateloup) C. Agardh (1824), p. 158.
Dasya simpliciuscula C. Agardh (1828), p. 122.

Thalli consisting of erect axes growing singly or in tufts from a discoid rhizoidal holdfast bearing adventitious rhizoidal side branches that spread over substratum and give rise to further erect axes; erect axes 2-5 (-7) cm high, simple or with 1-3 basal lateral branches, in larger specimens bearing simple lateral branches above, clothed except at base with pseudolaterals, producing a cylindrical outline, tapering gradually from 190-570 μm near the base to 50-75 μm near apices, corticated throughout except at extreme tips; colour brownish- to purplish-red; texture of main axes rigid, pseudolaterals soft but not flaccid.

Main axes composed of 5 periaxial cells, forming elongate primary rhizoidal filaments that resemble the periaxial cells, linking to them by secondary pit connections; secondary rhizoidal cortical filaments 6-16 μm wide, with cells 3-13 diameters long, developing and forming numerous secondary pit connections terminally and laterally with neighbouring cells; cortex becoming 1-2 layers thick and giving rise to adventitious monosiphonous filaments and polysiphonous branches in lower parts of thallus; periaxial cells linked to basal cells of pseudolaterals in vertical files by secondary pit connections; *pseudolaterals* borne on successive axial segments in a regular 2/5 clockwise spiral, tapering gradually when mature from 40-56 μm at the base to 10-16 μm near tips, with 4-5 (-7) dichotomies forming 45° angles and with 1 (2-3) naked segments between branch-bearing segments, with the basal cell unbranched, and the suprabasal cell and every alternate segment branched, terminating in filaments up to 15 segments long, the tips of which are deciduous in older plants and often replaced by adventitious terminal filaments; basal cells becoming elongated and wholly covered by cortical filaments in older axes, suprabasal cells remaining emergent and surrounded by a collar of rhizoidal cells; adventitious branching of pseudolaterals rare or absent.

Male and female plants unknown in the British Isles. *Tetrasporangial stichidia* abundant on third and fourth-order branches of pseudolaterals, occasional on adventitious monosiphonous branches, borne on 1-4-celled pedicels, elongate with sharply pointed tips, 480-1350 μm long x 95-120 μm wide, consisting of 15-30 (-35) tetrasporangium-bearing segments, increasing in number as the tips develop further; 5 periaxial cells and 5 tetrasporangia per segment, each stalk cell bearing 3 cover cells which may be undivided or divided and branched, covering about 1/3 of the tetrasporangium; tetrasporangia spherical, 36-52 μm diameter.

Epiphytic on smaller perennial algae, epilithic on bedrock or pebbles, rarely epizoic on barnacles, in shaded lower-shore pools on moderately exposed coasts, on vertical and horizontal surfaces at extreme low water on extremely sheltered shores, and subtidal to

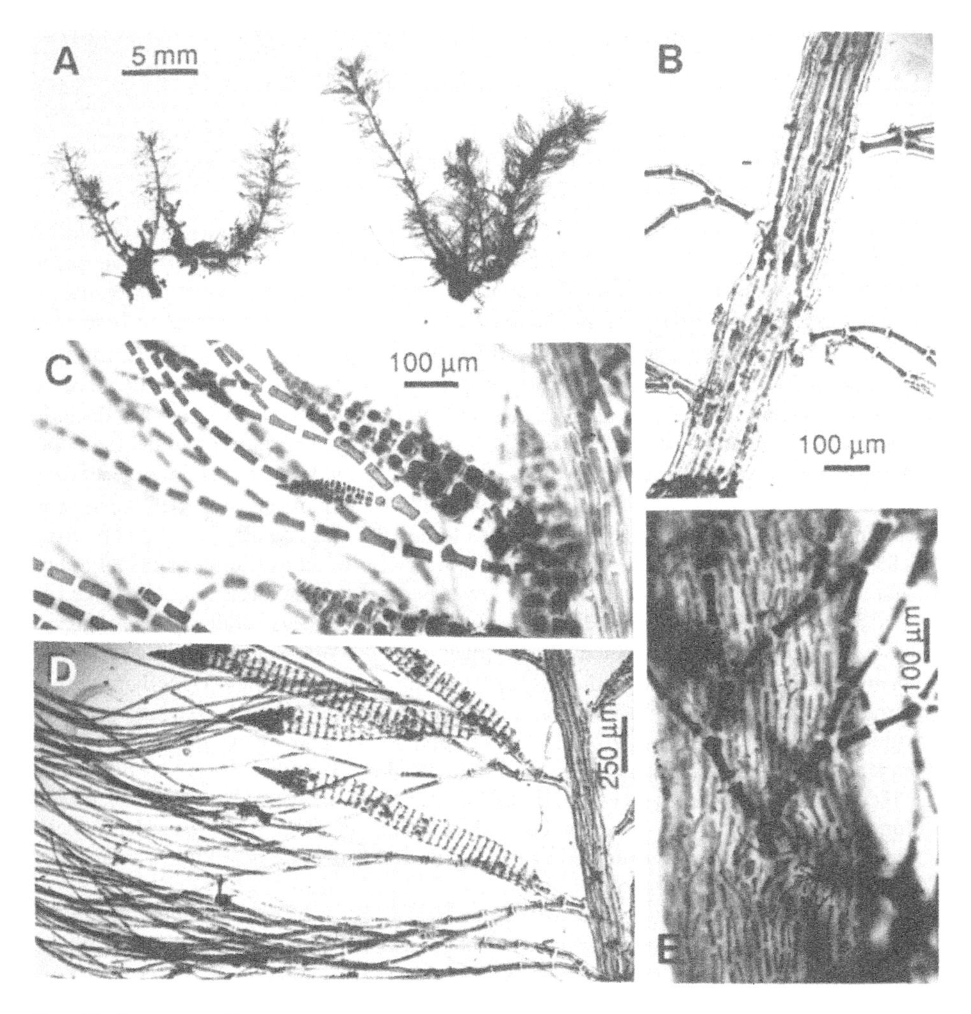

Fig. 85. *Dasya ocellata*

(A) Habit of typical plants with unbranched main axes (Devon, Oct.). (B) Young, lightly corticated axis bearing pseudolaterals that arise singly from unbranched, partly immersed basal cells and branch from the suprabasal cell but not the next cell (Cornwall, July). (C) Young tetrasporangial stichidia (Pembroke, Oct.). (D) Elongate old tetrasporangial stichidia borne on narrow pseudolaterals (as B). (E) Pseudolateral with suprabasal cell partly immersed by heavy cortication, showing branching pattern (as C).

20m depth at sheltered and exposed sites; usually rare, occurring as isolated individuals, occasionally abundant in favourable habitats.

South and south-west coasts: S. Devon, Cornwall, Pembroke; Dublin, Wicklow, Waterford, Cork and Kerry; Channel Isles. Records from Scotland are based on misidentifications of *Brongniartella byssoides*.

British Isles to Morocco and Canaries; Mediterranean; Caribbean (Taylor, 1960; Ardré, 1970).

Plants have been collected in July-November, with tetrasporangia throughout this period. The life history of *D. ocellata* in the British Isles appears to involve only tetrasporophytes, although female plants have been reported from Portugal (Ardré, 1970) and males from the Mediterranean (Falkenberg, 1901).

Thalli show little variation other than in the degree of branching of main axes.

Dasya punicea Meneghini ex Zanardini (1841), p. 66.

Type: Museo Civico Venice? A specimen from the type locality (Venice) was listed by De Toni & Levi (1888).

Dasya cattloviae Harvey in Gatty (1872), p. 91.

Thalli 5-14 cm high, consisting of a single main axis, attached by a discoid rhizoidal holdfast up to 2 mm in diameter, tapering gradually from 0.7-1.2 mm near base to 0.2 mm near apices, heavily corticated, bearing branches irregularly alternately at irregular intervals; major laterals decreasing in length upwards; younger axes densely clothed with pigmented pseudolaterals, becoming denuded below when older except for a few broken remnants; colour red to reddish-brown, occasionally brown or purple; texture of main axes tough and cartilaginous, pseudolaterals soft and flaccid.

Main axes with 5 periaxial cells per segment, linked by longitudinal secondary pit connections, giving rise to cortication near the apices, composed of rhizoidal filaments 6-8 µm wide that cover periaxial cells of older axes with 1-2 or more layers of cortex; *pseudolaterals* borne on successive axial segments in a regular 2/5 spiral, tapering gradually when mature from 50-60 µm at the base to 12-20 µm near tips, branched pseudodichotomously 5-6 times at a narrow angle, the basal cell and/or suprabasal cell branched, with subsequent branching unequal on either sides of dichotomy, one side branched from the next segment, the other side with 1 (-2) unbranched segments before the next branch, often terminated when young by multicellular hairs 4-8 µm in diameter; adventitious monosiphonous laterals frequent; *plastids* elongate or reticulate, becoming filiform.

Male thalli unknown. *Cystocarps* sessile, urceolate, 850-1125 µm long and 500-725 µm in diameter, with a straight or curved beak-like ostiole 300 µm long, containing branched chains of ellipsoid carposporangia 35-50 x 30-35 µm. *Tetrasporangial stichidia* borne on pseudolaterals, initially conical, then lanceolate, on long pedicels, 425-725 µm

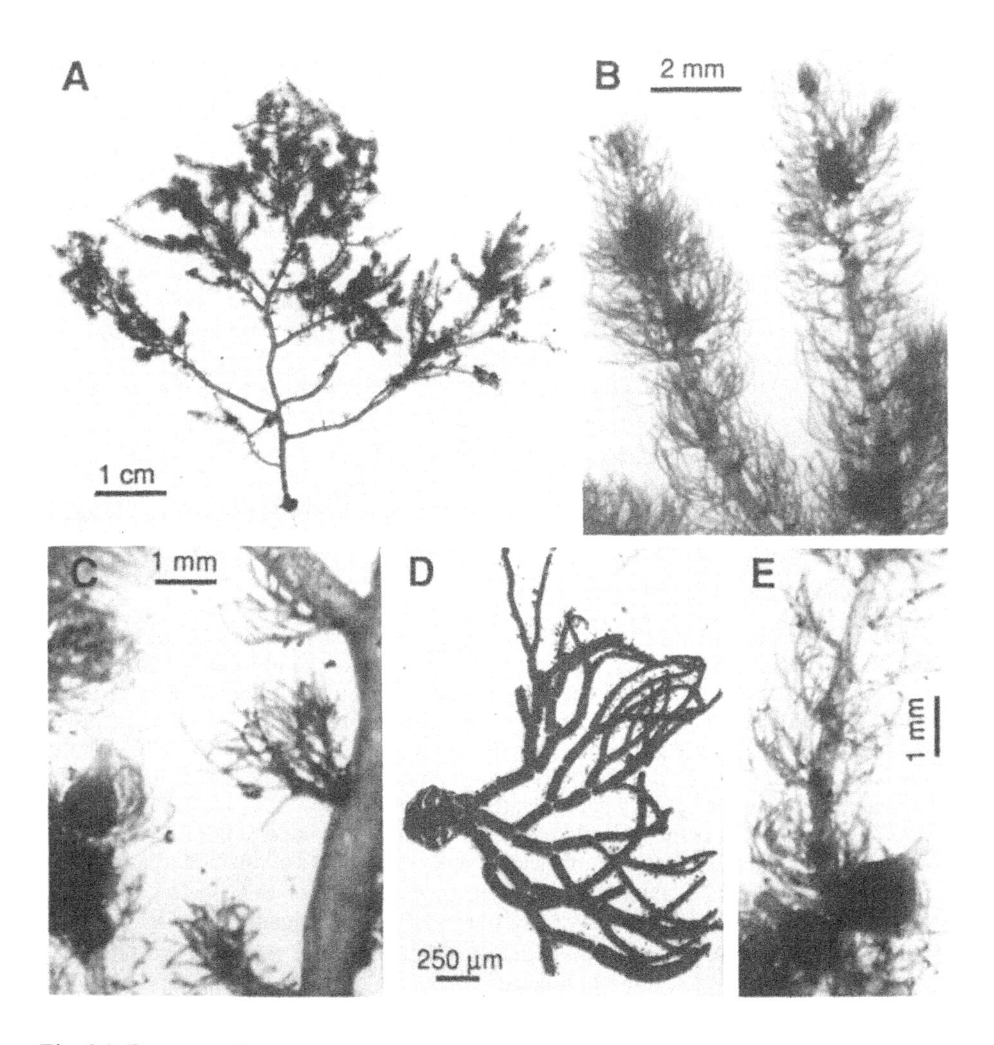

Fig. 86. *Dasya punicea*

(A) Habit of plants with naked lower main axes (A-E Dorset, July). (B) Dense arrangement of pseudolaterals. (C) Old, nearly naked axis with a few tufts of pseudolaterals, and a cystocarp. (D) T.S. of axis with pseudolateral, branched unequally on either sides of dichotomies, and an adventitious monosiphonous lateral borne on the basal cell. (E) Cystocarp with elongate ostiole.

long and 125-150 μm in diameter, consisting of up to 30 fertile segments terminated by 1-several sterile cells; each segment composed of 5 pericentral cells and bearing 5 tetrasporangia; tetrasporangia spherical, 35-40 μm in diameter.

On sandy or friable bedrock at 10-13 m depth on fairly sheltered coasts, growing as isolated plants amongst foliose algae; the great majority of British specimens were apparently collected as drift.

South coast of England (Sussex, Dorset, S. Devon); Channel Isles.

England to N. France; Mediterranean; Caribbean (Taylor, 1960; South & Tittley, 1986).

Plants have been collected only in July, August and October. Spermatangia have not been noted, cystocarps were observed in July and August, and tetrasporangia in July, August and October.

This species shows relatively little variation.

HETEROSIPHONIA Montagne, nom. cons.

HETEROSIPHONIA Montagne (1842), p. 4.

Type species: *H. berkeleyi* Montagne (1842), p. 5.

Thalli dorsiventrally organized, terete or compressed, polysiphonous, consisting of 1-several erect, alternately branched main axes arising from a discoid holdfast or prostrate axes; pseudolaterals separated by 2-9 internodal segments, monosiphonous, basally polysiphonous, or polysiphonous except at the tips, pigmented, persistent, 3-7 times subdichotomously branched; lateral axes arising from basal cells of pseudolaterals or replacing them; vegetative periaxial cells 4-12, cut off in alternating sequence, longitudinally connected by secondary pit connections, uncorticated or corticated by short or long descending rhizoidal filaments issuing from basal cells of pseudolaterals and lower ends of periaxial cells; adventitious monosiphonous filaments absent or present, borne on cortical cells.

Growth sympodial, initiated by oblique division of a uninucleate or multinucleate apical cell, the apical initial diverted laterally forming a pseudolateral or lateral branch, and the subapical cell continuing the main axis; pseudolaterals at first branched adaxially, subdichotomous at maturity; vegetative periaxial cells formed dorsiventrally in alternating sequence with the first cut off below the pseudolateral in nodal and internodal segments, the second towards the dorsal side, the third towards the ventral side, etc; in polysiphonous pseudolaterals the first periaxial cell abaxial, the second dorsal, the third ventral, etc.

Gametophytes dioecious. Spermatangial axes formed on monosiphonous portions of pseudolaterals; periaxial cells 4(-5) per segment, cut off in alternating sequence, dividing transversely and anticlinally to produce a single layer of spermatangial mother cells, each bearing 3-4 spermatangia. Procarps formed on segments below subdichotomies of a pseudolateral from the last (usually the fifth) periaxial cell on the adaxial side, consisting of a supporting cell, a 4-7 celled lateral sterile group, a 4-celled carpogonial branch and a 2-4 celled basal sterile group; prefertilization pericarp weakly developed before fertilization, formed by division of the third and fourth periaxial cells; fertilized carpogonium cutting off 2 connecting cells, one of which fuses with the auxiliary cell; sterile groups each usually dividing once after fertilization and their nuclei enlarging; gonimoblasts monopodially and unilaterally branched, bearing carposporangia in branched chains; fusion cell conspicuous, incorporating the supporting cell, auxiliary cell, central cell, and innermost gonimoblast cells; pericarp composed of about 14 filaments derived from the third and fourth periaxial cells; cystocarps flask-shaped, 3-5 cell layers thick, ostiolate, with or without a prominent beak. Tetrasporangia borne in whorls, 4-6 per segment, in modified axes (stichidia), either directly on the pseudolaterals, or on special monosiphonous or polysiphonous stalks; periaxial cells 5-7, cut off in alternating sequence, dividing longitudinally to produce 2 presporangial cover cells and then transversely into a stalk cell and a tetrasporangium; cover cells each dividing once forming 4 cells that cover the mature tetasporangium.

References: Rosenberg (1933), Parsons (1975).

One species in the British Isles.

Heterosiphonia plumosa (Ellis) Batters (1902), p. 83.

Lectotype: Ellis (1768), pl. 18, fig. C, probably based on material from Sussex.

Conferva plumosa Ellis (1768), p. 424.
Conferva coccinea Hudson (1778), p. 603.
Dasya coccinea (Hudson) C. Agardh (1828), p. 119.

Thalli consisting of erect axes 10-30 cm high growing singly or in tufts from a discoid holdfast up to 1 cm in diameter, composed of densely interwoven stolon-like prostrate axes that form matted rhizoids and can give rise to further erect axes; main axes distinct, 0.8-1.2 mm in diameter, compressed when young, becoming terete, branched in one plane irregularly alternate-distichously to 3-4 orders of branching, with all orders of branching bearing an alternate-distichous array of branched pigmented pseudolaterals, except where old axes have become denuded basally; bright red to dark red in colour, with a rigid texture and cartilaginous old axes.

Main axes composed of a wide axial cell and 9-10 periaxial cells, 2 of which are larger

---

Fig. 87. *Heterosiphonia plumosa*

(A) Habit (A-C Donegal, Dec.). (B) Main polysiphonous axis with alternate main branching. (C) Alternate arrangement of pseudolaterals with polysiphonous bases and monosiphonous ultimate filaments. (D) Cystocarps (Cornwall, Sep.). (E) Tetrasporangial stichidia (arrows) among dense pseudolaterals (as D).

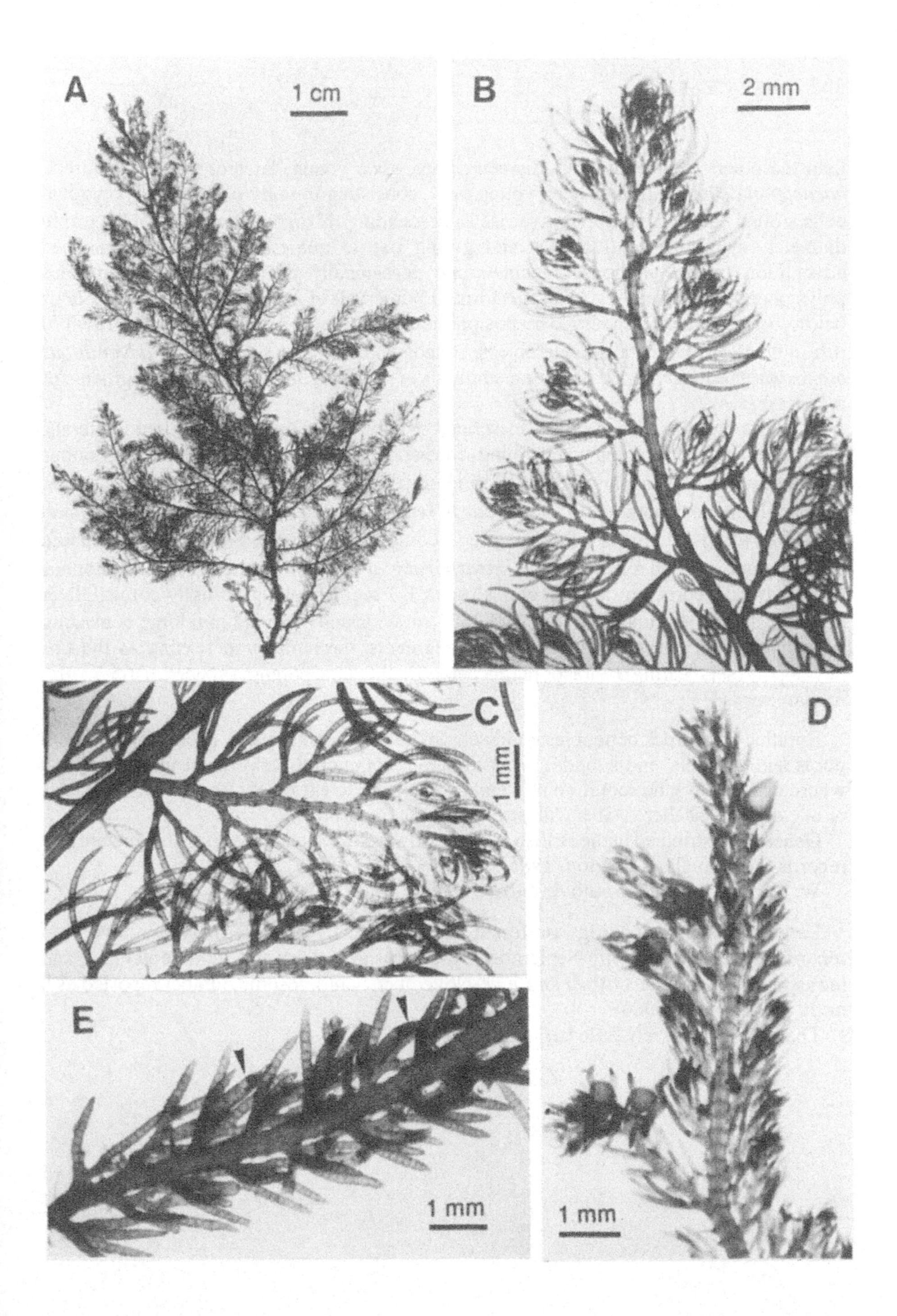
A
1 cm
B
2 mm
C
1 mm
D
E
1 mm
1 mm

than the others; segments 0.7-1 diameter long when young, later obscured by cortex; *cortication* developing rapidly on young axes, consisting initially of a ring of polygonal cells around each segment that give rise to descending rhizoidal filaments 25-40 μm in diameter, increasing in thickness and giving rise to numerous simple or branched adventitious monosiphonous branchlets and occasionally when older to adventitious polysiphonous laterals; *pseudolaterals* formed at intervals of 2-3 segments, polysiphonous below, bearing alternate incurved monosiphonous branchlets 12-16 cells long and 130-170 μm in diameter, with short acute apices, monosiphonous towards the tips; *adventitious monosiphonous branches* developing singly or in pairs on some unbranched segments of major axes.

Spermatangial branches borne in series on polysiphonous bases of pseudolaterals, resembling conical stichidia when young, consisting when mature of short polysiphonous stalk, cylindrical spermatangial axis bearing ellipsoid spermatangia 2-3 μm in diameter, and sterile monosiphonous tip. *Cystocarps* sessile, urceolate, 900-1000 μm long and 650-750 μm in diameter, thick-walled, containing branched chains of ellipsoid carposporangia 35-60 x 30-45 μm. *Tetrasporangial stichidia* formed on polysiphonous pseudolaterals, borne on polysiphonous stalks 1-3 segments long, initially conical, later cylindrical, with short pointed tips, 200-250 μm wide and up to 1.2 mm long, consisting initially of c. 12 tetrasporangium-bearing segments, the number increasing as the tips develop further, bearing 6 tetrasporangia per segment; tetrasporangia spherical, 65-75 μm in diameter.

Epilithic on bedrock or pebbles, rarely epiphytic on perennial algae, in deep lower-shore pools and channels, and subtidal to 33 m depth, common in kelp forest underflora and where exposed to sand scour, on moderately sheltered to extremely wave-exposed coasts or at extremely sheltered sites with strong current exposure.

Generally distributed in the British Isles, northwards to Shetland, although there are few records for E. England (Norton, 1985).

W. Norway to Spain (South & Tittley, 1986).

Large plants and sporelings are found throughout the year. Spermatangia have been recorded in January and July-September, cystocarps in January and June-September, and tetrasporangia in January-April and June-November. Chromosome number reported as n or $2n = 44$ (Magne, 1964).

Thalli show relatively little variation.

RHODOMELACEAE Areschoug

RHODOMELACEAE Areschoug (1847), p. 260 [as Rhodomeleae].

Germination bipolar by unequal division of spore into small primary rhizoid initial and large upper cell that divides transversely and longitudinally. Primary axis initially erect, attached by rhizoids or a rhizoidal disc; secondary axes either erect or prostrate. Growth by oblique or transverse division of large apical cell containing a prominent nucleus, if oblique resulting segment usually producing a lateral branch, if transverse the segment remaining naked; size of lateral initial and type of branch correlated with steepness of plane of division; primary branching exogenous with lateral initial issuing from high side of axial segment prior to first periaxial cell, or endogenous with lateral initial arising from distal end of axial cell after a full complement of periaxial cells have been cut off. Lateral branches developing into indeterminate or determinate axes (short shoots, phylloides, or trichoblasts); short shoots polysiphonous, cylindrical to compressed and either simple or branched up to 5-6 orders; phylloides polysiphonous, flattened and leaf-like; trichoblasts monosiphonous and either simple or branched to 5(-6) orders, the branching alternate-distichous to pseudodichotomous; normal and adventitious branches issuing from basal cells of trichoblasts; adventitious branches arising from trichoblast scar cells, periaxial cells or cortical cells. Branch symmetry either radial or dorsiventral, if radial, either spiral or alternate-distichous (divergence = 1/2). Each axial cell cutting off 4-24 periaxial cells close to apex, the first directly underneath the branch initial in branch-bearing segments or directly below a branch in naked segments, the second to the right or left of first, the third next to first on opposite side, the fourth adjacent to second, etc., in an alternating (rhodomelacean) sequence; in radial thalli second periaxial cell cut off in direction of spiral of lateral branches, to right if spiral right-handed, to left if left-handed*, with the remaining periaxial cells cut off in alternating sequence; in dorsiventral thalli the second periaxial cell cut off towards dorsal surface, to left if nearest branch above diverges to right, to right if nearest branch above diverges to left; lateral branches either maintaining their position directly above first periaxial cell or shifting to

* Using the physicists' rule, if the thumb is oriented in the direction of the apex the spiral follows the curl of the fingers of the right hand in a right-handed spiral, or the fingers of the left hand in a left-handed spiral.

position between the first and third periaxial cells; periaxial cells in successive segments either orthostichous or staggered in zig-zag fashion, either elongating together with the axial cell and becoming the same length or dividing transversely into 2-4(-6) tiers of cells alongside elongating axial cell; thalli ecorticate or weakly to strongly corticated, if corticated the periaxial cells partly or wholly covered by cortical filaments corresponding either to determinate lateral filaments or to descending rhizoidal filaments; cortical filaments emanating from anterior and posterior ends of periaxial cells and producing 2-6 (-many) orders of quaternately to subdichotomously branched filaments or rhizoidal in character, arising singly or in pairs from lower side of periaxial cells after periaxial cell elongation, and either overgrowing or descending between the periaxial cells below; secondary pit connections formed longitudinally between periaxial cells and sometimes also cortical cells.

Spermatangia formed on simple or compound polysiphonous axes (spermatangial axes) on modified trichoblasts or, less often, on ordinary polysiphonous axes, if on trichoblasts the spermatangial axis borne on a monosiphonous stalk two or more cells long, and either naked or subtended by a monosiphonous lateral filament; segments of spermatangial axes typically bearing 4(-5) periaxial cells and cortical filaments that tend to lie in a single layer; spermatangial mother cells formed from surface cortical cells, each bearing 2-3(-4) spermatangia; spermatangia separate or confluent in a common matrix, if confluent then producing multiple crops in which recently-formed spermatia replace spermatia that have been released; spermatangial axes ending bluntly or surmounted by a monosiphonous hair, simple or compound, sometimes ending in inflated sterile cells or united into plate-like structures fringed by inflated sterile terminal cells.

The female reproductive system of the Rhodomelaceae follows one of two different developmental patterns: one diagnostic for the subfamily Bostrychioideae, the other the subfamily Rhodomeloideae containing the rest of the tribes of the Rhodomelaceae. These are described separately as follows:

BOSTRYCHIOIDEAE: procarps formed near apex in 2-6(-many) successive segments on unspecialized polysiphonous branches; fertile segments each containing (4-)5(-6) periaxial cells and 1-4 procarps, with the first periaxial cell usually sterile; procarp consisting of a supporting cell (= fertile periaxial cell), a lateral sterile group-1 composed of 2-6 cells, and a (3-)4-celled carpogonial branch; second sterile group absent; pericarp absent prior to fertilization. Supporting cell dividing transversely after fertilization, cutting off an auxiliary cell; carpogonium dividing twice cutting off a capping cell to the outside and a connecting cell facing the auxiliary cell; each cell of sterile group-1 dividing once, doubling number of sterile cells; auxiliary cell undivided or dividing several times transversely to produce row of 2-4(-8) primary gonimoblast cells; gonimoblasts initially branched monopodially and bearing terminal clavate carposporangia, replaced by sympodial branching from subterminal cells with the terminal cells maturing successively into carposporangia; pericarp consisting of cortical filaments initiated on all sides of the fertilized procarp, the terminal cells transforming into apical initials of up to 14 pericarpic axes that converge to form an ostiole; pericarp segments cutting off 2(-3) periaxial cells towards the surface, each of which forms a cortical filament, corticated even in species in which the vegetative axes are ecorticate; mature sterile group-1 either resembling ordinary cortical filaments, or the terminal cells greatly enlarged, filling the cystocarp cavity and containing enlarged nuclei.

RHODOMELOIDEAE : procarps borne on polysiphonous lateral branches terminated by branched or unbranched monosiphonous filaments (trichoblasts), usually with 1 procarp per branch on suprabasal segment; exceptionally 2-3(-4) procarps per branch; fertile segment bearing (4-)5 periaxial cells, the last-formed, adaxial periaxial cell (usually the fifth) producing the procarp; supporting cell (= fertile periaxial cell) bearing a lateral sterile group-1 composed of 2(3-6) cells, a (3-)4-celled carpogonial branch, and a basal sterile group-2 composed of 1(2-4) cells (absent in tribe Lophothalieae); a prefertilization pericarp developing primarily from the two lateral periaxial cells (usually 3 and 4), composed of branched cortical filaments and forming valvate structures on either side that nearly enclose the procarp; trichogyne initially directed anteriorly, parallel to trichoblast. Supporting cell enlarging and cutting off auxiliary cell transversely or, in exceptional cases, functioning directly as an auxiliary cell; fertilized carpogonium cutting off a distal capping cell and a posteriolateral connecting cell that lies between the auxiliary cell and central cell; cells of sterile groups 1 and 2 each dividing once, doubling the number of sterile group cells; diploid nucleus expanding within auxiliary cell after fusion with connecting cell, the auxiliary cell then dividing transversely cutting off a single gonimoblast initial; postfertilization pericarp arising from terminal cells of prefertilization pericarp and converting into apical initials of up to 16(-18) axial filaments which converge and form the ostiole; each axial cell of a pericarpic filament normally cutting off 2 (-3) periaxial cells towards the outside which may remain ecorticate or form a cortex; gonimoblasts initially monopodially branched and bearing terminal carposporangia, afterwards the branching sympodial with successive terminal cells maturing into carposporangia; supporting cell, auxiliary cell, inner gonimoblast cells, central cell and inner sterile group cells fusing to form a large central fusion cell; mature carposporangia released through the terminal ostiole.

Tetrasporangia produced on main axes, polysiphonous lateral branches, or regenerated axes born on trichoblasts. Each fertile segment bearing 1 or 2(3-5) tetrasporangia; fertile periaxial cell cutting off 2(3) cover cells which remain unbranched, or in exceptional cases form a cortex, followed by transverse division cutting off a single tetrasporangium; a third, postsporangial cover cell may be formed. Tetrasporangia dividing into four tetrahedrally arranged tetraspores that are ordinarily released as a unit through a pore between paired cover cells.

References: Scagel (1953), Kylin (1956), Hommersand (1963), Parsons (1975).

Subfamily BOSTRYCHIOIDEAE Hommersand (1963), p. 334.

Tribe BOSTRYCHIEAE Falkenberg (1901), p. 504.

BOSTRYCHIA Montagne, nom. cons.

BOSTRYCHIA Montagne (1842), p. 39.

Lectotype species: *B. scorpioides* Montagne ex Kützing (1849), p. 893 [see Schmitz (1889), p. 449].

Thalli dorsiventral, often with the tips curved towards the ventral side, consisting of prostrate indeterminate axes attached to the substratum either by clusters of rhizoids derived from periaxial or cortical cells, or by rhizoidal haptera borne terminally on special branches, and bearing 0-3(-4) orders of creeping or erect alternate or irregularly arranged determinate lateral branches; polysiphonous throughout, or the ultimate branches partly or entirely monosiphonous; periaxial cells 4-9, each dividing transversely into two tiers of cells with the basal cell retaining the pit connection to the axial cell; uncorticated or corticated by up to 5 layers of cortical cells.

Branching exogenous by oblique division of the apical cell, rarely endogenous from axial cells adjacent to branches bearing haptera; trichoblasts absent; periaxial cells formed in alternating sequence with the first cut off dorsally towards the abaxial side, the second dorsal towards the adaxial side and the rest lateral or ventral; cortical filaments (when present) ternately or dichotomously branched, the cortication of the basal tier cell completed prior to cortication of the second tier cell.

Gametophytes monoecious or dioecious. Spermatangia superficial on polysiphonous branches or transformed monosiphononous filaments 5-25 segments long, straight or coiled, and blunt or terminated by a sterile tip; spermatangial mother cells each producing 1-3(-4) spermatangia. Procarps formed in series near the apex on ordinary determinate branches containing (4-)5 periaxial cells per segment, any periaxial cell except the first potentially fertile and often with two or more procarps per segment; each procarp consisting of a single sterile group and (3-)4-celled carpogonial branch; prefertilization pericarps absent; fertilized carpogonium cutting off a distal and a posteriolateral connecting cell, the latter fusing directly with the auxiliary cell; cells of the sterile group each dividing once after fertilization, sometimes also enlarging apically; auxiliary cell cutting off a gonimoblast initial which may divide transversely 3-7 times producing a row of primary gonimoblast cells that branch subdichotomously and bear clavate carposporangia terminally, thereafter the gonimoblasts sympodially branched from subterminal cells below the differentiating carposporangia; cystocarp subterminal, ovoid to globular, ostiolate; pericarp initiated after fertilization, composed of 6-14 axial filaments arising from periaxial cells surrounding the procarp, corticated as in vegetative axes; fusion cell absent. Tetrasporangia borne in ordinary polysiphonous branches or specialized terminal filaments (stichidia) up to 25 segments long and either blunt or terminated by a sterile tip; fertile segments containing 4-5 periaxial cells, with most or all bearing tetrasporangia; tetrasporangia protected by 2-3 cover cells and their tier cells, or by branched cortical filaments like those in vegetative axes.

References: Hommersand (1963), King & Puttock (1989).

One species in the British Isles.

Bostrychia scorpioides (Hudson) Montagne ex Kützing (1849), p. 893.

Holotype: Dillenius in Ray (1724), pl. 2, fig. 6, based on a specimen in OXF (see Irvine & Dixon, 1982).

Fucus scorpioides Hudson (1762), p. 471.
Fucus amphibius Hudson (1778), p. 590.

Thalli forming dense turfs 2-6 cm high, either loosely embedded in mud or attached to substrata by haptera formed by prostrate parts of axes; erect axes terete, branched alternate-distichously to 4 orders in a complanate arrangement, with a triangular outline, at wide branch angles up to 90°; main axes 0.3-0.45 mm in diameter, bearing first-order laterals that decrease in length upwards but are often reduced to broken stumps; mature ultimate laterals 60-110 μm in diameter, straight and spine-like, tapering gradually to pointed apices; tips of all orders of branching tightly inrolled on lower side of thallus; peg-like haptera developing at branch nodes on prostrate thalli, consisting of fused groups of multicellular rhizoidal filaments 10-15 μm in diameter; thalli dull brownish-purple, bleaching to dull yellow, main axes tough and flexible.

Main axes growing from apical cells 12-14 μm in diameter, fully corticated but with the internal structure visible in young axes, consisting of thick-walled axial cells, 8-15 μm in diameter and 75-100 μm long, each surrounded by 2 tiers of 6 cells c. 7 μm in diameter, bearing the periaxial cells in a median position, with a second ring of derivative cells cut off above them; *cortex* of young axes consisting of a single layer of round to oval cortical cells 8-14 μm in diameter embedded in wall material, later increasing to 3-4 cells deep; *medullary cells* formed in older axes, consisting of 1-2 layers of elongate cells surrounding periaxial cells; all orders of branching including ultimate laterals fully corticate; *plastids* discoid in outer cortical cells.

Plants dioecious. *Spermatangial sori* formed on last 1-2 orders of branching, continuous over cylindrical branchlets 130-200 μm in diameter, up to 1.2 mm long, consisting of a layer of spermatangial mother cells bearing cylindrical spermatangia 10-13.5 μm x 2.5-4 μm. *Cystocarps* spherical, 350-450 μm in diameter, with small, indistinct ostiole;

carposporangia clavate, 100-120 x 20-25 μm. *Tetrasporangial stichidia* cylindrical, c. 250 μm in diameter, normally terminal on reproductive branchlets, but occasionally subterminal because apex resumes vegetative growth (Prud'homme van Reine & Sluiman, 1980); tetrasporangia borne 1-4 per fertile segment, spherical, 80-100 μm in diameter; tetraspores released in tetrads, rounding up to 40-80 μm in diameter (Prud'homme van Reine & Sluiman, 1980).

Growing on mud, plants, and stone and wooden structures near extreme high water, forming dense stands among flowering plants in salt marshes, at wave-sheltered sites, frequently in estuaries and other low-salinity situations.

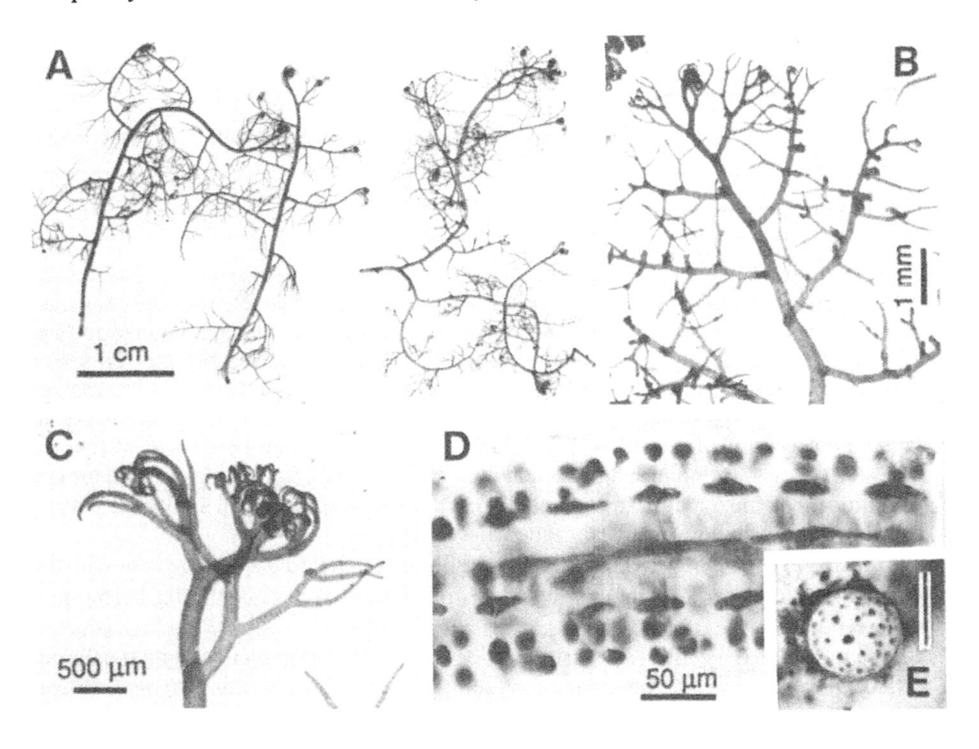

Fig. 88. *Bostrychia scorpioides*

(A) Habit of plants removed from turf (Down, Oct.). (B) Detail of thallus showing alternate arrangement of major laterals (B-E Pembroke, undated). (C) Tightly inrolled filaments at apex. (D) Axial structure visible through cortex, with 2 tiers of cells surrounding each axial cell, bearing cortical cells. (E) T.S. of axis showing axial cell and 6 periaxial cells surrounded by cortex (scale = 100 μm).

Coasts of England, Wales and Scotland, northwards to Orkney; Ireland; Channel Isles.

British Isles to Morocco; Mediterranean; S. Africa, S.E. Australia, New Zealand and South America (Prud'homme van Reine & Sluiman, 1980, including records of *B. harveyi* Montagne, *B. australasica* Sonder, *B. distans* Harvey in J. D. Hooker and *B. vulgaris* Haufe).

Large plants occur throughout the year, but reproductive structures are rarely present; vegetative fragmentation appears to be the main form of reproduction. Specialized reproductive structures are formed only in late summer: spermatangia have been recorded in August; cystocarps in July-September; and tetrasporangia in August-September. All records of gametangia in the British Isles are from the south coast of England in the years 1882-1888 (Prud'homme van Reine & Sluiman, 1980). High temperatures are required for tetrasporangial development (Prud'homme van Reine & Sluiman, 1980). Non-reproductive field-collected plants from the Netherlands placed under culture conditions formed tetrasporangial stichidia within 10 days at 20°C, under 18 h days. Fewer stichidia developed at 15°C, and at 12°C the only stichidia formed were sterile. Released spores germinated at 15°C and 20°C but the germlings failed to reach maturity in culture. Prud'homme van Reine & Sluiman suggested that although European *B. scorpioides* might potentially undergo a dioecious *Polysiphonia*-type life history, this does not normally take place in the field. Environmental factors, particularly the lack of sufficiently long periods at high temperatures, might prevent completion of the life history in N.W. Europe. In southern Europe, however, climate was unlikely to disrupt a sexual life history, and other causes required investigation. Two possible explanations of the lack of gametophytes were that haploid gametophytes were of low viability, or that tetrasporophytes gave rise to diploid tetraspores which recycled the tetrasporophytic phase.

B. scorpioides shows little morphological variation, and is easily recognizable in the British Isles as the only member of this distinctive genus.

Subfamily RHODOMELOIDEAE Hommersand (1963), p. 335.

Tribe RHODOMELEAE (Areschoug) Schmitz (1889), p. 446.

RHODOMELA C. Agardh, nom. cons.

RHODOMELA C. Agardh (1822), p. 368.

Lectotype species: *R. subfusca* (Woodward) C. Agardh [= *R. confervoides* (Hudson) Silva (1952), p. 269].

Thalli perennial, terete, consisting of one to several erect main axes up to 30 (-45) cm high from a discoid base and lateral stolons; spirally branched at 3-5 mm intervals, with up to six orders of progressively shorter, more slender branches; vegetative trichoblasts present, faintly pigmented when young, colourless and 5-6 times pseudodichotomously branched at maturity, deciduous; adventitious branches mostly axillary; periaxial cells 6 (-7), divided transversely into several tiers of cells, with the uppermost cell retaining the pit connection to the axial cell; cortication 2-3 cell layers thick above, becoming 6-8 (-10) layers below.

Growth exogenous, initiated by slightly oblique division of the apical cell, each segmental cell either producing a vegetative trichoblast or a lateral branch in a 1/4 to 2/7 right-handed or left-handed spiral; periaxial cells formed in rhodomelacean sequence with the first cut off directly below the branch initial, periaxial and cortical cells dividing transversely and periclinally, forming successive layers of descending cortical files; adventitious branches arising from superficial cortical cells.

Reproductive structures on ordinary and/or adventitious branches. Gametophytes dioecious. Spermatangia superficial on polysiphonous branches or modified trichoblasts, the cortex 2 cell layers thick; spermatangial mother cells each cutting off 2 spermatangia by oblique divisions. Procarps originating from the fifth periaxial cell of the suprabasal segment of a fertile trichoblast, consisting of a 2-celled lateral sterile group, a 4-celled carpogonial branch and a 1-celled basal sterile group, pre-fertilization pericarp present, derived primarily from the second and third periaxial cells; fertilized carpogonium cutting off a distal and a posteriolateral connecting cell, the latter fusing with an extension from the inner side of the auxiliary cell; cells of sterile groups each dividing once after fertilization; auxiliary cell cutting off a single, sickle-shaped gonimoblast initial that branches subdichotomously and bears clavate carposporangia terminally, thereafter the gonimoblasts sympodially branched from subterminal cells below the differentiating carposporangia; fusion cell conspicuous, incorporating the central cell, supporting cell, auxiliary cell and most inner gonimoblast cells; cystocarp globose, ovoid or urceolate, surrounded by a pericarp 2-3 cell layers thick with a terminal ostiole. Tetrasporangia formed in opposite pairs from the third and fourth periaxial cells in penultimate and/or ultimate branches, tetrahedrally divided, protected by two conspicuous cover cells.

References: Kylin (1914, 1934), Rosenvinge (1923-24), Masuda (1982).

The North Atlantic members of this genus show a high degree of morphological variability, which has led to widely differing views of species taxonomy. Three species, *R. confervoides*, *R. lycopodioides* and *R. virgata* Kjellman, have been reported from N.W. Europe, but Rosenvinge (1923-24) considered that the last two should be regarded as varietal forms of *R. confervoides* (as *R. subfusca*). Rosenvinge investigated the formation of adventitious branching, previously considered to be diagnostic of *R. lycopodioides*, and showed that it also occurred in *R. confervoides*. The most recent checklists for the British Isles (e. g. South & Tittley, 1986) include both *R. confervoides* and *R. lycopodioides*. These are distinguished mainly by overall habit, but as noted by Harvey (1846), "specimens are sometimes found which have an intermediate character". Price & Tittley (1978) suggested that the taxonomic status of *R. lycopodioides* had yet to be established.

Rosenvinge (1923-24) discussed the morphological and phenological features considered by Kjellman to be diagnostic of *R. virgata* from the Arctic. The most important of these were (1) the formation of short-lived reproductive branchlets in winter on main axes of the previous year's growth, in contrast to the formation in *R. confervoides*, during spring, of reproductive branchlets that later become vegetative; and (2) gradual decrease of cell size towards the outer cortex in *R. virgata* vs. the distinct difference in *R. confervoides* between the large-celled medulla and deeply-pigmented, small-celled cortex. From a comparative study of these features in Danish *Rhodomela* populations, Rosenvinge concluded that the apparent differences between *R. virgata* and *R. confervoides* represented responses to ecological conditions, the '*R. virgata*' form being characteristic of plants from deeper water. Intermediate morphologies were observed, in particular the presence of both short-lived and persistent reproductive branchlets on the same plants. Kornmann & Sahling (1978) likewise reported that *R. virgata* grew in the sublittoral at Helgoland, whereas *R. confervoides* occurred in the lower intertidal. They reported that *R. virgata* could be distinguished from *R. confervoides* by the coarser, tougher thalli and the formation, on main axes, of reproductive branchlets, that were shed after spore release, in contrast to reproduction at apices in *R. confervoides*. Although *R. virgata* has not been reported for the British Isles, Hiscock (1986) described plants from eastern England that were closely similar to those from Helgoland identified as this species. She suggested that a gradation might exist between *R. lycopodioides*, *R. confervoides* and *R. virgata*. After examination of a range of specimens collected subtidally and intertidally from different types of habitat in the British Isles, we concur with Rosenvinge in concluding that plants corresponding to *R. virgata* and *R. confervoides* cannot be regarded as distinct species. Considerable morphological overlap occurs and the most significant feature, the formation of short-lived vs. persistent reproductive branchlets, is variable within individual thalli.

We recognize the occurrence of *R. lycopodioides* on northern shores of the British Isles, largely on the basis of habitat and general morphology, but particular specimens can be difficult to identify. These conclusions are tentative, and a complete taxonomic revision of the North Atlantic members of this genus is required.

KEY TO SPECIES

Epilithic and on mobile substrata; second-order laterals of varying lengths, adventitious branches few .... *R. confervoides*

Growing on kelp stipes; second-order and subsequent laterals more or less equal in length, adventitious branches abundant, obscuring primary branches .... *R. lycopodioides*

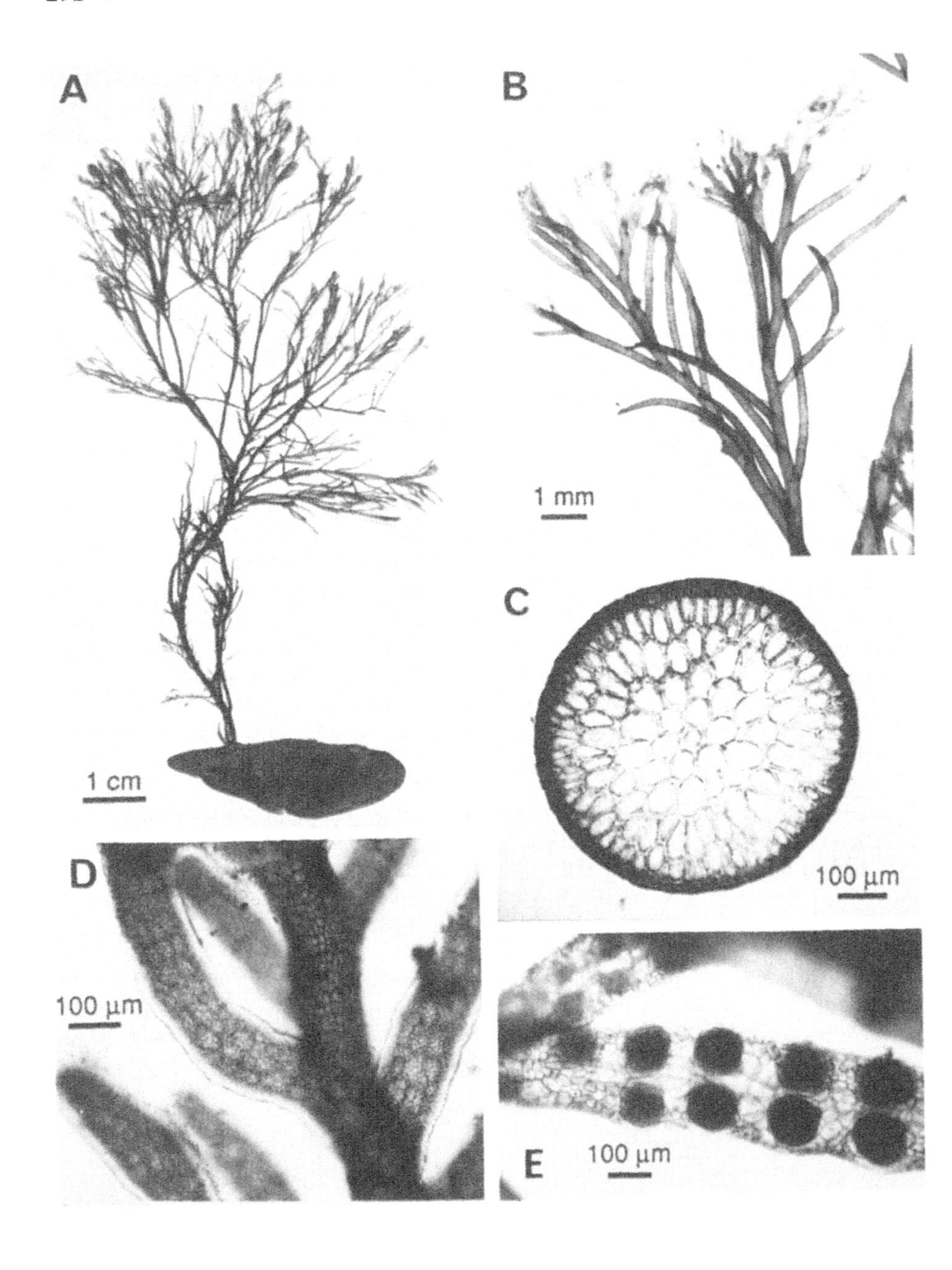
A
1 cm
B
1 mm
C
100 µm
D
100 µm
E
100 µm

Rhodomela confervoides (Hudson) Silva (1952), p. 268.

Neotype*: BM-K, ex Herb. E. Forster. Cornwall (Penzance), 1802.

Fucus confervoides Hudson (1762), p. 474.
Fucus subfuscus Woodward (1791), p. 131.
Fucus variabilis Goodenough & Woodward (1797), p. 220 (see Harvey, 1850, pl. 264).
Rhodomela subfusca (Woodward) C. Agardh (1822), p. 378.

Thalli attached by solid discoid holdfast up to 6 mm in diameter, 7-30 (50) cm high, straggly and sparsely branched to bushy and densely branched, linear to irregularly pyramidal in shape, consisting of terete erect axes arising singly or in dense tufts; *main axes* distinct, expanding upwards from the holdfast to a maximum diameter of 1-1.6 mm, bearing a spiral arrangement of branches; *laterals* up to 1.5 mm in diameter and similar in length to main axes, branched again spirally to a further 2-4 orders of branching, more densely towards main apices, which are surrounded and overtopped; last-order branchlets initially incurved, later straight or bent outwards, constricted near point of insertion, with a maximum diameter just above this, tapering gradually towards the blunt apices, often broken or damaged in older parts of thalli; *adventitious branching* usually sparse, mainly in axils of broken laterals; thalli brownish-red, darkening with age, old axes tough and cartilaginous, young axes crisp and fairly rigid when fresh, soft when liquid-preserved.

Main axes with apical cells 8-12 μm in diameter, sometimes bearing a spiral arrangement of trichoblasts up to 500 μm long, simple or branched pseudodichotomously 1-2 times, trichoblasts and branches formed in a 1/4-2/7 spiral (Rosenvinge, 1923-24), consisting in TS of axial cell 35-60 μm in diameter, 6 (-7) polygonal periaxial cells, 2-3 layers of thick-walled, radially elongated, pseudoparenchymatous cells up to 130 μm long x 90 μm wide, and 1-2 layers of cortical cells, the outermost layer rectangular to polygonal in surface view, 12-40 μm long x 5-25 μm wide, with articulations visible through cortex; *cortical thickening* developing in older axes as an opaque cortical layer up to 50-250 μm thick, often asymmetric around axis, composed of densely packed, little-pigmented radiating filaments of cells 25-60 μm long x 13-20 μm wide; *plastids* discoid.

Fig. 89. *Rhodomela confervoides*

(A) Habit of plant on stone (A-E Down, Nov.). (B) Apices with spermatangial sori. (C) T.S. of axis showing axial cell and 5 periaxial cells surrounded by pseudoparenchymatous medulla and pigmented cortex. (D) Spermatangial sori on young axes. (E) Branch with paired tetrasporangia visible through cortex.

---

* A neotype has been selected among historical specimens from Cornwall, rather than Yorkshire, the other locality cited by Hudson, to avoid confusion with other members of the genus, as *R. confervoides* appears to be the only species present in Cornwall.

Spermatangial sori developing on young branchlets at apices and on adventitious reproductive branchlets, continuous or patchy over 1-2 orders of branching, consisting of a layer of conical or cylindrical spermatangial mother cells 12-15 µm long x 4-6 µm wide, bearing ellipsoid spermatangia 4-6 µm x 3-4 µm; released spermatia slightly ovoid, 4.5-6 µm long. *Cystocarps* spherical to slightly urceolate, 300-375 µm high x 325-475 µm wide, with narrow ostiole c. 50 µm in diameter; carposporangia clavate, 100-125 x 30-50 µm. *Tetrasporangia* borne in last 2 orders of branching near apices and in adventitious reproductive branchlets, in slightly compressed branchlets 125-225 µm wide and 125 µm thick, formed in pairs in each segment, spherical, 110-150 µm in diameter; released tetraspores 50-75 µm in diameter.

Growing on bedrock, boulders, cobbles and other mobile substrata, rarely epiphytic on larger algae, including stipes of *Laminaria hyperborea*, in lower-shore pools and damp places and subtidal to 27 m, particularly abundant on sand-covered or sand-scoured bedrock, at sites ranging widely in environmental conditions, from sheltered to extremely wave-exposed, with or without exposure to strong tidal currents, often common in tidal rapids.

Generally distributed in the British Isles (Norton, 1985).

Spitzbergen to N. Spain; Labrador to New Jersey (South & Tittley, 1986). A circumboreal distribution has been reported (Lüning, 1990), but this includes records of *R. lycopodioides*.

Thalli are perennial, but marked changes of appearance take place through the year. Harvey (1850, pl. 264) reported that plants were clothed with ramuli in summer but that these drop off before winter, leaving rough broken stumps of branches. However, Dunn in Blackler (1974) described active growth on surviving plants in November at St Andrews, Fife, and we have likewise observed abundant young branches from November onwards at a sheltered site in Ireland; Masuda (1982) found that at Roscoff, France, plants bore well-developed branches in December. There have also been conflicting reports regarding the seasonality of reproduction in *R. confervoides*. Harvey (1850) suggested that tetrasporangia were found in winter only in specialized lateral branchlets, but were borne in terminal branches in summer, and Dunn in Blackler (1974) reported that tetrasporangia and cystocarps were borne in lateral branchlets from January onwards. We have observed reproduction in adventitious branchlets in specimens collected at various localities in the British Isles from January to June. At sheltered sites in Ireland and Wales, however, reproductive structures develop in terminal branches from late November to March and may persist until June. It seems likely that reproductive structures are formed in terminal branches initially, if these branches have not been eroded, and adventitious reproductive branchlets develop subsequently on the same individuals. If, however, all young branches have been lost, which is more frequent from late winter onwards, reproduction takes place only in adventitious branchlets. Spermatangia recorded in November-May; cystocarps in January, March, May and June, and tetrasporangia in November-June.

Released tetraspores of an isolate from Roscoff grown at 5-14°C, in 14 h days, gave rise to dioecious plants that formed gametangia after 2-3 months, when 4 cm high (Masuda, 1982). Mature cystocarps were present 1 month later; cultured carpospores developed into fertile tetrasporophytes within 3-5 months at 5-14°C, in 14 h days. Reproductive structures were not formed at 18°C by either gametophytes or tetrasporophytes. At Roscoff, *R. confervoides* reproduces in winter to spring, when the water temperature is 9-14°C, which correlates well with inhibition of reproduction at higher temperatures. The chromosome number is $n = 32$ (Magne, 1964).

This species shows considerable variation in branching pattern and axis diameter, and in the position of reproductive structures. Plants with narrower axes and long ultimate branchlets appear to be characteristic of very sheltered and reduced-salinity habitats, while the most robust thalli are found in the shallow subtidal zone at moderately wave-exposed sites. Coarse, sparsely branched thalli are sometimes referred to *R. virgata* or *R. confervoides* var. *virgata* (see generic notes). Plants with this morphology typically form small adventitious reproductive branchlets directly from main axes, and the branchlets are shed after reproduction. However, thalli otherwise corresponding to typical *R. confervoides*, bearing terminal tetrasporangial branchlets, may also form adventitious deciduous reproductive branchlets, and no consistent distinctions between these forms could be detected.

R. confervoides, including plants morphologically similar to *R. virgata*, usually bears parasitic thalli of *Harveyella mirabilis* (Reinsch) Schmitz & Reinke in Reinke (see Vol. 1, 2A).

Robust plants resemble *Cystoclonium purpureum* (see Vol. 1, 1), but differ by the discoid holdfast in contrast to the branching rhizoidal holdfast of *C. purpureum*, and by the pseudoparenchymatous rather than filamentous medulla. Some forms of *Gracilaria verrucosa* (Hudson) Papenfuss (see Vol. 1, 1) could be confused with *R. confervoides* but *G. verrucosa* lacks an obvious single apical cell and the distinctive axial structure in TS. Whereas axes of *R. confervoides* have a central axial cell surrounded by 6 periaxial cells, axes of *G. verrucosa* have no recognizable periaxial cells. In addition, the reproductive structures of *G. verrucosa* differ markedly in that cystocarps are nodular and spermatangial pits and tetrasporangia are scattered and inconspicuous, whereas cystocarps of *R. confervoides* are urn-shaped and spermatangia and tetrasporangia are borne conspicuously on apical branchlets. Very fine plants of *R. confervoides* are similar to some species of *Polysiphonia*, but are fully corticate at the apices rather than ecorticate or with cortication developing at some distance from the apical cell as in *Polysiphonia* species.

Rhodomela lycopodioides (Linnaeus) C. Agardh (1822), p. 377.

Lectotype: LINN 1274.20. Scotland.

Fucus lycopodioides Linnaeus (1767), p. 717.
Rhodomela subfusca var. *lycopodioides* (Linnaeus) Gobi (1878), p. 24.

Thalli 1-50 (60) cm high, consisting of dense tufts of erect axes with a bottle-brush appearance, arising from solid discoid holdfast up to 8 mm in diameter; *main axes* distinct, expanding upwards from the holdfast to a maximum diameter of 0.6-1.5 mm, naked below, densely clothed with laterals of approximately equal length, borne in a spiral arrangement; *laterals* either remaining short, obscured by adventitious branching, or becoming similar in length and diameter to main axes, clothed with a spiral arrangement of short second-order laterals bearing 1-3 further orders of short branches in an irregularly spiral to distichous or secund arrangement; ultimate branchlets 0.1-0.3 mm in diameter, usually straight, with reflexed tips; *adventitious laterals* abundant, developing from main axes and first-order laterals and obscuring the primary branching pattern, some axillary but mostly arising randomly on major axes, simple or branched to 1-2 orders; thalli dark brownish-red, old axes tough and cartilaginous, young axes crisp and fairly rigid when fresh, soft when liquid-preserved.

Main axes with apical cells 12-14 μm in diameter, apparently lacking trichoblasts, consisting in TS of the axial cell 40-60 μm in diameter, (5-)6 polygonal periaxial cells, 1-2 layers of thick-walled, pseudoparenchymatous cells, and 1-2 layers of cortical cells, the outermost layer rectangular to polygonal in surface view, 18-40 μm long x 9-18 μm wide; articulations scarcely visible through cortex; *cortical thickening* developing on older axes as an opaque layer 50-250 μm thick, often asymmetrical around axis, composed of densely

Fig. 90. *Rhodomela lycopodioides*

(A) Habit of plant from kelp stipe, with numerous long axes densely clothed with short adventitious laterals (A-B Sutherland, Aug.). (B) Detail of axis densely clothed with sparsely branched, incurved laterals.

packed, little-pigmented radiating filaments of cells 30-60 μm long x 20-25 μm wide, and a thick pigmented cuticular layer; *plastids* discoid.

Plants dioecious. [Spermatangial and cystocarpic material not available.] *Cystocarps* spherical to slightly urceolate with narrow ostiole (Harvey, 1846, pl. 50). *Tetrasporangia* developing in last 2 orders of branching near apices and in adventitious reproductive branchlets, formed in pairs in each segment of slightly compressed tetrasporangial branchlets 130-150 μm wide, spherical, c.150 μm in diameter.

Epiphytic on stipes of *Laminaria hyperborea*, occasionally epilithic on bedrock, in lower intertidal pools and from near extreme low water to 5 m depth, common both in moderately sheltered sealochs and in shallow wave-exposed kelp forest.

Northern coasts of the British Isles, southwards to N. Ireland, S.W. Scotland and Yorkshire (Norton, 1985). Older reports from south coasts apparently resulted from taxonomic confusion with *R. confervoides*.

Spitzbergen to British Isles; Iceland, Greenland and Arctic Canada (South & Tittley, 1986). Records of *R. lycopodioides* from the N.E. Pacific require reassessment (Scagel et al. 1989).

Thalli are perennial. Harvey (1846, pl. 50) reported that plants were clothed with branching ramuli in summer but that these drop off in winter, leaving rough broken stumps of branches. Spermatangia have been recorded in June; there are no dated records of cystocarps; and tetrasporangia have been recorded in January and June-August.

There is comparatively little morphological variation except seasonal changes.

Parasitic thalli of *Harveyella mirabilis*, which are common on *Rhodomela confervoides*, have not been observed in collections of *R. lycopodioides* from the British Isles.

R. lycopodioides is unlikely to be confused with algae other than species of *Rhodomela* (see under *R. confervoides* for comparison with similar genera).

ODONTHALIA Lyngbye, nom. cons.

ODONTHALIA Lyngbye (1819), p. 9.

Type species: *O. dentata* (Linnaeus) Lyngbye (1819), p. 9.

Thalli perennial, terete at base, compressed to flattened above, consisting of one to several erect main axes up to 25 cm high arising from an expanded basal disc, alternate-distichously branched with up to 6 orders of progressively shorter branches, the branches distinct or partly confluent with the main axis, and with or without a central midrib and lateral wings; vegetative trichoblasts absent; adventitious branches borne in branch axils or laterally along margins; periaxial cells 6 in main axes (sometimes with 4 periaxial and 2 pseudoperiaxial cells), divided transversely into 2-3 tier cells, with the upper cell retaining the pit connection to the axial cell; cortication 6-20 (-30) cell layers thick.

Growth exogenous, initiated by a steeply oblique division of the apical cell, the lateral initials originating mostly from every third segment, alternately to one side and then the other; periaxial cells formed in rhodomelacean sequence with the first cut off abaxially, directly below the branch initial, with the next 4 lateral on the flattened surfaces, and the last adaxial; periaxial and cortical cells dividing transversely 1-2 times and periclinally forming successive layers of outwardly growing, descending cortical files; adventitious branches originating from superficial cortical cells.

Reproductive structures borne on ordinary and/or adventitious branches; gametophytes dioecious. Spermatangia superficial below apices or on marginal branches; spermatangial mother cells each cutting off 2 spermatangia by oblique divisions. Procarps on short polysiphonous branches or, more rarely, on fertile trichoblasts, formed from the fifth periaxial cell on suprabasal and sometimes more distal segments, often with two or more procarps on a single branchlet; pre- and postfertilization stages developing as in *Rhodomela*; cystocarps globose to ovoid with a thick pericarp, naked or subtended by a spine or branchlet. Tetrasporangia formed in opposite pairs from the third and fourth periaxial cells in successive segments on unmodified or specialized penultimate and/or ultimate branches, tetrahedrally divided, protected by two conspicuous cover cells.

References: Kylin (1934), Masuda (1982).

One species in the British Isles.

Odonthalia dentata (Linnaeus) Lyngbye (1819), p. 9.

Holotype: LINN 1274.72. Atlantic Ocean.

Fucus dentatus Linnaeus (1767), p. 718.

Thalli attached by solid discoid holdfast 5-13 mm in diameter, composed of fused radiating rhizomes; *erect axes* arising singly or in dense tufts with a variable overall shape, from linear to irregularly pyramidal or flabellate, 5-30 (40) cm high and up to 25 cm broad, each consisting of compressed, stipe-like denuded midrib 3-5 mm in width, with remnants of eroded laminae, increasing in length with age of plant up to 10 cm, irregularly branched to 2-4 orders of branching, bearing branches of various ages on old axes, complanate when young, 2-4 mm wide and 40-100 μm thick with midrib 150-350 μm thick, branched alternate-distichously; *laterals* remaining unbranched and wing-like with abaxial and adaxial margins curving outwards, or bearing a further 1-3 orders of similar alternate-distichous laterals; ultimate laterals wing-like, with acute apices, the youngest

---

Fig. 91. *Odonthalia dentata*

(A) Habit of mature thallus (Donegal, May). (B) Densely branched female branchlets with clustered cystocarps, borne in axils of main axis (B & D Down, May). (C) Alternately branched female axis bearing urceolate cystocarps with broad ostioles and basal spurs (Down, Feb.). (D) Tetrasporangial branchlets with paired tetrasporangia in each segment.

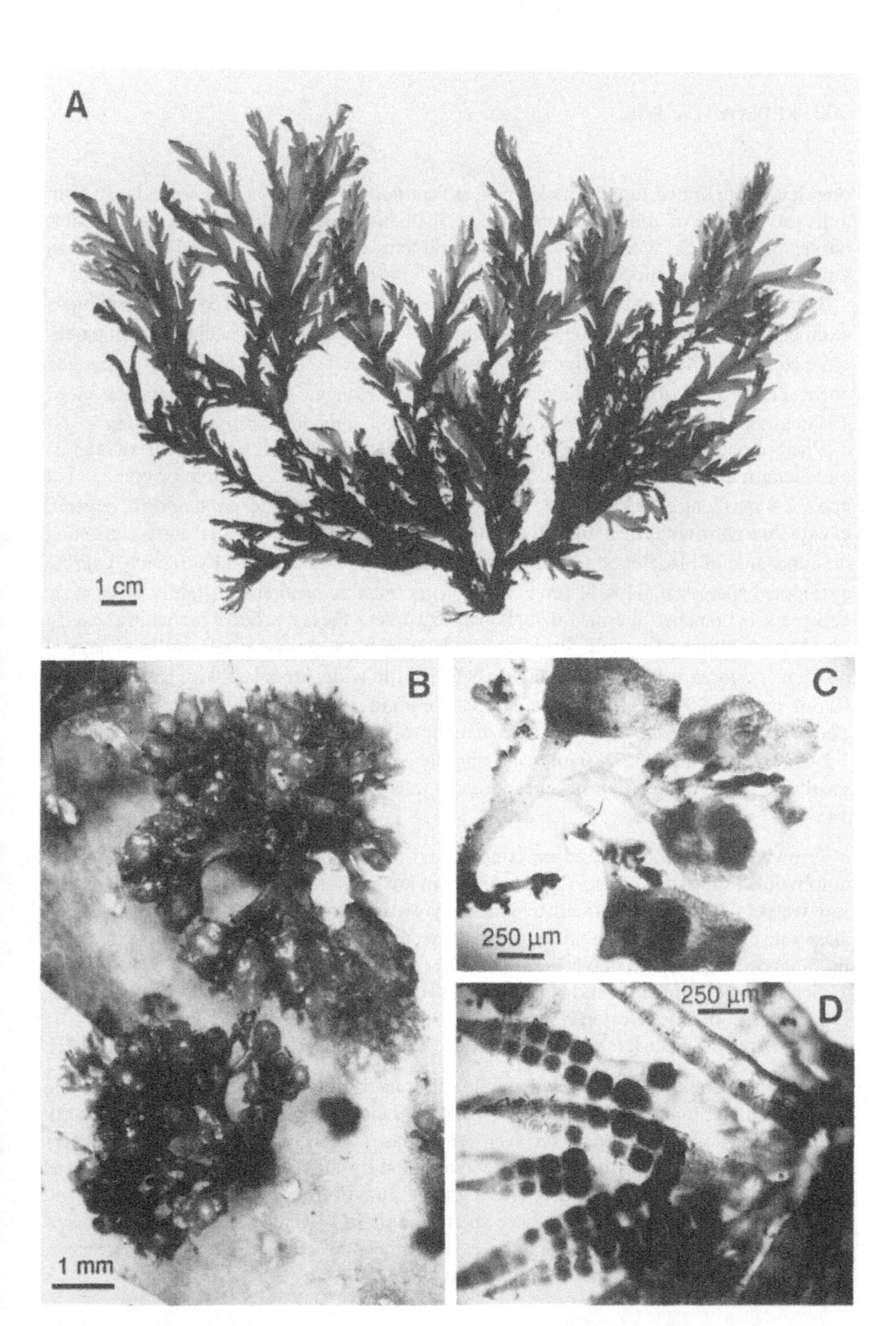
A
1 cm
B
1 mm
C
250 µm
250 µm
D

ones level with or overtopping main apex; *adventitious branches* arising secondarily from axils and margins of young and mature axes; thalli brownish-red to deep purple in colour, darkening with age, with a glossy surface, old axes extremely tough and cartilaginous, young axes soft and slippery.

Main axes with apical cells 12-14 µm in diameter, consisting in TS of a conspicuous axial cell c. 60 µm in diameter, little-pigmented medulla of thick-walled pseudoparenchymatous cells 40-60 µm wide x 12-35 µm high, and 1-2 layers of deeply pigmented cortical cells c. 15 µm high, the outer cortical layer elongate-polygonal in surface view, 15-35 µm x 15-20 µm; *plastids* reticulate and beaded in medulla, discoid in cortex.

Plants dioecious; all reproductive structures formed in dense clusters in axils of laterals and along margins of main axes. *Spermatangial bladelets* ovate, simple or branched at apex, 2-4 mm long x 1-2 mm wide and c. 50 µm thick, with a thickened midrib, covered except for a narrow sterile margin and apices by continuous spermatangial sorus consisting on either side of bladelet of a layer of spermatangial mother cells 8-11 µm wide, bearing cylindrical spermatangia 8-11 µm x 3.5-5.5 µm. *Female branchlets* slightly to strongly compressed, branched alternate-distichously to 3 orders, the last order of branching bearing an alternate series of simple fertile or sterile branchlets, procarps adaxial in subapical position; *cystocarps* 600-850 µm high x 500-650 µm wide, urceolate with basal spur and flared ostiole 300-600 µm wide; carposporangia clavate, 210-400 x 50-100 µm. *Tetrasporangial branchlets* alternate-distichously branched, the ultimate branchlets 1.2-2.2 mm long and 250-320 µm wide, slightly compressed with acute apices, becoming swollen after forming a pair of tetrasporangia in each segment; tetrasporangia spherical, 145-190 µm in diameter.

Growing on bedrock, boulders, cobbles and other mobile substrata, less frequently epiphytic on stipes of *Laminaria hyperborea*, in lower-shore pools and from near extreme low water to 20 m, at moderately to extremely wave-exposed sites and at wave-sheltered sites with exposure to strong tidal currents; most abundant in shallow kelp forest, with the biomass decreasing with depth towards the lower limit of kelp.

Scotland, N.E. England and N.E. Ireland, southwards to Wigtown, Isle of Man, Mayo, Down and Yorkshire (Norton, 1985).

Spitzbergen to British Isles; Arctic Canada (Ellesmere Is.) to Nova Scotia (Kain, 1982).

Thalli are perennial, but those on mobile substrata do not survive the winter. At the Isle of Man, a high proportion of plants was lost after their first year, but subsequent mortality was low and some individuals survived for at least 4 years, and possibly up to 9 years (Kain, 1984). New growth began in February, and by July maximum plant weight was reached and growth stopped; there was a gradual loss of fronds throughout summer and autumn. Isle of Man populations bore spermatangia in November-February, cystocarps in December-June, and tetrasporangia in November-June (Kain, 1982); in Scotland, cystocarps and tetrasporangia occur until August.

There is comparatively little morphological variation; *O. dentata* is among the most easily recognizable red algae in the British Isles and is unlikely to be misidentified.

Tribe BRONGNIARTELLEAE Parsons (1975), p. 691.

BRONGNIARTELLA Bory

BRONGNIARTELLA Bory (1822), p. 516.

Type species: *B. elegans* Bory (1822), p. 516 [= *B. byssoides* (Goodenough & Woodward) Schmitz (1893a), p. 217].

Thalli radially organized, terete, polysiphonous, consisting of one to several erect main axes up to 50 cm high arising from short prostrate axes attached to the substratum by unicellular rhizoids usually bearing digitate adhesion discs; lateral axes spirally to irregularly arranged, up to 25 cm long; vegetative trichoblasts one per segment, pigmented, 2-6 times subdichotomously branched, persistent or the tips deciduous; adventitious branches rare or absent; periaxial cells 7, fewer in prostrate branches and weakly developed laterals; cortication absent.

Growth monopodial, initiated by slightly oblique division of the apical cell, each segment bearing a trichoblast in a tight (1/7) right-handed spiral; lateral branches originating from basal cell of trichoblast on the left-hand side; periaxial cells cut off in alternating sequence, with the first either to the right or left of the trichoblast in different species, linked by secondary pit connections to periaxial cells in the segment below.

Gametophytes dioecious. Spermatangial axes cylindrical, clustered spirally on short lateral axis, or borne terminally on a trichoblast branch; periaxial cells 5 (-6) per segment, dividing transversely and anticlinally to form a single fertile layer; spermatangia 4-5 per spermatangial mother cell. Procarps originating from the fifth periaxial cell on the second basal segment of a fertile trichoblast, consisting of a 2-celled lateral sterile group, a 4-celled carpogonial branch (a 1-celled basal lateral present in some), and a 1-celled basal sterile group; prefertilization pericarp present, derived primarily from the third and fourth periaxial cells; fertilized carpogonium cutting off a distal and a posteriolateral connecting cell, the latter fusing with the inner side of the auxiliary cell; cells of the sterile groups each dividing once after fertilization and their nuclei enlarging; auxiliary cell cutting off a single primary gonimoblast initial which divides transversely and gives rise initially to monopodially branched gonimoblast filaments bearing clavate, terminal carposporangia, thereafter sympodially branched from subterminal cells below the differentiating carposporangia; fusion cell large, incorporating the supporting cell, auxiliary cell and inner gonimoblast cells; cystocarp flask-shaped, 2-layered, ostiolate, with a slight to prominent beak. Tetrasporangia borne in short, stichidia-like lateral branches with 6 or 7 periaxial cells per segment, formed singly in a tight spiral from the second periaxial cell, tetrahedrally divided, covered by 2 conspicuous to greatly swollen presporangial cover cells.

References: Rosenberg (1933), Parsons (1980).

One species in the British Isles.

Brongniartella byssoides (Goodenough & Woodward) Schmitz (1893a), p. 217.

Lectotype: BM-K, ex Herb. Goodenough. Hampshire (Christchurch), 1794.

Fucus byssoides Goodenough & Woodward (1797), p. 229.
Conferva byssoides (Goodenough & Woodward) J. E. Smith (1799), pl. 547.
Polysiphonia byssoides (Goodenough & Woodward) Greville (1824), p. 309.

Thalli 7-30 cm high, composed of dense cylindrical to irregularly pyramidal tufts of erect axes attached by tangled prostrate axes, with further prostrate axes developing from bases of erect axes; *erect axes* consisting of numerous main axes 0.2-0.6 mm in diameter, bearing a spiral arrangement of first-order laterals that are initially short and even-sized, giving the thallus a cylindrical outline, but continue to grow so that the mature thallus is pyramidal in shape; first-order laterals bearing a further 2-3 orders of spirally-arranged laterals, all axes except the denuded bases of old thalli clothed with incurved, pigmented trichoblasts that resemble monosiphonous branchlets; young thalli bright red, soft but not delicate, mature plants dark brownish-red, bleaching to light brown, with tough, flexible old axes.

Prostrate axes cylindrical, ecorticate, 60-100 µm in diameter, composed of axial and 5-6 periaxial cells, with segments 0.5-1 diameter long, attached by numerous rhizoids 30-50 µm in diameter terminating in discoid pads, lacking trichoblasts but branching frequently to form further prostrate axes; *erect axes* 200-600 µm in diameter, with 7 straight or slightly spiral periaxial cells; segments 1-3 diameters long, each bearing a pigmented trichoblast, the basal cell of which gives rise to a polysiphonous lateral of unlimited growth, and itself becomes polysiphonous (see Parsons, 1980, for details); *trichoblasts* branched pseudodichotomously 1-3 times at a wide angle, up to 1.5 mm long and 70 µm in diameter, the last branches either short or prolonged into multicellular filaments tapering to 5 µm in diameter; *polysiphonous laterals* similar in structure to main axes and giving rise to several further orders of spiral branching in the same arrangement; *plastids* numerous, ribbon-like.

Plants dioecious. *Spermatangial branchlets* borne in dense spiral clusters on short

---

Fig. 92. *Brongniartella byssoides*

(A) Habit (Donegal, May). (B) Apex of main axis bearing spirally arranged laterals, densely clothed with pigmented trichoblasts (Galway, Feb.). (C) Ecorticate polysiphonous axis with pigmented trichoblasts (Donegal, Dec.). (D) Densely clustered spermatangial branchlets (D-E Mayo, May). (E) Tetrasporangia (t) borne in unspecialized laterals.

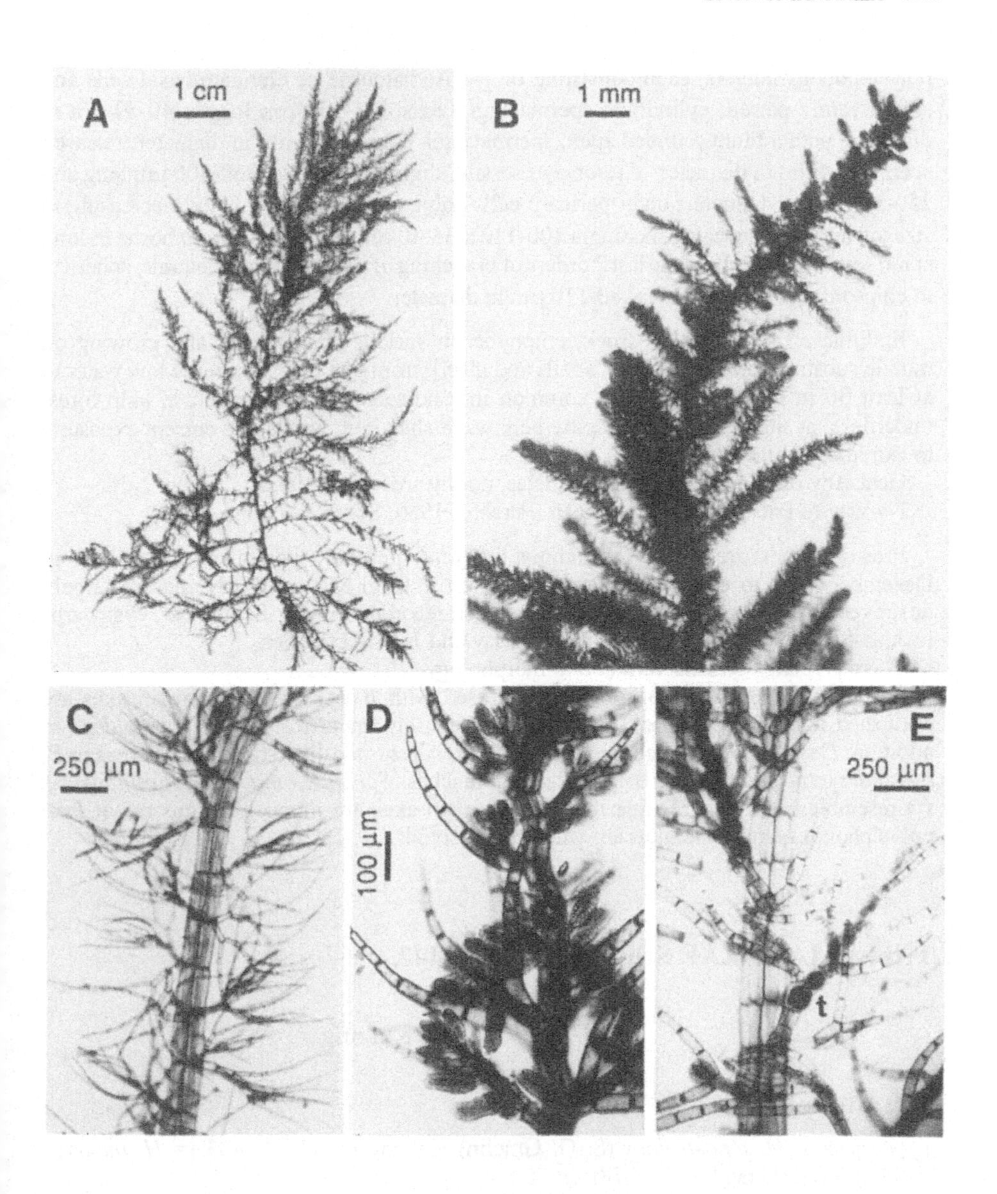
A
1 cm
B
1 mm
C
250 µm
D
100 µm
E
250 µm
t

polysiphonous laterals, each consisting of 1-2 isodiametric or elongate basal cells and single, rarely paired, cylindrical spermatangial axis 125-150 µm long x 40-50 µm in diameter, with a blunt rounded apex; spermatangia ellipsoid, 4 µm in diameter; released spermatia 4 µm in diameter. *Cystocarps* sessile, slightly urceolate, 350-400 µm long and 250-350 µm in diameter; outer pericarp cells polygonal, in straight rows, decreasing in size towards ostiole; carposporangia 100-110 x 35-40 µm. *Tetrasporangia* borne in long spiral series of up to 10 in the last 2 orders of branching of polysiphonous laterals, spherical to ellipsoid, 90-120 µm long x 80-110 µm in diameter.

Epilithic on horizontal bedrock, epiphytic on various small algae, and growing on mobile substrata such as pebbles, shells and maerl, from just below extreme low water to at least 30 m depth, particularly common in sand-scoured habitats and in kelp forest underflora, at sites ranging from extremely wave-sheltered, with some current exposure, to extremely wave-exposed.

Generally distributed in the British Isles, northwards to Shetland.

Norway to Portugal; Mediterranean (Parsons, 1980; South & Tittley, 1986).

Prostrate axes are found throughout the year, probably perennial, giving rise in December-April to erect axes that become fertile from May onwards and are largely destroyed by October. Spermatangia have been recorded in May-September, cystocarps in June-September, and tetrasporangia in May and July-September.

There is relatively little variation in morphology.

Species of *Dasya* (q. v.) are often confused with *B. byssoides*, but main axes are corticated to some degree in *Dasya* spp. and entirely uncorticated in *B. byssoides*. In addition, *Dasya* spp. form tetrasporangia in specialized pod-like stichidia, whereas in *B. byssoides* tetrasporangia are borne in lateral branches. *Sphondylothamnion multifidum* (q. v.) resembles *B. byssoides* in habit, but the main axes are monosiphonous rather than polysiphonous, and branchlets are whorled not spiral.

Tribe AMANSIEAE Schmitz (1889), p. 447.

HALOPITHYS Kützing

HALOPITHYS Kützing (1843), p. 433.

Type species: *H. pinastroides* (S. G. Gmelin) Kützing (1843), p. 433 [= *H. incurvus* (Hudson) Batters (1902), p. 78].

Thalli dorsiventrally organized, perennial, terete, consisting of one to several erect main axes up to 30 cm high from a discoid holdfast and 4-5 orders of progressively shorter, pectinate branches terminating in attenuated, incurved or helically inrolled tips; primary branches produced alternately to one side in pairs from successive segments at regular or irregular intervals; main axes and major lateral branches clothed with simple, mostly adventitious lateral branches up to 1 cm long; vegetative trichoblasts present or absent, formed in a row along the dorsal midline, deciduous; periaxial cells 5, elongating to the same length as the axial cell; cortication composed of progressively smaller isodiametric or angular cells, 2-3 cell layers thick above, increasing in thickness towards the base.

Growth initiated by slightly oblique division of the apical cell in helicoid succession, with each segmental cell potentially capable of producing a trichoblast on the dorsal side; periaxial cells formed in alternating sequence, the first dorsal towards the abaxial side, the second dorsal towards the adaxial side, the third abaxial towards the ventral side, the fourth adaxial towards the ventral side and the fifth ventral; branching endogenous, the laterals initiated alternately in pairs from successive axial cells at more or less regular intervals, emerging between the 3rd and 5th and the 4th and 5th periaxial cell respectively, and deflected inwardly towards the ventral surface; periaxial cells cutting off 3 cortical initials (2 anterior and 1 posterior) which divide obliquely and periclinally to produce a tightly compact cortex of ternately to subdichotomously branched cortical filaments; adventitious branches originating from superficial cortical cells.

Gametophytes dioecious. Spermatangial axes formed in series at the tips of branches on the dorsal side, subglobose, borne on the suprabasal cell of a modified trichoblast; spermatangial mother cells and spermatangia developing as in *Polysiphonia*. Procarps formed from the fifth (ventral) periaxial cell on the second basal segment of a modified trichoblast in series on the dorsal side at the tips of branches, prefertilization and postfertilization development as in *Polysiphonia*; 1-2 cystocarps maturing in a fertile branch, ovoid to slightly urceolate with a thick pericarp. Tetrasporangia formed in pairs from the third and fourth periaxial cells in successive segments on a fertile branch, covered by 2 presporangial and 1 postsporangial cover cells.

References: Falkenberg (1901), Kylin (1956).

One species in the British Isles.

Halopithys incurvus (Hudson) Batters (1902), p. 78.

Lectotype: BM-K, ex herb. Pulteney. England, unlocalized.

Fucus incurvus Hudson (1762), p. 470.
Fucus pinastroides S. G. Gmelin (1768), p. 127.
Rytiphloea pinastroides (S. G. Gmelin) C. Agardh (1817), p. 25.
Halopithys pinastroides (S. G. Gmelin) Kützing (1843), p. 433.

Thalli 10-30 cm high, consisting of dense tufts of terete erect axes, attached by solid discoid holdfast up to 1 cm in diameter; erect axes bushy and densely branched, with a flat-topped, obconical to irregular outline; *main axes* 0.8-1 mm in diameter, bearing an irregular arrangement of indeterminate laterals, and densely covered below with short laterals; *determinate laterals* simple, up to 0.5-0.9 mm in diameter, increasing in length towards

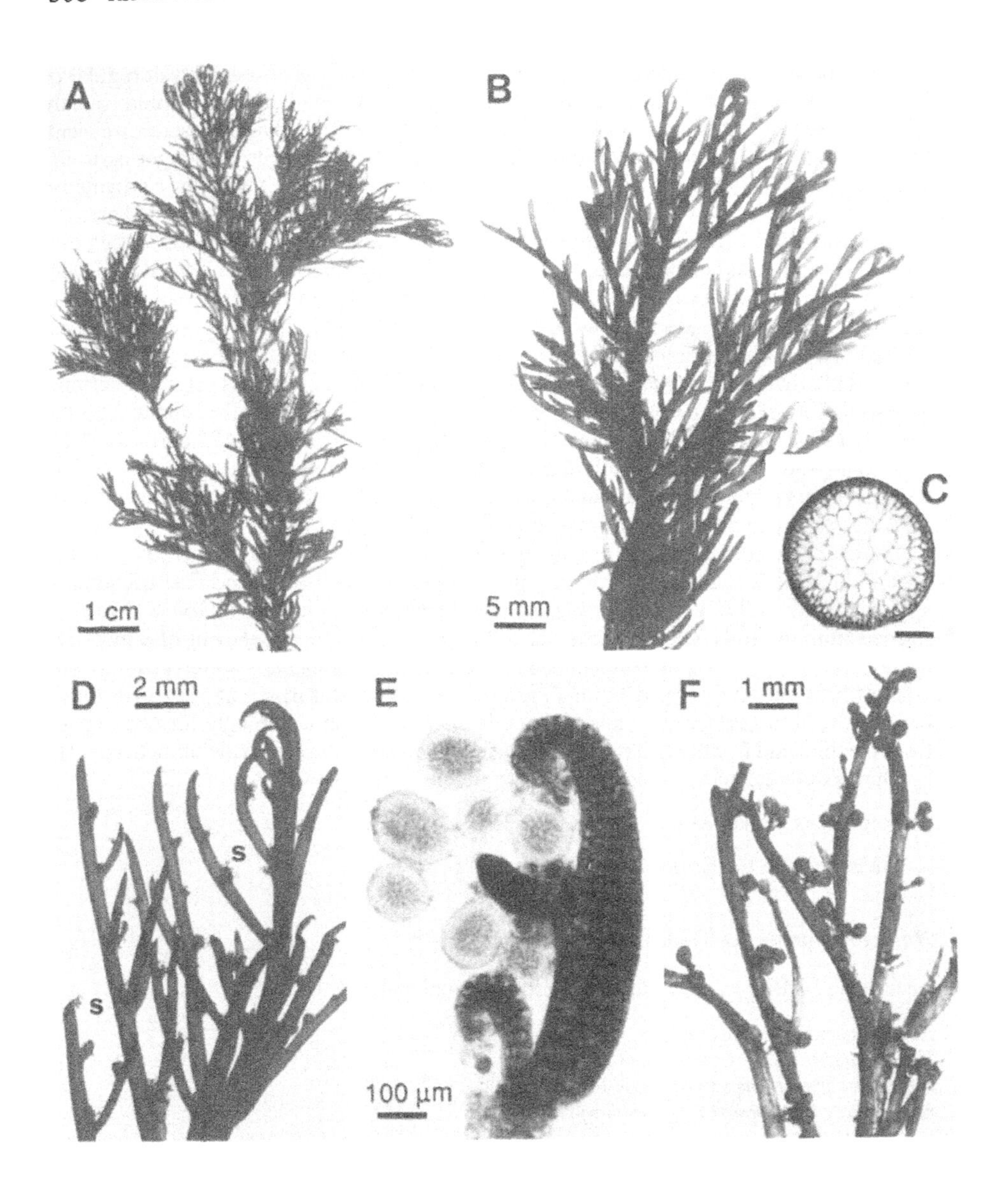
A
B
C
1 cm
5 mm
D
2 mm
E
F
1 mm
s
s
100 µm

plant apex but often reduced to broken stumps near holdfast, straight, with inrolled apices, bearing regularly-spaced pairs of straight, erect adaxial branchlets 0.5-0.6 mm in diameter, with tightly inrolled tips; a further order of branching sometimes present, consisting of pairs of adaxial branchlets; *adventitious laterals* abundant on older axes; bright red to dark brownish-red in colour, old axes tough and cartilaginous, young axes succulent and fairly brittle.

Main axes with apical cells 12-14 µm in diameter, consisting in TS of a polygonal axial cell 125-150 µm in diameter, 5 polygonal periaxial cells equal to or larger than axial cell, 3-4 layers of thick-walled pseudoparenchymatous cells decreasing in size outwards, and 1-2 layers of cortical cells, the outermost layer polygonal in surface view, 18-35 µm in diameter; *segments* 0.2-0.3 diameters long, visible through cortex of young axes; *cortex* increasing in thickness on older axes, up to 250 µm thick, often asymmetrical around axis, composed of radiating filaments of thick-walled cells 30-70 µm long x 25-20 µm wide; *trichoblasts* abaxial, occasionally present on lateral apices, up to 400 µm long and branched 2-4 times; *plastids* reticulate to beaded in medulla and inner cortex, discoid in outer cortical cells.

Plants dioecious; reproductive structures formed on small, tightly inrolled branchlets that develop in pairs or small groups at intervals on last-order laterals and axillary to them. *Spermatangial branchlets* developing in series on convex adaxial face of reproductive branchlets and on short lateral branchlets formed adaxially on these branchlets, one per segment, consisting of 1-2-celled monosiphonous stalk and spherical to ovoid spermatangial axis 125-225 µm long x 125-150, covered with spermatangial mother cells bearing ellipsoid spermatangia 6-8 µm x 4-5 µm. *Procarps* formed on convex adaxial face of reproductive branchlets; *cystocarps* developing singly or paired on branchlet, ovoid to slightly urceolate, 525-775 µm high x 525-625 µm wide, with an ostiole c. 125 µm in diameter; carposporangia pyriform, 100-130 x 40-65 µm. *Tetrasporangia* spherical, formed in pairs in each segment of reproductive branchlets.

Growing on bedrock and pebbles in lower-shore pools and damp places and from near extreme low water to 13 m depth at moderately wave-sheltered sites on open coasts; sometimes becoming abundant in favourable habitats.

North Cornwall and south coast of England from Cornwall to Kent, most common in

---

Fig. 93. *Halopitys incurvus*

(A) Habit (A-C Devon, Nov.). (B) Detail of axes, bearing paired laterals with tightly inrolled apices and a secund arrangement of last-order branches. (C) T.S. of main axis with large axial cell surrounded by 6 periaxial cells and pseudoparenchymatous medulla (scale = 250 µm). (D) Detail of male thallus with spermatangial branchlets (s) borne in tufts on last-order laterals (D-E Dorset, Nov.). (E) Stalked ellipsoid spermatangial axes. (F) Tufts of cystocarps (Devon, Aug., in BM).

Dorset (Norton, 1985); verified from one site in Kerry (Guiry, 1978); Channel Isles. Records from outside this area may be of drift material.

England to Mauritania; Mediterranean (Ardré, 1970).

Thalli perennial; new axes arise in winter and spring from surviving parts of mature plants. Spermatangia recorded for August-November and cystocarps in January-March. No dated collections of tetrasporangial plants have been located; although Harvey (1847, pl. 85) illustrated tetrasporangia, the source of his material was not indicated. Reproductive plants generally comprise only a small proportion of the population at one time; cystocarpic and tetrasporangial plants were not observed during the present study.

This species shows little morphological variation.

H. incurvus is among the more easily recognizable members of this family. The only possible confusion is with robust plants of *Rhodomela confervoides* (q. v.), but the young laterals of *H. incurvus* are tightly inrolled, whereas in *R. confervoides* they are only slightly curved. The axial cell of *H. incurvus* is large, more-or-less equal in size to the periaxial cells, but in *R. confervoides* it is much smaller than the periaxials.

Tribe POLYSIPHONIEAE Schmitz (1889), p. 447.

POLYSIPHONIA Greville, nom. cons.

POLYSIPHONIA Greville (1823-4), pl. 90.

Type species: *P. urceolata* (Dillwyn) Greville 1824, p. 309, typ. cons. [= *P. stricta* (Dillwyn) Greville (1824), p. 309].

Thalli radially organized, polysiphonous, consisting of erect or decumbent axes arising from a discoid or fibrous holdfast, or composed of prostrate axes attached by unicellular or multicellular, simple or digitate rhizoids and giving rise to erect indeterminate and/or determinate axes; vegetative trichoblasts present or absent, nearly colourless, 2-4 (-5) times pseudodichotomously branched, deciduous, except the basal cell (scar cell), spirally arranged, one per segment or with naked intervening segments, or infrequent and irregularly disposed, or absent; indeterminate axes terete, exogenous, either replacing trichoblasts or borne on basal cells of trichoblasts, or endogenous, or adventitious; determinate axes, likewise, replacing trichoblasts, borne on basal cells of trichoblasts, endogenous, adventitious, or absent; periaxial cells 4-24, elongating to the same length as the axial cells and connected longitudinally by secondary pit connections, straight or spirally twisted, sometimes dividing longitudinally to form an equal number of pseudoperiaxial cells; cortication absent, basal, or moderate to extensive, composed of outwardly growing, descending filaments.

Growth monopodial, by steeply to slightly inclined divisions of the apical cell; lateral initials cut off 1-several segments below the apex, usually in a right-handed spiral;

vegetative periaxial cells formed in alternating sequence with the first cut off directly below the trichoblast or branch initial in nodal and internodal segments, the second in the direction of the spiral (usually to the right), the third in the opposite direction (usually to the left), etc., and with the lateral branch displaced secondarily, usually to the left, arising between the first and third periaxial cells; the first periaxial cell typically abaxial in basal segments of side branches; conjunctor cells and initials of cortical filaments cut off from the basal corners of periaxial cell; rhizoids mostly originating from periaxial cells and either separated by a septum, or the cytoplasm continuous with the bearing cell; adventitious indeterminate or determinate branches originating from scar cells, periaxial cells, or cortical cells, or endogenous.

Gametophytes monoecious or dioecious, sometimes with mixed phases. Spermatangial axes borne on the second-basal cell of a modified trichoblast, either terminal or subtended by a trichoblast branch; fertile periaxial cells 4 per segment, dividing anticlinally and obliquely forming 2 (-3) layers, with the surface layer fertile; spermatangia 2-3 per spermatangial mother cell. Procarps originating from the fifth periaxial cell on the second-basal segment of a modified trichoblast, consisting of a 2-celled lateral sterile group, a (3-) 4-celled carpogonial branch, and a 1-celled basal sterile group; prefertilization pericarp present, derived primarily from the third and fourth periaxial cells in the fertile segment; fertilized carpogonium cutting off a distal and a posteriolateral connecting cell, the latter fusing with an extension from the inner side of the auxiliary cell; cells of the sterile groups each dividing once after fertilization, with sterile group-1 becoming 4-celled and sterile group-2 becoming 2-celled, and with their nuclei enlarging; auxiliary cell cutting off a single gonimoblast initial apically that branches subdichotomously and bears terminal clavate carposporangia, thereafter the gonimoblasts sympodially branched from subterminal cells below the differentiating carposporangia; fusion cell conspicuous, incorporating the central cell, supporting cell, auxiliary cell and most inner gonimoblast cells; cystocarp globose to flask-shaped with a central ostiole and with or without a protruding beak, surrounded by a pericarp composed of 12-14 axial filaments, each bearing 2 periaxial cells and sometimes also cortical cells. Tetrasporangia on main axes or side branches, usually 1 per segment from the third periaxial cell, and either arranged spirally or in straight series, depending on the presence or absence of trichoblast analogues; tetrahedrally divided when mature, completely covered by 2 lateral presporangial cover cells and sometimes a third, basal postsporangial cover cell.

References: Kylin (1923), Womersley (1979), Hommersand & Fredericq (1990).

Important characters for identification of *Polysiphonia* species include the number of periaxial cells, which is easily determined in transverse section; whether or not cortical filaments develop secondarily from the periaxial cells and partially or entirely cover them; whether young branches develop in the axils of trichoblasts or replace trichoblasts; whether rhizoids remain in open connection with periaxial cells or are cut off from them; the formation of tetrasporangia in straight or spiral rows in fertile axes; and the size and shape of spermatangial axes and cystocarps.

Of the 24 species placed in *Polysiphonia* in previous British checklists (South & Tittley, 1986), *P. fruticulosa* is treated here as a representative of the genus *Boergeseniella*; British records of *P. rhunensis* Bornet & Thuret appear to belong to *P. fibrata*; *P. richardsonii*, *P. spinulosa* and *P. violacea* sensu Harvey are considered to be conspecific with *P. fibrillosa*; *P. spiralis* is treated as a synonym of *P. stricta* (= *P. urceolata*); *P. nigrescens* is replaced by the older name *P. fucoides*; and British records of *P. insidiosa* appear to be based on the recently reported *P. harveyi*. We have been unable to verify records of *P. sanguinea* C. Agardh (1827, p. 638), while *P. devoniensis* is described as a new species from the British Isles.

KEY TO SPECIES

1 Periaxial (= pericentral) cells 4 2
   Periaxial cells 5 or more 12
2 Cortication lacking 3
   Cortication present near base at least 7
3 Rhizoids conspicuous and abundant on prostrate axes, remaining in open connection with periaxial cells (see Fig. 94C).................... 4
   Rhizoids cut off from periaxial cells by cross-walls (see Fig. 101D) 6
4 Plants forming extensive mats; trichoblasts abundant, branched 2-4 times; spermatangial axes borne on branched fertile trichoblasts; cystocarps globose; tetrasporangia in spiral series .... *P. devoniensis*
   Plants growing as distinct tufts; trichoblasts sparse, branched 0-2 times; spermatangial axes replacing trichoblasts; cystocarps urceolate (rarely globose); tetrasporangia in straight series 5
5 Thalli forming rounded cushions 2-5.5 cm high, axes ≤75 μm wide; spermatangial axes lacking sterile terminal cells .... *P. atlantica*
   Thalli forming dense erect tufts 5-25 cm high, axes ≥(50-)100 μm wide; spermatangial axes terminating in 3-5 sterile cells .... *P. stricta*
6 Prostrate axes extensive, matted, forming numerous erect axes; branching axillary to trichoblasts (see Fig. 102F); plastids small, discoid to beaded, covering all walls of periaxial cells .... *P. fibrata*
   Holdfast discoid, secondary prostrate axes sparse or absent, erect axes solitary; plastids large, ribbon-like, absent on outer walls of periaxial cells that therefore appear transparent .... *P. harveyi*
7 Branching axillary to trichoblasts (see Fig. 102F), trichoblasts conspicuous and persistent 8
   Branches replacing trichoblasts, trichoblasts conspicuous or not 9
8 Holdfast discoid, sometimes with secondary prostrate axes; thalli solitary, with distinct main axis; cystocarps ovoid or globular with narrow ostiole.................... *P. fibrillosa*
   Prostrate axes extensive; erect axes numerous, tufted; cystocarps shallow, with wide ostiole .... *P. fibrata*
9 Young branches spindle-shaped, markedly constricted basally .... *P. elongata*

Young branches not constricted basally 10
10 Periaxial cells appearing transparent, plastids absent on outer walls .... *P. harveyi*
Periaxial cells appearing pigmented, plastids present on outer walls 11
11 Cortication developing on axes 1-2 orders of branching from apex; all axes more-or-less rigid .... *P. elongata* (low salinity form)
Cortication developing on axes ≥4 orders of branching from apex; young axes delicate and flaccid; bisporangia common .... *P. elongella*
12 Periaxial cells 5-8 13
Periaxial cells ≥9 19
13 Cortication absent 14
Cortication present near base at least 17
14 Periaxial cells 6 .... *P. denudata* (sterile thalli)
Periaxial cells 7-8 15
15 Branching dichotomous, young axes paired and incurved, resembling *Ceramium*; gametangia and tetrasporangia unknown, specialized propagules formed at apices .... *P. furcellata*
Branching spiral on main axes, apices straight or corkscrew-like but not incurved; gametangia and tetrasporangia common 16
16 Main axes ≥200 µm wide, apices spiral, corkscrew-like .... *P. nigra*
Main axes ≤40 µm wide, apices straight .... *P. foetidissima*
17 Axes consisting largely of 4 periaxial cells, only some segments appearing to have 5 periaxials (due to cortication) .... *P. fibrillosa*
Axes consisting of 5 or more periaxial cells throughout 18
18 Main axes dichotomously branched at a wide angle; holdfast discoid, lacking prostrate axes; periaxial cells 5-7 .... *P. denudata*
Main axes straight, bearing laterals spirally; prostrate axes well-developed; periaxial cells 6-8 .... *P. brodiaei*
19 Cortication lacking 20
Cortication present on lower main axes .... *P. fucoides*
20 Segments ≤0.5 diameters long in mature axes; branching pseudodichotomous; epiphytic on fucoids, especially *Ascophyllum nodosum* .... *P. lanosa*
Segments >1 diameter long in mature axes 21
21 Periaxial cells >20 .... *P. opaca*
Periaxial cells ≤20 22
22 Major axes ≤40 µm wide .... *P. foetidissima*
Major axes ≥200 µm wide 23
23 Laterals more-or-less equal in length to major axes .... *P. fucoides*
Laterals short, arranged spirally or alternately on major axes 24
24 Thalli erect; branching axillary to trichoblasts; laterals borne in a spiral arrangement .... *P. nigra*
Thalli broad and spreading; branches replacing trichoblasts; laterals borne distichously in part of thallus at least 25

25 Major axes ≤300 µm wide, periaxial cells not inflated .... *P. simulans*
Major axes ≥350 µm wide, periaxial cells inflated ....*P. subulifera*

Polysiphonia atlantica Kapraun & J. Norris (1982), p. 226.

Lectotype: TCD. Clare (Miltown Malbay), 1831, *Harvey*.

Polysiphonia macrocarpa Harvey in Mackay (1836), p. 206, non *Polysiphonia macrocarpa* (C. Agardh) Sprengel (1827), p. 350.

Thalli forming dense cushion-like tufts 0.5-3 cm high and 2-4 cm wide, consisting of interwoven prostrate axes and numerous erect axes branched to 5 or more orders; lower axes often bearing series of short reflexed branchlets; upper branches equal in length to main axes, dull red to brownish-red in colour, with an extremely soft and flaccid texture.

Axes ecorticate, consisting of small axial cell and 4 periaxial cells; *prostrate axes* curved downwards at tips, with large and conspicuous apical cells 12-14 µm in diameter, dividing transversely, the first-formed periaxial cell ventral in position, branching endogenously near curved tip of apex, 60-90 µm in diameter when mature, with segments 0.5-2.5 diameters long; *rhizoids* formed medially by periaxial cells and remaining in open connection with them, 20-30 µm in diameter and up to 1 mm long; periaxial cells in vicinity of rhizoids sometimes dividing transversely; *erect axes* curved adaxially when young, later straight or reflexed, increasing in diameter from domed apical cells 8-12 µm wide, to 40-50 µm in diameter just behind apices and reaching a maximum of 60-75 µm, with segments (0.5)1-3 diameters long; primary laterals endogenous, emerging adaxially near apex; adventitious laterals originating at a distance from apex, endogenous, never arising from scar cells or periaxial cells, with only two periaxial cells maturing normally in basal segment; exogenous branches and trichoblasts occasional to common on gametophytes, rare or absent on tetrasporophytes, with all periaxial cells maturing in basal segment; *trichoblasts* formed in 1/4 spiral divergence, 7-12 µm in diameter and up to 375 µm long, branched 1-2 times; *plastids* in the form of angular plates or becoming beaded.

Plants dioecious. *Spermatangial branchlets* and occasional interpolated trichoblasts initiated in a 1/4 right-handed spiral, but deflected secondarily and appearing to form a left-handed spiral, borne in dense clusters at apices, 115-145 µm long when mature,

---

Fig. 94. *Polysiphonia atlantica*

(A) Habit (A & E-I Donegal, July). (B) Prostrate axis with numerous young erect axes, attached by rhizoids (B-D Devon, Mar.). (C) Rhizoid showing cytoplasm continuous with periaxial cell. (D) Apex growing from large domed apical cell. (E) Axis with long, delicate trichoblasts. (F) Tufts of spermatangial branchlets at apices. (G) Incurved spermatangial branchlets lacking terminal sterile cells. (H) Slightly urceolate cystocarp. (I) Tetrasporangia borne in long straight series.

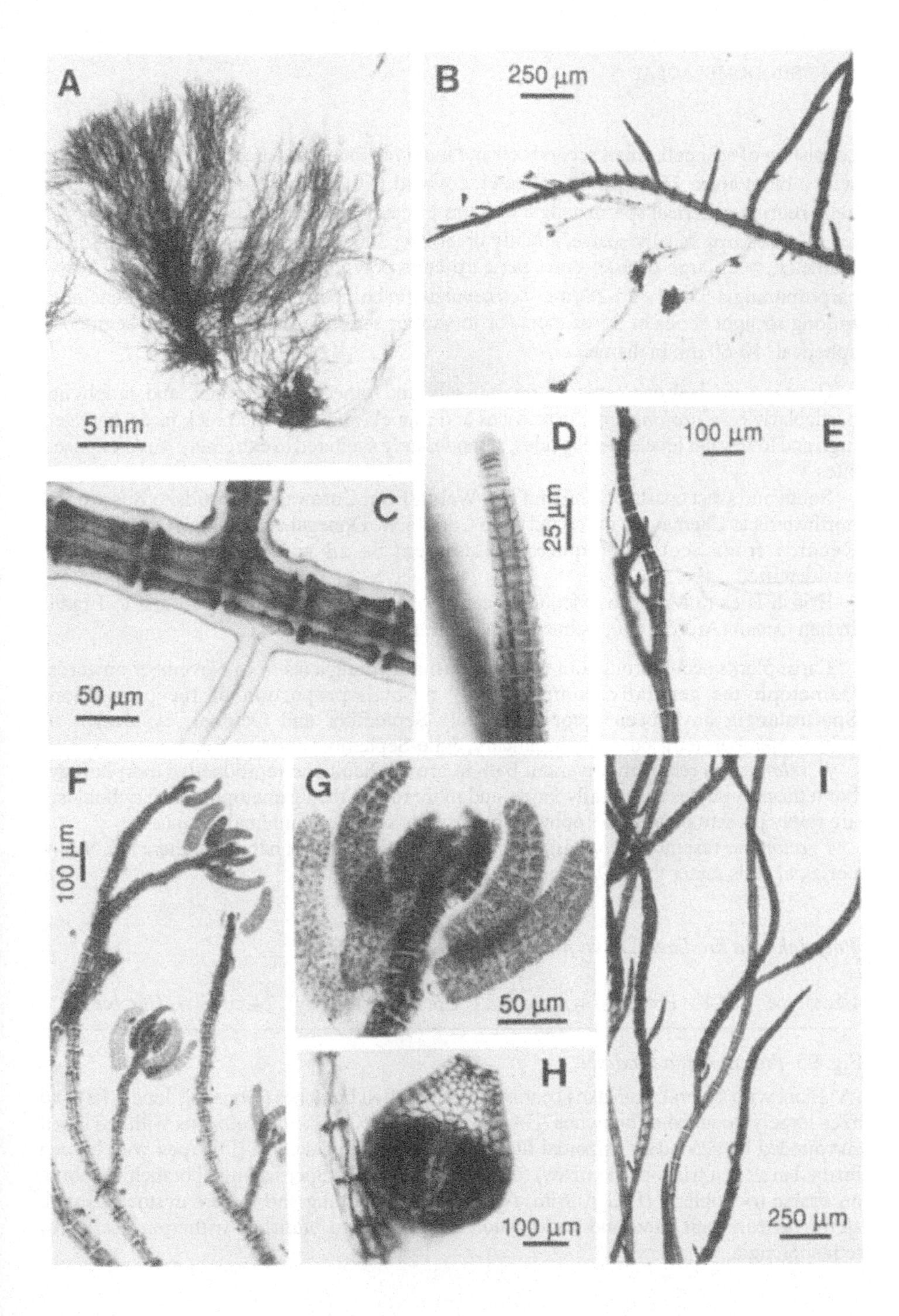
A
5 mm
B
250 µm
C
50 µm
D
25 µm
E
100 µm
F
100 µm
G
50 µm
H
100 µm
I
250 µm

consisting of scar cell, small second cell and incurved, banana-shaped, spermatangial axis with a blunt apex, 28-40 µm in diameter, covered with a layer of spermatangial mother cells bearing spherical spermatangia c. 2 µm in diameter. *Procarps* numerous on upper axes; *cystocarps* usually sparse, slightly urceolate, 300-360 µm high and 260-330 µm in diameter, with large ostiole; outer pericarp cells polygonal, arranged in vertical rows; carposporangia 55-80 x 20-28 µm. *Tetrasporangia* borne in last two orders of branching, in long straight series of 25 or more, or in shorter series separated by sterile segments, spherical, 50-60 µm in diameter.

Growing on bedrock, epizoic on mussels and other invertebrates, and epiphytic, particularly on *Corallina* spp., in crevices and runnels and on shaded rock faces, between high and low water levels of neap tides, at moderately sheltered to extremely wave-exposed sites.

South and west coasts of England and Wales, from Cornwall eastwards to Sussex and northwards to Caernavon; in Ireland from Cork to W. Donegal and Dublin; Channel Isles. Records from Scotland require reassessment as all specimens examined were misidentified.

British Isles to Morocco; Mediterranean; W. Atlantic from North Carolina to Brazil; Indian Ocean (Ardré, 1970; Schneider & Searles, 1991).

Large plants occur throughout the year, forming young axes from November onwards. Gametophytes generally comprise only a small proportion of the population. Spermatangia have been recorded in July-September and October; cystocarps in July-September; tetrasporangia in January, June-September and October.

P. atlantica is relatively invariant both in growth habit and reproductive morphology, but tetrasporophytes are usually larger and more robust than gametophytes. Trichoblasts are rarely present on tetrasporophytes, but usually common on female plants.

P. atlantica resembles *Lophosiphonia reptabunda* (q. v.) in habit, but there are only 4 periaxial cells rather than 10 or more as in *L. reptabunda*.

Polysiphonia brodiaei (Dillwyn) Sprengel (1827), p. 349.

Lectotype: BM-K. Probable syntype: LD. Cork (Bantry Bay), 24 vi 1807, *Hutchins*.

Fig. 95. *Polysiphonia brodiaei*

(A) Plant with several main axes bearing densely tufted branches (Donegal, June). (B) Old axes largely denuded of branches (B-C Cork, Nov.). (C) T.S. of main axis with axial cell surrounded by secondary rhizoidal filaments and 7 periaxial cells. (D) Apex with branch formed in axil of trichoblast (arrow) (Cornwall, Sep.). (E) Spermatangial branchlets borne on sparse trichoblasts (E-G Antrim, Apr.). (F) Developing and mature cystocarps with spiral pericarp cell rows and wide ostioles. (G) Tufted branches with spiral series of tetrasporangia.

A
B
2 cm
2 cm
C
250 µm
D
100 µm
E
100 µm
F
250 µm
G
250 µm

Conferva brodiaei Dillwyn (1809), pl. 107.

Thalli 3-36 cm high, consisting of dense tufts of one to many main axes and numerous smaller axes, attached by a solid mass 1-3 mm in diameter of aggregated rhizoids, and a mat of much-branched, interwoven prostrate axes that develop from base of young erect axes, spreading to 1.5 cm in diameter and giving rise to further erect axes; *main axes* 0.6-0.9 mm in diameter, remaining distinct and sometimes forming a few major laterals up to 0.6 mm in diameter, bearing a spiral arrangement of profusely branched tufts of narrower laterals at a narrow angle, resulting in a narrow cylindrical or pyramidal outline to the branch tips, and becoming partially or entirely denuded at the base when older; dark brownish-or purplish-red in colour, bleaching to straw-yellow; old axes fairly tough and flexible; young laterals extremely soft and flaccid.

Prostrate axes growing from apical cells 13-14 µm in diameter, lacking trichoblasts, typically with 6 periaxial cells, becoming corticated and increasing in diameter from 50 µm near apices to 0.6 mm; *rhizoids* 25-80 µm in diameter, formed by small cells cut off from posterior end of periaxial cells; *erect axes* growing from apical cells 7 µm in diameter, with 6-8 periaxial cells, rapidly becoming heavily corticated and forming 1-2 layers of thick-walled rhizoidal filaments between axial and periaxial cells; cortex initiated 3 orders of branching from apices, forming one cell layer over periaxial cells and 3-4 layers in the grooves between them; segments 0.5-2 diameters long; *trichoblasts* numerous, borne in an approximately 1/6 spiral divergence, 5-8 µm in diameter and up to 400 µm long, branched 1-2 times, composed of uninucleate cells, deciduous, leaving conspicuous scar cells; *branches* formed in axils of trichoblasts, up to 5 or more orders developing on main axes and major laterals, increasing in density towards apices; ultimate branchlets 50 µm in diameter with segments 0.7-1.5 diameters long; *plastids* discoid.

Plants dioecious; all reproductive structures borne abundantly on terminal 1-3 orders of branching. *Spermatangial axes* borne at the first dichotomy of fertile trichoblasts, 160-280 µm long and 50-75 µm in diameter, cylindrical to slightly conical, with a rounded or pointed apex, lacking sterile terminal cells when mature, covered with a layer of spermatangial mother cells bearing spherical spermatangia 3-4 µm in diameter. *Cystocarps* narrow with wide ostiole when developing, urceolate when mature, 425-500 µm high and 325-500 µm in diameter; outer pericarp cells irregularly diamond-shaped, formed in irregularly spiral rows, larger around ostiole; ostiole 75-125 µm wide; carposporangia clavate, 80-115 x 20-40 µm. *Tetrasporangia* formed in long spiral series of which only 2-5 are mature, spherical when mature, 70-80 µm in diameter.

Growing on bedrock, mussels and limpets, and epiphytic on crustose corallines and various smaller algae, in pools from mid-intertidal to extreme low water and submerged on floating structures such as marina pontoons, less frequently subtidal on stones and *Laminaria hyperborea* stipes and fronds, to 8 m depth, occurring intertidally at moderately to extremely wave-exposed sites, subtidally at sheltered sites with some current exposure.

Generally distributed in the British Isles.

Norway to Portugal; Mediterranean; Atlantic Canada; Pacific North America from California to Washington; Australia and New Zealand (Kapraun & Rueness, 1983; Adams, 1991). Pacific populations are usually found near harbours and may have been spread by shipping (Womersley, 1979; Adams, 1991).

Seasonal behaviour varies with habitat. Rock-pool populations become conspicuous in February-April, reproducing when very small, and reach their maximum size in June-July after which they die back or become heavily epiphytized. On floating structures and in sheltered subtidal habitats, large fertile plants can be found throughout the year. Spermatangia have been recorded for March-October and December, cystocarps in April-October and December, and tetrasporangia in March-October and December. Chromosome number is $n = 29\text{-}31$ (Magne, 1964).

Morphology varies greatly according to environment. Plants growing on pontoons or in lower-shore pools are elongate and flaccid, with a cylindrical outline to the laterals borne on each main axis; thalli in upper-shore pools are densely tufted and pyramidal.

Polysiphonia ceramiaeformis P. Crouan & H. Crouan (1867), p. 158.

Lectotype: PC 'Algues Marines du Finistère', no. 305. France (baie de Laninon, Brest), undated, *Crouan*.

Thalli 2.5-5.5 cm high and 1.5-5 cm broad, consisting of dense irregularly rounded tufts of one to many much-branched erect axes attached by tangled prostrate axes; erect axes 0.3-0.5 mm in diameter, lacking distinct main axes and forming corymbose apices; bright red in colour, bleaching to straw-yellow, with a succulent texture.

All axes ecorticate, with 10-12 periaxial cells; *prostrate axes* becoming erect at the tips, lacking trichoblasts, increasing to 290-430 μm in diameter, with segments 0.4-0.9 diameters long; *rhizoids* numerous, cut off from periaxial cells; *erect axes* growing from apical cells 10-12 μm in diameter, increasing to 300-500 μm in diameter, with segments 1.3-2 diameters long and straight or spirally twisted transparent periaxial cells; axial cell comprising 1/7-1/5 of axis diameter; branches and trichoblasts formed in a spiral divergence of 1 in 11-13; *trichoblasts* frequent on females and tetrasporophytes, up to 600 μm long, branched 2-4 times, decreasing from 20-25 to 6 μm diameter, the basal cells multinucleate and apical cells uninucleate, leaving conspicuous scar cells when shed; *branches* developing at intervals of 2-9 segments, replacing trichoblasts, in an alternate or pseudodichotomous arrangement, to 9 orders of branching, borne at acute angles; ultimate branches forcipate; *adventitious branches* developing occasionally from lower main axes; *plastids* irregularly elongate to convoluted.

Plants dioecious. *Spermatangial branches* consisting of scar cell, an elongate cell and the cylindrical spermatangial axis 200-330 μm long and 30-65 μm in diameter, with a rounded apex lacking sterile terminal cells, covered with a layer of spermatangial mother cells bearing spherical spermatangia 3-4 μm in diameter. *Cystocarps* ovoid to slightly pyriform, 310-380 μm high and 240-290 μm in diameter; outer pericarp cells polygonal,

A
1 cm
B
1 mm
C
100 µm
D
E
50 µm
100 µm
F
G
100 µm

in straight rows, decreasing in size towards ostiole; carposporangia clavate, 100-125 x 35-50 μm. *Tetrasporangia* formed in last 4 orders of branching, in spiral series of 1-9, often interrupted by sterile segments, ellipsoid, 90-100 μm long x 75-80 μm wide.

On silty bedrock and blades of *Ulva* sp. in pools at low water of neap tides at sheltered to very sheltered sites.

Known with certainty only from a few sites in one area of Dorset, where its occurrence may be restricted to a narrow range of habitats. A report from Norfolk requires confirmation.

British Isles to N.W. France; Mediterranean (Lauret, 1970). Previous confusion with *Polysiphonia furcellata* precludes further assessment of its distribution.

Populations are apparently ephemeral and occur very sporadically; all collections have been made in April, some including plants with spermatangia, cystocarps and tetrasporangia.

There appears to be little morphological variation in the few known samples from the British Isles. Batten (1923) considered that *P. ceramiaeformis* represented "a young form of *P. furcellata*". However, Lauret (1970) showed that Mediterranean populations of these species differed in several important morphological characters, and these distinctions have been confirmed for British material.

Polysiphonia denudata (Dillwyn) Greville ex Harvey in W. J. Hooker (1833), p. 332.

Lectotype: BM-K. Syntypes: BM-K. Hampshire (Southampton), *Biddulph*.

Conferva denudata Dillwyn (1809), p. 85, pl. G.
Hutchinsia variegata C. Agardh (1824), p. 153.
Polysiphonia variegata (C. Agardh) Zanardini (1841), p. 60.

Thalli 3-25 cm high, consisting of a single much-branched erect axis, with a broad and irregularly rounded outline, attached by a disc 1-3 mm in diameter composed of aggregated rhizoids; main axes 0.6-0.7 mm in diameter near holdfast, branching pseudodichotomously repeatedly at an angle of 70-90° and gradually decreasing in diameter towards apices where younger axes are very fine and profusely branched; thalli purplish-red, older axes cartilaginous, young branches delicate and flaccid.

Erect axes growing from apical cells 7-8 μm in diameter, c. 50 μm in diameter when

---

Fig. 96. *Polysiphonia ceramiaeformis*

(A) Habit, showing densely tufted branches (A-G Dorset, Apr.). (B) Detail of axes, showing irregularly alternate branching and forcipate apices. (C) Apex with incurved tips and sparse inconspicuous trichoblasts. (D) Spermatangial branchlets, lacking sterile tips (herbarium specimen, 1884, in BM). (E) T.S. of axis with 12 periaxial cells and small axial cell (scale = 100 μm). (F) Mature cystocarp with narrow ostiole. (G) Tetrasporangia.

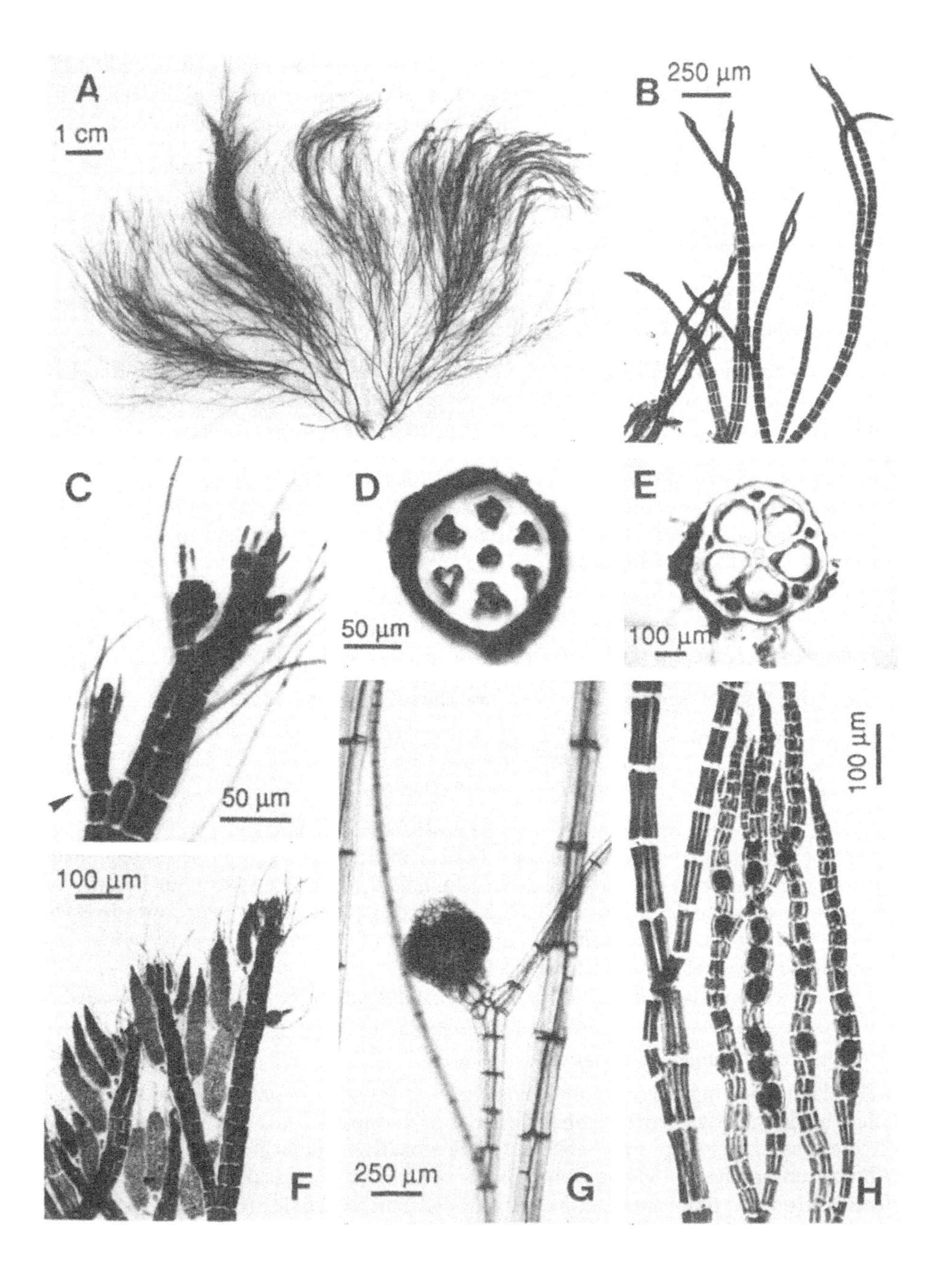
A
1 cm
B
250 µm
C
50 µm
D
50 µm
E
100 µm
100 µm
F
250 µm
G
100 µm
H

young, with 5-7 periaxial cells and segments 2-3 diameters long; cortication initiated several orders of branching from apices, growing downwards in grooves between periaxial cells and only covering axis entirely near base; segments near base of main axes 0.5-0.6 diameters long; *trichoblasts* numerous, borne on every segment in a 1/5 to 1/6 spiral divergence, 5-15 µm in diameter and up to 400 µm long, simple on female gametophytes, branched 1-2 times on males and tetrasporophytes, composed of uninucleate cells, deciduous leaving conspicuous scar cells; *branches* formed in axils of trichoblasts, typically at intervals of 4-6 segments near apices; up to 6 or more orders developing at irregular intervals, both branches of a dichotomy usually lying in one plane; *adventitious branches* later formed sparsely in various planes, *plastids* irregularly-shaped, round to elongate or slightly convoluted.

Plants dioecious; all reproductive structures borne on terminal 1-3 orders of branching. *Spermatangial axes* formed at the first dichotomy of fertile trichoblasts, conical to cylindrical with a rounded or pointed apex, 160-430 µm long and 50-90 µm in diameter, with or without 1-4 elongate or isodiametric sterile terminal cells when mature, covered with a layer of spermatangial mother cells bearing spherical spermatangia 2-3 µm in diameter. *Cystocarps* narrow when young, angled towards bearing axis by curvature of the wide stalk, barrel-shaped when mature, broader at the equator, 475-550 µm high and 240-450 µm in diameter, with ostiole 175 µm wide; outer pericarp cells irregularly polygonal, formed in vertical rows, decreasing in size towards ostiole; carposporangia 90-130 x 40-65 µm. *Tetrasporangia* formed in spiral series, distorting branch into a spiral, only 1-5 per series mature at one time, spherical to elliptical, 70-100 µm long x 50-100 µm in diameter.

Growing on bedrock and stones, on large algae such as *Chorda filum*, and on artificial structures such as pontoons in yacht marinas, at extreme low water and submerged on floating structures, at extremely sheltered sites with moderate to strong current exposure; usually occurring as isolated plants except in marinas where this species is sometimes abundant.

South coast of England from Cornwall eastwards to Hampshire; reported occurrence in Ireland requires confirmation. Reports from northern coasts represent misidentifications of other *Polysiphonia* species such as *P. brodiaei*.

Fig. 97. *Polysiphonia denudata*

(A) Habit, showing pseudodichotomously branched main axes and numerous branches (Devon, Nov.). (B) Axes of sterile thallus lacking cortication (Cornwall, undated). (C) Apex with branch formed in axil of trichoblast (arrow) (as A). (D) T.S. of axis of sterile thallus, with 6 periaxial cells and no cortex (as B). (E) T.S. of axis of female thallus, with 5 periaxial cells and developing cortication (as A). (F) Spermatangial branchlets, with long sterile tips (Devon, Oct.). (G) Mature cystocarp with wide ostiole (as A). (H) Spirally arranged tetrasporangia (Hampshire, Nov.).

Netherlands to Portugal and W. Africa; Mediterranean; W. Atlantic from Maine to Caribbean and Brazil; ?Pacific islands (Schneider & Searles, 1991).

Mature thalli have been collected from April to November. Spermatangia recorded in May and November; cystocarps in April, June and September-November; and tetrasporangia in April, June, August and October-November. The main axes of large plants usually bear numerous epiphytes. An isolate from Texas followed a *Polysiphonia*-type life history in culture, tetraspore tetrads developing into 2 male and 2 female gametophytes (Edwards, 1970). This isolate grew well at 19-30°C, forming reproductive structures within 2-4 weeks under different conditions including 8 h and 16 h days, and completed the life history within 6 weeks. Chromosome number of North Carolina populations is $n = 30$ (Kapraun, 1978a).

Lauret (1970) noted a dimorphism of reproductive and non-reproductive plants in Mediterranean populations of *P. denudata*. Reproductive thalli had 5-6 periaxial cells and were corticated, with abundant trichoblasts, whereas sterile plants, which could become as large as fertile ones, had 6-7 periaxial cells and lacked cortication. The majority of recent collections of *P. denudata* in the British Isles are fertile, consist of 5-6 periaxial cells, and have extensive cortication. Some non-reproductive specimens have 6 periaxial cells and lack cortication and trichoblasts; Batten (1923) reported free-floating sterile plants lacking cortication. It therefore seems likely that dimorphism also occurs in the British Isles. This point is significant because the type collection consists of large sterile thalli with 6 periaxial cells and no cortication and thus differs from the majority of large specimens of this species. In the Mediterranean and North Carolina, spermatangial axes terminate in 1-3 sterile cells. In England, some marina populations form sterile tips while plants from other habitats lack them.

Polysiphonia devoniensis Maggs & Hommersand, sp. nov.

Thalli forming tufts 1.2-5 cm high, consisting of matted prostrate axes and numerous erect axes, resembling *Polysiphonia scopulorum* Harvey (1855, p. 540) and *P. rudis* Harvey & J. D. Hooker (1845, p. 183), differing from them in the position of spermatangial axes, which are borne singly at the first dichotomy of long fertile trichoblasts, whereas in these

---

Fig. 98. *Polysiphonia devoniensis*

(A) Habit of densely tufted plant (A, D-E & G-I holotype and isotype material, Devon, Aug.). (B) Habit of finely branched plant (Dorset, June). (C) Axis with numerous rhizoids and branches (Devon, June). (D) Apex with long, much-branched trichoblasts. (E) Elongate spermatangial branchlets distributed along younger axes (holotype slide, in BM). (F) Prostrate axis showing rhizoids connected to periaxial cells (as C). (G) Spermatangial axis with long sterile tip cells (holotype slide, in BM). (H) Mature cystocarp with regularly polygonal pericarp cells (holotype slide, in BM). (I) Branches distorted into spiral by tetrasporangia.

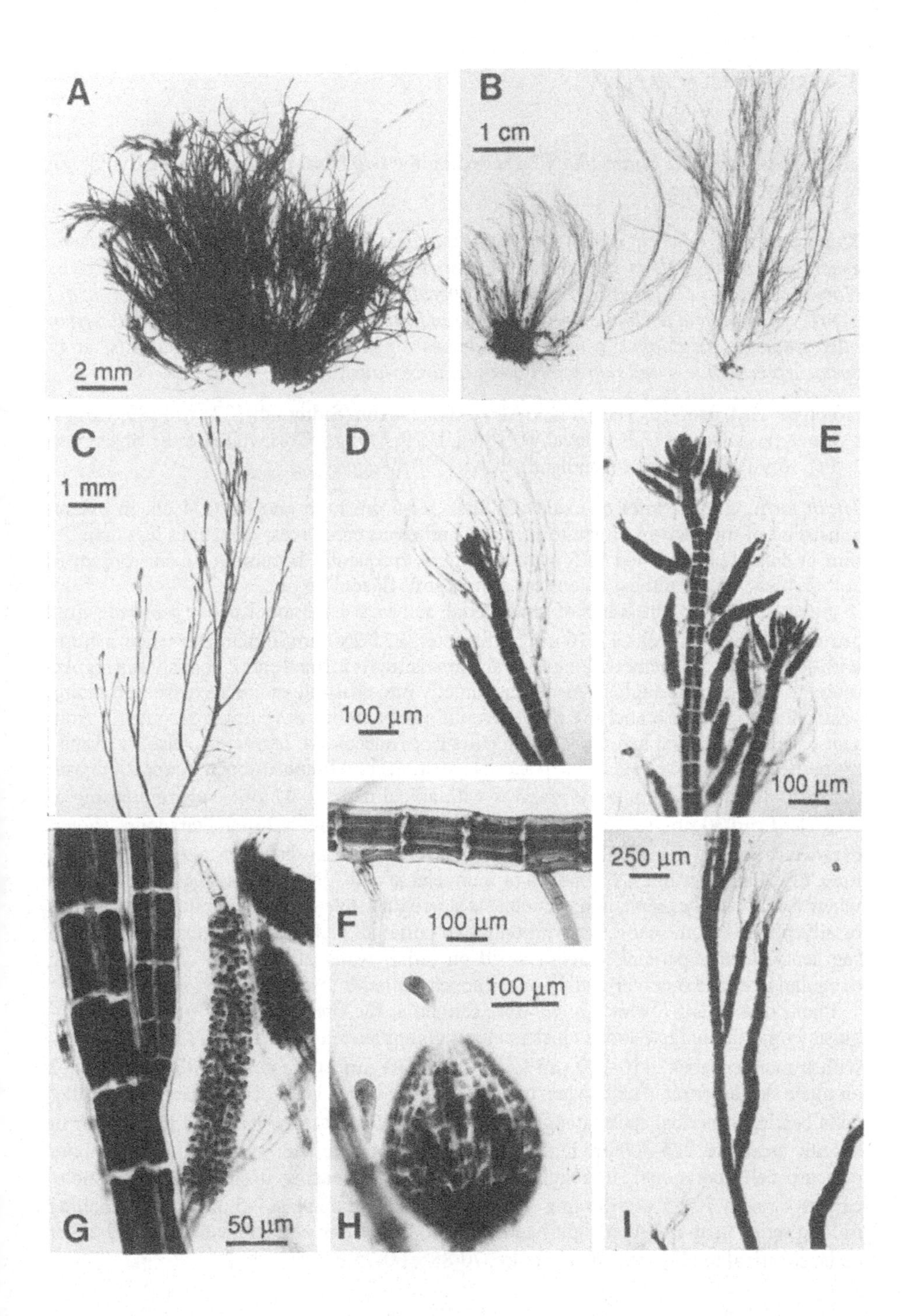
A
2 mm
B
1 cm
C
1 mm
D
100 µm
E
100 µm
F
100 µm
G
50 µm
H
100 µm
I
250 µm

other species they are borne singly or paired on the two basal cells, and in the ovoid, not urceolate, cystocarps.

Thalli caespites 1.2-5 cm altos formantes, ex axibus implicatis prostratisque et plurimis erectis compositi. Affinis Polysiphoniae scopulorum *Harvey (1855, p. 540) et* P.rudis *Harvey & Hooker (1845, p. 183), e quibus differunt positione axium spermatangiorum qui in prima dichotomia trichoblastorum longorum fecundorumque singulatim portati, sed in aliis speciebus in duabus cellulis basalaribus singulatim aut binatim portatis, et in cystocarpiis ovoideis, sed non in cystocarpiis urceolatis.*

Holotype: BM. Isotypes: BM, PC, GALW, US. Devon (Sidmouth), 22 viii 1989, *Maggs*. Paratypes: Devon (S.E. Clovelly), 18 vii 1979, *Maggs*; Cornwall (Lower Sharpnose Pt), 16 vii 1979; Dorset (Studland), 10 vi 1990, *Maggs*.

Thalli forming erect tufts or extensive turfs 1.2-5 cm high and up to 4 cm in extent, consisting of interwoven prostrate axes and numerous erect axes; erect axes less than 0.3 mm in diameter, branched to 5 or more orders, frequently becoming repent; brown in colour, bleaching to almost colourless, with a soft, flaccid texture.

Axes ecorticate, consisting of small axial cell and 4 periaxial cells; *prostrate axes* growing from apical cells 12-16 μm in diameter, 225-280 μm in diameter when mature, with segments 0.5-1 diameter long; lateral branch initials formed endogenously at irregular intervals and variable angles, developing directly into prostrate or erect axes, or remaining vestigial; exogenous branching rare; adventitious branches developing frequently from scar cells and vestigial laterals, often in pairs from successive segments; *rhizoids* formed by periaxial cells in a median or posterior position, remaining in open connection with them, 20-80 μm in diameter; *erect axes* with apical cells 10-12 μm wide, increasing in diameter from 30 μm near apices to 50-280 μm, segments 0.5-2 diameters long; *trichoblasts* formed in a 1/4 spiral, few to abundant, 10-12 μm in diameter and up to 600 μm long, branched 1-4 times, composed of uninucleate cells, leaving conspicuous scar cells when shed; *branches* replacing trichoblasts at irregular intervals, exogenous, endogenous or adventitious from scar cells, with only two periaxial cells maturing normally in basal segment; *plastids* parietal, shield-shaped on radial walls of periaxial cells, becoming irregularly beaded on outer walls so axes appear rather transparent.

Plants dioecious. *Spermatangial axes* formed at the first dichotomy of fertile trichoblasts borne in elongate loose clusters along young axes, cylindrical to slightly conical, with a rounded apex, 110-200 μm long and 25-30 μm in diameter, with or without 2 elongate sterile terminal cells when mature, covered with a layer of spermatangial mother cells bearing spherical spermatangia 4 μm in diameter. *Cystocarps* ovoid to globular or slightly urceolate, 275-300 μm high and 210-250 μm in diameter, with large ostiole; outer pericarp cells polygonal, in regular vertical rows decreasing in size towards ostiole; carposporangia 70-95 x 25-35 μm. *Tetrasporangia* borne in last 2-3 orders of branching in long series in an irregular right-handed spiral, formed by periaxial cell-3, with 3 cover cells, spherical to ellipsoid when mature, 70-85 x 60-75 μm.

Growing on hard or friable bedrock, often through sand cover or where exposed to sand scour, from low water level of neap tides to 4 m depth, at moderately sheltered to extremely wave-exposed sites, generally occurring in abundance when present.

Dorset, N. and S. Devon, and N.E. Cornwall; common and widely distributed in appropriate habitats within this area, but not observed outside it.

Plants have been observed only from June-August. Gametophytes generally form only a small proportion of the population. Spermatangia, procarps and cystocarps recorded in August; tetrasporangia in June-August.

This species shows relatively little morphological variation, but details of branching differ between gametophytes and tetrasporophytes. In males, branches are exogenous near apices, supplemented by adventitious branching from scar cells towards the base; in females, branching is endogenous, rarely exogenous from basal cells of trichoblasts, or adventitious from scar cells; and in tetrasporophytes, branches are exogenous, replacing trichoblasts, or adventitious from scar cells towards the base.

P. devoniensis shows similarities in habit and vegetative structure to *Polysiphonia scopulorum* Harvey (1855, p. 540) from Australia and *P. rudis* Harvey & J. D. Hooker (1845, p. 183) from New Zealand (see Womersley, 1979), and it may be closely related to them. However, *P. devoniensis* differs reproductively. Spermatangial axes replace one branch at the first dichotomy of long fertile trichoblasts, whereas in *P. scopulorum* and *P. rudis* they are paired, replacing both branches, or are single and terminal on the two basal cells. Although the morphology of fertile male trichoblasts varies in some *Polysiphonia* species, such as *P. fucoides*, this variation is normally seen within individual thalli. No specimens of *P. devoniensis* have been found with paired spermatangial axes, and there are no reports of *P. scopulorum* with spermatangial axes borne singly at the first dichotomy of fertile trichoblasts. Cystocarps are ovoid in *P. devoniensis* but urceolate in *P. scopulorum*; tetrasporangia distort branches into a definite spiral in *P. devoniensis* but this occurs only to a small degree in *P. scopulorum*. *P. scopulorum* is also known from Europe (Ardré, 1970, as *Lophosiphonia*) and comparison of our material of *P. devoniensis* with specimens of *P. scopulorum* from France (Biarritz, 28 vii 1870, *Bornet*, LD 39420) confirms the distinctness of these entities. The Biarritz plants have much shorter axial segments, shorter lateral branches with shorter series of tetrasporangia, and are generally much more regular in appearance than *P. devoniensis*.

Our specimens of *P. devoniensis* also resemble closely a small species of *Polysiphonia* reported from Portugal, tentatively identified as *P. funebris*, which was originally described from the Mediterranean (Ardré, 1970; F. Ardré, pers. comm.). However, all known Mediterranean and Atlantic collections of *P. funebris* are non-reproductive, lacking important taxonomic characters. Thus, although we recognize the possibility that *P. devoniensis* may be conspecific with *P. funebris*, or at least with the specimens from Portugal, we prefer to propose it as a new species until further research on *P. funebris* can be carried out.

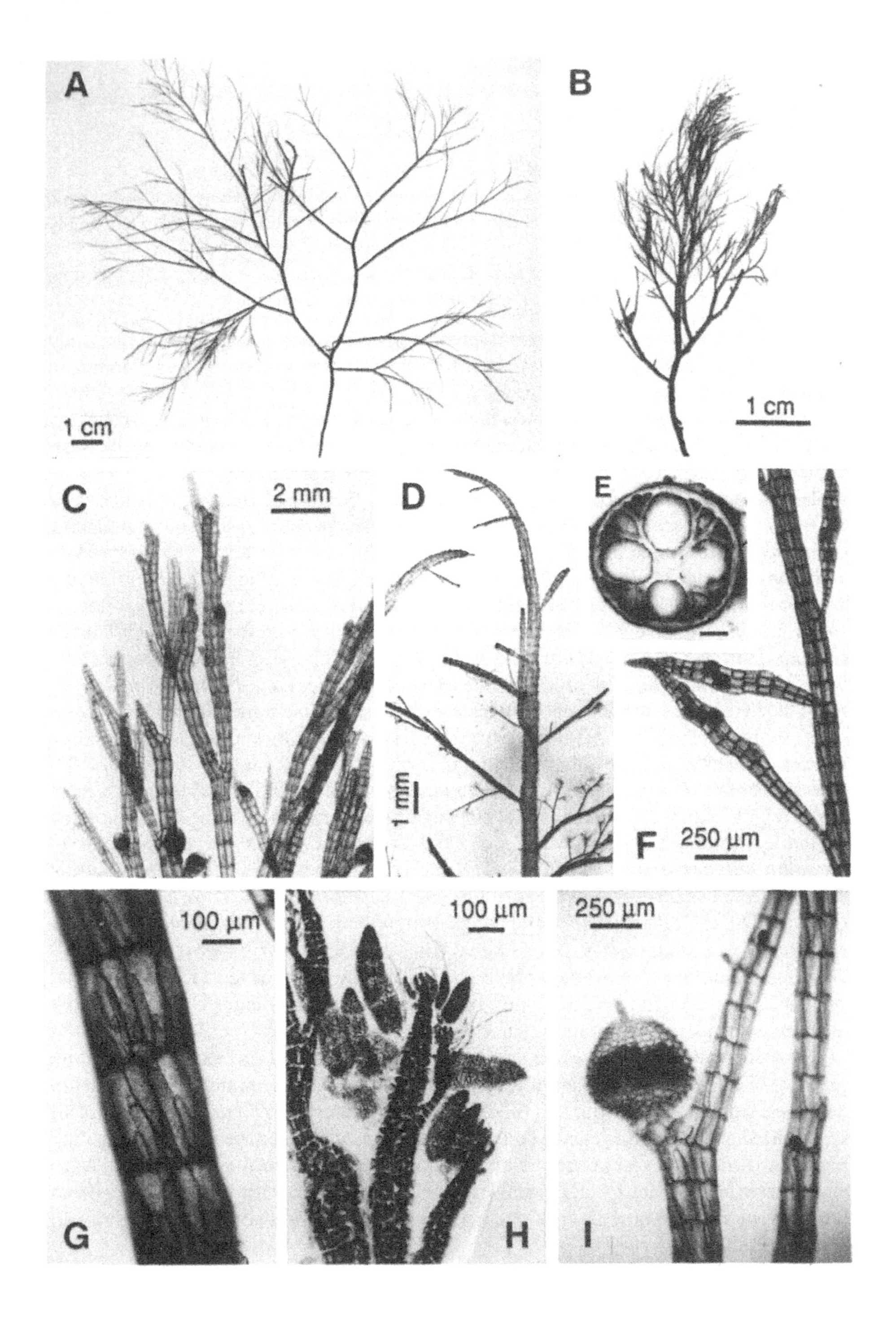
A
B
1 cm
1 cm
C
2 mm
D
1 mm
E
F
250 µm
G
100 µm
H
100 µm
I
250 µm

Small plants of *P. devoniensis* may resemble *Lophosiphonia reptabunda* (q. v.) in habit, but there are only 4 periaxial cells rather than 10 or more periaxials as in *L. reptabunda.*

Polysiphonia elongata (Hudson) Sprengel (1827), p. 349.

Lectotype: BM-K. England, unlocalized (see Irvine & Dixon, 1982).

Conferva elongata Hudson (1762), p. 484.

Thalli 5-30 cm high, consisting of a single branched erect axis with a narrow to broad outline, attached by a holdfast; holdfast consisting initially of a disc 1-2 mm in diameter composed of downgrowing cortical filaments, subsequently increasing in size by forming short radiating, heavily corticated, prostrate axes 0.3-0.7 mm wide that terminate in clusters of rhizoids 40-130 μm in diameter; *main axis* 0.7-2.4 mm in diameter, unbranched for basal 1-2 cm, either remaining distinct throughout thallus, bearing laterals at irregular intervals, or branching pseudodichotomously one to several times, gradually decreasing in diameter towards apices, usually densely branched to 4 or more orders but sometimes entirely denuded; *adventitious branches* occurring irregularly among primary branches; young branches bright red, succulent and easily damaged; old axes dark brownish-red, fairly tough and flexible.

Erect axes with 4 periaxial cells, spirally arranged; segments 0.5-0.7 diameters long when young, 0.3-0.5 diameters long in main axes; cortication developing rapidly, initially growing downwards in grooves between periaxial cells, increasing to 3-5 layers in thickness on major axes; *trichoblasts* usually sparse, formed in a 1/4 spiral divergence, longer on male gametophytes, up to 500 μm in length, branched 0-3 times, composed of uninucleate cells, deciduous leaving small scar cells; *branches* replacing trichoblasts at irregular intervals of 1-7 segments, spindle-shaped when young because apices and bases are markedly attenuated; *plastids* irregularly-shaped, round to elongate or slightly convoluted.

Plants dioecious. *Spermatangial trichoblasts* borne on every segment in a 1/4 spiral divergence at the apices of young branches and on densely branched adventitious laterals; spermatangial axes borne singly on basal cell of fertile trichoblasts or replacing one or both

---

Fig. 99. *Polysiphonia elongata*

(A) Plant with open branching (Down, Nov.). (B) Densely branched plant (Cork, Dec.). (C) Typical young axes showing rapid development of cortication (Down, May). (D) Detail of thallus with unusual narrow branches lacking attenuate bases (Sutherland, Aug.). (E) T.S. of axis with 4 periaxial cells and cortication (Donegal, June). (F) Last-order branches with strongly attenuate bases and spiral tetrasporangia (Donegal, Dec.). (G) Spiral periaxial cells with spiral cortication (as A). (H) Spermatangial branchlets lacking sterile tips (Glamorgan, Mar.). (I) Mature cystocarp with wide ostiole (as F) .

branches at first, and sometimes second, dichotomy, 330-500 µm long and 85-130 µm in diameter, cylindrical to slightly conical, initially incurved, straight when mature, with a rounded apex lacking sterile terminal cells, covered with a layer of spermatangial mother cells bearing spherical spermatangia 4 µm in diameter. *Cystocarps* present in abundance on last 3 orders of branching, ovoid when developing, ovoid to slightly urceolate when mature, 725-950 µm high and 675-775 µm in diameter, with ostiole 125 µm wide; outer pericarp cells polygonal in straight, regular rows, sometimes slightly larger around ostiole; carposporangia 125-200 x 40-75 µm. *Tetrasporangia* formed in last 2 orders of branching in long spiral series of which only 1-5 are mature, distorting branch into a spiral, tetrasporangia spherical, 70-110 µm in diameter.

Growing on bedrock, stones, shells, maerl, artificial structures such as ropes and marina pontoons, and various algae, occasionally on *Laminaria hyperborea* stipes, in lower-shore pools and from extreme low water to 27 m depth, most abundant in pools and the shallow subtidal zone, at extremely wave-sheltered sites with strong tidal currents and on open coasts with slight to strong wave-exposure.

Generally distributed in the British Isles, northwards to Shetland.

Norway to Portugal; Mediterranean; W. Atlantic from Prince Edward Is. to New England (Kapraun & Rueness, 1983).

Thalli perennial (Batten, 1923); new growth formed by surviving axes from November onwards. Large fertile plants can be found throughout the year. Spermatangia have been recorded for February-October, and are most abundant in March and April; cystocarps recorded in February and April-December; and tetrasporangia in February and April-December. Chromosome number is $n = 36$ (Austin, 1959).

Specimens of this species collected on open coasts generally show little morphological variation, the two most characteristic features being the attenuated bases of branches, and the rapid development of cortication. Rosenvinge (1923-24) described Danish populations that differed in branch morphology and cortication. In *P. elongata* forma *schuebelerii* (Foslie) Rosenvinge and forma *baltica* Rosenvinge, branches are not attenuate at the base, cortication is much less dense, and axes are narrower, with longer segments. Reduced-salinity growth forms of *P. elongata* are occasionally found in Scottish and Irish sealochs, and resemble those described by Rosenvinge. Individual thalli may exhibit this type of morphology in basal portions and then develop into a typical growth form; Rosenvinge's forms of *P. elongata* undoubtedly represent a response to environmental conditions.

Denuded heavily corticated axes superficially resemble algae with terete thalli such as *Gracilaria verrucosa*, but can be identified by the four periaxial cells seen in TS.

Polysiphonia elongella Harvey in W. J. Hooker (1833), p. 334.

Lectotype: TCD. Probable syntypes: BM, LD. Dorset (Sidmouth), *Griffiths & Cutler*. Lectotype undated, probably collected in September 1831.

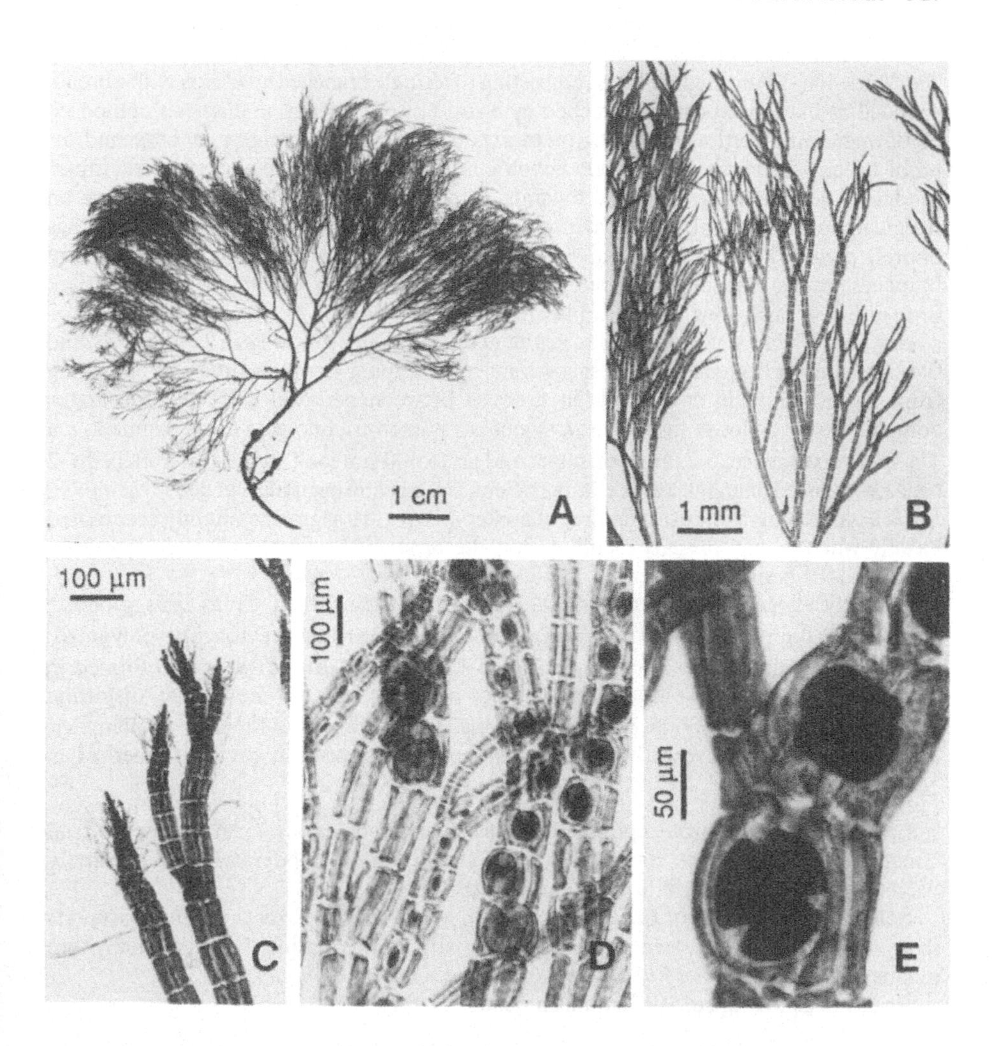

Fig. 100. *Polysiphonia elongella*

(A) Habit (Cork, Dec.). (B) Young axes showing formation of several orders of branching before development of cortication (Inverness, May). (C) Apex showing branches non-axillary to trichoblasts (Cork, Aug.). (D) Developing and mature bisporangia (as B). (E) Tetrasporangia (as C).

Thalli 2.5-10 (-13) cm high, solitary, consisting of a single branched erect axis with a broad, rounded or flat-topped outline, attached by a solid disc 0.7-1 mm in diameter composed of downgrowing cortical filaments; main axis 0.3-0.6 mm in diameter, unbranched for basal 1-3 cm, then branching pseudodichotomously at a wide angle one to several times, gradually decreasing in diameter towards apices; major axes bearing laterals in an irregularly spiral pattern, usually densely branched to 4 or more orders but sometimes entirely denuded; smaller adventitious branches occurring occasionally among primary branches; young branches bright red, flaccid, delicate and easily damaged, old axes dark brownish-red, fairly tough and flexible.

Erect axes with 4 periaxial cells; segments 1.5-3 diameters long in young axes and 0.4-0.5 diameters long in main axes; cortication developing 4 or more orders of branching from apices, growing downwards in grooves between periaxial cells and forming a complete cover on lower main axes; *trichoblasts* numerous, borne on every segment in a 1/4 spiral divergence, 6-7 μm in diameter and up to 400 μm long, simple or branched 1-2 times, composed of uninucleate cells, persistent, leaving conspicuous scar cells when shed; *branches* replacing trichoblasts at irregular intervals of 1-10 segments, slightly constricted basally when young; *plastids* round to elongate or slightly convoluted, often sparse on outer walls of periaxial cells.

Probably dioecious. [Spermatangial material not available.] *Cystocarps* globose, 300-500 μm high and 300-400 μm in diameter; outer pericarp cells rectangular-polygonal; ostiole c. 75 μm diameter; carposporangia 55-100 x 30-50 μm. *Bisporangia* formed in last 3 orders of branching in long spiral series of which only 1-4 are mature, distorting branch into a spiral, obliquely divided, ovoid to spherical, 70-125 μm in diameter; *tetrasporangia* formed sparsely amongst bisporangia, tetrahedrally divided, spherical, c. 100 μm in diameter.

Epiphytic on larger algae, epilithic on bedrock and pebbles, also growing on artificial substrata, from extreme low water to 13 m depth, at moderately to extremely wave-sheltered sites with little or no current exposure.

South and west coasts of Britain and Ireland, northwards to Ross & Cromarty; reports from Shetland and eastern Scotland requiring confirmation because many are misidentifications of other *Polysiphonia* species such as *P. elongata*.

British Isles to France; Mediterranean (Batten, 1923).

Mature reproductive plants have been collected from April onwards. Harvey (1848, pl. 146) reported that all branchlets are lost by October, the plants consisting during winter only of major axes; new branches are formed by surviving thalli in spring. The great majority of thalli are bisporangial, with a few tetrasporangia. Spermatangia have been recorded in January, February, July and August, and cystocarps in April, May, July and September. Bisporangia mixed with a few tetrasporangia have been observed in April, May and July-October.

This species shows little variation in morphology other than seasonal changes.

Polysiphonia fibrata (Dillwyn) Harvey in W. J. Hooker (1833), p. 226.

Lectotype: BM-K. Moray (Forres), undated, *Brodie*.

Conferva fibrata Dillwyn (1809), p. 84, pl. G.

Thalli forming dense tufts 0.5-10 (20) cm high and up to 4 cm wide, consisting of a solid mat of interwoven repent axes with numerous much-branched, very fine erect axes; main axes often bearing regular series of tufts of lateral branches; dull brownish-red in colour, with an extremely soft and flaccid texture, decaying rapidly after collection.

Erect axes repent basally, curving upwards towards apices and becoming erect, growing from apical cells 7-8 µm in diameter, with 4 periaxial cells, c. 50 µm in diameter when young, increasing to 200-250 µm in diameter in main axes; segments 1-1.5 diameters long when young, increasing to 2-5.5 diameters long; cortication developing only near bases of main axes or absent; repent axes 125-250 µm in diameter, with segments 0.5-1 diameter long; *rhizoids* formed abundantly, often paired, cut off from posterior end of periaxial cells, 20-60 µm in diameter; *trichoblasts* numerous, borne on every segment in a 1/4 spiral divergence, 8 µm in diameter and up to 600 µm long, branched 1-4 times, composed of uninucleate cells, deciduous leaving conspicuous scar cells; *branches* formed in axils of trichoblasts at intervals of 1-7 segments, up to 5 or more orders developing at irregular intervals; *plastids* small, discoid or beaded.

Plants dioecious; all reproductive structures borne abundantly on terminal 1-3 orders of branching. *Spermatangial axes* formed at the first dichotomy of fertile trichoblasts, 100-175 µm long and 25-40 µm in diameter, cylindrical with a blunt rounded apex, lacking sterile terminal cells when mature, covered with a layer of spermatangial mother cells bearing spherical spermatangia 2-3 µm in diameter. *Cystocarps* globular to ovoid when developing, angled away from bearing axis, distinctly stalked when mature, subglobular, 300-385 µm high and 265-360 µm in diameter, with a wide ostiole, sometimes slightly flared; outer pericarp cells rectangular to polygonal, formed in vertical rows, much larger around the ostiole; carposporangia 50-80 x 20-30 µm. *Tetrasporangia* formed in spiral series, distorting branch into a spiral, only 1-5 mature per series, spherical to elliptical, 60-80 µm long x 60-70 µm in diameter.

Growing on bedrock, epizoic on limpets, mussels and other invertebrates, and epiphytic, particularly on *Corallina* spp., in pools from mid-tide level to near extreme low water and occasionally subtidal to 20 m depth, tolerant of sand cover, at moderately to extremely wave-exposed sites.

Widely distributed in the British Isles.

British Isles to Spain (South & Tittley, 1986). The closely similar species *P. sertularioides* (Grateloup) J. Agardh is widespread in the Mediterranean (Lauret, 1967).

Basal portions of thalli are probably perennial; young erect axes occur in April, and plants are reproductive from April or May until November. Spermatangia have been

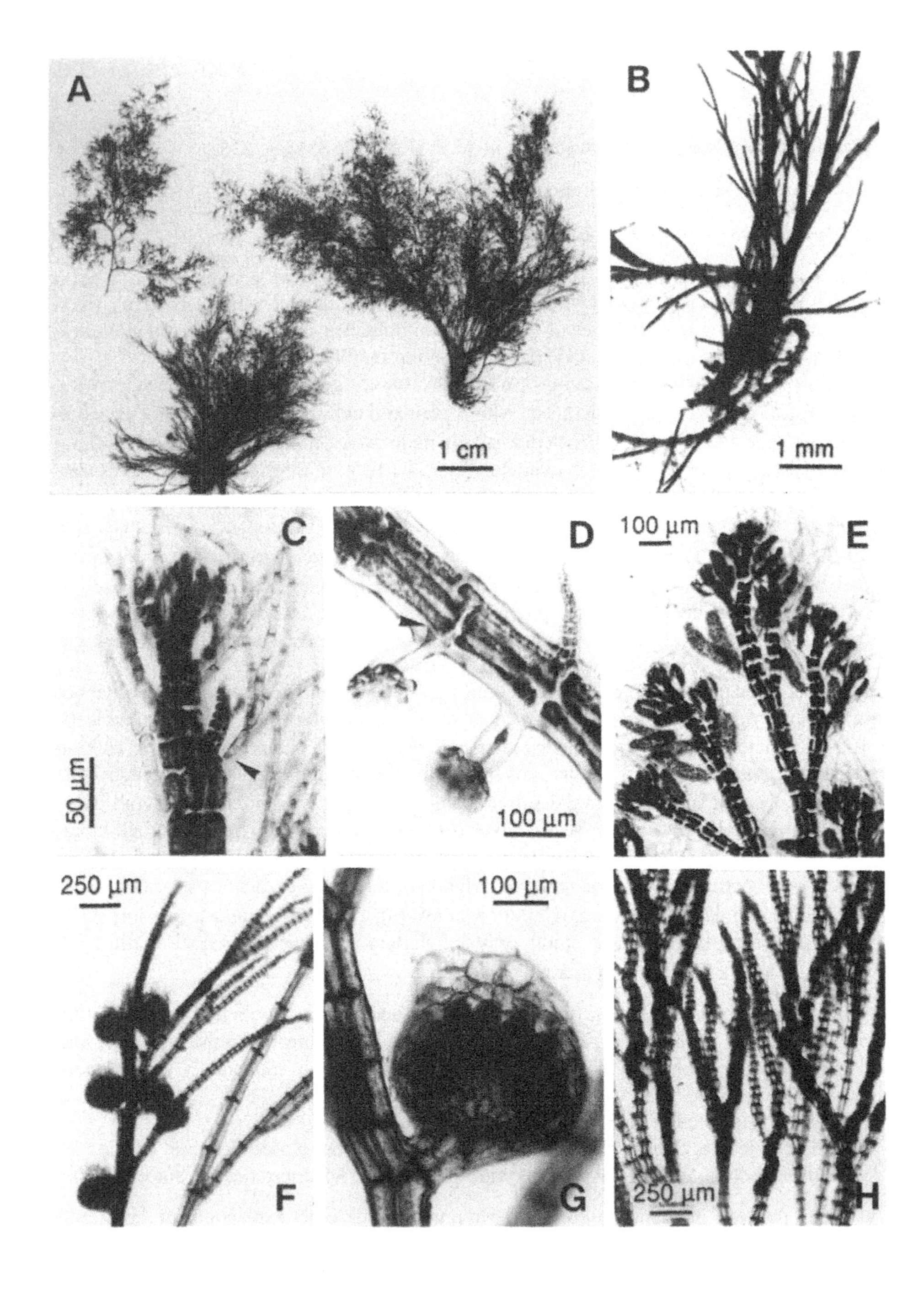
A
B
1 cm
1 mm
C
D
E
100 µm
50 µm
100 µm
250 µm
100 µm
F
G
250 µm
H

recorded in May-November; cystocarps in May-November; and tetrasporangia in April-September.

Axis diameter and the degree of cortication are variable. Subtidal plants have narrow ecorticate axes whereas some specimens growing on exposed coasts form robust, fairly heavily corticated lower main axes.

Several species from the N.E. Atlantic and the Mediterranean appear to be closely related to *P. fibrata*. *P. rhunensis* Thuret in Thuret & Bornet (1878, p. 84), described from Normandy, France, has been reported from the south coast of England (Batten, 1923), but British specimens attributed to this species are indistinguishable from *P. fibrata*. *P. orthocarpa* Rosenvinge (1923-24, p. 412) from Denmark is closely similar to *P. rhunensis*, differing principally in the shape of cystocarps. Some subtidally-collected specimens from the British Isles resemble *P. orthocarpa*; although cystocarpic material has not been observed, we consider these plants to be a lax form of *P. fibrata*. *P. sertularioides* from the Mediterranean is difficult to distinguish from finely branched specimens of *P. fibrata*, and the shape of cystocarps shown by Lauret (1967) appears identical to those of *P. fibrata*. *P. sanguinea* C. Agardh (1827, p. 638), described from the Adriatic (Venice), also has similarities to *P. fibrata*, and most British records are referable to *P. fibrata*. This group of taxa may represent a cluster of closely related species or a single, rather variable, species.

Polysiphonia fibrillosa (Dillwyn) Sprengel (1827), p. 349.

Lectotype: BM-K (Kapraun & Rueness, 1983, fig. 60). Probable syntypes: LD 40372. Sussex (Brighton), July 1807, *Borrer*.

Conferva fibrillosa Dillwyn (1809), p. 86, pl. G.
Polysiphonia spinulosa Greville (1824), p. 90.
Polysiphonia carmichaeliana Harvey in W. J. Hooker (1833), p. 328.
Polysiphonia richardsonii W. J. Hooker ex Harvey in W. J. Hooker (1833), p. 333.
Polysiphonia griffithsiana Harvey (1841), p. 91.
Polysiphonia violacea sensu Harvey (1848), pl. 209.
Polysiphonia grevillii Harvey (1849b), p. 86.
Polysiphonia violacea f. *fibrillosa* (Dillwyn) Areschoug (1850), p. 52.
Polysiphonia violacea var. *griffithsiana* Batten (1923), p. 302.

Thalli 1.5-25 cm high, consisting initially of a single much-branched erect axis with a

---

Fig. 101. *Polysiphonia fibrata*

(A) Thalli with numerous main axes, one separated out at top left, bearing densely tufted branches (A & C-G Clare, June). (B) Base of thallus showing erect axes and some that become secondarily prostrate (Donegal, July). (C) Apex with branch formed in axil of trichoblast (arrow). (D) Rhizoids cut off from periaxial cells (arrow). (E) Spermatangial branchlets on trichoblasts, lacking sterile tips. (F) Cystocarps with wide ostioles. (G) Shallow, cup-like cystocarp with wide ostiole surrounded by enlarged, transparent pericarp cells. (H) Tufted branches with spiral series of tetrasporangia (as B).

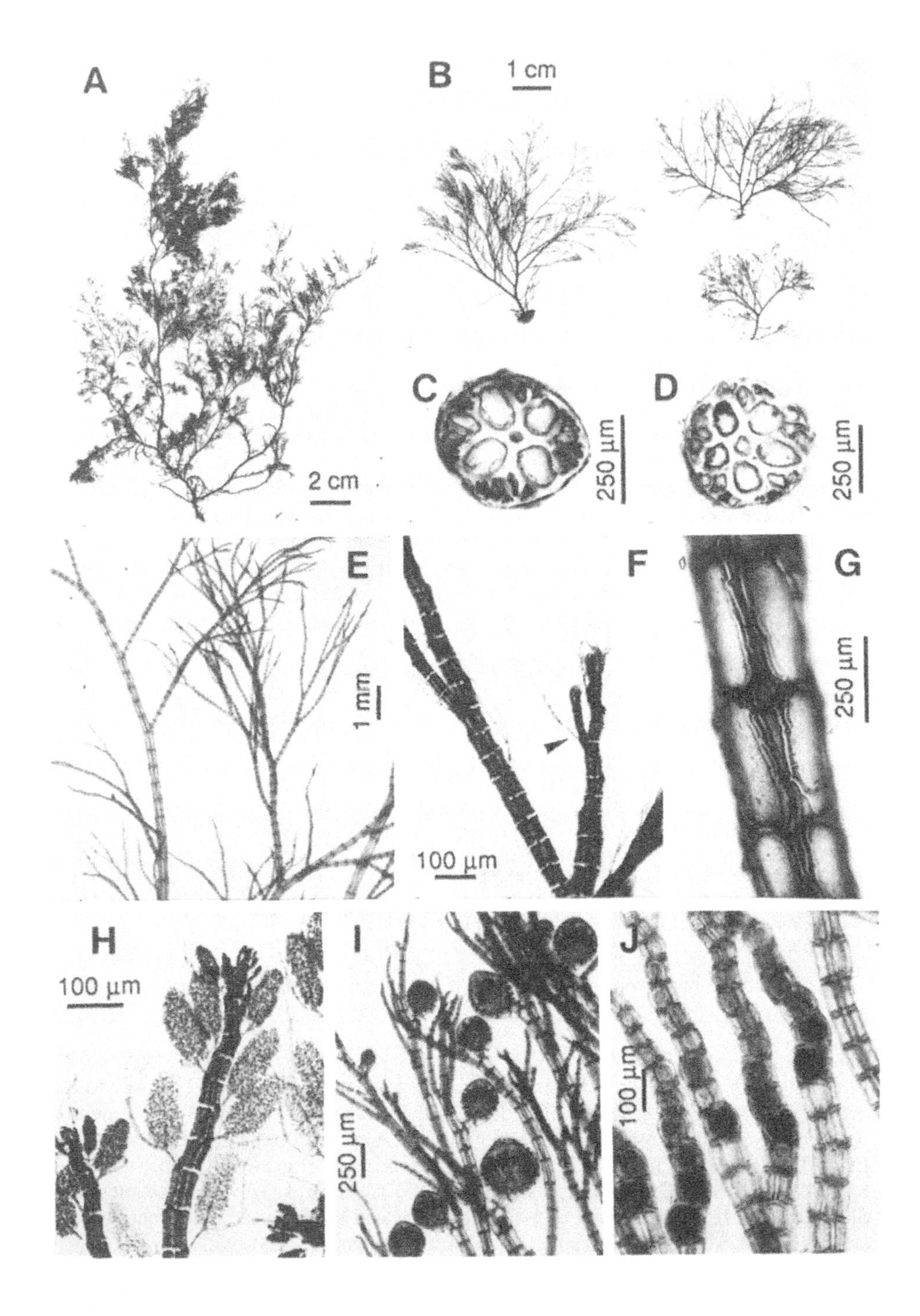
A
B
1 cm
2 cm
C
250 µm
D
250 µm
E
1 mm
F
100 µm
G
250 µm
H
100 µm
I
250 µm
J
100 µm

linear to rounded or very irregular outline, attached by a discoid holdfast 0.5-3 mm in diameter composed of aggregated cortical filaments and clusters of rhizoids; repent axes and further erect axes developing secondarily from disc and base of erect axes; main axes 0.3-0.8 mm in diameter, either remaining distinct and bearing spirally arranged laterals, or branching, often pseudodichotomously, from near the holdfast; brownish- or purplish-red in colour, bleaching to straw-yellow, old axes fairly tough and flexible, young laterals extremely soft and flaccid.

Erect axes repent basally, curving upwards towards apices and becoming erect, growing from apical cells 8-9 μm in diameter, 30-60 μm in diameter when young, consisting of small axial cell and 4, often spiral, periaxial cells; segments 1-2 diameters long, 0.5-2 diameters long in major axes; cortication initiated 2-4 orders of branching from apices, growing downwards in grooves between periaxial cells, covering axis entirely only in robust plants; *rhizoids* cut off from periaxial cells, 20-35 μm in diameter; *trichoblasts* numerous, borne on every segment in a 1/4 spiral divergence, 8-18 μm in diameter and up to 600 μm long, branched 2-4 times, composed of uninucleate cells, deciduous, leaving conspicuous scar cells; *branches* formed in axils of trichoblasts at intervals of 1-5 segments; main axes and major laterals branched to 5 or more orders; ultimate laterals spine-like or cylindrical; adventitious branches developing abundantly from scar cells on young and old axes, often on every segment; *plastids* numerous, discoid to beaded.

Plants dioecious; all reproductive structures borne abundantly on terminal 1-3 orders of branching. *Spermatangial axes* formed at the first dichotomy of fertile trichoblasts, 100-150 μm long and 50-55 μm in diameter, ellipsoid to conical-cylindrical with a blunt rounded apex, lacking sterile terminal cells when mature, covered with a layer of spermatangial mother cells bearing ellipsoid spermatangia 5 x 4 μm. *Cystocarps* ovoid when developing, with a distinct stalk, globular to barrel-shaped when mature, sometimes broader at the equator, 430-480 μm high and 330-500 μm in diameter, with an ostiole 1/3-1/7 of cystocarp diameter; outer pericarp cells polygonal or rectangular, formed in vertical rows, decreasing in size towards ostiole, sometimes with enlarged cells around ostiole when mature; carposporangia 70-80 x 30-50 μm. *Tetrasporangia* formed in spiral

---

Fig. 102. *Polysiphonia fibrillosa*

(A) Thallus with single main axis branched to several orders, typical of the "*fibrillosa*" form, collected from kelp stipe (Donegal, May). (B) Thalli with single main axes below, branched pseudodichotomously above, typical of the "*violacea*" form (B-E, G & I-J Down, Nov.). (C, D) T.S. through axis of one plant, showing 4 periaxial cells (in C) and appearing to have 5 periaxial cells in D, surrounded by cortex. (E) Detail of branching. (F) Apex with branch formed in axil of trichoblast (arrow) (Clare, July). (G) Axis showing cortical filaments growing downwards in grooves between periaxial cells. (H) Ellipsoid spermatangial axes on trichoblasts, lacking sterile tips (Sutherland, Aug.). (I) Globular cystocarps with narrow ostioles. (J) Spiral series of tetrasporangia.

series, distorting branch slightly into a spiral, only 1-3 per series mature at once, spherical, 70-80 µm in diameter.

Growing on a wide variety of substrata including bedrock, stones, algae such as *Laminaria hyperborea* stipes, *Zostera* leaves, and artificial structures, in lower-shore pools and from extreme low water to at least 10 m depth, most abundant just below low-water level, in a wide range of environmental conditions, from extremely wave-sheltered with current exposure, to wave-exposed open-coast sites.

Generally distributed in the British Isles, northwards to Shetland.

N. Norway to France; Baltic (Kapraun & Rueness, 1983).

Mature thalli can be found throughout the year, apparently existing as a series of short-lived overlapping generations although holdfasts possibly perennate. Spermatangia recorded in February and April-December; cystocarps in February and April-November; and tetrasporangia in February-December. An isolate from Denmark followed a *Polysiphonia*-type life history in culture (Koch, 1986), except that females occasionally formed spermatangial axes on a few branches.

This species shows great morphological variability in overall size, branching pattern and the form of last-order branchlets, as Rosenvinge (1923-24) observed in Denmark. A range of growth forms can be found in single populations, but some may be related to environmental conditions. Many of the growth forms have been described as separate taxa. We have examined type material of the taxa listed in synonymy above, and consider that all represent *P. fibrillosa*. The original collection of *P. richardsonii* was reported to have 5 periaxial cells; some specimens of *P. fibrillosa* give the appearance of 5 periaxials in some segments, but this probably results from enlargement of a cortical filament. The name *P. fibrillosa* has generally been used in the British Isles for large specimens with a single main axis, whereas plants with branching main axes were more commonly identified as *P. violacea*. Rosenvinge (1923-24) considered that these two species could not be distinguished and should be be merged. Kapraun & Rueness (1983) showed that type material of *P. violacea* (Roth) Sprengel (1827, p. 348) was identifiable as *Polysiphonia fucoides* [as *P. nigrescens* (Hudson) Greville], and *P. fibrillosa* appears to be the oldest available name. Koch's (1986) report that Danish isolates of *P. fibrillosa* and *P. violacea* were intersterile is based on a misidentification of *P. harveyi* (q. v.) as *P. fibrillosa*.

Polysiphonia foetidissima Cocks ex Bornet (1892), p. 314 *.

Lectotype: PC, 'Algarum Fasciculi', no. 29. Probable syntypes distributed as this exsiccatum. Cornwall (Plymouth), undated (?1855), *Cocks*.

---

* *Polysiphonia stuposa* Zanardini ex Kützing (1864, p. 18) predates *P. foetidissima*, with which it may be conspecific, although Bornet believed them to be separate; we have examined type material of this taxon (L 941. 242. 226), a fragment of the holotype specimen in MEL, which appears to be very similar to *P. foetidissima*.

Thalli forming dense tufts 6-11 cm high, consisting of interwoven prostrate axes and numerous much-branched, very fine erect axes; dull brownish-red in colour (herbarium material), with an extremely soft and flaccid texture.

Erect axes growing from apical cells 7-8 μm in diameter, with 7-8 periaxial cells, ecorticate, 20-40 μm in diameter; segments 1-2 diameters long; *rhizoids* 25-55 μm in diameter, cut off from periaxial cells; *trichoblasts* numerous, apparently borne on every segment, 5-8 μm in diameter, branched 1-2 times, deciduous leaving conspicuous scar cells; *branches* formed in axils of trichoblasts at irregular intervals, straight, slightly to markedly constricted basally. [Few details of morphology are available due to the lack of recent collections.]

No details of gametangial material available. *Tetrasporangia* formed singly or in series, greatly distorting branch, spherical to elliptical, 55-80 μm long x 40-70 μm in diameter.

No details of habitat in the British Isles are available; Cocks' specimens were collected on the west side of Plymouth Sound at sites with little wave-exposure and fairly strong tidal currents.

S.E. Cornwall. The only verified records of this species in the British Isles are the original collections made by Cocks. Its occurrence at Plymouth may have been ephemeral

Fig. 103. *Polysiphonia foetidissima*

(A) Habit of probable syntype herbarium specimen (A-C Cornwall, undated, in BM). (B) Axes lacking cortication. (C) T.S. through axis showing 8 periaxial cells (arrows) around axial cell.

as there were no subsequent reports and recent searches at the original sites were unsuccessful. Reports from other sites in Cornwall and from Sussex (Batters, 1902) do not appear to be supported by specimens.

British Isles to N.W. France (Normandy, in C), S.W. France, Portugal; Mediterranean; Bermuda (Bornet, 1892; Ardré, 1970; Taylor, 1960).

All British specimens were collected in autumn and were tetrasporangial.

Batten (1923) reported that specimens from Swanage were corticated, whereas material in the Cocks exsiccatum is ecorticate, as are populations at Biarritz (Bornet, 1892).

Polysiphonia fucoides (Hudson) Greville (1824), p. 308.

Neotype: BM-K, in Herb. Lightfoot. Undated, unlocalized specimen annotated "named repeatedly by Hudson *C. fucoides*"*

Conferva fucoides Hudson (1762), p. 485.
Conferva nigrescens Hudson (1778), p. 602.
Ceramium violaceum Roth (1797), p. 150 (see Kapraun & Rueness, 1983).
Polysiphonia violacea (Roth) Sprengel (1827), p. 348.
Polysiphonia nigrescens (Hudson) Greville ex Harvey in W. J. Hooker (1833), p. 332.
Polysiphonia atropurpurea Moore in Harvey (1841), p. 89.

Thalli 3-20 (30) cm high, consisting of dense cylindrical to irregularly rounded tufts attached by tangled prostrate axes compacted by densely matted rhizoids; tufts complanate or cylindrical, composed of few to many main axes 0.3-0.9 mm in diameter; branching pattern highly variable; main axes usually remaining distinct, bearing much-branched laterals in an irregularly spiral to regularly alternate-distichous arrangement at narrow to wide branch angles, typically with corymbose apices, becoming denuded at the base of

---

Fig. 104. *Polysiphonia fucoides*

(A) Irregularly branched thallus (A, D & F Antrim, Nov.). (B) Thallus with regularly alternate complanate branching (Donegal, Apr.). (C) Prostrate axes attached by numerous matted rhizoids (Cornwall, Mar.). (D) Apices with typical fan-like branching. (E) Axes of different orders showing development of cortication (as C). (F) T.S. of ecorticate axis with 17 periaxial cells around large axial cell (scale = 100 μm). (G) T.S. of corticated axis (scale = 250 μm) (Pembroke, Dec.). (H) Spermatangial axes borne in clusters at apices (Down, Nov.). (I) Slightly urceolate cystocarps with wide ostioles (Glamorgan, June). (J) spirally arranged tetrasporangia (as I).

---

* This neotype has been selected from specimens discussed by Irvine & Dixon (1982).

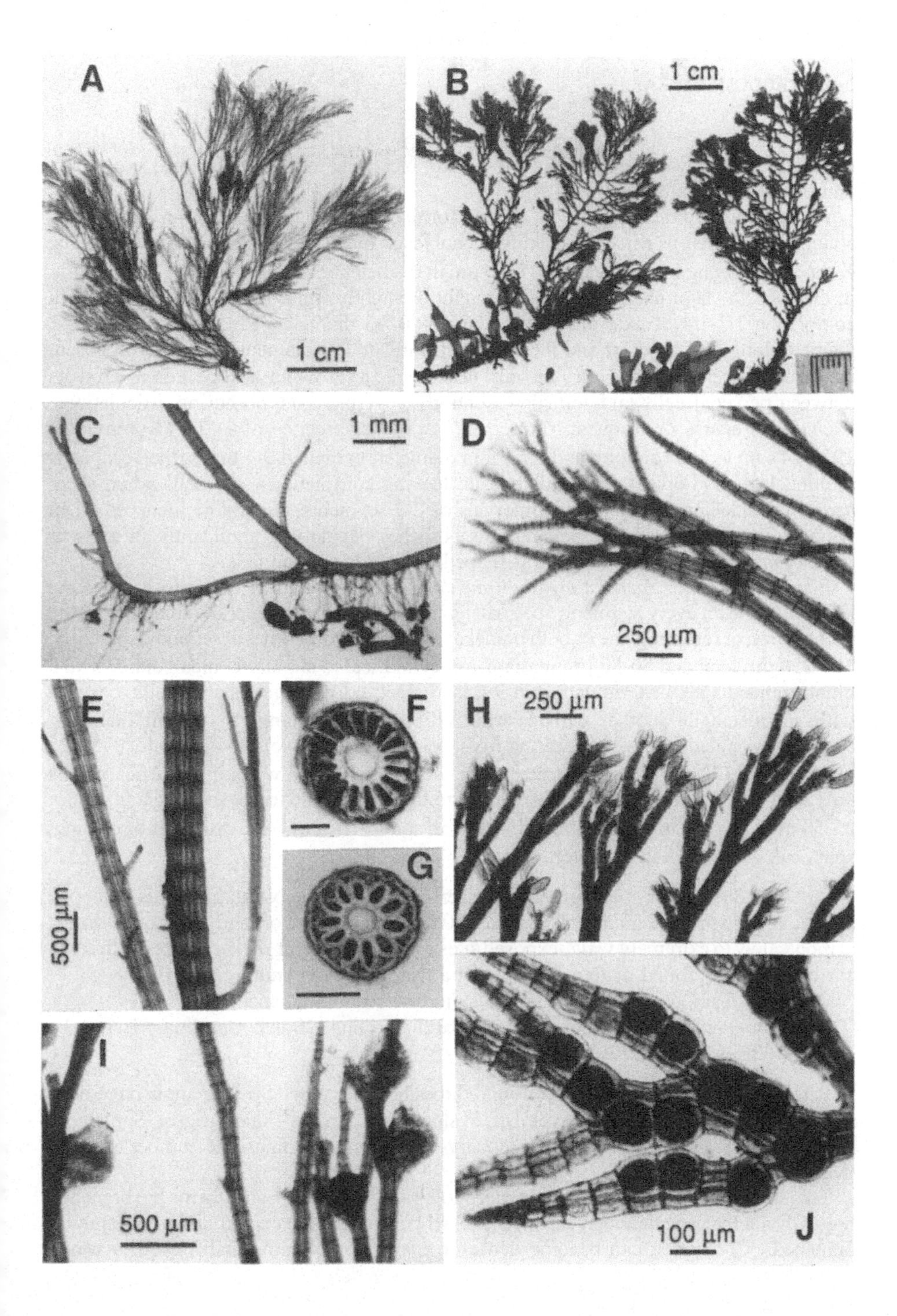
A
1 cm
B
1 cm
C
1 mm
D
250 µm
E
500 µm
F
G
H
250 µm
I
500 µm
100 µm
J

old thalli; dark brownish-red to black in colour, bleaching to light brown; old axes tough and flexible in texture.

Prostrate axes branching frequently to form erect axes and further prostrate axes, attached by numerous rhizoids, lacking trichoblasts; *rhizoids* cut off from periaxial cells; *erect axes* growing from apical cells 8-10 μm in diameter, increasing to up to 500-750 μm in diameter in major axes, with 11-21 straight or slightly spiral periaxial cells; axial cell comprising 1/3-1/4 of axis diameter; segments 0.7-3 diameters long in young axes, 1 diameter long near holdfast and increasing upwards to 2-3 diameters; cortication absent or formed either on major axes or only near holdfast, growing downwards in grooves between periaxial cells and sometimes entirely covering axis, obscuring articulations; *trichoblasts* sparse or frequent, formed in a spiral divergence of c. 2/5 (Rosenvinge, 1923-24), up to 400 μm long and 25 μm in diameter, branched 0-4 times, the basal cells multinucleate and apical cells uninucleate, leaving conspicuous scar cells when shed; *branches* replacing trichoblasts at intervals of 2-3 segments, straight or incurved when young, sometimes forcipate; adventitious branches developing abundantly in axils of primary branches; *plastids* irregularly-shaped, discoid to convoluted.

Plants dioecious. *Spermatangial branchlets* densely tufted at the apices of young laterals, borne on every segment; spermatangial axes terminal on suprabasal cell of fertile trichoblasts, or replacing one or both branches at first and sometimes at second dichotomy, 160-250 μm long and 50-80 μm in diameter, cylindrical to slightly conical, either with a blunt rounded apex or terminating in 3-6 isodiametric to elongate sterile cells, covered with spermatangial mother cells bearing ellipsoid spermatangia 4 μm in diameter. *Cystocarps* angled away from bearing axes, on thick stalks, ovoid to urceolate, 375-525 μm long and 300-425 μm in diameter, with ostiole 75-100 μm wide; outer pericarp cells polygonal, in straight rows, decreasing in size towards ostiole; carposporangia 90-130 x 25-50 μm. *Tetrasporangia* formed singly or in spiral series in last 2 orders of branching, spherical, 75-100 μm in diameter.

Growing on bedrock, pebbles, shells, maerl and artificial substrata, in pools and open situations from the mid-intertidal to at least 20 m depth, tolerant of sand cover on bedrock, at moderately to extremely wave-exposed sites and, subtidally, at sheltered sites with some current exposure; found in abundance under a wide range of conditions.

Generally distributed in the British Isles, northwards to Shetland.

Russia to Portugal; Newfoundland to North Carolina, ?South Carolina, ?Bermuda; Alaska (Schneider & Searles, 1991).

Perennial; large mature thalli are found throughout the year. Spermatangia have been recorded in January-July and November, cystocarps in January-June, August, September and November, and tetrasporangia in January-November. Chromosome number is $n = 30$ (Magne, 1964).

This species shows a wide range of morphological variation. Some forms are branched regularly alternate-distichously, in one plane, whereas others are very irregularly or spirally branched. Cortication can become dense on main axes of some thalli, even in wave-sheltered environments, but other thalli remain ecorticate. Axis diameter is also highly variable: narrow axes are often found in sealochs and may be associated with reduced salinity.

The branching pattern of some forms of *P. fucoides* resembles *Boergeseniella thuyoides*, but only the main axes of *P. fucoides* are corticated, whereas *B. thuyoides* axes are entirely corticated from just behind the apices.

Polysiphonia furcellata (C. Agardh) Harvey in W. J. Hooker (1833), p. 332.

Lectotype: LD 40907. France (Brittanny), undated, *Bonnemaison*.

Hutchinsia furcellata C. Agardh (1828), p. 91.
Polysiphonia forcipata J. Agardh (1842), p. 127.
Polysiphonia furcellata var. *forcipata* J. Agardh (1852), p. 1025.

Thalli 1-10 cm high and 1.5-10 cm broad, consisting of dense irregularly rounded tufts of numerous pseudodichotomously branched erect axes with corymbose apices, attached by branching repent axes; bright red or brownish-red in colour, bleaching to straw-yellow, with a flaccid texture.

Axes growing from apical cells 10-12 µm in diameter, ecorticate, with 7-8 straight or spiral periaxial cells, enlarging to 240 µm in diameter; segments 2-4 diameters long; *rhizoids* numerous on repent axes, cut off from periaxial cells, 15-70 µm in diameter; *trichoblasts* formed with a spiral divergence of 1 in 7, sometimes sparse or absent, lacking on some segments, up to 1.5 mm long, branched 2-4 times, 15-25 µm in diameter, the basal cells multinucleate and apical cells uninucleate, leaving conspicuous scar cells when shed; *branches* replacing trichoblasts, either at regular intervals of 4-5 segments so that all branches tend to lie in one plane, or irregularly at intervals of 1-15 segments; mature axes branched alternately or pseudodichotomously at a wide angle, to 10 or more orders of branching; ultimate branches forcipate or divaricate, 70-80 µm in diameter, with segments 0.7 diameters long; adventitious branches absent; *plastids* irregularly-shaped, elliptical to ribbon-like.

Gametangial and tetrasporangial plants unknown. *Propagules* developing at apices, consisting of swollen, densely-pigmented and starch-filled branchlets 1-1.8 mm long, 110-125 µm in diameter, simple or branched once, subtended by little-pigmented axes that function as abscission zones.

Loose-lying or attached secondarily to pebbles, invertebrates, maerl and other algae at moderately to extremely wave-sheltered sites with moderate current exposure, from extreme low water to 10 m depth.

South coast of England, eastwards to Sussex; known from only a few sites elsewhere, in Wales (Pembroke), Isle of Man, W. Scotland (Hebrides, Ross & Cromarty) and Ireland (Cork, Galway, Donegal, Antrim); Channel Isles.

British Isles to Canaries; Mediterranean (Ardré, 1970).

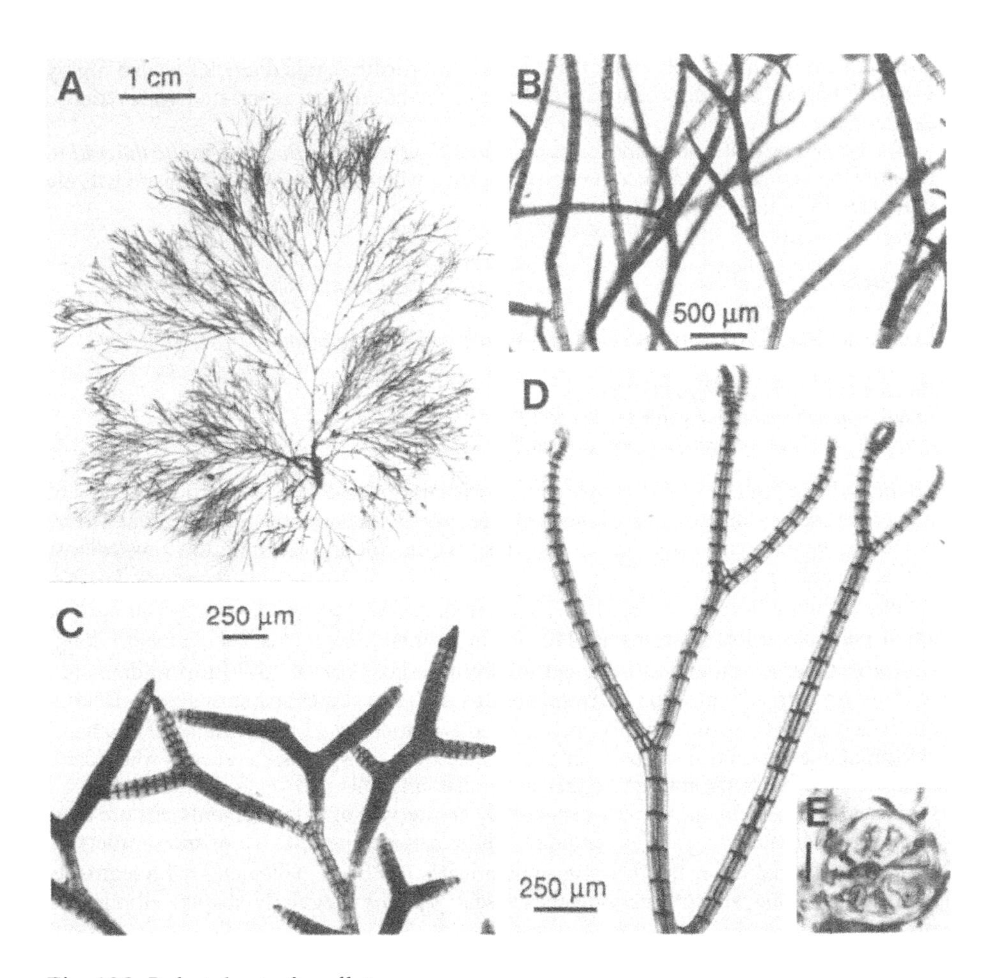

Fig. 105. *Polysiphonia furcellata*

(A) Habit (A & C-D Galway, June). (B) Wide-angled branching of mature, ecorticate axes (Donegal, Dec.). (C) Swollen propagules with dense contents borne at branch tips. (D) Young branches with incurved, forcipate tips. (E) T.S. through axis showing 8 periaxial cells around small axial cell (as B).

Thalli are present throughout the year, but are reduced to creeping axes in winter. In the British Isles, reproduction occurs by fragmentation and the formation of specialized propagules, which have been observed in June, July and September. Lauret (1970) likewise observed only vegetative reproduction in French Mediterranean populations of this species. Depictions of tetrasporangial branchlets (e. g. Harvey, 1846, pl. 7) appear to result from a misinterpretation of the swollen, starch-filled propagules. Reports of tetrasporangia and gametangia from Brittany and Portugal (Feldmann, 1954; Ardré, 1970) may involve *P. ceramiaeformis* (q. v.), which was previously considered conspecific with *P. furcellata.*

The most striking variation in morphology is in the form of apices. These are sometimes strongly hooked inwards, resembling *Ceramium* species and corresponding to 'var. *forcipata*', and sometimes divergent. Both forms can occur on the same individual.

Polysiphonia harveyi Bailey (1848), p. 38.

Lectotype: TCD. Co-types: BM-K. USA (Stonington, Connecticut), July 1847, *Bailey.*

Polysiphonia havanensis δ *insidiosa* J. Agardh (1863), p. 960.
Polysiphonia insidiosa (J. Agardh) P. Crouan & H. Crouan (1867), p. 156, non *Polysiphonia insidiosa* Greville ex J. Agardh (1851), p. 926

Thalli 0.5-10 cm high, solitary, consisting of dense tufts of much-branched erect axes with a broad, rounded or flat-topped outline, attached either by a solid disc composed of downgrowing cortical filaments, or (when growing on *Codium*) by a dense cluster of rhizoids; prostrate axes and further erect axes developing secondarily from disc and base of erect axes; main axis 0.2-0.6 mm in diameter, typically unbranched for basal 0.5-1 cm, then branching irregularly or pseudodichotomously at a wide angle several times, gradually decreasing in diameter towards apices, bearing laterals in an irregularly spiral pattern, usually densely branched to 4 or more orders; young branches brownish-red in colour bleaching to yellow, succulent to flaccid, old axes dark brownish-red, becoming fairly tough and flexible.

Erect axes growing from apical cells 8-10 μm in diameter, with 4 straight or spiral periaxial cells; segments 0.7 diameters long in young axes, 0.6-1 diameter long in main axes; cortication developing on all major axes or only near holdfast, or absent, growing downwards in grooves between periaxial cells, sometimes entirely covering lower main axes; *trichoblasts* sparse to abundant, usually borne on every segment in a 1/4 spiral divergence, 6-10 μm in diameter and up to 1 mm long, simple or branched 1-3 times, composed of uninucleate cells, leaving conspicuous scar cells when shed; *branches* replacing trichoblasts at irregular intervals of 1-5 segments, not constricted basally when young; adventitious branches formed abundantly by scar cells on young and older axes, typically on every segment and often paired; *plastids* elongate to ribbon-like or convoluted, lying only on radial walls of periaxial cells so the outer walls appear transparent, with conspicuous multiple nuclei.

Plants dioecious. *Spermatangial axes* formed at the first dichotomy of fertile

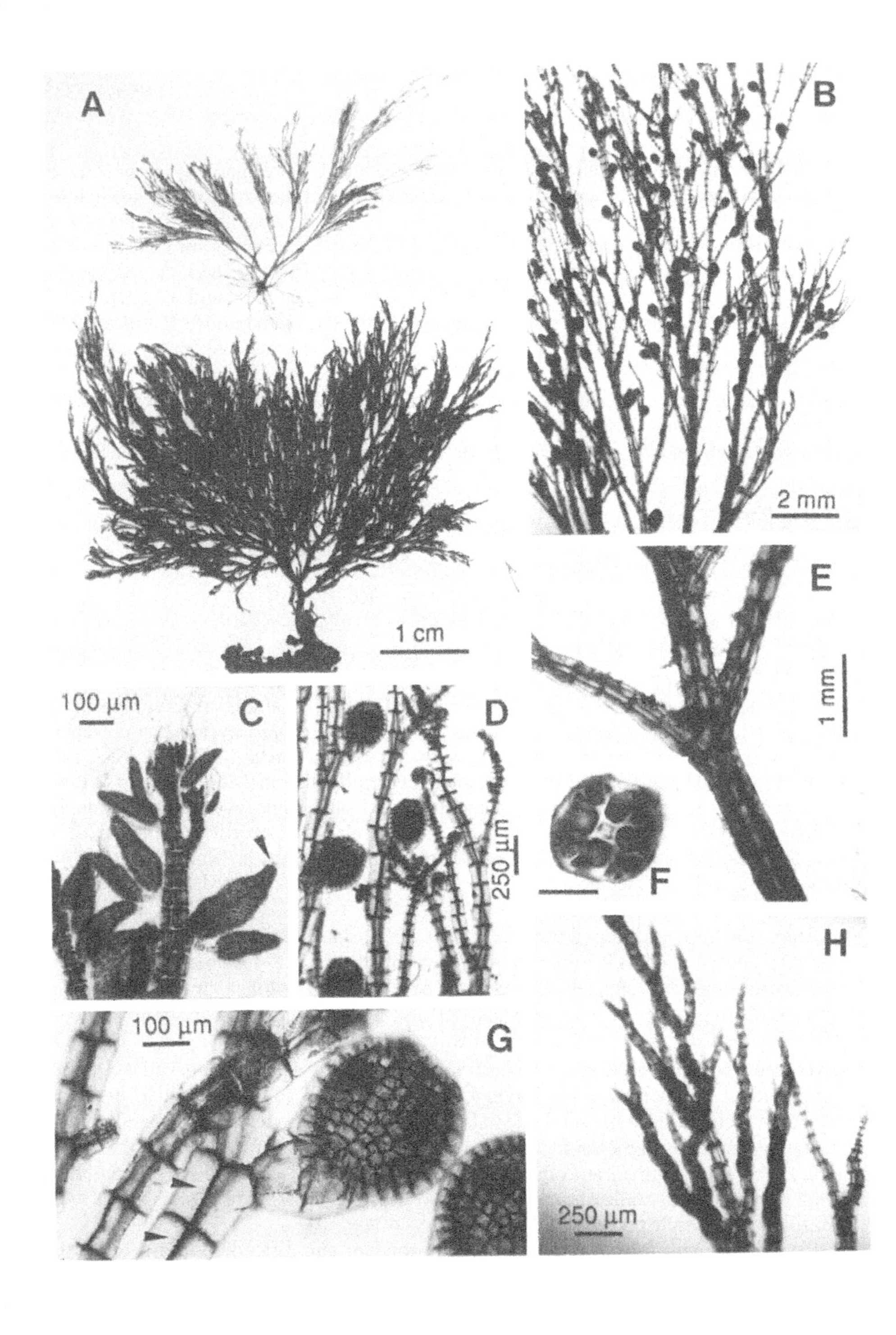
A
B
2 mm
1 cm
E
1 mm
100 µm
C
D
250 µm
F
H
100 µm
G
250 µm

trichoblasts, sometimes also replacing one branch at second dichotomy, rarely terminal on suprabasal cell of trichoblast, 150-350 µm long and 50-75 µm in diameter, cylindrical, with 1-2 isodiametric sterile terminal cells when mature, covered with spermatangial mother cells bearing ellipsoid spermatangia. *Cystocarps* borne on wide stalks, angled inwards towards axis, globular, 475-600 µm high and 350-425 µm in diameter, with ostiole c. 125 µm wide; outer pericarp cells polygonal, in straight rows, decreasing in size towards ostiole; carposporangia clavate, 40-115 x 30-40 µm. *Tetrasporangia* formed in last 3 orders of branching, in long spiral series, often interrupted by sterile segments, ellipsoid when mature, 80-100 µm long x 75-80 µm wide.

Epiphytic on a wide variety of algae including *Codium* spp., *Chorda filum* and *Chondrus crispus*, on *Zostera* spp. and growing on artificial substrata such as ropes, in pools from high water of neap tides to extreme low water, rarely subtidal to 3 m depth, at extremely wave-sheltered sites with some current exposure and on open, moderately wave-exposed coasts, sometimes occurring in great abundance.

South and east coasts of England (Cornwall to Hampshire and Essex); Pembroke, S.W. and S.E. Scotland (Wigtown and Fife); Ireland (Kerry, Galway, Donegal, Antrim, Down, Dublin); Channel Isles.

Norway (Rueness, pers. comm.), Denmark (Koch, 1986, as *P. fibrillosa*), Helgoland (Kornmann & Sahling, 1978, as *P. violacea*) and N.W. France; east coast of North America from Newfoundland to South Carolina.

Mature plants and sporelings occur throughout the year, and are most abundant from March to November. Bases of large thalli can perennate and become heavily epiphytized; the majority of individuals appear to be ephemeral. Spermatangia have been observed in January-April and June-November, cystocarps in January and March-November and tetrasporangia in January-November. An isolate from Denmark (Koch, 1986, as *P. fibrillosa*) grew well and reproduced at 4-22°C. It followed a *Polysiphonia*-type life history for three generations in culture, except that vegetative propagules occasionally developed in abundance from modified male trichoblasts and gave rise to further male plants. Chromosome number of Irish plants is $n = 29 \pm 1$.

Size and general appearance is very variable in British populations of *P. harveyi*. Male

---

Fig. 106. *Polysiphonia harveyi*

(A) Delicate plant above (Cornwall, Mar.) and robust plant below (Down, Nov.). (B) Branching of female plant with cystocarps (B, D & H Down, July). (C) Spermatangial axes on trichoblasts, with sterile tip cells (arrow) (Donegal, June). (D) Globular cystocarps. (E) Base of main axis, showing cortication (E-F Dorset, May). (F) T.S. of axis with 4 periaxial cells and cortication (scale = 250 µm). (G) Axis showing convoluted plastids only on radial walls of periaxial cells, with globular cystocarp (Devon, Nov.). (H) Spiral series of tetrasporangia.

thalli are usually smaller than females and tetrasporophytes, sometimes reproducing when only 0.5 cm high. In the Danish isolate, males were much smaller than female thalli (Koch, 1986). The degree of cortication and density of branching vary greatly: large bushy plants usually develop cortication on main axes, while delicate males, and some populations in sheltered environments, are ecorticate. Koch (1986) noted that cortication was much reduced in culture compared to field-collected thalli, and that plants grown at 4°C were characterized by strongly twisted periaxial cells.

P. harveyi has only recently been recognized in the British Isles (Maggs & Hommersand, 1990), but *Polysiphonia insidiosa* (J. Agardh) P. Crouan & H. Crouan, which was first collected in France in or before 1832 (specimen in LD), was reported from England by Batten (1923). *P. insidiosa* differs from *P. harveyi* principally in its lack of cortication. In view of the variation in cortication observed in the British Isles, we consider that this character cannot be used to discriminate between these species; *P. insidiosa* appears to be conspecific with *P. harveyi*. Although most of Batten's records of *P. insidiosa* were misidentifications, specimens collected at Weymouth by Cotton in 1908 (in BM) represent the first known British collection of *P. harveyi*. In the British Isles, *P. harveyi* may be an introduction from the W. Atlantic, but further comparisons are necessary with closely-related Pacific species such as *P. strictissima* J. D. Hooker & Harvey (1845b, p. 538) from New Zealand. *P. strictissima* shows a high degree of variability, including several forms previously recognized as separate species (Adams, 1991). *P. ferulacea* Suhr ex J. Agardh (1863, p. 980) closely resembles *P. harveyi*, and the formation of male propagules in both species (Kapraun, 1977b; Koch, 1986) emphasizes the similarities between them.

Two varieties of *P. harveyi* have been reported from the eastern USA. *P. harveyi* var. *olneyi* (Harvey) Collins, based on *P. olneyi* Harvey (1853, p. 40) is characterized by narrow flaccid axes and spermatangial axes with sterile terminal cells (Kapraun 1977c, 1978a). Chromosome number is $n = 28$ (Kapraun, 1978a). However, the lectotype of *P. olneyi* (TCD) has branches axillary to trichoblasts, and *P. olneyi* may be conspecific with *P. fibrillosa*. *P. harveyi* var. *arietina* (Bailey) Harvey is a rigid, hooked, seasonal growth form of *P. harveyi* (Taylor, 1957), which has a chromosome number in North Carolina of $n = 32$ (Kapraun, 1978a); it has not been observed in the British Isles.

Polysiphonia lanosa (Linnaeus) Tandy (1931), p. 226.

Holotype: LINN 1274.23. Iceland, undated.

Fucus lanosus Linnaeus (1767), p. 718.
Ceramium fastigiatum Roth (1800b), p. 463 (see Tandy, 1931).
Vertebrata fastigiata (Roth) S. F. Gray (1821), p. 338.
Polysiphonia fastigiata (Roth) Greville (1824), p. 308.
Vertebrata lanosa (Linnaeus) Christensen (1967), p. 93.

Thalli forming dense spherical tufts 3-7.5 cm in diameter on fucoids, attached by rhizoids penetrating into host; erect axes branching pseudodichotomously repeatedly, with corymbose apices; prostrate axes formed secondarily, creeping over host; dark brownish-red in colour, with a tough, rigid texture.

Axes growing from conspicuous apical cells 13-16 μm in diameter, ecorticate, with 12-24 straight or slightly spiral periaxial cells, 70-150 μm in diameter when young, increasing to 300-500 μm in diameter; axial cells comprising about 1/3 of total diameter; segments 0.4-0.5 diameters long throughout thallus; *rhizoids* either cut off by periaxial cells or remaining continuous with them (Rawlence & Taylor, 1970); *trichoblasts* absent; *branches* developing at irregular intervals, e. g. 9-13 segments, at narrow angles in young axes, wider in older axes; *plastids* inconspicuous.

Plants dioecious. *Spermatangial branches* developing in dense tufts at apices, consisting of scar cell, a single elongate suprabasal cell and cylindrical, rarely branched, spermatangial axis 200-310 μm long and 40-80 μm in diameter, with a rounded apex lacking sterile terminal cells, covered with spermatangial mother cells bearing ovoid spermatangia 3-4 μm in diameter. *Cystocarps* ovoid, 330-380 μm high and 260-330 μm in diameter, with narrow ostiole; outer pericarp cells polygonal, in irregular rows, decreasing in size towards ostiole; carposporangia clavate, 100-130 x 25-30 μm. *Tetrasporangia* formed in last 2 orders of branching, in spiral series often interrupted by sterile segments, distorting branches when mature, spherical, 90-110 μm in diameter; released tetraspores 85 μm in mean diameter (Pearson & Evans, 1990).

Growing on the fucoid algae *Ascophyllum nodosum*, *Fucus vesiculosus* and *Fucus serratus,* reportedly also very rarely on stones, at moderately exposed to extremely wave-sheltered sites, sometimes with moderate current exposure, throughout the intertidal vertical ranges of these fucoids from near high water of neap tides to extreme low water. Spore germination occurs preferentially on *A. nodosum*, apparently as a result of the larger number of protected niches offered by its growth form; settlement on *F. vesiculosus* is more or less confined to wound sites (Pearson & Evans, 1990; Lining & Garbary, 1992). *P. lanosa* is tolerant of reduced salinity: brackish-water ecotypes of *P. lanosa* showed high rates of photosynthesis at salinities down to 0 ppt, whereas marine ecotypes failed to photosythesize below 10 ppt (Lüning, 1990).

Generally distributed in the British Isles (Norton, 1985).

Norway and Iceland to Spain; Greenland and Newfoundland to New England (South & Tittley, 1986).

Large mature thalli are present throughout the year. Spermatangia have been recorded in February-July and December; cystocarps in January, April-June, and August-October. In N. Wales, peak tetraspore release occurred in August and by September the remaining reproductive structures appeared to be necrotic (Pearson & Evans, 1990). Peak spore settlement also occurred in August and small unbranched plants were attached by primary rhizoids in September. A high rate of loss took place from September to January, and surviving plants were branched and easily visible to the naked eye in May. Young

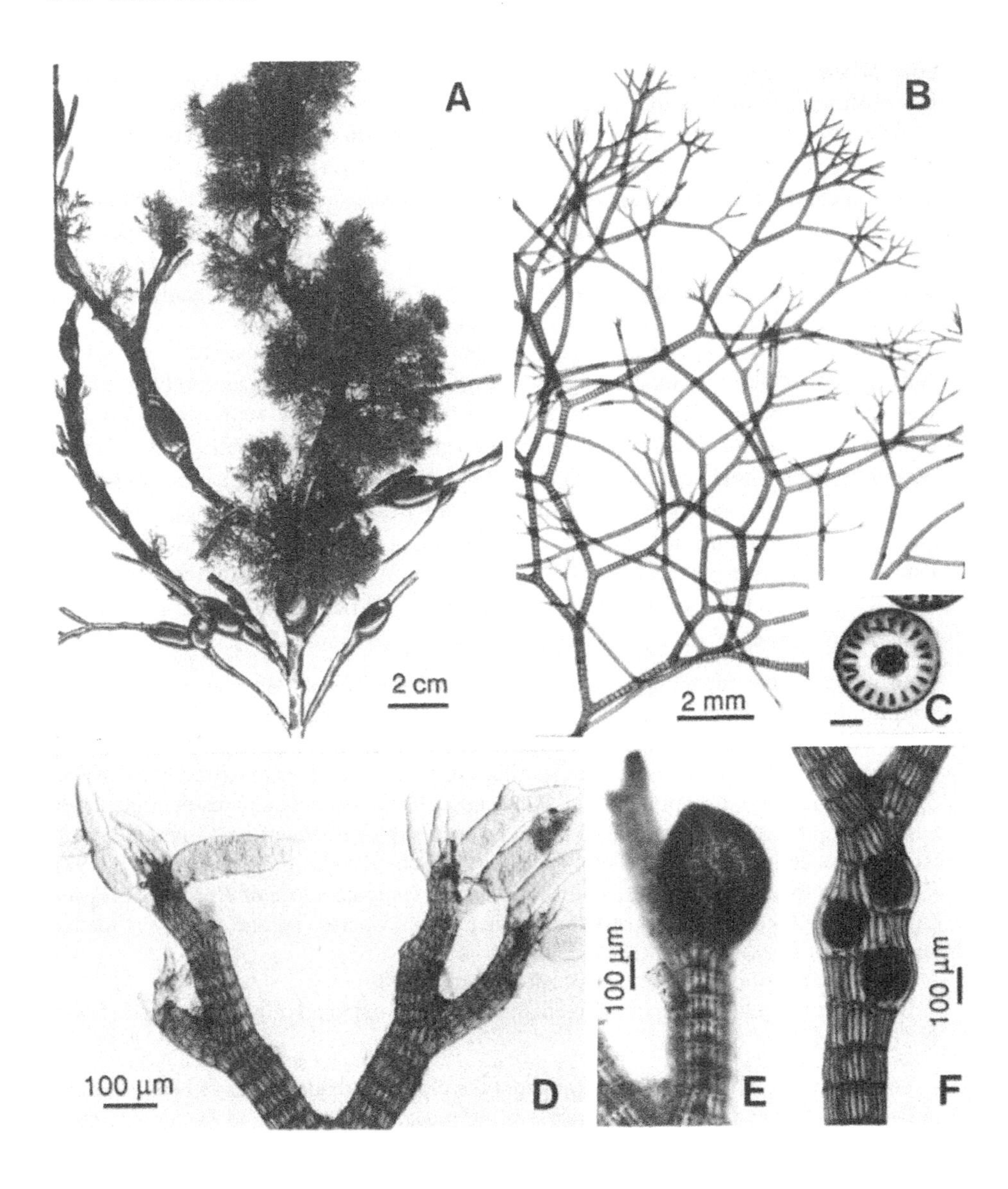
A
B
C
D
E
F
2 cm
2 mm
100 µm
100 µm
100 µm

sporelings form uniseriate filaments, the basal cell of which gives rise to a primary penetrating rhizoid (Rawlence & Taylor, 1970).

This species shows very little morphological variation.

P. lanosa usually bears numerous thalli of the parasite *Choreocolax polysiphoniae* Reinsch (see Vol. 1, 2A).

Polysiphonia nigra (Hudson) Batters (1902), p. 81.

Neotype*: BM. Durham (Marsden), 12 vi 1971, *P. Edwards*.

Conferva nigra Hudson (1762), p. 481.
Conferva badia Dillwyn (1809), p. 85, pl. G (see Harvey, 1848, pl. 172).
Conferva atrorubescens Dillwyn (1809), pl. 70.
Polysiphonia atrorubescens (Dillwyn) Greville (1824), p. 308.
Polysiphonia badia (Dillwyn) Sprengel (1827), p. 350.

Thalli 3-20 (30) cm high, consisting of dense cylindrical to irregularly rounded tufts attached by extensive tangled prostrate axes; erect axes much-branched, with numerous major axes 0.2-0.3 mm in diameter, bearing short laterals or clusters of laterals in an irregularly spiral arrangement at acute angles; young axes bright red, with a soft, succulent texture; older axes brownish-red, bleaching to pale brown.

All axes ecorticate, with (8) 9-13 (14) spirally-twisted periaxial cells; *prostrate axes* branching frequently to form erect axes and further prostrate axes, attached by numerous rhizoids, lacking trichoblasts; *rhizoids* cut off from periaxial cells, 30-75 µm in diameter; *erect axes* growing from apical cells 8-9 µm in diameter, increasing to 200-300 µm in diameter in major axes; axial cell comprising 1/3-2/5 of total diameter; segments 0.3-0.5 diameters long near holdfast, increasing to 2-3 diameters in middle of plant; *trichoblasts* sparse or frequent, formed in a spiral divergence of 1 in 3-6 (Rosenvinge, 1923-24), up to 800 µm long, branched 1-4 times, decreasing from 20-25 to 10 µm in diameter, the basal cells multinucleate and apical cells uninucleate, leaving conspicuous scar cells when shed; *branches* developing in axils of trichoblasts at intervals of 3-8 segments, constricted basally

---

Fig. 107. *Polysiphonia lanosa*

(A) Thalli epiphytic on *Ascophylllum nodosum* (Down, June). (B) Pseudodichotomous wide-angled branching, showing short segments throughout thallus (Pembroke, Dec.). (C) T.S. of axis with numerous periaxial cells around large axial cell (scale = 100 µm) (Ross, Nov.). (D) Spermatangial branchlets, lacking trichoblasts (as B). (E) Cystocarp with narrow ostiole (as C). (F) Spiral series of tetrasporangia (as B).

---

* Irvine & Dixon (1982) lectotypified Hudson's description, but this is no longer valid (ICBN, 1988, Art. 9.1).

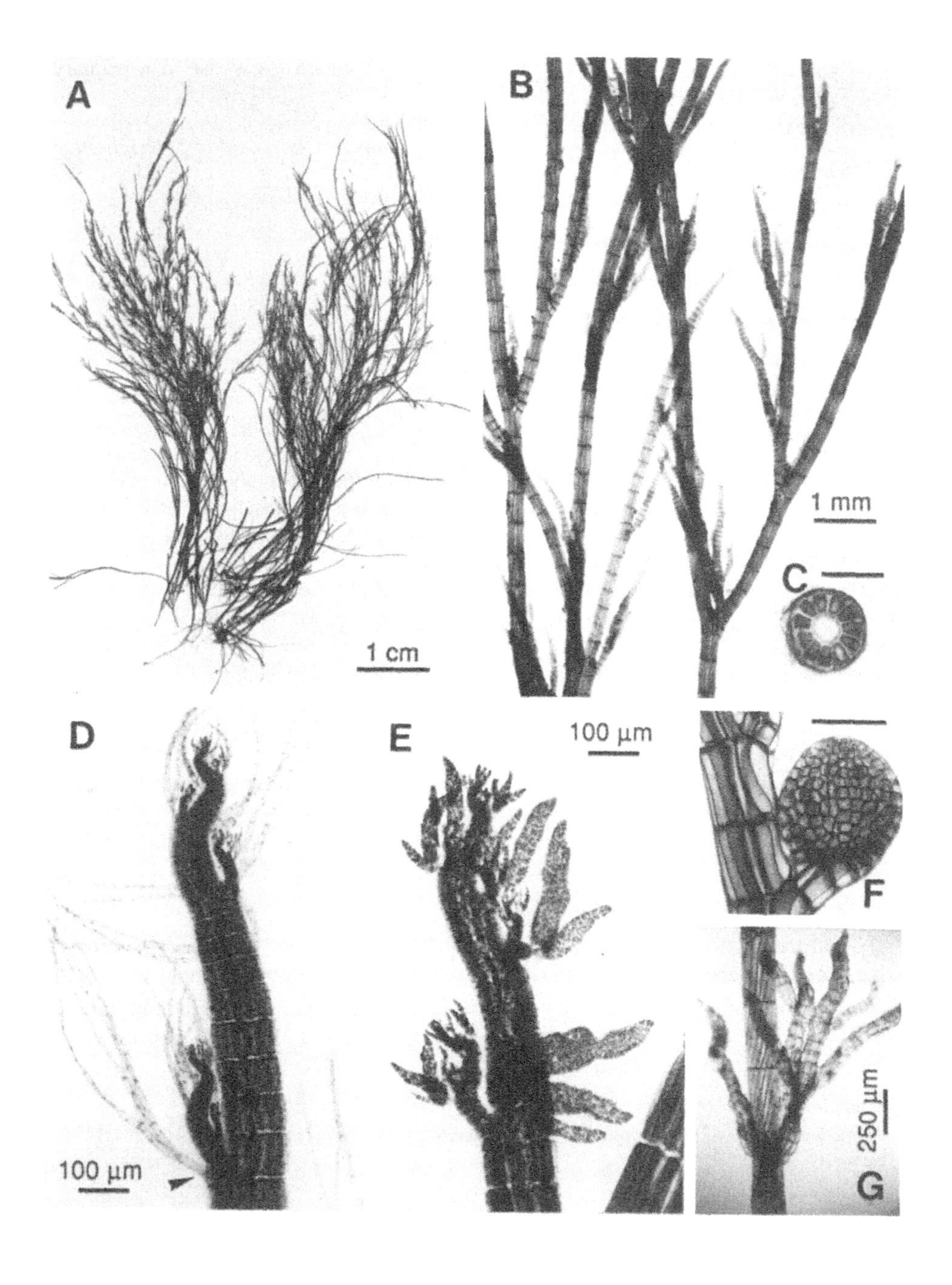
A
B
1 mm
C
1 cm
D
E
100 µm
F
100 µm
250 µm
G

when young, with conspicuously spirally twisted apices and periaxial cells and segments 0.4-0.5 (1.2) diameters long; adventitious branches developing abundantly in axils of primary branches; *plastids* irregularly-shaped, discoid to convoluted.

Plants dioecious. *Spermatangial branchlets* densely tufted at the apices of young laterals, borne on every segment; spermatangial axes replacing one or both branches of fertile trichoblasts at first dichotomy, 200-330 µm long and 50-75 µm in diameter, cylindrical to slightly conical, incurved, with a rounded apex lacking sterile terminal cells; spermatangia ellipsoid, 3 µm in diameter. *Cystocarps* angled towards bearing axes and tightly pressed to them, globular, 400-500 µm in diameter, with ostiole 80 µm wide; outer pericarp cells polygonal, in straight rows, decreasing in size towards ostiole; carposporangia 75-100 x 35-50 µm. *Tetrasporangia* formed in short lateral branches, in long spiral series of which 1-5 are mature, tightly packed when mature, spherical to ellipsoid, 100-125 µm long x 100-110 µm wide.

On bedrock, pebbles, shells, maerl and other mobile substrata, in pools and damp places in the lower intertidal and from extreme low water to at least 20 m depth, particularly abundant in the intertidal, growing through sand cover on bedrock, at moderately to extremely wave-exposed sites and, subtidally, at sheltered sites with strong current exposure.

Generally distributed in the British Isles, northwards to Shetland.

Norway to Portugal; Newfoundland to New Jersey (South & Tittley, 1986).

Large mature thalli can be found throughout the year, and prostrate axes are probably perennial. Gametophytes are much less frequent than tetrasporophytes. Spermatangia have been recorded in March and May, cystocarps in April-June and August, and tetrasporangia in February-September.

There is relatively little morphological variation.

Polysiphonia opaca (C. Agardh) Moris & De Notaris (1839), p. 264.

Possible lectotypes: LD 41663-41666. Adriatic Sea.

Hutchinsia opaca C. Agardh (1824), p. 148.

Fig. 108. *Polysiphonia nigra*

(A) Tufted thalli consisting of numerous erect axes (Pembroke, Dec.). (B) Main axes bearing tufted short laterals with attenuate bases (Cornwall, Mar.). (C) T.S. of axis with 11 periaxial cells around large axial cell (scale = 250 µm) (as A). (D) Apex, showing development of spirally twisted branches in axils of trichoblasts (arrow) (D-E Inverness, May). (E) Spermatangial axes borne singly or in pairs (as B). (F) Cystocarp with narrow ostiole (Antrim, Apr.). (G) Main axis with spirally arranged periaxial cells, bearing twisted branches with spirally arranged tetrasporangia (as A).

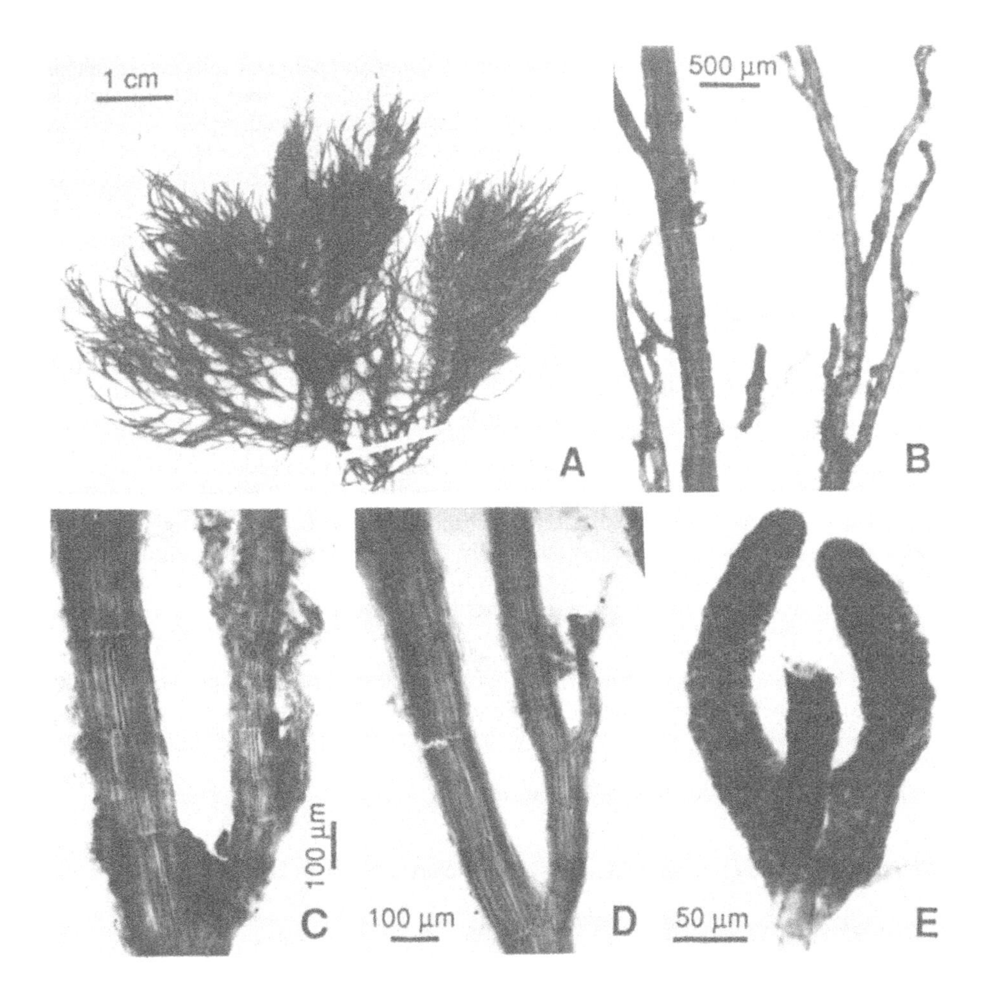

Fig. 109. *Polysiphonia opaca*

(A) Habit of herbarium specimen (A-E "?Jersey", undated, in BM). (B) Axes of different ages. (C) Mature ecorticate axis and young branch with spiral apex. (D) Mature axes. (E) Three spermatangial axes borne on one basal cell, lacking sterile tip cells.

Thalli up to 5.5 cm high and 5 cm broad, consisting of dense irregularly rounded tufts of erect axes up to 0.3 mm in diameter; erect axes branching frequently at acute angles; dark brown in colour (herbarium material).

Erect axes growing from apical cells c. 12 μm wide, with twisted apices, increasing in diameter from 40 to 300 μm, ecorticate, with c. 24 straight periaxial cells; axial cell comprising c. 1/3 of axis diameter; segments 0.8-1.5 diameters long; *rhizoids* cut off from periaxial cells of lower main axes, c. 50 μm in diameter, terminating in discoid pads; *trichoblasts* abundant, up to 1 mm long, branched several times, to 20-25 μm in diameter, leaving conspicuous scar cells when shed; *branches* replacing trichoblasts at intervals of 3-6 segments, spirally twisted into a corkscrew shape when young.

Plants dioecious. *Spermatangial axes* borne on suprabasal cell of fertile trichoblasts, replacing both branches at first dichotomy, cylindrical to slightly conical, 130-270 μm long and 30-45 μm in diameter, with a rounded apex lacking sterile terminal cells. Cystocarpic and tetrasporangial material not available. [Few details of morphology are available due to the lack of recent British collections. See Lauret (1970) for a complete description of Mediterranean plants.]

No details of habitat in the British Isles are available. In N. France, *P. opaca* grows on rock and coralline algae at mid-tide level (Crouan & Crouan, 1867).

The only herbarium material that we have located (BM, ex Herb. Batters) is labelled '?Jersey' (undated), and seems unlikely to be the specimen reported from Petit Port, Guernsey by Batters (1896). Specimens from Devon and Cornwall attributed to *P. opaca* are misidentifications of *P. nigra* or *Lophosiphonia reptabunda*.

British Isles to N. Spain; Mediterranean; West Indies, Canaries and E. Australia (South & Tittley, 1986; Athanasiadis, 1987).

No data on seasonality are available for the British Isles. Crouan & Crouan (1867) recorded *P. opaca* in autumn in N. France; it occurs throughout the year in the Mediterranean (Lauret, 1970).

The only species in the British Isles with which *P. opaca* is likely to be confused is *Lophosiphonia reptabunda*, but *L. reptabunda* has a turf-like habit, with erect axes less than 3 cm high, whereas *P. opaca* thalli grow singly and are up to 5 cm or more high; axes are only up to 150 μm in diameter in *L. reptabunda* in contrast to up to 300 μm in *P. opaca*.

Polysiphonia simulans Harvey (1849b), p. 89.

Lectotype: BM-K, ex Herb. Harvey. Devon (Tor Abbey), 20 v 1831, *Griffiths*.

Thalli 2-8 cm high and 2-10 cm broad, consisting of dense irregularly rounded tufts of numerous much-branched erect axes attached by tangled prostrate axes; distinct main axes lacking; erect axes 0.2-0.3 mm in diameter, branched irregularly alternately to several orders at narrow angles, with the ultimate branchlets usually short and spine-like, borne

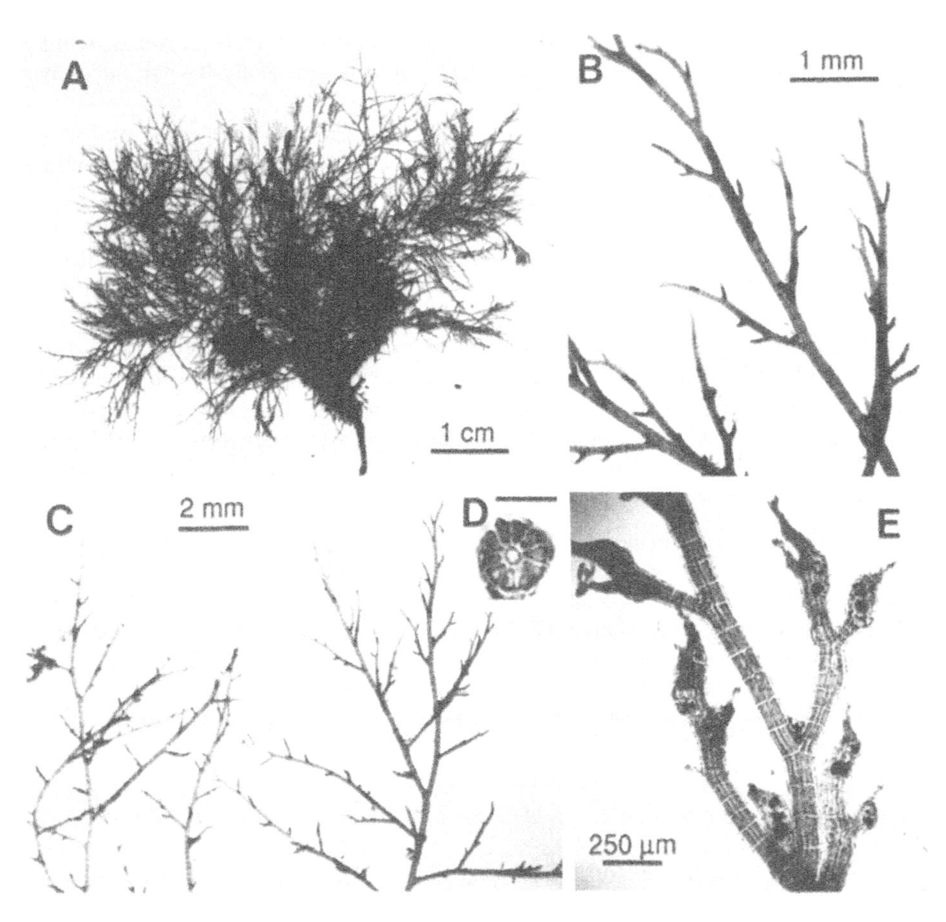

Fig. 110. *Polysiphonia simulans*

(A) Habit of epiphytic thallus (A-C Cornwall, Oct.). (B) Branching pattern, showing incurved short branches. (C) Main axes bearing irregularly alternate arrangement of short laterals. (D) T.S. of ecorticate axis with 10 periaxial cells around small axial cell (scale = 100 µm) (D-E Pembroke, Dec.). (E) Tetrasporangia distorting short laterals into spirals.

in an alternate-distichous arrangement, or becoming spiral; bright red in colour, with a crisp, fairly rigid texture.

Axes growing from apical cells 10-13 μm in diameter, increasing to 250-300 μm in diameter, ecorticate, with 10-13 straight or spiral periaxial cells; axial cell comprising 1/5 of axis diameter; segments 0.8-1.3 diameters long; trichoblasts and branches borne in a spiral divergence of c. 1 in 12, interspersed with unbranched segments; *rhizoids* cut off from posterior end of periaxial cells, 20-40 μm in diameter; *trichoblasts* sparse, up to 750 μm long, branched 3-4 times, decreasing from 30-40 μm to 6 μm diameter, the basal cells multinucleate and apical cells uninucleate, leaving conspicuous scar cells when shed; *branches* replacing trichoblasts at intervals of 4-9 segments, incurved or slightly forcipate when young, with conspicuously spirally-twisted apices; adventitious branches developing occasionally from scar cells; *plastids* irregularly-shaped, discoid, beaded or convoluted.

Gametophytes not known with certainty. *Tetrasporangia* formed in last 2 orders of branching, in spiral series of 1-4, often interrupted by sterile segments, spherical, 50-70 μm in diameter.

Epiphytic on various algae such as *Chondrus crispus* and *Laminaria* blades, in pools at low water of neap tides and at extreme low water, at moderately to extremely wave-sheltered sites, with or without exposure to currents. Although rarely found, *P. simulans* occasionally occurs in large quantities.

South coast of England from Cornwall eastwards to Dorset; Pembroke; Cork, Kerry; Channel Isles. Records from outside this area (e. g. Orkney) appear to be based on misidentifications of *P. fucoides*.

British Isles to N.W. France (South & Tittley, 1986).

Large mature plants have been collected from April to October and young thalli observed in December. The great majority of plants appear to be tetrasporophytes; tetrasporangia have been recorded in April-July. In September and October all thalli were non-reproductive. The only known example of spermatangia is on a tetrasporophyte collected in May. Cystocarps have not been observed with certainty, although reported by several authors. Many herbarium specimens are misidentified, and old herbarium material can be difficult to distinguish from that of *P. fucoides*.

There is little morphological variation other than in the regularity of the alternate short branchlets, which are sometimes obviously distichous, but may become spiral. *P. simulans* can bear a superficial resemblance to *Boergeseniella fruticulosa* (q. v.), from which it can be distinguished by the lack of cortication in contrast to the complete cortication in *B. fruticulosa*.

Polysiphonia stricta (Dillwyn) Greville (1824), p. 309.

Lectotype: BM-K. Glamorgan (Swansea), undated, *Dillwyn*.

Conferva stricta Dillwyn (1804), pl. 40.
?*Ceramium strictum* Roth (1806), p. 130.

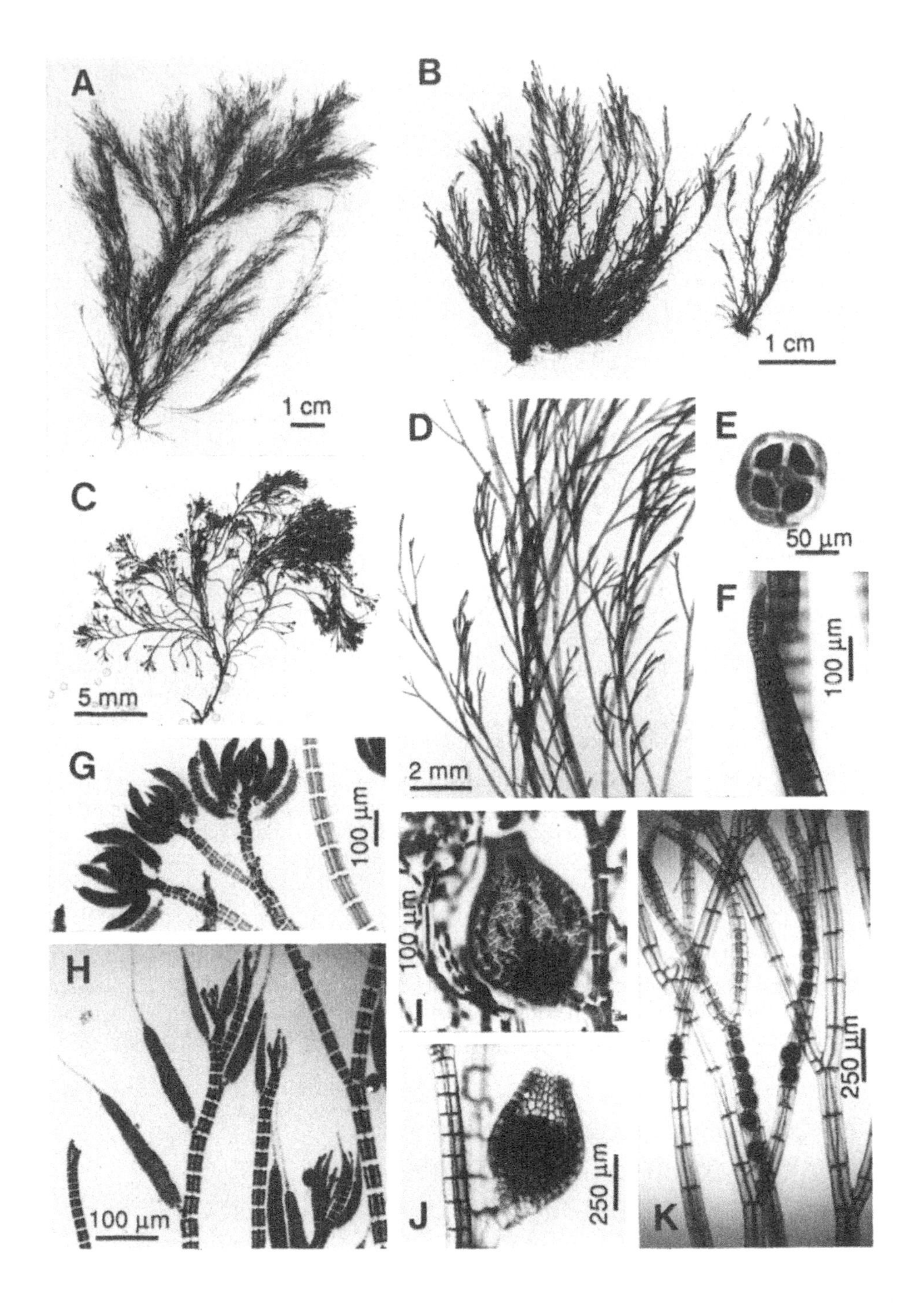
A
1 cm
B
1 cm
C
5 mm
D
2 mm
E
50 µm
F
100 µm
G
100 µm
H
100 µm
I
100 µm
J
250 µm
K
250 µm

Conferva urceolata Lightfoot ex Dillwyn (1809), p. 82, pl. G.
Conferva patens Dillwyn (1809), p. 83, pl. G.
Polysiphonia urceolata (Lightfoot ex Dillwyn) Greville (1824), p. 309.
Polysiphonia spiralis Batten (1923), p. 287.

Thalli forming dense tufts that vary from cushion-like to narrow and elongate, 2-25 cm high and 2-10 cm wide, consisting of interwoven prostrate axes and numerous erect sparsely to much-branched axes; lower axes sometimes bearing series of short reflexed branchlets; bright red to dark brown in colour, ranging in texture from soft and flaccid to harsh and rigid. Axes ecorticate, consisting of small axial cell and 4 periaxial cells; *prostrate axes* much-branched, branching endogenous, giving rise to erect axes and further prostrate axes, mature prostrate axes 50-85 μm in diameter, segments 0.6-1.5 diameters long, lacking trichoblasts; *rhizoids* formed singly or in pairs in a median position on little-pigmented periaxial cells, remaining in open connection with them, 30-45 μm in diameter, simple or branched and terminating in simple or compound lobed to digitate discoid pads; *erect axes* growing from domed apical cells 9-12 μm wide, increasing to 50-300 μm in diameter; segments 1-3 (-6) diameters long; periaxial cells straight or spiral; *trichoblasts* absent, sparse or frequent, not borne on every segment, up to 10 μm in diameter and 500 μm long, simple or branched 1-3 times, composed of uninucleate cells; *branches* replacing trichoblasts if present, formed in a 1/4 spiral at intervals of 1-5 segments, typically every 4 segments; several orders of branching formed on major laterals, in an irregularly spiral to dichotomous pattern; *plastids* discoid.

Plants dioecious. *Spermatangial branchlets* densely clustered at apices; spermatangial axes terminal on isodiametric to elongate suprabasal cell, incurved (rarely straight), 175-360 μm long and 40-80 μm wide, covered with spermatangial mother cells bearing ovoid spermatangia 3 μm in diameter, and terminating in 3-7 isodiametric to elongate sterile cells. *Cystocarps* borne on wide stalks, angled inwards so that they lie parallel to axis, usually distinctly urceolate, occasionally globular, 325-725 μm high and 250-540 μm in diameter, typically with projecting ostiole 100-150 μm long; outer pericarp cells polygonal, in regular vertical rows, usually decreasing in size towards ostiole, rarely with

---

Fig. 111. *Polysiphonia stricta*

(A) Habit of large plant of the "*stricta*" form from subtidal kelp stipe (Donegal, May). (B) Small densely tufted thalli of the "*patens*" form, collected on limpet shell (Antrim, Feb.). (C) Estuarine thallus resembling *P. subtilissima* (Devon, July). (D) Detail of branching of typical subtidal plant (Down, Dec.). (E) T.S. of ecorticate axis with 4 periaxial cells (Donegal, June). (F) Apex of lateral branch with large apical cell (Down, Nov.). (G) Clustered incurved spermatangial axes with short sterile tips (Cornwall, July). (H) Elongate straight spermatangial axes with long sterile tips, in the "*subtilissima*" form (H-I as C). (I) Urceolate cystocarp in the "*subtilissima*" form. (J) Slightly urceolate cystocarp in a typical form (Mayo, May). (K) Long straight series of tetrasporangia (Donegal, May).

a ring of larger cells around ostiole; carposporangia clavate, 125-200 x 35-50 µm. *Tetrasporangia* borne in last 2-3 orders of branching, in straight series of variable length, sometimes distorting axis into a pod-like shape, spherical, 40-125 µm in diameter.

On bedrock, pebbles, shells and other unstable and artificial substrata, epizoic on limpets, mussels and other invertebrates, epiphytic on various algae including *Laminaria hyperborea* stipes, in pools and damp places from low water of neap tides to extreme low water and subtidal to at least 20 m, at sites ranging from extremely sheltered to extremely wave-exposed, with negligible to strong current exposure; certain morphological forms are common in particular habitats.

Generally distributed in the British Isles.

Russia to Spain; Mediterranean; W. Atlantic from Labrador to North Carolina; Argentina (Schneider & Searles, 1991).

Large plants occur throughout the year in marinas and sheltered places. Spermatangia have been recorded in February-August and October-December; cystocarps in February-October; tetrasporangia in February-August and October-December. Gametangia and tetrasporangia occasionally occur on the same thalli. A Norwegian isolate identified as *P. urceolata* grew well at 10-20°C but failed to reproduce in culture (Kapraun, 1979). Chromosome number in N. Carolina populations is $n = 30$ (see Cole, 1990).

This species, as circumscribed here, shows a wide range of morphological variation. However, preliminary studies indicate that a species complex is present in the British Isles. The name most commonly used for this group is *P. urceolata*, but until species limits in this complex can be characterized, we have used the oldest available name, *P. stricta*. Numerous morphological forms have been recognized at different taxonomic levels (Kapraun & Rueness, 1983), but a genetic interpretation of the complex has not yet been attempted. In the British Isles, some populations resemble *P. subtilissima* Montagne (type locality: French Guiana, Central America), which is widely distributed in the tropics (Womersley, 1979). The most characteristic features are the very narrow vegetative axes (<100 µm in diameter) and straight spermatangial axes with elongate sterile tips; further research is necessary to determine whether they are environmentally influenced, perhaps resulting from growth in low-salinity conditions. Womersley noted that *P. subtilissima* appeared to form part of a complex of several closely related species including *P. urceolata*. Another distinctive form, originally described as *Conferva patens*, is found most commonly on kelp stipes in the British Isles. It is characterized by wide, rigid axes, recurved laterals, lack of trichoblasts, strongly urceolate cystocarps and incurved spermatangial axes with short sterile tips. In marinas, and lower-shore pools, populations correspond better with *Conferva stricta*, with softer, narrower axes, more abundant trichoblasts and spermatangial axes terminating in elongate tips.

Polysiphonia subulifera (C. Agardh) Harvey (1834), p. 301.

Lectotype: LD 41607. Italy (Venice), undated.

Hutchinsia subulifera C. Agardh (1827), p. 638.

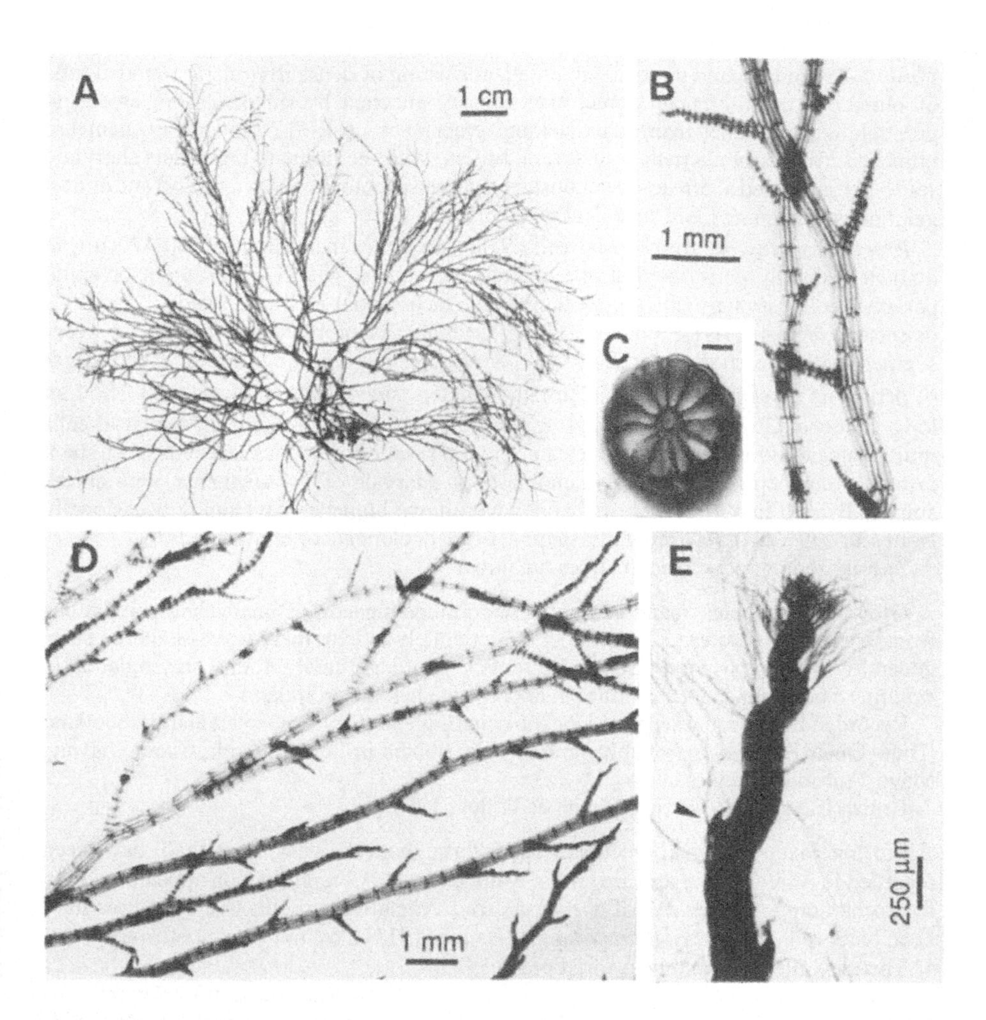

Fig. 112. *Polysiphonia subulifera*

(A) Habit (A-E Galway, June). (B) Detail of ecorticate axes with short spine-like branches. (C) T.S. of axis, showing 12 periaxial cells, with inflated outer walls, around small axial cell. (D) Branching pattern. (E) Apex, showing development of branches in axils of long trichoblasts (arrow).

Thalli 5-20 cm high and up to 20 cm broad, consisting of dense irregularly rounded tufts of numerous much-branched erect axes loosely attached by rhizoids from erect and prostrate axes; distinct main axes lacking; erect axes up to 0.3-0.5 mm in diameter, branched irregularly alternately to several orders, with the ultimate branchlets short and spine-like, arranged alternate-distichously; bright red in colour; young axes soft and almost gelatinous, becoming more succulent when older.

Erect axes growing from apical cells 9-11 µm wide, increasing to 360-470 µm in diameter, slightly constricted at articulations, ecorticate, with 12-13 straight or spiral periaxial cells, strongly convex on their outer faces; axial cell comprising 1/5 of axis diameter; segments 1-1.5 diameters long; trichoblasts or branches formed on every segment with a spiral divergence of 1 in 6-8; *rhizoids* cut off along the whole cell length of periaxials in repent axes, 40-100 µm in diameter; *trichoblasts* abundant, up to 1.5 mm long, branched 3 times, decreasing from 25-40 µm to 6 µm in diameter, the basal cells multinucleate and apical cells uninucleate, leaving large, conspicuous scar cells when shed; *branches* developing in axils of trichoblasts at intervals of 3-4 segments, with apices spirally twisted into a corkscrew shape; adventitious branches developing occasionally from scar cells; *plastids* irregularly-shaped, beaded, elongate or convoluted.

Specialized reproductive structures unknown.

Growing on pebbles, maerl and crustose coralline algae or as unattached populations from extreme low water to 20 m depth at moderately to extremely wave-sheltered sites, generally with exposure to strong currents. Although rarely found in England, *P. subulifera* occurs in large quantities in appropriate habitats in Ireland.

Recorded from widely separated localities in England (Devon, Dorset) and W. Scotland (Bute, Outer Hebrides); probably generally distributed in Ireland (Cork, Clare, Galway, Mayo, Antrim); Channel Isles.

British Isles to N.W. France (South & Tittley, 1986).

Prostrate axes and small erect axes occur throughout the year; large thalli have been recorded in May-October and are probably ephemeral. All reproduction appears to be by fragmentation as neither sexual nor specialized vegetative reproductive structures have been observed. Reports of tetrasporangia (Batten, 1923) have not been confirmed.

There is little morphological variation.

P. subulifera superficially resembles *Boergeseniella fruticulosa* (q. v.), from which it can be distinguished by the lack of cortication in contrast to the fully-corticate axes of *B. fruticulosa*.

BOERGESENIELLA Kylin

BOERGESENIELLA Kylin (1956), p. 507.

Type species: *B. fruticulosa* (Wulfen) Kylin (1956), p. 507.

Thalli radially organized, polysiphonous, differentiated into prostrate and erect axes; prostrate axes lacking trichoblasts, bearing spirally or irregularly arranged prostrate branches of limited or unlimited growth attached to the substratum by clusters of rhizoids, and branches that develop into erect axes; erect axes branched alternate-distichously or in a 2/5 spiral, or the branching sometimes subdichotomous due to secondary torsion; trichoblasts initiated in a 1/5 right-handed spiral from every segment, replaced by branches at intervals of 3-7 segments; every segment bearing a trichoblast or branch initial, or with 1-several naked segments interpolated between branch-bearing segments, then not disturbing the 1/5 phyllotaxy of the lateral initials; lateral initials developing into branches of limited growth (short shoots) or of unlimited growth; periaxial cells 8-12; cortication developing near the apex, wholly or only partly covering the periaxial cells; adventitious branches developing in the axils of older branches.

Growth by slightly oblique division of the apical cell in a tight right-handed spiral; trichoblasts, lateral branches and periaxial cells developing as in *Polysiphonia*, except for the specificity in the arrangement of the branch initials.

Gametophytes dioecious. Spermatangial axes replacing one or both branches of a fertile trichoblast at the level of the first dichotomy, developing as in *Polysiphonia.* Procarps originating from the fifth periaxial cell on the suprabasal segment of a modified trichoblast; pre- and postfertilization stages developing as in *Polysiphonia*; cystocarps sessile, ovoid to slightly urceolate, lateral or appearing axile, situated between the axis and a lateral branch, the fusion cell conspicuous, highly branched, and candelabra-like. Tetrasporangia originating from the third periaxial cell, borne in spiral series in much-branched laterals, covered by 2 presporangial cover cells.

References: Kylin (1956), Öztig (1959).

We follow Kylin in recognizing the genus *Boergeseniella*, characterized chiefly by the arrangement of trichoblasts and branches. The two British representatives of this genus, *B. fruticulosa* and *B. thuyoides,* were placed in previous British checklists in the genera *Polysiphonia* and *Pterosiphonia,* respectively.

KEY TO SPECIES

Branch angles of main axes >60°; laterals formed at intervals of (3) 4-7 segments; cortex of mature axes with a ring of large (periaxial) cells in each segment, separated by smaller cortical cells; periaxial cells 11-12.. *B. fruticulosa*

Branch angles of main axes <30°; laterals formed at intervals of 3 segments; large and small cortical cells intermixed, periaxial cells not distinguishable in cortex of mature axes; periaxial cells 8-10 .... *B. thuyoides*

Boergeseniella fruticulosa (Wulfen) Kylin (1956), p. 507.

Lectotype: Wulfen (1789), pl. 16, fig. 1.

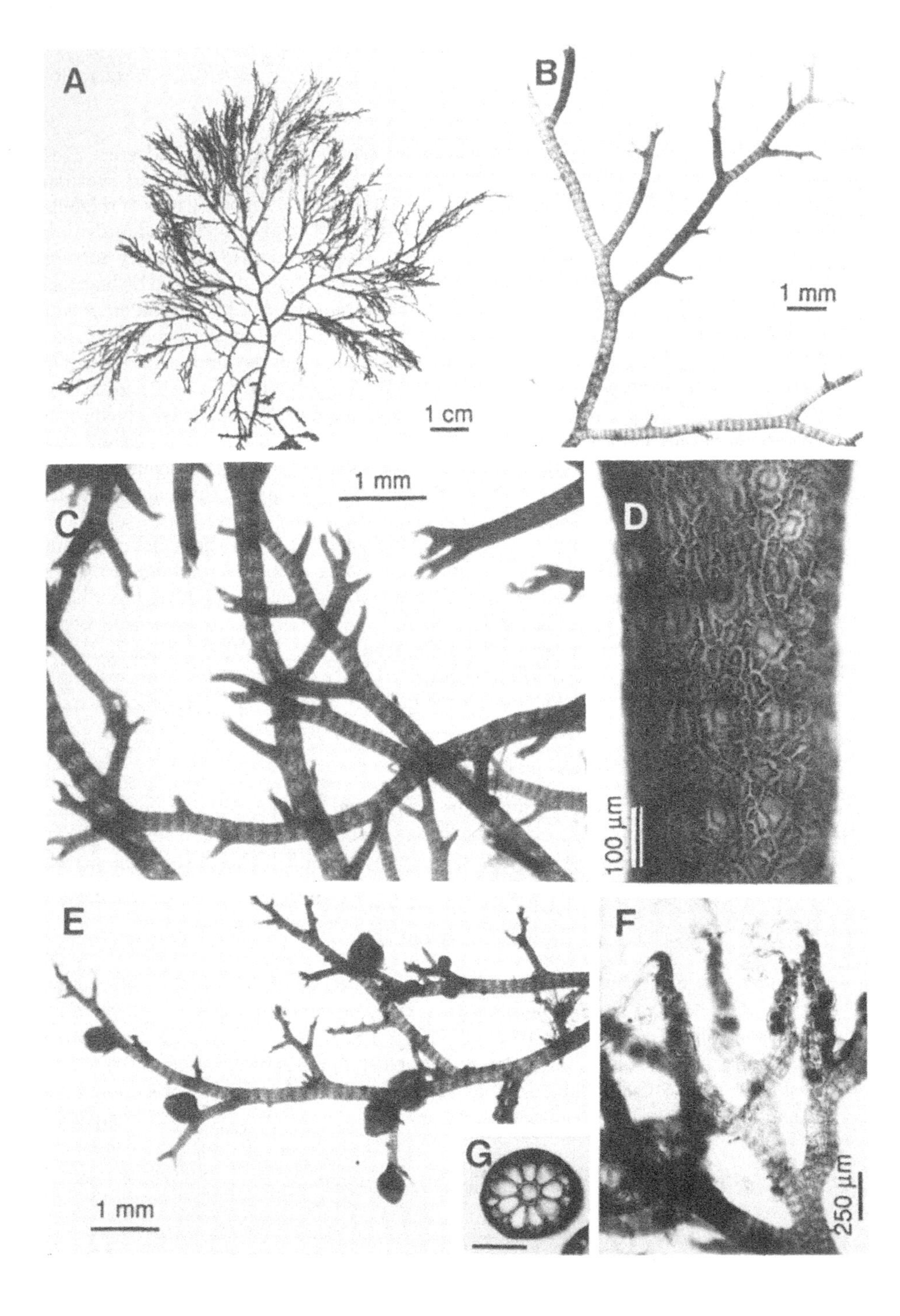
A
B
1 mm
1 cm
1 mm
C
D
100 µm
E
F
G
250 µm
1 mm

Fucus fruticulosus Wulfen (1789), p. 159.
Polysiphonia fruticulosa (Wulfen) Sprengel (1827), p. 350.
Polysiphonia martensiana Kützing (1843), p. 432.
Rytiphloea fruticulosa (Wulfen) Harvey (1849a), pl. 220.
Boergeseniella martensiana (Kützing) Ardré (1970), p. 198.

Thalli 5-15 cm high, consisting of spreading tufts of erect axes attached by tangled prostrate axes; tufts composed of few to many erect axes 0.3-0.9 mm in diameter, branched pseudodichotomously below, alternately in younger parts, at a wide angle; laterals borne in an irregularly spiral or regularly alternate-distichous, complanate arrangement; brownish-red in colour, bleaching to straw-coloured, with a harsh, rigid texture.

Prostrate axes branching frequently to form erect axes and further prostrate axes, attached by clusters of rhizoids 30-90 μm in diameter; *erect axes* growing from apical cells c. 10 μm in diameter, with 11-12 periaxial cells, increasing to 350-900 μm in diameter in major axes; axial cell comprising c. 1/6 of axis diameter; segments 0.4-0.6 diameter long; *cortication* developing at apices, only partly covering periaxial cells when young so that each segment has a ring of larger periaxial cells separated by transverse and longitudinal bands of smaller cortical cells, obscuring articulations in older axes; *trichoblasts* abundant or absent, formed in a spiral divergence of 1/5, up to 2 mm long and 45 μm in diameter, branched 3-5 times, the basal cells multinucleate and apical cells uninucleate; *branches* replacing trichoblasts at intervals of 3-7 segments in a 2/5 spiral divergence (Öztig, 1959), incurved when young, sometimes forcipate; adventitious branching developing in axils of older laterals; *plastids* discoid to irregularly elongate.

Spermatangial branchlets densely tufted at the apices of young laterals; spermatangial axes replacing one branch of fertile trichoblast at first dichotomy, cylindrical, incurved, terminating in several elongate sterile cells. *Cystocarps* sessile, angled away from bearing axes, slightly urceolate or pyriform, 600-725 μm long and 600-725 μm in diameter, with ostiole 75 μm wide; outer pericarp cells polygonal, irregularly arranged, decreasing in size towards ostiole; carposporangia 150-200 x 70-80 μm. *Tetrasporangia* formed in spiral series in much-branched laterals, distorting branch into a corkscrew shape, spherical, 75-95 μm in diameter.

Epiphytic on various algae, particularly *Corallina* spp., *Cystoseira* spp. and maerl,

---

Fig. 113. *Boergeseniella fruticulosa*

(A) Habit of thallus with widely spreading branches (Galway, Feb.). (B) Detail of branching, showing wide angles and irregularly alternate short laterals (B-C & G Cork, Nov.). (C) Major axes with irregularly alternate short laterals; note banded appearance of cortex. (D) Cortex showing small cells arranged around rows of larger (periaxial) cells (Cornwall, Sep.). (E) Axes bearing cystocarps (Donegal, Oct.). (F) Tetrasporangia formed in spiral series in much-branched laterals (Donegal, July). G. T.S. of axis with 9 periaxial cells (scale = 250 μm).

epilithic on bedrock, occasionally growing on pebbles, dead shells and other mobile substrata, in pools and damp places between high and low water levels of neap tides and subtidal to at least 7 m depth, tolerant of sand cover on bedrock, at moderately to extremely wave-sheltered sites, often with some current exposure, and in sheltered situations on wave-exposed shores.

Common on southern and western coasts of the British Isles, northwards to Shetland, rare in E. Scotland and very rarely recorded from E. England.

British Isles to Morocco and Canaries; Mediterranean (Athanasiadis, 1987).

Large thalli are found throughout the year in sheltered locations. At a site in Dublin, spermatangia occurred in May-December, cystocarps in May and July-December, and tetrasporangia in May-December; large plants were sparse from December until May (H. Parkes, pers. comm.). Elsewhere, cystocarps have also been noted in April, and tetrasporangia in January and April. Gametophytes are much less frequent than tetrasporophytes. Trichoblasts are generally absent in winter and abundant from May-August.

There is little morphological variation, except that some thalli show a more regularly distichous branching pattern, with laterals developing every (3) 4 segments, whereas other thalli are less regularly branched, with 4-7 segments between branches. Mediterranean populations differ from those in the Atlantic, having only 3 segments between branches (Öztig, 1959) and 9-10 rather than 10-11 periaxial cells. Ardré (1970) treated Mediterranean and Atlantic forms as separate species, and recognized the Atlantic one as *Boergeseniella martensiana* (type locality: Biarritz, France).

Boergeseniella thuyoides (Harvey in Mackay) Kylin (1956), p. 508.

Lectotype: TCD. Clare (Miltown Malbay), 1831, *Harvey**.

Polysiphonia thuyoides Harvey in Mackay (1836), p. 205.

Fig. 114. *Boergeseniella thuyoides*

(A) Habit of thallus with erect main axes (A, C-D & F Antrim, Jan.). (B) Regularly alternate branching with narrow branch angles (Antrim, June). (C) T.S. of axis with 10 periaxial cells (scale = 250 μm). (D) Base of plant, showing prostrate axes bearing erect axes. (E) Cortex consisting entirely of small cells (Dublin, May). (F) Apex showing regularly alternate branching. (G) Clusters of spermatangial axes (G-I Cornwall, Mar.). (H) Cystocarps. (I) Tetrasporangia.

---

* The protologue also referred to specimens collected at Portrush by Moore, but this collection is heterogeneous, including specimens of *Polysiphonia fucoides* (H. Parkes, pers. comm.)

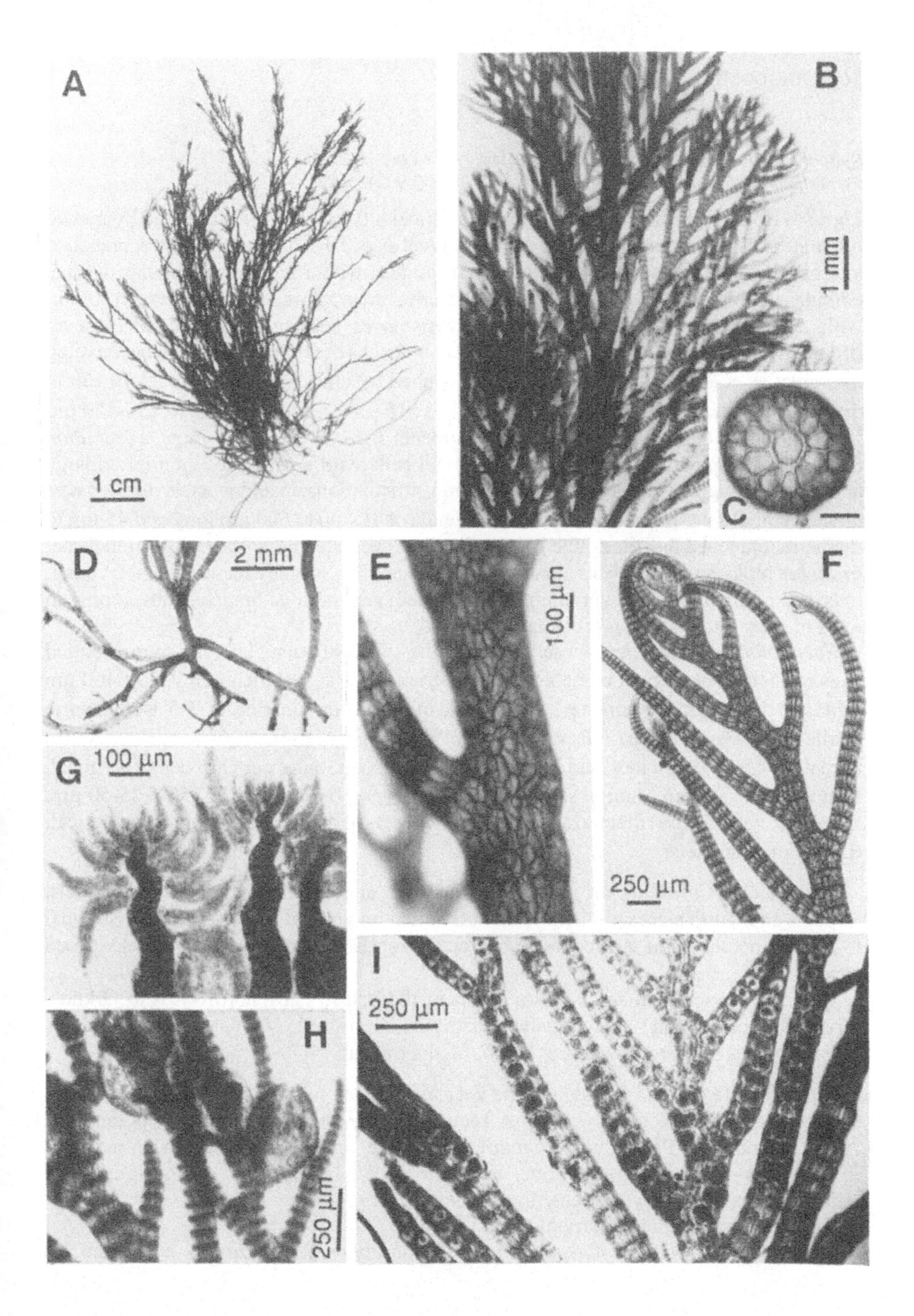
A
1 cm
B
1 mm
C
D
2 mm
E
100 µm
F
250 µm
G
100 µm
H
250 µm
I
250 µm

Rytiphlaea thuyoides (Harvey in Mackay) Harvey (1849a), pl. 221.
Pterosiphonia thuyoides (Harvey in Mackay) Batters (1902), p. 82.

Thalli 5-15 cm high, consisting of dense cylindrical tufts of erect axes attached by matted prostrate axes; tufts composed of few to many erect axes 0.3-0.5 mm in diameter, branched alternate-distichously to 1-3 orders at narrow angles, with a linear to triangular outline, complanate; older axes becoming denuded basally; brownish-red to dark brown in colour, with a flexible, cartilaginous texture. *Prostrate axes* bearing short branchlets in an irregularly spiral arrangement and also forming branches of unlimited growth that develop into erect axes and further prostrate axes, attached by clusters of rhizoids 20-70 µm in diameter; *erect axes* growing from apical cells c. 10 µm wide, increasing to 250-450 µm in diameter, with 8-10 periaxial cells; segments 0.5-0.6 diameters long; *cortication* developing at apices, rapidly covering periaxial cells with a pattern of intermixed large and small polygonal cortical cells, obscuring articulations in older axes; *trichoblasts* abundant or absent, formed in a spiral divergence of 1/5, up to 600 µm long and 45 µm in diameter, branched 3-4 times, the basal cells multinucleate and apical cells uninucleate; *branches* replacing trichoblasts at intervals of 3 segments, strongly incurved when young, overtopping main apices, later straight or reflexed; adventitious branches developing in axils of older laterals; *plastids* discoid to beaded.

Spermatangial branchlets densely tufted at the apices of young laterals; spermatangial axes replacing one or both branches of fertile trichoblast at first dichotomy, 300-460 µm long and 50-90 µm in diameter, cylindrical, incurved, terminating in 3-5 isodiametric sterile cells; spermatangia ellipsoid, 3-4 µm in diameter. *Cystocarps* sessile, ovoid to urceolate, 450-650 µm long and 350-450 µm in diameter; outer pericarp cells polygonal, in straight rows, decreasing in size towards ostiole; carposporangia 75-150 x 25-50 µm. *Tetrasporangia* formed in spiral series in last 1-2 orders of branching near apices, spherical, 65-85 µm in diameter.

Epiphytic on *Corallina* spp., epilithic on bedrock, and epizoic on limpets and mussels, in pools and damp places on the lower shore, often on vertical rock, very rarely subtidal to 2 m depth, tolerant of sand cover on bedrock, at moderately to extremely wave-exposed sites.

Common on southern and western coasts of the British Isles, northwards to Orkney, rare in E. Scotland and England; widely distributed in Ireland.

British Isles to Morocco (Ardré, 1970); Mediterranean.

Large thalli are found throughout the year in sheltered locations. At a site in Dublin, spermatangia and cystocarps were most frequent in May-December, and tetrasporangia in June-December (H. Parkes, pers. comm.). Elsewhere, spermatangia have been recorded in February-May and September, cystocarps in March, May and August, and tetrasporangia in January, March-July and September-October.

There is relatively little morphological variation.

The branching pattern of some forms of *Polysiphonia fucoides* (q. v.) resembles

Boergeseniella thuyoides, but only the main axes of *P. fucoides* are corticated, whereas *B. thuyoides* axes are entirely corticated.

Tribe PTEROSIPHONIEAE Falkenberg (1901), p. 261.

PTEROSIPHONIA Falkenberg

PTEROSIPHONIA Falkenberg in Schmitz & Falkenberg (1897), p. 443.

Type species: *P. cloiophylla* (C. Agardh) Falkenberg in Schmitz & Falkenberg (1897), p. 443.

Thalli radially organized, bilaterally symmetrical, polysiphonous, differentiated into prostrate and erect systems; indeterminate prostrate axes cylindrical to slightly compressed, attached by unicellular rhizoids or clusters of rhizoids often terminating in discoid pads, and bearing dwarfed lateral branches in alternate-distichous arrangement, some of which develop into either prostrate or erect indeterminate axes; erect axes subcylindrical to compressed, bearing both determinate and indeterminate lateral axes in alternate-distichous arrangement from every second or third segment; determinate axes consisting of 1-3 orders of branches in different species; lateral indeterminate axes replacing or developing terminally from determinate axes; vegetative trichoblasts mostly absent, when present resembling the trichoblasts of *Polysiphonia*; periaxial cells 4-20, elongating to the same length as the axial cells; cortication absent, or present throughout most of the thallus and several layers thick; lateral branches coalescing with the main axis over a length of 1.5 to 4.5 segments in different species, reinforcing bilateral symmetry; adventitious branches absent.

Growth monopodial, by oblique division of the apical cell; branch initials issuing alternately from subterminal cells every second or third segment (or further apart in prostrate axes); periaxial cells formed in alternating sequence with the first cut off directly below the branch initial in nodal and internodal segments, the second in the direction of radial symmetry (usually to the right), the third in the opposite direction (usually to the left), etc; and with lateral branches displaced, arising between the first and third periaxial cells; first periaxial cells abaxial in basal segments of lateral branches; cortication, when present, by 2 ascending and 1 descending cortical initial formed from periaxial cells and then ternately to dichotomously branched; rhizoids issuing from periaxial cells or cortical cells, separated by a septum.

Gametophytes dioecious. Spermatangial axes typically borne on the suprabasal cell of a modified trichoblast, either terminal or subtended by a trichoblast branch; densely clustered at apices of the last two orders of branching, sometimes overtopped by polysiphonous lateral branches bearing additional spermatangial branchlets; formed on every other segment in alternate-distichous arrangement, or on every segment in spiral arrangement; periaxial cells 4 per segment; spermatangia developing as in *Polysiphonia*. Procarps and young cystocarps formed on modified trichoblasts near apices of branches as in *Polysiphonia*, with the pericarp corticated or uncorticated. Tetrasporangia clustered in 3 (-4) orders of subcylindrical branchlets resulting mostly from resumed growth of determinate axes; tetrasporocytes one per segment, on the second or third periaxial cell, alternating to front and back sides on main axes and in straight series on ultimate branchlets, dividing tetrahedrally, covered by 2 lateral presporangial and a third, basal postsporangial cover cell.

References: Suneson (1940), Hommersand (1963), Ardré (1967).

Useful characters for identification of *Pterosiphonia* species include the presence or absence of cortication; numbers of periaxial cells (both in transverse section and in face view); and the arrangement and form of dwarf laterals on prostrate axes. In some species, these appear to be spirally arranged (although this may result from torsion of the prostrate axes), whereas in others they are distichous. The degree of coalescence of short laterals with main axes is difficult to determine without careful preparation, but the effect on the position of laterals relative to each other is readily seen.

In the most recent algal checklist for the British Isles (South & Tittley, 1986), three species of *Pterosiphonia* are listed: *P. complanata*, *P. pennata* and *P. thuyoides*. In our treatment, the position of *P. complanata* is unchanged, *P. thuyoides* is transferred to *Boergeseniella,* and *P. pennata* is recognized as one of a group of three species.

Dixon & Parkes (1968) showed that the first legitimate use of the epithet *pennata* in relation to the genus *Pterosiphonia* was that by C. Agardh (1824), whose type material was in Lund. However, this specimen is not of the species for which the name *P. pennata* is currently used. It has main axes 250-265 μm diameter, with 6-7 periaxial cells visible in face view (about 11 periaxials in total), and short laterals coalesced with main axes for 1.5 segments. In contrast, the alga currently known as *P. pennata* has 6-8 periaxials and main axes less than 250 μm wide (Ardré, 1967; 1970); it is here referred to *P. pinnatula* (Kützing) comb. nov. C. Agardh's collections of '*Hutchinsia pennata*' from various sites are heterogeneous, including specimens of both *P. pennata* and *P. pinnulata*; this situation was first appreciated by Kützing (1843), who described *Polysiphonia pinnulata* based on the smaller entity. A third species in the *P. pennata* group, *P. ardreana* sp. nov., was previously reported from Portugal and France as *Pterosiphonia spinifera* var. *robusta* Ardré (1967), p. 43, nom. illeg.

KEY TO SPECIES

1 Cortication covering all axes .... *P. complanata*
  Cortication absent 2
2 Erect thallus outline triangular, >10 mm wide; **prostrate axes** bearing a distichous arrangement of distichously branched dwarf laterals on every second segment (see Fig. 117B) .... *P. parasitica*

Erect thallus outline linear, 1-3 (-6) mm wide; **prostrate axes** bearing a spiral arrangement of simple dwarf laterals on every third segment (see Fig. 118C) 3

3 Main axes ≤125 µm wide, consisting of 6-8 periaxial cells; prostrate axes with 6 periaxial cells .... *P. pinnulata*

Main axes >200 µm wide, consisting of 9-12 periaxial cells; prostrate axes with 7-12 periaxial cells 4

4 Main axes 200-360 µm wide, with 5-6 periaxial cells visible in face view; short lateral branchlets coalesced with main axes for (1) 1.5-2.5 (typically 2) segments, the bases of successive laterals overlapping only slightly or not at all (see Figs 118B, D) .... *P. pennata*

Main axes 360-480 µm wide, with 7-10 periaxial cells visible in face view; short lateral branchlets coalesced with main axes for 2.5-3.5 segments, the bases of successive laterals overlapping (see Figs 115D, E) .... *P. ardreana*

Pterosiphonia ardreana Maggs & Hommersand, sp. nov.

Thalli composed of terete, ecorticate prostrate axes, with 9-12 periaxial cells, forming ecorticate, complanate, alternate-distichously branched erect axes 2-4 cm high, with 10-12 periaxial cells, tetrasporangial branchlets developing as branched cylindrical extensions to determinate laterals; similar to *P. spinifera* (Kützing) Norris & Aken (1985, p. 62) from Peru and to *P. pennata* from Europe, differing principally in the greater degree of coalescence of laterals with main axes, 2.5-3 segments in *P. ardreana* compared to 1.5-2 segments in *P. spinifera* and *P. pennata*.

Thalli ex axibus teretis ecorticatis prostratis compositi; cellulae 9-12 periaxiales formant axes 2-4 cm altos erectos distiche alternatimque ramosos et complanatos ecorticatosque cellulis 10-12 periaxialibus. Ramuli tetrasporangiorum evolventes ut extensiones ramosae cylindricae lateralium determinatorum. Affinis P. spinifera *(Kützing) Norris & Aken (1985, p. 62) de Peruvia et* P. pennata *de Europa, e quibus differt lateralibus axes principales multo coalescentioribus habentibus, et segmentis 2.5-3 (ex comparatione segmentis 1.5-2 in aliis).*

Holotype: BM. Isotypes: PC. Cornwall (Nerope Rocks, Padstow), 30 vi 1977, *K. Hiscock*.

Pterosiphonia spinifera var. *robusta* Ardré (1967), p. 43, nom. illeg.

Thalli forming small tufts or extensive turfs of numerous erect axes 2-4 cm high arising from interwoven prostrate axes; erect axes bearing a complanate, alternate-distichous arrangement of short lateral branchlets and occasional indeterminate branches, with a linear-lanceolate or triangular outline due to branching of main axes, 2-4 mm wide when main axes are unbranched, up to 15 mm in width if branched; colour brownish-red, texture rigid.

All axes ecorticate; *prostrate axes* terete, 190-215 µm in diameter, attached by

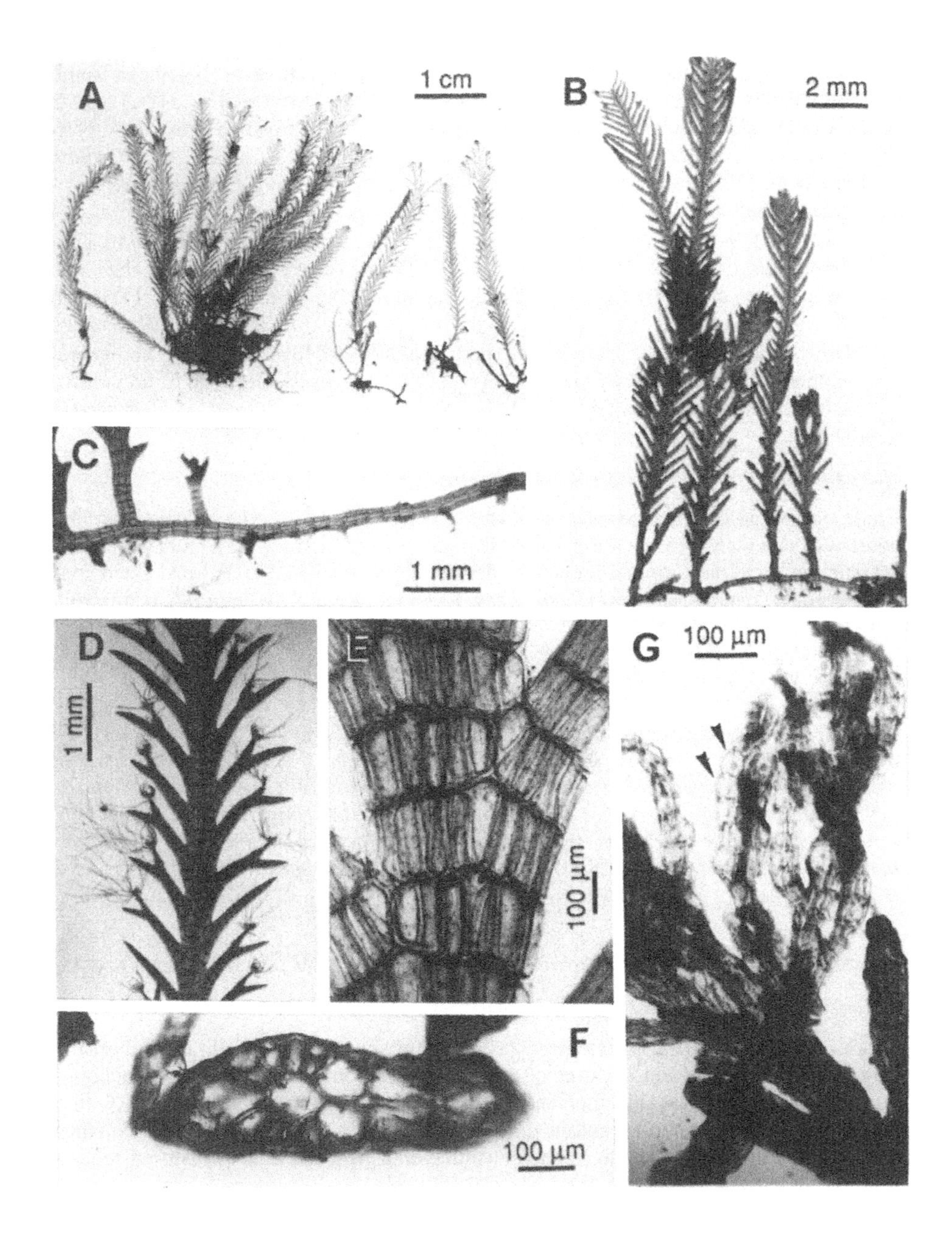
A
1 cm
B
2 mm
C
1 mm
D
1 mm
E
100 µm
G
100 µm
F
100 µm

unicellular rhizoids terminating in multicellular discoid pads, consisting of axial and 9-12 periaxial cells; segments 0.7 diameters long; branches formed by every third segment in an apparently spiral arrangement (that may result from torsion of axes), either remaining simple and spur-like, or developing into further prostrate axes or erect axes; *erect axes* terete below, becoming compressed, 360-480 μm wide, with 10-12 periaxial cells; segments 0.3-0.5 diameters long; every other segment branched; *determinate branches* borne in an alternate-distichous pattern, coalesced with main axis for 2.5-3 axial segments, increasing in length upwards to 21-30 segments, with 10 periaxial cells; lowermost branches simple, initially incurved, sometimes becoming straight, mostly forming 1-4 short spine-like laterals from alternate segments in a distichous arrangement, or secondarily developing into indeterminate axes; *trichoblasts* borne around apices in an alternate-distichous pattern, replacing branchlets at first, then formed on every segment, to 1.8 mm long, branched 4-6 times, composed of uninucleate cells; *plastids* irregularly reticulate.

Gametangia unknown. Tetrasporangial branchlets developing as cylindrical extensions to determinate lateral branches, branched to 3 orders, 50-80 μm in diameter, becoming swollen and forming tetrasporangia in long spiral series; tetrasporangia 1 per segment, with 3 cover cells, distorting branchlet into a corkscrew shape when mature, spherical, c. 100 μm in diameter.

Epilithic on sandy bedrock and boulders among *Rhodochorton floridulum* (Dillwyn) Nägeli in lower-shore pools and on sand-scoured vertical and horizontal bedrock from 1 m above extreme low water to 20 m depth; on crustose corallines and among *Alaria esculenta* plants at extremely wave-exposed sites; and on pebbles at 14 m at a moderately sheltered site.

Widespread in N. Cornwall (Cornish sites in Hiscock & Maggs, 1984, fig. 19, as *P. pennata*); otherwise known from only a few sites in Pembroke and Clare.

British Isles to Portugal; Mediterranean (Ardré, 1970 and pers. comm.).

Large plants have been observed between March and November; probably perennial. Tetrasporangia were observed in March in the British Isles and in Portugal (Ardré, 1967). Trichoblasts were present in June and July, and absent in August, September and March. An isolate from Cornwall (29 iii 90) formed tetrasporangia (80 μm diameter) in culture at 15°C, 16:8 h light: dark.

Fig. 115. *Pterosiphonia ardreana*

(A) Habit of holotype (Cornwall, June). (B) Prostrate axis with several erect axes (B-C & F Cornwall, Mar.). (C) Prostrate axis showing spiral arrangement of erect axis initials. (D) Complanate erect axis with regularly alternate short laterals bearing trichoblasts (D-E from type collection, Cornwall, June). (E) Stained axis showing coalescence of 2.5 segments between main axis and lateral. (F) T.S. of complanate erect axis. (G) Tetrasporangia (arrows) in much-branched fertile extensions of short laterals (Cornwall, Mar. 1954, in BM).

The degree of coalescence of short laterals with main axes increases during development of an axis, from about 2.5 segments to 2.5-3 segments. Cultured plants show a reduction in coalescence from 3 segments in the parent field-collected thalli to 1.5-2.5 segments in young erect axes; the number of periaxial cells tended to decrease by 1-2. Trichoblasts were not formed in culture.

Ardré (1967) described plants from Portugal and France, previously identified as *Pterosiphonia pennata*, as *Pterosiphonia spinifera* var. *robusta* (due to an error, this appeared in Ardré (1967), p. 43, as *Pterosiphonia spinifera* var. *spinifera* [F. Ardré, pers. comm.]). This name was illegitimate as no type was designated. We consider that the plants involved represent a separate species from *P. pennata* and the closely similar (possibly conspecific) species *P. spinifera*, differing principally in the greater degree of coalescence of lateral branchlets with the main axes, 2.5-3 segments in *P. ardreana* compared to 1.5-2 segments in *P. spinifera* and to 1.5-2.5 segments in *P. pennata*. *P. ardreana* and *P. pennata* occupy different habitats in the British Isles and exhibit phenological differences; they also remain distinct in culture.

Pterosiphonia complanata (Clemente) Falkenberg (1901), p. 265

Lectotype: MA (algae) 1464 (see Cremades, 1993, fig. 4b). Spain (Tarifa).

Fucus complanatus Clemente (1807), p. 316.
Fucus cristatus γ *articulatus* Turner (1808), pl. 23, fig. H.
Polysiphonia cristata Harvey in Mackay (1836), p. 205.
Polysiphonia complanata (Clemente) J. Agardh (1863), p. 933.

Thalli 3-10 cm high, consisting of flabellate to cylindrical tufts of erect axes attached by tangled prostrate axes; tufts composed of few to many compressed erect axes 0.3-0.9 mm in diameter; branching complanate, alternate-distichous to 5-6 orders, with a linear to triangular outline as lengths of laterals increase upwards, and flat-topped apices; branch angles narrow; older axes becoming denuded basally; bright red to brownish-red in colour, with a flexible, cartilaginous texture.

All axes consisting of 5 periaxial cells surrounded by a cortex that increases in thickness with age; *prostrate axes* subcylindrical, 450-600 μm wide and 350 μm thick, heavily corticated, attached by single rhizoids 50-125 μm in diameter and by peg-like haptera composed of intertwined rhizoids, both terminating in discoid pads, bearing an alternate-distichous arrangement of dwarf, distichously branched laterals every 3-4

Fig. 116. *Pterosiphonia complanata*

(A) Habit of unusually narrow thalli showing erect axes arising from extensive prostrate axes (Cornwall, Mar.). (B) Detail of branching from every third segment, with tetrasporangia borne near apices (B-E Cornwall, July). (C) Cortex, showing axial segments visible through complete cortication. (D) T.S. of complanate erect axis, with 5 periaxial cells. (E) Tetrasporangia borne in last 2 orders of branching.

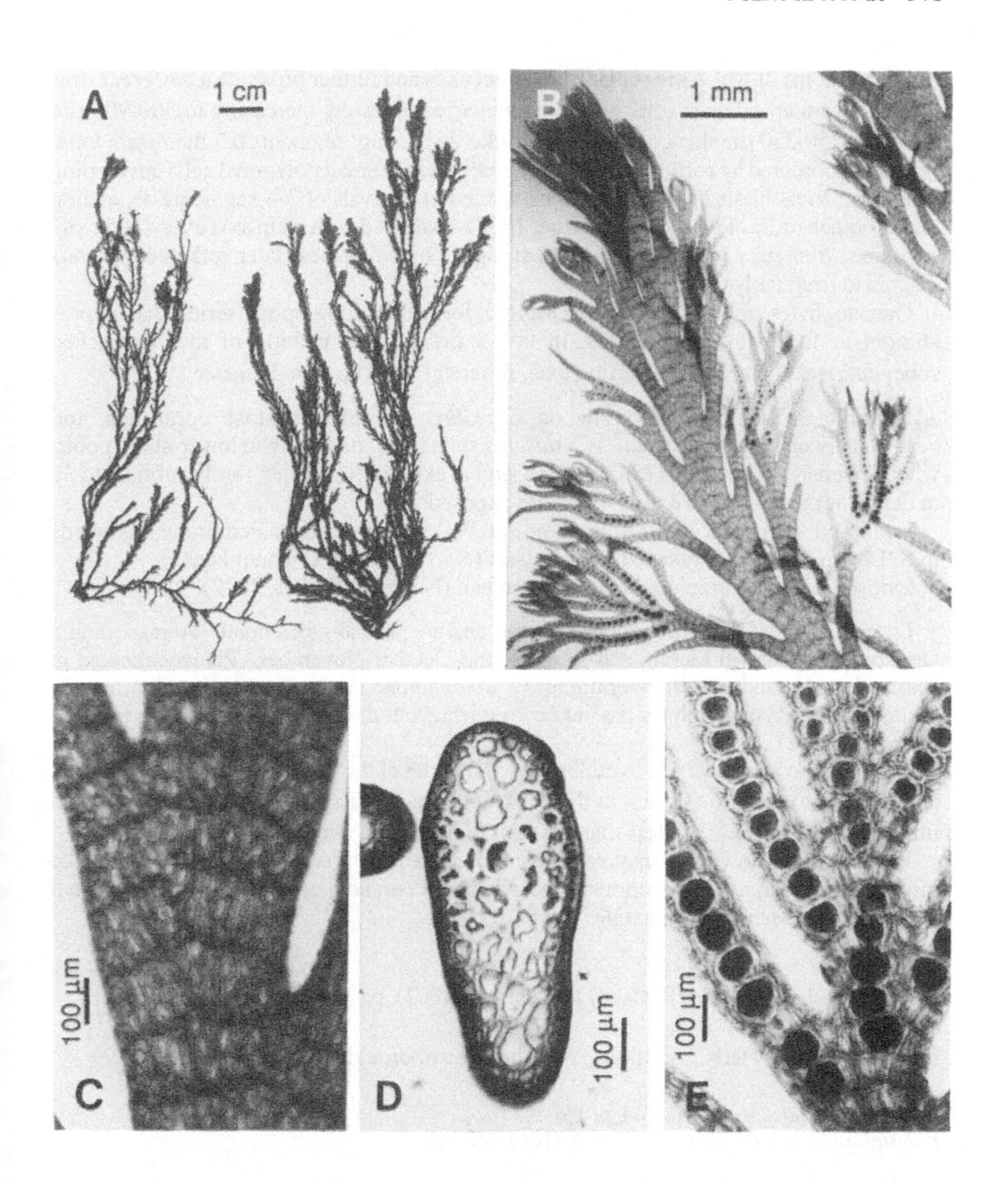
A
1 cm
B
1 mm
100 µm
C
D
100 µm
100 µm
E

segments, some of which are replaced by erect axes and further prostrate axes; *erect axes* growing from apical cells c. 12 µm in diameter, compressed, increasing to 300-900 µm wide and 250-350 µm thick, with a midrib-like thickening; segments 0.3 diameters long, becoming obscured by cortication in older axes; cortex of small polygonal cells developing near apex; trichoblasts lacking; *branches* formed at intervals of 3-4 segments depending on the branch order of the axis (see Ardré, 1967), coalesced with main axis over 4-4.5 axial segments, triangular when young, with straight pointed apices, later reflexed; *plastids* discoid to irregularly elongate.

Gametophytes unknown. *Tetrasporangia* formed in long spiral series that appear straight in face view of the axis, in last 3 orders of branching of much-branched subcylindrical laterals, not distorting axis, spherical, 65-95 µm in diameter.

Epilithic on bedrock, epiphytic on *Corallina* spp. and crustose corallines, and occasionally on large algae such as *Cystoseira* spp., most commonly in lower-shore pools, where extensive populations can be found, and at extreme low water, rarely subtidal to 30 m depth, at moderately to extremely wave-exposed sites.

South and south-west coasts of England and Wales, northwards to Pembroke, eastwards to S. Devon, south and west coasts of Ireland (Norton, 1985); Channel Isles.

British Isles to Mauritania; W. Mediterranean (Norton & Parkes, 1972).

Large thalli are found throughout the year, and are probably perennial. Tetrasporangia have been recorded in March, July, August and October-November. Plants collected in March bore abundant tetrasporangia; *P. complanata* appears to be predominantly winter-fertile. Gametophytes are unknown throughout the range of this species (Norton & Parkes, 1972).

Thalli vary considerably in overall shape and width of the axes. Broad flabellate plants tend to have axes up to 900 µm in diameter, whereas the main axes of linear plants with little-developed laterals are less than 500 µm in maximum width.

Plants with narrow axes may resemble *Boergeseniella thuyoides* (q. v.), but can be distinguished in transverse section by the number of periaxial cells. *P. complanata* has 5 periaxial cells whereas *B. thuyoides* has 8-10.

Pterosiphonia parasitica (Hudson) Falkenberg (1901), p. 265

Neotype: BM-K, Herb. Lightfoot. Yorkshire (Scarborough), *Frankland*.

Conferva parasitica Hudson (1762), p. 486.
Polysiphonia parasitica (Hudson) Harvey (1848), pl. 147.

Thalli either entirely prostrate, to 5 cm wide, or consisting of prostrate axes bearing tufts of erect axes 2-7 cm high and wide that may become repent at apices; *erect thalli* complanate, irregularly triangular or pyramidal in shape, with a distinct main axis 0.2-0.3 mm diameter, bearing 3-4 orders of alternate-distichously branched laterals, decreasing in

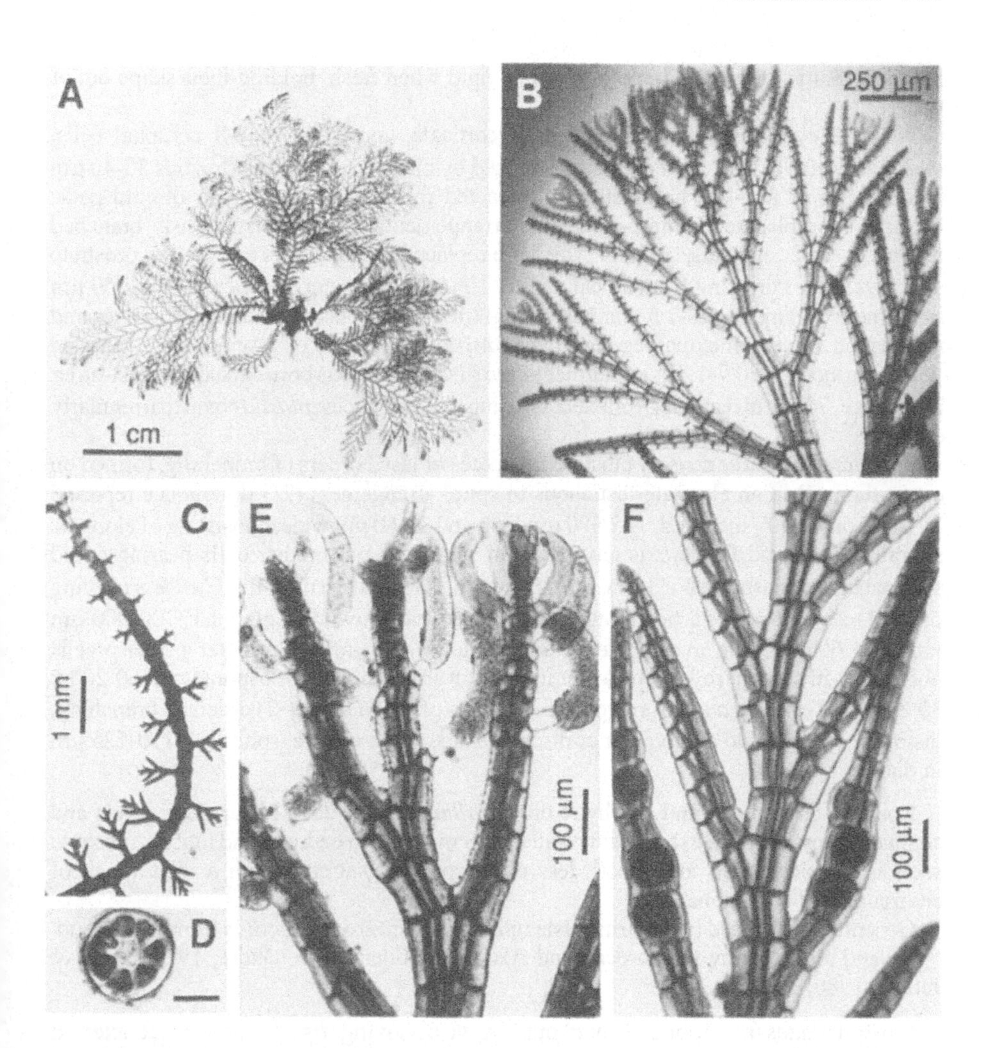

Fig. 117. *Pterosiphonia parasitica*

(A) Habit of typical pyramidal plant (Donegal, Apr.). (B) Apex, showing regularly alternate branching on every second axial segment (Donegal, May). (C) Prostrate axis with bilateral arrangement of branched dwarf laterals (Donegal, July). (D) T.S. of erect axis, with 7 periaxial cells (scale bar = 100 µm) (Galway, June). (E) Spermatangial axes clustered at apices (Down, May). (F) Tetrasporangia in last-orders laterals (Mayo, May).

length upwards; bright red in colour; plants rigid when fresh, holding their shape out of water.

All axes cylindrical to subcylindrical, ecorticate, consisting of 7-8 periaxial cells; *prostrate axes* 90-200 µm in diameter, attached by clusters of separate rhizoids 30-40 µm in diameter, or peg-like holdfasts of compacted rhizoids terminating in discoid pads, bearing a complanate alternate-distichous arrangement of dwarf, distichously branched laterals every 2 segments, some of which are replaced by erect axes and further prostrate axes; *erect axes* growing from apical cells c. 10 µm in diameter, increasing to c. 250 µm in diameter; segments 0.8-1.5 diameters long; trichoblasts absent in tetrasporophytes and non-reproductive gametophytes; *branches* formed at intervals of 2 segments, coalesced with main axis over 0.7-1.3 axial segments, the first on each axis borne abaxially; last-order branches c. 70 µm in diameter, reflexed, tapering to blunt apices; *plastids* oval to irregularly elongate.

Spermatangial axes densely clustered at apices of last 2 orders of branching, formed on every segment in an alternate-distichous to spiral arrangement (2/5 divergence reported by Suneson, 1940), incurved, 175-300 µm long and 35-60 µm wide, consisting of elongate suprabasal cell and fertile axis covered with spermatangial mother cells bearing ovoid spermatangia measuring 5 x 3 µm, terminating in 1-3 short sterile cells. *Cystocarps* lying parallel to bearing axes when mature, on narrow stalks, ovoid to globular, 700-900 µm long and 600-775 µm in diameter, with ostiole 50-100 µm wide; outer pericarp cells polygonal, in straight rows decreasing in size towards ostiole; carposporangia 160-200 x 30-80 µm. *Tetrasporangia* formed in spiral series of 1-6 in last 1 (-2) orders of branching, distorting laterals into a curved or corkscrew shape when mature, spherical, 110-125 µm in diameter.

Epilithic on bedrock and epiphytic on *Corallina* spp. in deep lower-shore pools and channels, growing from extreme low water to 33 m depth on bedrock and mobile substrata such as pebbles, shells and maerl, less often epiphytic, at sites with a wide range of environmental conditions.

Generally distributed in the British Isles although there are few records from E. England.

Iceland and Norway to Morocco and Azores; Mediterranean (Ardré, 1970; South & Tittley, 1986).

Prostrate axes are found throughout the year, giving rise to new erect axes in February-April; on open coasts these reach their maximum size from June onwards and are reduced to main axes by September. Spermatangia have been recorded in May-June; cystocarps in May-August; and tetrasporangia in April-September. Tetrasporophytes are usually predominant; male gametophytes are rarely observed.

There is relatively little variation in morphology, except that plants on extremely wave-exposed coasts may be almost entirely prostrate.

Pterosiphonia pennata (C. Agardh) Falkenberg (1901), p. 263.

Lectotype: LD 39271 (holotype lost; see Dixon & Parkes, 1968). Unlocalized.

Hutchinsia pennata C. Agardh (1824), p. 146.
Ceramium pennatum Roth (1800a), p. 171, pro parte, non *Conferva pennata* Hudson (1762, p. 486) (see Dixon & Parkes, 1968).
Polysiphonia pennata (C. Agardh) J. Agardh (1842), p. 141.

Thalli forming small tufts or loose, open turfs 1-3 cm high, composed of numerous erect axes arising from interwoven prostrate axes; erect thalli complanate, with a pinnate arrangement of short lateral branchlets and occasional long branches giving a linear-lanceolate or triangular outline, usually 1.5-2 mm wide, up to 6 mm wide when main axes are branched; brownish-red in colour, with a rigid texture.

All axes ecorticate; *prostrate axes* terete, 190-215 μm diameter, attached by unicellular rhizoids terminating in multicellular discoid pads, with 8-10 periaxial cells; segments 0.6-1.0 diameters long; *branches* formed by every third segment in an apparent spiral (that may result from torsion of axis), either remaining simple and spur-like, or developing into further indeterminate prostrate axes or erect axes; *erect axes* terete below, becoming slightly compressed, 190-360 μm wide, with 9-11 periaxial cells; segments 0.4-0.5 diameters long, every other one branched; *determinate branches* borne in an alternate-distichous pattern, coalesced with main axis over 1.5-2.5 axial segments, composed of 8-10 periaxial cells, short and spur-like below, increasing upwards to 19-23 segments in length, mostly simple, initially incurved, becoming straight or reflexed, tapering gradually to an acute point; some branches forming 2-4 short laterals from alternate segments in a distichous arrangement, or developing secondarily into indeterminate axes; *trichoblasts* absent or formed near apices on every segment, initially in an alternate-distichous arrangement, then spiral, up to 960 μm long and 25 μm in diameter, branched 4-5 times, composed of uninucleate cells; *plastids* discoid to irregularly reticulate.

Gametangia and tetrasporangia unknown in the British Isles.

Epilithic on muddy bedrock and pebbles, and epiphytic on crustose corallines and maerl, usually growing amongst other turf-forming algae, from extreme low water to 10 m depth, at moderately exposed to extremely wave-sheltered sites, often with moderate to strong current exposure.

Recorded from scattered localities in south-west Britain and Ireland: N. Devon (Lundy), Glamorgan; Cork, Clare, Galway.

British Isles to Atlantic France and ?Spain; Mediterranean. Reported to be widely distributed in the Atlantic and Pacific Oceans (Schneider & Searles, 1991), but its distribution is difficult to assess at present due to confusion between a group of morphologically similar species.

Probably perennial, mature plants found throughout the year. Specialized reproductive structures unknown in European populations; plants spreading vegetatively by extensive

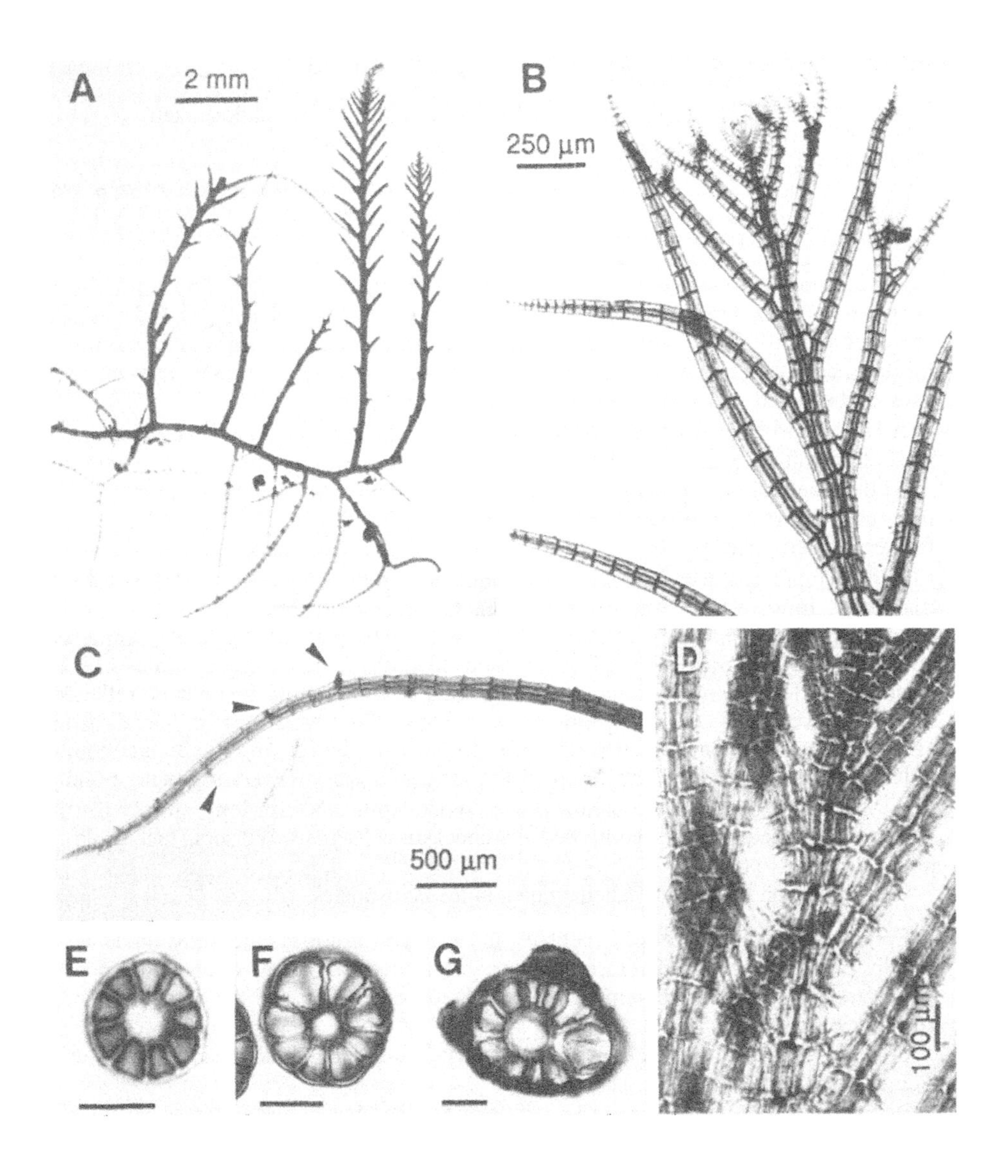
A
2 mm
B
250 µm
C
500 µm
D
100 µm
E
F
G

growth, followed by fragmentation, of prostrate axes. Trichoblasts were present in June and July but not in February. An isolate from Galway (3 ii 90) did not form tetrasporangia or gametangia in culture.

The degree of coalescence of short laterals with main axes increases during development of an axis, from about 1.5 segments to 2-2.5 segments. Cultured plants showed a reduction in coalescence from 2 segments in the parent field-collected thalli to 0.75-1 segment in young erect axes; the number of periaxial cells also tended to decrease by 1-2. Trichoblasts were not observed in culture.

Pterosiphonia pinnulata (Kützing) Maggs & Hommersand comb. nov.

Lectotype: Kützing (1863), pl. 23 a-d* Italy (Genoa).

Polysiphonia pinnulata Kützing (1843), p. 416.

Thalli forming dense turfs, up to 10 cm in extent, consisting of entangled prostrate axes and numerous erect axes; erect axes 1-3 cm high and 0.9-3 mm wide, with a complanate branching pattern and linear to narrowly triangular outline; colour dull purplish-brown; texture soft but not flaccid.

All axes ecorticate; *prostrate axes* terete, 60-100 µm in diameter, attached by unicellular rhizoids terminating in multicellular discoid pads, with 6 periaxial cells; segments 0.6-1.7 diameters long, every third one forming a lateral branch in an apparent spiral (that may result from torsion of axes); *branches* remaining as simple and spur-like, or developing into further prostrate axes or erect axes; *erect axes* terete below, becoming slightly compressed, 70-125 µm wide, with 6-7 (-8) periaxial cells; segments 1-1.8 diameters long; lowermost 2-7 segments unbranched, then every other segment branched; *determinate branches* short, borne in an alternate-distichous pattern, coalesced with main axis for 0.5-1.5 axial segments, 20-26 segments in length, usually simple, initially incurved,

Fig. 118. *Pterosiphonia pennata*

(A) Habit of plant with branched prostrate axes and several erect axes (A & C Galway, Apr.). (B) Apex, showing regularly alternate branching on every second axial segment (Galway, June). (C) Prostrate axis with spiral arrangement of simple dwarf laterals (arrows). (D) Erect axis showing coalescence of laterals with main axis over 2 axial segments (D & G Devon, July). (E) T.S. of prostrate axis, with 9 periaxial cells (scale bar = 100 µm) (E-F Clare, Aug.). (F) T.S. of erect axis from below lowermost laterals, showing 8 periaxial cells (scale bar = 100 µm). (G) T.S. of compressed erect axis with 10 periaxial cells (scale bar = 100 µm).

* Type material is apparently missing.

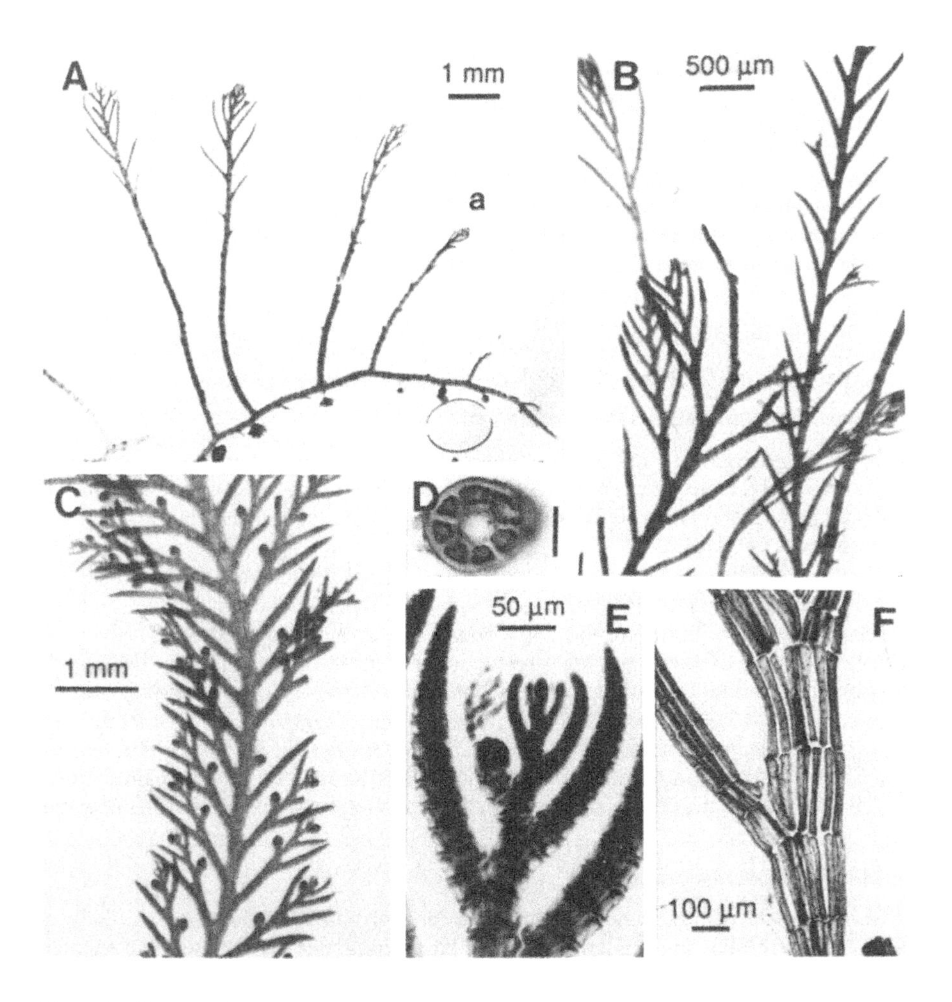

Fig. 119. *Pterosiphonia pinnulata*

(A) Part of plant removed from algal turf, consisting of a prostrate axis bearing small erect axes (A-F Pembroke, Mar.). (B) Young erect axes with regularly alternate branching. (C) Mature erect axis with numerous unfertilized procarps. (D) T.S. of erect axis with 7 periaxial cells (scale bar = 50 µm). (E) Apex of axis with procarp. (F) Erect axis squashed slightly to show axial cell and 7 periaxial cells, with coalescence of about 0.7 axial segments between main axis and lateral.

becoming straight or reflexed, tapering gradually to an acute point; *trichoblasts* formed on females, up to 260 µm long and 4 µm wide, branched 3-4 times, with uninucleate cells; *plastids* discoid to irregularly lobed and reticulate.

Male plants unknown in the British Isles. *Procarps* with pericarps 50-75 µm in diameter; fertilized procarps not observed. *Tetrasporangial branchlets* developing as cylindrical extensions to determinate lateral branches, branched irregularly alternately, 75-100 µm in diameter, becoming swollen and forming tetrasporangia in long spiral series; tetrasporangia 1 per segment, distorting branchlet into a corkscrew shape when mature, spherical, 55-70 µm in diameter.

Growing on muddy pebbles, boulders and dead oyster shells at extreme low water in a wave-sheltered inlet with strong tidal currents, and in moderately sheltered lower-shore pools.

Collected at one site in Devon and at one site in Pembroke (Milford Haven), where all plants were female. As this species has not been found in similar habitats elsewhere, and Milford Haven is a port with an oyster farm, it is likely that this population has been introduced.

British Isles, S.W. France, Portugal, S. Spain; Mediterranean; California (Abbott & Hollenberg, 1976), Japan (Masuda, 1973).

Mature female plants were observed in February and March, and showed no evidence of fertilization; tetrasporangia occurred in November. A Japanese isolate followed a *Polysiphonia*-type life history in culture (Masuda, 1973, as *P. pennata*).

No data on form variation is available for the British Isles. In fertile female plants, the simple short branches resume growth, forming a further order of branching, an irregularly alternate series of sterile branchlets and female branchlets. Specimens collected in southern Europe are usually much smaller than the British plants, rarely more than 1 cm high.

Tribe HERPOSIPHONIEAE Schmitz & Falkenberg (1897), p. 457.

LOPHOSIPHONIA Falkenberg

LOPHOSIPHONIA Falkenberg in Schmitz & Falkenberg (1897), p. 459.

Type species: *L. obscura* (C. Agardh) Falkenberg in Schmitz & Falkenberg (1897), p. 459, nom. illeg. [= *L. subadunca* (Kützing) Falkenberg (1901), p. 496].

Thalli polysiphonous, dorsiventrally organized, terete, ecorticate, composed of creeping indeterminate axes attached to substratum by unicellular rhizoids and erect determinate axes that may be simple or 1-2(-3) times branched; trichoblasts absent or rare on prostrate axes, and either present or absent on erect axes, if present, long, tufted and persistent, or short, inconspicuous and often deciduous; periaxial cells 4, 6-7, or 10-20 in different species; adventitious branches rare, frequent, or abundant.

Growth by successive slightly oblique divisions of a dome-shaped apical cell in straight series, causing the tip to curve downwards or forwards; the first periaxial cell dorsal or abaxial in axes lacking trichoblasts, the rest cut off in alternating sequence and elongating to the same length as the axial cell; all polysiphonous branches endogenous, originating from the upper ends of axial cells after a full complement of periaxial cells has been cut off; lateral and ventrolateral initials usually developing into prostrate axes; dorsal and dorsolateral initials usually developing into erect determinate axes; trichoblasts exogenous, either produced in a zig-zag on the abaxial side or spiral, commonly shifting from unilateral to spiral arrangement in the same axis, most abundant on erect, fertile determinate axes; rhizoids derived from periaxial cells, either cut off or remaining in open connection with them in different species.

Gametophytes dioecious. Spermatangial axes solitary, in pairs, or clustered in threes or fours on suprabasal cells of modified trichoblasts; spermatangial mother cells and spermatangia formed as in *Polysiphonia*. Procarps and cystocarps developing on trichoblasts at the tips of determinate axes, developing (where known) as in *Polysiphonia*. Tetrasporangia borne in simple or branched erect axes, one per segment on the second periaxial cell to the right of the trichoblast or its scar cell, or in straight series to the right of the nearest trichoblast or scar cell above; arranged in rows, zig-zags, or spirals as determined by the arrangement of the trichoblasts, covered by 2 presporangial cover cells.

Reference: Falkenberg (1901).

Only one species is included here. Although *L. subadunca* was reported to occur in Ireland in a mixed turf with *L. reptabunda* (Cullinane, 1970), we have been unable to confirm its presence in the British Isles.

Lophosiphonia reptabunda (Suhr ex Kützing) Kylin (1956), p. 359.

Holotype: L 955. 62. 97. Undated, unlocalized (? subtropical Africa).

Polysiphonia reptabunda Suhr ex Kützing (1843), p. 417.
Polysiphonia obscura sensu J. Agardh (1842), p. 123, non *Hutchinsia obscura* C. Agardh (1828), p. 108.

Thalli forming dense mats 0.5-2 cm thick, consisting of intertwined prostrate axes and numerous erect axes, often interconnected by rhizoids; erect axes simple or branched up to 4 orders; thalli dark brown in colour, with a fairly rigid texture.

All axes consisting of axial cell and 10-19 periaxial cells; *prostrate axes* curved downwards at tips, with large and conspicuous apical cells c. 15 µm in diameter, 70-150 µm in diameter when mature, composed of segments 0.8-1 diameter long, branching frequently at irregular intervals, often in pairs on opposite sides of the dorsal midline on successive or alternate segments; branches tapered to base with reduced numbers (10-13) of periaxial cells, and with the basal cell embedded within the bearing axis and covered by a single abaxial periaxial cell; *rhizoids* frequent, often paired on same segment, cut off medially or near posterior end of periaxial cells, 25-45 µm in diameter, becoming greatly elongate, occasionally branched, with digitate or discoid holdfast pads; *erect axes* growing from apical cells 12-15 µm in diameter, curved adaxially, up to 50 segments long before

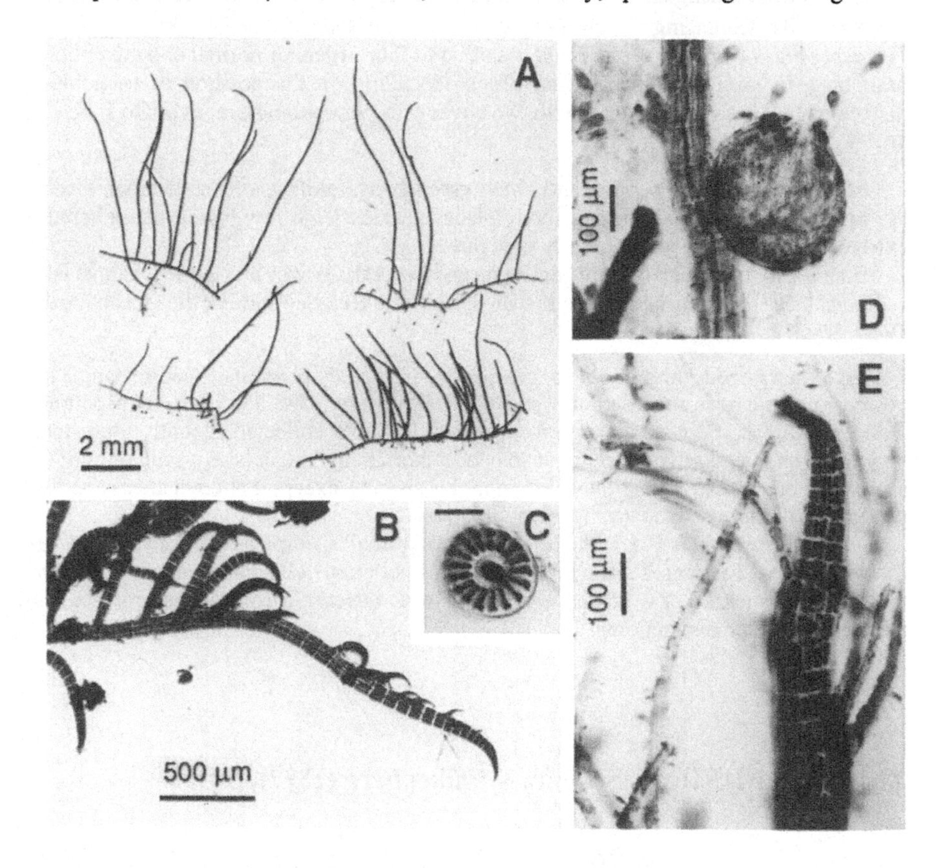

Fig. 120. *Lophosiphonia reptabunda*

(A) Pieces of thallus removed from algal turf, with several prostrate axes bearing erect axes (Dorset, Nov.). (B) Apex of prostrate axis with numerous forwardly-curved erect axes (B-C & E Dorset, Aug.). (C) T.S. of erect axis with 17 periaxial cells (scale bar = 50 µm). (D) Cystocarp (Dorset, Aug. 1883, in BM). (E) Apex of axis with trichoblasts and spirally arranged tetrasporangia.

initiating trichoblasts or branching, 70-100 μm in diameter, with segments 0.7-1 diameter long; *trichoblasts* initially abaxial, becoming spiral on every segment in a 1/4 divergence, up to 40 μm in diameter and 2.1 mm long, branched 3-4 times, borne in conspicuous tufts; *plastids* discoid.

Spermatangial axes strongly incurved, with blunt apices (Buffham, 1888). *Cystocarps* angled towards bearing axis, globular, 250-310 μm in diameter, with irregularly arranged pericarp cells decreasing in size towards ostiole; carposporangia 75-100 x 25-40 μm. *Tetrasporangia* formed in erect branches of up to four orders of normal and adventitious branching, in long straight or spiral series depending on the position of trichoblasts, distorting the axes when mature, with two cover cells per sporangium, ovoid to spherical, 70-75 x 50-75 μm.

Growing on bedrock, pebbles and *Fucus vesiculosus* holdfasts on shaded rock faces in the upper intertidal zone and in muddy places near extreme low water, at sheltered to extremely sheltered sites; apparently very rare.

South coast of England (Cornwall, Devon and Dorset); Galway and Cork; Channel Isles.

British Isles to Portugal; Mediterranean; Black Sea, Red Sea and Pacific Ocean (Ardré, 1970; South & Tittley, 1986).

Large plants occur throughout the year, and are probably perennial. Spermatangia and cystocarps have been recorded only in August (Buffham, 1888 and BM); tetrasporangia noted in August, September and November. All plants collected recently were tetrasporangial and vegetative perennation may be important in British Isles populations of this species. There is relatively little variation in this species, except that erect axes can either be simple or bear a secund arrangement of laterals.

L. obscura resembles *Polysiphonia atlantica* and small specimens of *P. devoniensis* and *P. stricta* (q. v.) in habit, but the large number of periaxial cells distinguishes *L. obscura* from these species, all of which have 4 periaxials. *L. obscura* is sometimes confused with *P. opaca* (q. v. for distinctions).

Tribe CHONDRIEAE Schmitz & Falkenberg (1897), p. 432.

CHONDRIA C. Agardh, nom. cons.

CHONDRIA C. Agardh (1817), p. 443.

Type species: *C. tenuissima* (Goodenough & Woodward) C. Agardh (1817), p. 18 [= *C. capillaris* (Hudson) M. Wynne (1991), p. 317].

Thalli consisting of one to several terete to compressed or flattened erect main axes to

30(-45) cm high from a discoid holdfast and lateral stolons bearing haptera and often forming secondary erect axes, radially organized, composed of a central axis surrounded by inflated inner cortical cells resembling medulla and covered by a smooth outer cortical layer; branching irregularly radial to 3-4 orders, with each branch initially constricted at the base and either tapering at the tip or the tip blunt or recessed forming an apical pit; trichoblasts colourless, borne on exposed tips or emerging from apical depressions; adventitious branches formed in axils of pre-existing branches.

Growth by successive slightly oblique divisions of a dome-shaped apical cell, with each segmental cell cutting off a trichoblast initial from its high side and 5 periaxial cells in alternating sequence; trichoblasts arranged in a 2/5 right- or left-handed spiral, 5-6 times alternate-distichously to subdichotomously branched, deciduous except the basal cell which persists as circular scar cell; successive periaxial cells in staggered arrays, each normally cutting off 2 basal conjunctor cells that link to the periaxial cells below forming secondary pit connections; cortical filaments quadri-, tri-, and dichotomously branched up to 5 orders of filaments with all cells, including surface cells, forming secondary pit connections with the cells below; descending rhizoidal filaments formed from lower ends of periaxial and inner cortical cells and growing through the intercellular spaces; all cells elongating and expanding more or less simultaneously, with the periaxial cells becoming as long as the axial cells; lenticular or band-like thickenings developing in walls of periaxial and inner cortical cells in some species; indeterminate branches originating from the basal cells of trichoblasts or scar cells.

Gametophytes dioecious. Spermatangia developing on single or paired, discoid to reniform or bilobed plate-like axes formed laterally on the suprabasal cells of modified trichoblasts, each composed of up to 5 orders of axial filaments lying in a plane terminated by a 1-3 layered margin of inflated sterile cells; fertile segments cutting off initials on either side that branch in one plane forming a layer of spermatangial mother cells, each bearing 2-3 spermatangia. Procarps initiated from the fifth periaxial cell on suprabasal cells of trichoblasts, consisting of a lateral sterile group containing 6-12 cells, a 4-celled carpogonial branch, and a basal sterile group containing 2-8 cells; perfertilization pericarp well-developed, consisting of 9-12 filaments derived mostly from the third and fourth periaxial cells; auxiliary cell cut off after fertilization or the supporting cell functioning directly as an auxiliary cell; carpogonium evidently cutting off 2 connecting cells, one of which unites with the auxiliary cell; cells of the lateral and basal sterile groups each dividing once; gonimoblast initial bearing monopodially branched gonimoblast filaments and obovoid carposporangia formed terminally in pairs; fusion cell large, composed of the central cell, supporting cell, inner sterile cells, auxiliary cell and inner gonimoblast cells; pericarp globose to urceolate with a simple or protruding ostiole, consisting of over 18 axial filaments, with each segment bearing 2 periaxial cells and a layer of cortical cells; cystocarp subtended by a spur formed from the tip of the fertile trichoblast, or the spur indistinct or lacking. Tetrasporangia formed in whorls of 1-3(-4) beneath the outer cortical layer, initiated singly from the distal ends of elongated periaxial cells, adaxial, covered by

2 presporangial cover cells and subtended by an abaxial cortical filament, released individually through a pore formed by separation of the paired, superficial cover cells.

References: Kylin (1928), Gordon-Mills (1987), Gordon-Mills & Womersley (1987).

KEY TO SPECIES

1 Apices attenuate, terminating in an acute point, usually surrounded by trichoblasts .... *C. capillaris*
Apices obtuse, terminating in a shallow depression from which tufts of trichoblasts may protrude 2
2 Young axes show vivid turquoise iridescence when alive; diamond pattern lacking; wart-like aborted branch initials borne in a spiral on all axes (see Fig. 122B), some developing into rhizoidal holdfasts .... *C. coerulescens*
Not iridescent; diamond pattern visible through cortex of young axes (see Fig. 123B); aborted branch initials absent .... *C. dasyphylla*

Chondria capillaris (Hudson) M. Wynne (1991), p. 317.

Neotype: BM-K, ex Herb. Goodenough. Dorset (Weymouth) (see Wynne, 1991; Gordon Mills, 1987).

Ulva capillaris Hudson (1778), p. 571, non *Fucus capillaris* Hudson (1778), p. 591.
Fucus tenuissimus Withering (1796), p. 117 (see Wynne, 1991).
Chondria tenuissima (Withering) C. Agardh (1817), p. 18.

Thalli consisting of cylindrical erect axes arising singly or in dense tufts from solid discoid holdfast 2-4 mm in diameter; erect tufts 5-25 cm high, cylindrical to irregularly rounded in shape, with distinct main axes bearing laterals at irregular intervals in a spiral pattern, to 3-4 orders of branching; main axes 0.8-1.2 mm in diameter; ultimate laterals 0.25 mm in diameter; axillary branching abundant; laterals near holdfast growing out horizontally, curving downwards and reattaching by secondary rhizoidal holdfasts c. 0.5 mm in diameter; colour dull brownish-red, bleaching to pale yellow; plants fairly tough and cartilaginous in texture, decaying rapidly after collection.

Apices attenuate, terminating in conspicuous apical cell; *trichoblasts* borne in a spiral around young axes, each up to 15 μm in diameter, branched 5-6 times, leaving pale scar cells when shed that give a pitted appearance to young axes; *main axes* consisting in TS of axial cell and 5 rounded periaxial cells c. 125 μm in diameter, 2-3 layers of subcortical cells, frequently with lenticular wall thickenings, and a single layer of cortex; rhizoidal filaments formed in medulla of older axes; *cortical cells* elongate in surface view, 35-100 x 6-25 μm, with conspicuous secondary pit connections; *branches* borne in an irregularly spiral arrangement, spindle-shaped when young, those of last 2 orders markedly constricted basally; ultimate laterals 200-250 μm in diameter; *plastids* ribbon-like, radiating outwards in cortical cells.

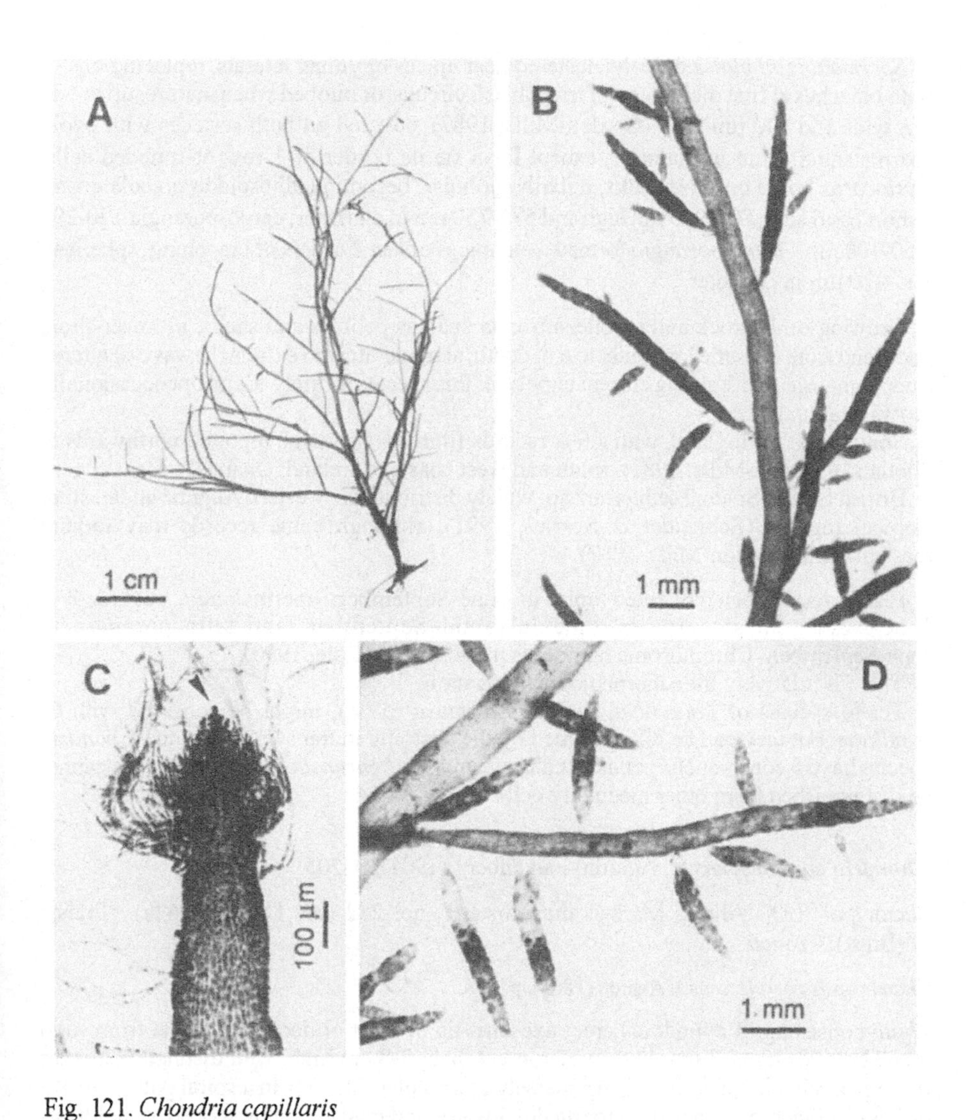

Fig. 121. *Chondria capillaris*

(A) Habit (Galway, June). (B) Axes bearing spindle-shaped young laterals (B & D Dorset, June). (C) Attenuate apex showing protruding apical cell (arrow) surrounded by trichoblasts (Cork, Nov.). (D) Tetrasporangia borne in spindle-shaped laterals.

Spermatangial plates densely clustered near apices of young laterals, replacing one or both branches at first dichotomy of trichoblast, circular or bilobed when mature, up to 500 μm wide and 330 μm long (Gordon-Mills, 1987), covered on both surfaces with ovoid spermatangia 4 μm in diameter, except for a sterile border of 1 row of rounded cells. *Cystocarps* borne on short stalks, initially globular, becoming ellipsoid to urceolate with a short basal spur, 800-950 μm high and 575-750 μm in diameter; carposporangia 170-290 x 50-100 μm. *Tetrasporangia* formed near apices of last 2 orders of branching, spherical, 100-120 μm in diameter.

Growing on bedrock and mobile substrata such as pebbles and shells, in lower-shore pools and from extreme low water to 6 m depth, at moderately to extremely wave-sheltered sites, sometimes with strong current exposure; this species is rarely found but occasionally grows in abundance.

South coast of England, with a few records from W. Scotland, reported northwards to Shetland (Gordon-Mills, 1987); south and west coasts of Ireland; Channel Isles.

British Isles to Spain; Mediterranean; widely distributed in western Atlantic and in other tropical regions (Schneider & Searles, 1991), although some records may require reassessment (Gordon-Mills, 1987).

Plants have been collected only in June-September; spermatangia recorded in June-August, cystocarps in June-July and September, and tetrasporangia in June-September. Chromosome number is n = c. 25 (see Cole, 1990).

There is relatively little morphological variation.

Terete species of *Laurencia*, such as *L. obtusa* (q. v.), might be confused with *C. capillaris*, but they can be differentiated by the anatomy in transverse section: *Chondria* species have 5 conspicuous periaxial cells, whereas in *Laurencia* the periaxial cells cannot be distinguished from other medullary cells.

Chondria coerulescens (J. Agardh) Falkenberg (1901), p. 205.

Lectotype: LD, 'Algues Marines du Finistère', no. 282 (see Dixon, 1962a). France (Brest), *Crouan*.

Chondriopsis coerulescens J. Agardh (1863), p. 808.

Thalli consisting of cylindrical erect axes arising in erect or decumbent tufts from solid lobed holdfast up to 7 mm in diameter; erect thalli 3-8 cm high, with distinct main axes 0.4-0.5 mm in diameter, branching sparsely at irregular intervals in a spiral pattern to 1-3 orders of branching; branches curving downwards and reattaching by secondary holdfasts that may form stolon-like outgrowths; dark purplish-brown with a striking turquoise iridescence when alive, flexible and cartilaginous in texture.

Apices obtuse, terminating in circular depression 75-100 μm in diameter, sometimes with protruding trichoblasts up to 10 μm in diameter, branched 2-4 times; *erect axes* consisting in TS of rounded axial cell and 5 periaxial cells c. 100 μm in diameter,

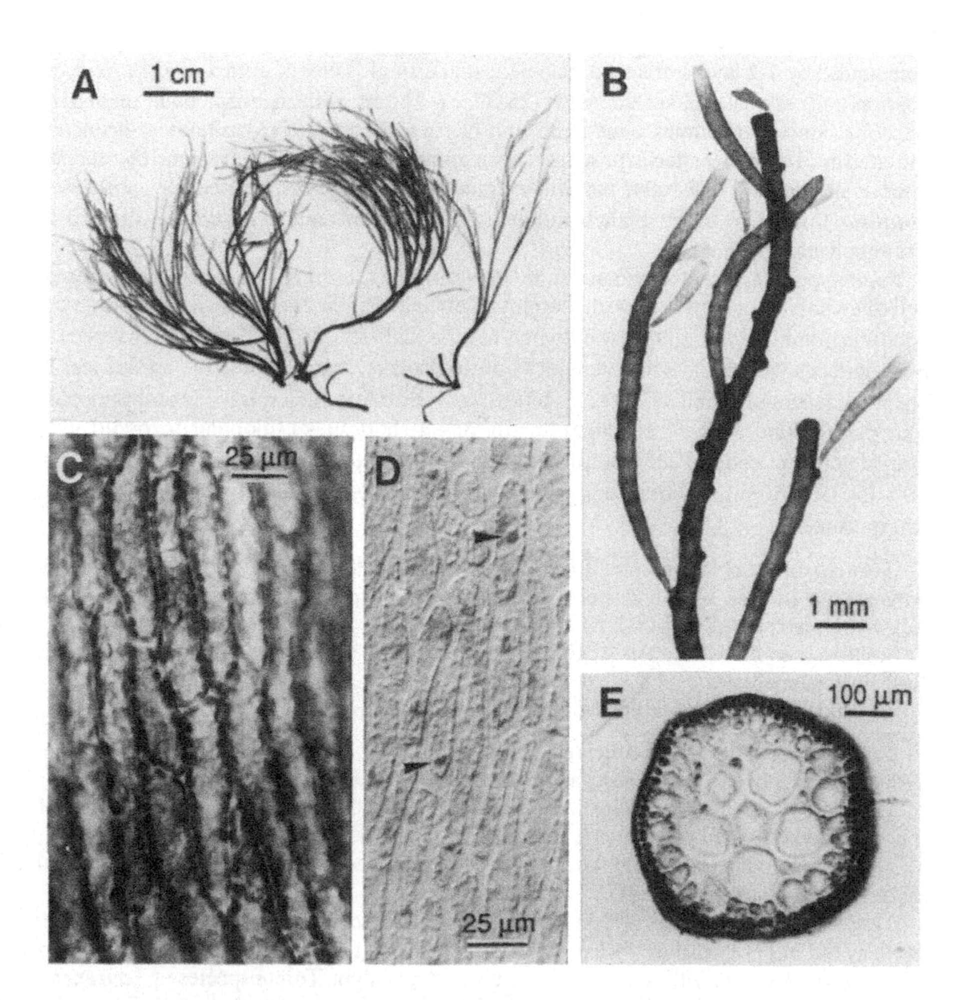

Fig. 122. *Chondria coerulescens*

(A) Habit (A & D Hampshire, undated). (B) Young axes showing clavate young laterals and spirally-arranged knob-like initials of undeveloped laterals (B-C & E Cornwall, Oct.). (C) Cortex with elongate outer cortical cells. (D) Cortex showing spherical hyaline bodies (arrows) in outer cortical cells. (E) T.S. of erect axis with 5 periaxial cells readily distinguishable from other medullary cells.

surrounded by 1-2 layers of subcortical cells and a single layer of cortex 12-25 µm deep; *cortical cells* elongate in surface view, 25-85 x 6-16 µm, with inconspicuous secondary pit connections, sometimes containing one or more spherical brown bodies; *branches* initiated in a 1/5 spiral pattern, most initials remaining remain wart-like, replaced by clavate laterals at irregular intervals; mature laterals markedly constricted basally; *rhizoidal holdfasts* formed by some branch initials, c. 300 µm in diameter; *plastids* discoid to irregularly elongate.

Spermatangial plates formed in a spiral on young axes, terminal on elongate suprabasal cell of trichoblast or borne at first dichotomy of branched trichoblast, initially heart-shaped, becoming circular or slightly lobed when mature, 250-525 µm in diameter, covered on both surfaces with ovoid spermatangia 4 µm in diameter, except for a 1-2 -celled sterile border of rectangular cells 13-30 x12-16 µm, soon shed leaving a spiral of non-pigmented scar cells around axes. *Cystocarps* sessile or on short stalks, globular to ovoid with non-protuberant ostiole, 725-800 µm high and 650-800 µm in diameter; carposporangia 100-150 x 30-50 µm. Tetrasporangia formed near apices of laterals, spherical, 130-150 µm in diameter.

Growing on pebbles in mud from extreme low water to 4 m depth, at extremely wave-sheltered sites, usually in inlets with moderate to strong current exposure; although *C. coerulescens* is rarely found, it can be abundant in favourable habitats.

South and south-east coasts of England: S. Cornwall, S. Devon, Hampshire, Sussex, Suffolk; Channel Isles.

England to Morocco; Mediterranean (Ardré, 1970).

Large plants have been collected in May-October, bearing spermatangia in June, cystocarps in July and September, and tetrasporangia in June and September.

Morphology shows relatively little variation except that some thalli consist only of inconspicuous isolated erect axes whereas others form dense tufts.

The characteristic iridescence results from the presence in the cortex of hemispherical aggregations of small spheres (G. Feldmann, 1970).

C. scintillans G. Feldmann (1964, p. 45), another iridescent species, was described from Brittany but has not yet been found in the British Isles. It is much larger and more robust than *C. coerulescens*, with rigid erect axes up to 20 cm high. Terete species of *Laurencia*, such as *L. obtusa* (q. v.) might be confused with *C. coerulescens*, but they can be separated by the anatomy in transverse section, as *Chondria* species have 5 conspicuous periaxial cells, whereas in *Laurencia* the periaxial cells cannot be distinguished.

Chondria dasyphylla (Woodward) C. Agardh (1822), p. 350.

Neotype: BM-K. Essex (Yarmouth), undated, *Turner*.

Fucus dasyphyllus Woodward (1794), p. 293.
Laurencia dasyphylla (Woodward) Greville (1830), p. 112.

Fig. 123. *Chondria dasyphylla*

(A) Habit (Galway, June). (B) Detail of thallus bearing numerous cystocarps, showing diamond patterning visible through cortex, and obtuse apices of laterals (B-D Donegal, July). (C) Optical L.S. of young axes showing regular arrangement of periaxial cells and trichoblasts protruding from the obtuse apex. (D) Spermatangial plates at tip of young axis. (E) Tetrasporangia borne in clavate laterals (Dorset, June).

Thalli consisting of cylindrical erect axes arising singly or in dense tufts from solid discoid holdfast up to 5.5 mm in diameter; erect thalli 6-21 cm high, irregularly rounded or pyramidal in shape, with distinct main axes, bearing laterals at irregular intervals in a spiral pattern, to 3-4 orders of branching, usually decreasing in length upwards; main axes 0.7-1.5 mm in diameter; ultimate laterals 0.5 mm in diameter; axillary branching abundant; laterals near holdfast growing out horizontally, curving downwards and reattaching by secondary rhizoidal holdfasts c. 0.5 mm in diameter; colour bright purplish-red, bleaching to pale yellow; old axes tough and cartilaginous, younger ones with a softer, fairly delicate texture when fresh and decaying rapidly after collection.

Apices obtuse, terminating in circular depression, sometimes with protruding trichoblasts up to 20 μm in diameter, branched 5-6 times; *main axes* with a conspicuous diamond pattern visible through the cortex of young axes, consisting in TS of axial cell, 5 rounded periaxial cells c. 100 μm in diameter, 3-5 layers of subcortical cells, frequently with lenticular wall thickenings, intermixed with rhizoidal filaments in older axes, and a single layer of cortex; *cortical cells* elongate in surface view, 30-150 x 14-32 μm, with conspicuous secondary pit connections; *branches* borne in an irregularly spiral arrangement every 2-8 segments, clavate when young, markedly constricted basally in last 2 orders of branching; ultimate laterals 250-550 μm in diameter; *plastids* ribbon-like, reticulate, arranged radially in young cells.

Spermatangial plates densely clustered near apices of young laterals, terminal on elongate suprabasal cell of trichoblast or borne at first dichotomy of branched trichoblast, circular or bilobed when mature, 400-600 μm wide and 300-400 μm long, covered on both surfaces with ovoid spermatangia 4 μm in diameter, except for a 1-celled sterile border of rectangular cells 20-35 x 20-30 μm, soon shed leaving a spiral of non-pigmented scar cells around axes. *Cystocarps* borne on short stalks, initially globular, becoming ellipsoid to slightly urceolate, 900-1075 μm high and 650-725 μm in diameter; carposporangia 150-300 x 75-110 μm. *Tetrasporangia* formed near apices of last 2 orders of branching, spherical, 150-200 μm in diameter.

Epilithic on bedrock, more frequently on mobile substrata such as pebbles, shells and maerl, tolerant of sand cover, in mid- to lower-shore pools and from extreme low water to 20 m depth, occasional on wave-exposed bedrock lacking kelp forest but rarely under kelp canopy, common at wave-sheltered sites with moderate to strong current exposure; particularly abundant in *Zostera marina* beds.

Widely distributed in the British Isles, reported northwards to Shetland (Gordon-Mills, 1987), with few records from east coasts of Scotland and N. England; Channel Isles.

Sweden to Spain; Mediterranean; widely distributed in tropical and temperate regions (Schneider & Searles, 1991), although some records may require reassessment (Gordon-Mills, 1987).

Regenerating holdfasts and small erect axes occur throughout the year on mobile substrata, mature thalli developing from April onwards; sporelings were noted in

June-September. Spermatangia recorded in April-October, cystocarps in May-August and October, and tetrasporangia in April and June-October. Chromosome number is $n = 31$ (see Cole, 1990).

There is relatively little morphological variation.

Terete species of *Laurencia*, such as *L. obtusa* (q. v.), might be confused with *C. dasyphylla*, but they can be separated by the anatomy in transverse section, as *Chondria* species have 5 conspicuous periaxial cells, whereas in *Laurencia* the periaxial cells cannot be distinguished.

Tribe LAURENCIEAE Schmitz (1889), p. 447.

LAURENCIA Lamouroux, nom. cons.

LAURENCIA Lamouroux (1813), p. 41.

Type species: *L. obtusa* (Hudson) Lamouroux (1813), p. 42 [see Schmitz (1889), p. 447].

Thalli consisting of one to several terete to compressed or flattened erect main axes to 30 cm high from a discoid holdfast and lateral stolons bearing haptera and secondary erect axes, radially organized, composed of inflated inner cortical cells resembling medulla and a smooth outer cortical layer; texture soft and fleshy to firm and cartilaginous; branching irregularly radial, alternate-distichous, subdichotomous or whorled up to 4(-5) orders, with the branches tapering towards the base and the tips blunt or swollen with prominent apical pits; trichoblasts colourless, borne in the apical depressions.

Growth by successive oblique divisions of a 3-sided apical cell situated in an apical depression, each segmental cell cutting off a trichoblast initial from the high side and two lateral periaxial cells, one on either side; trichoblasts arranged in a 2/5 right- or left-handed spiral, 5-6 times alternate-distichously to subdichotomously branched, deciduous except the basal cell; axial cells remaining short, the central axis soon obscured; periaxial cells bearing 5(-6) orders of cortical fialaments, with the outermost cells forming the surface cortical layer; basal cells of trichoblasts, periaxial cells and inner cortical cells expanding radially a short distance below the apex, and interconnecting by secondary pit connections; surface cells either connected longitudinally by secondary pit connections, or secondary pit connections rare or absent in the surface layer; rhizoidal filaments evidently present, growing between the cortical filaments; indeterminate branches originating from the basal cells of trichoblasts.

Gametophytes dioecious. Spermatangial axes in obconical to cup-shaped terminal receptacles or in deep pockets along the sides of fertile branches, borne laterally on suprabasal cells of modified trichoblasts, either naked or subtended by a trichoblast filament and terminated by 1-3 obovoid to obconical sterile cells; each fertile segment cutting off 4 periaxial cells bearing quadri-, tri-, and dichotomous filaments and forming a single layer of spermatangial mother cells, each bearing 2-3(-4) spermatangia. Procarps formed in apical depressions from the fourth (adaxial) periaxial cell on the suprabasal cell of fertile trichoblasts, consisting of a lateral sterile group containing 4-6 cells, a 4-celled carpogonial branch, and a basal sterile group containing 2-3 cells, displaced progressively towards the margin of the apical depression and outer surface of the fertile branch; prefertilization pericarp derived primarily from the two lateral periaxial cells in the fertile segment; auxiliary cell cut off after fertilization or the supporting cell functioning directly as an auxiliary cell; cells of the lateral and basal sterile groups each dividing once; gonimoblast filaments initially monopodially branched, shifting to sympodial branching during carposporangial formation; carposporangia terminal, clavate; pericarp 4-6 cells thick, ostiolate, formed by axial filaments in which each segment bears 2 periaxial cells and 2 or more layers of cortical cells; fusion cell large, incorporating the supporting cell, auxiliary cell, central cell, inner pericarp cells and inner gonimoblast cells; cystocarps lateral on the sides of fertile branches. Tetrasporangia initiated singly at the distal ends of periaxial cells while still inside the apical depression and displaced apically and lying outside the pit prior to cleavage; either adaxial or abaxial and covered by two presporangial cover cells and a cortical filament; mature tetrasporangia either oriented parallel to the axis of growth and the tip undisturbed, or at right angles to it, causing the tip to swell or become crescent-shaped.

References: Kylin (1923, 1928), Saito (1967, 1969, 1982), Saito & Womersley (1974).

Useful features for identification of *Laurencia* species include the holdfast, which should be collected if possible, and details of the outer cortical layer, such as secondary pit connections visible in surface view, the presence or absence of spherical hyaline bodies called 'corps en cerise' (although these are ephemeral in fixed material), and the relative size and pigmentation of the outermost and adjacent cortical layers.

Four species of *Laurencia* are listed in recent British Isles checklists (South & Tittley, 1986). One of these, *L. platycephala*, is typified by a specimen of *L. hybrida*. Magne (1980) reported that the lectotype of *L. platycephala* lacked secondary pit connections in the outer cortex. We have confirmed this observation, and conclude that this specimen is *L. hybrida*. The species currently known as *L. platycephala* appears instead to be *L. truncata*, described from the Mediterranean. In addition, we have included two further species, *L. osmunda* and *L. pyramidalis*. *L. osmunda* has been the subject of taxonomic confusion since shortly after its initial description. Although Stackhouse (1801, p. 46, pl. 11) appreciated the diagnostic significance of its discoid holdfast without stolons, in contrast to the stoloniferous holdfast of *L. pinnatifida*, subsequent confusion with *L. hybrida* obscured this distinction. *Laurencia pyramidalis* has been treated as a variety of *L. obtusa* since its proposal, although J. Agardh (1852) noted that it could well represent a separate species. In the British Isles, it is readily separable from *L. obtusa* by its morphology, phenology, ecology and distribution.

Nam (pers. comm.) has proposed that all British species other than *L. obtusa* and *L. pyramidalis* should be removed from *Laurencia* on the basis of fundamental anatomical differences, and has suggested that the genus *Osmundea* Stackhouse may be available for these species.

KEY TO SPECIES

1 Holdfast discoid, lacking stolons (see Fig. 124D) 2
  Holdfast stoloniferous or tangled (see Fig. 127A) 4
2 Secondary pit connections present in outer cortex (see Fig. 129B); main axes compressed, laterals compressed to terete; spermatangial receptacles terminal, cup-like; thalli usually epiphytic .... *L. truncata*
  Secondary pit connections absent in outer cortex; thalli very rarely epiphytic 3
3 Main axes terete or slightly compressed, apices with circular pit; spermatangial receptacles terminal, open and cup-like .... *L. hybrida*
  Main axes flattened, apices with terminal groove; spermatangial receptacles lateral, deep urn-shaped cavities .... *L. osmunda*
4 Secondary pit connections absent in outer cortex; holdfast stolon-like, axes compressed or rarely terete, apices with terminal groove; spermatangial receptacles lateral, deep urn-shaped cavities; plants forming extensive turfs on rock .... *L. pinnatifida*
  Secondary pit connections and spherical hyaline bodies present in outer cortex; holdfast tangled, axes terete, apices with circular depression; spermatangial receptacles terminal; plants epiphytic 5
5 Thalli rigid, brittle, colour reddish-orange to bright orange; in shallow mid-shore pools and subtidal, 4-12 (-16) cm high .... *L. obtusa*
  Thalli soft, flexible, colour purplish-brown; at extreme low water, 12-20 cm high .... *L. pyramidalis*

Laurencia hybrida (A. P. de Candolle) Lenormand ex Duby (1830), p. 951.

?Holotype: PC, ex Herb. Montagne. France (?Gravure, Calvados) (F. Ardré, pers. comm.).

Fucus hybridus A. P. de Candolle (1815), p. 30.
Fucus pinnatifidus var. γ *angustus* Turner (1802), p. 268.
Laurencia caespitosa Lamouroux (1813), p. 43.
Laurencia platycephala Kützing (1865), p.23.

Thalli attached by discoid holdfast up to 15 mm in diameter, lacking stolons; erect axes arising in dense clumps 2-8.5 (-15) cm high, with distinct main axes bearing distichous or spirally arranged, irregularly alternate to opposite but not whorled lateral branches, producing a pyramidal outline 0.5-5 cm wide; *main axes* cylindrical to slightly compressed, 0.5-1.2 mm wide x 0.5-0.8 mm thick, denuded below in larger thalli by erosion of lower laterals; *first-order laterals* slightly compressed, bearing a further 1-2 orders of branching in same arrangement; *ultimate branchlets* terete, densely clustered; colour deep purplish-brown to bright green or yellowish; texture cartilaginous.

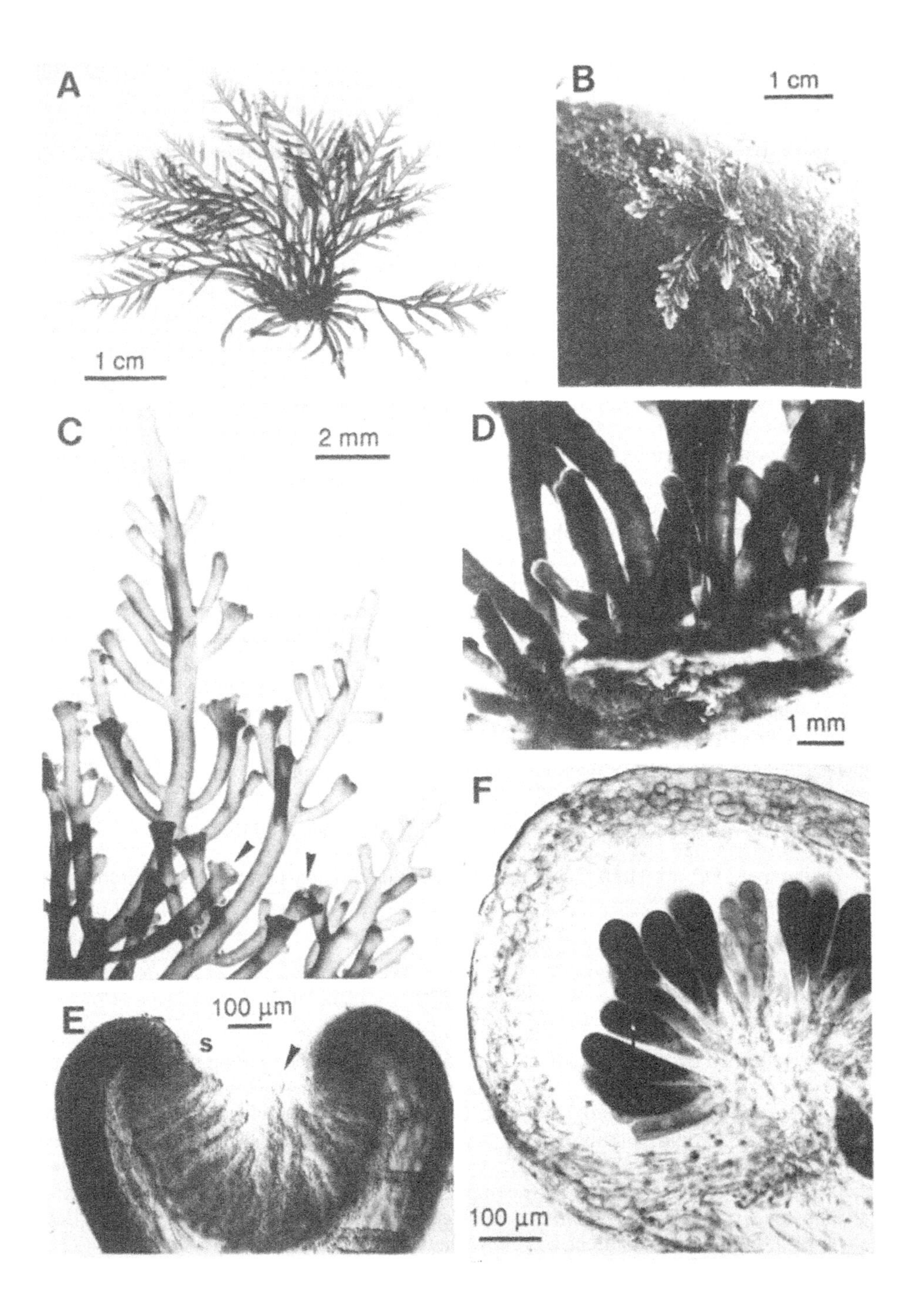
A
1 cm
B
1 cm
C
2 mm
D
1 mm
E
100 µm
s
F
100 µm

Axes growing from apices with circular terminal depression, consisting in TS of pseudoparenchymatous medulla of thin-walled cells 20-40 µm diameter, lacking lenticular wall thickenings, merging into 4-5 layers of subcortical cells, surrounded by a cortical layer 16-35 µm deep; *outer cortex* conspicuously less pigmented and smaller than inner cortex, with cells polygonal to elongate in surface view, 26-76 x 16-32 µm, lacking secondary pit connections,

Spermatangial receptacles terminating lateral branches that later renew growth on either side of the receptacle and form further terminal receptacles, developing as shallow cups, maturing to discoid plates 1-1.7 mm wide, containing numerous branched or unbranched spermatangial axes; spermatangial axes 250-400 µm long x 45-65 µm wide, terminating in groups of 3-20 spherical sterile cells containing bright yellow pigment, covered with spermatangial mother cells bearing densely packed ovoid spermatangia 7-10 x 5-6 µm; released spermatia spherical, 4-6 µm in diameter. *Cystocarps* sessile, ovoid with non-protuberant ostiole, 0.6-1 mm in diameter, containing clavate carposporangia 110-170 x 50-70 µm. *Tetrasporangia* formed in last 2 orders of branching, in terete branchlets, developing adaxially, 85-110 µm in diameter when mature.

Epilithic on bedrock and pebbles, epizoic on limpets, occasionally epiphytic on *Fucus* spp., in pools and on open shaded rock from upper mid-shore to lower shore, but never subtidal, occasionally forming dense turf in sandy pools, often mixed with *Laurencia pinnatifida*, at moderately sheltered to extremely wave-exposed sites.

Generally distributed in the British Isles, northwards to Shetland.

British Isles to S. Portugal (Albufeira, 23 iii 1989, *Maggs*). Reports from Scandinavia (e.g. Lyngbye, 1819, p. 40, as *Gelidium pinnatifidum* var. *angustum*) are based on other species of *Laurencia*.

Annual or perennial; at Clare Island (Mayo), sporelings appeared in September, developed during winter, reproduced in spring and disappeared in May (Cotton, 1912). In general, in the British Isles, small plants and young erect axes produced by perennating holdfasts are noticeable in October; spermatangia occur in November-June, cystocarps develop in April and mature in May-July, and tetrasporangia are present in December-July. Chromosome number is $n = 31$ (Austin, 1956).

Fig. 124. *Laurencia hybrida*

(A) Habit (Pembroke, Dec.). (B) Habit on rock, showing erect axes radiating from small holdfast (Down, Nov.). (C) Detail of male thallus with irregularly arranged laterals, some terminating in cup-like spermatangial receptacles (arrows) (as A). (D) Discoid holdfast bearing numerous erect axes of different ages (Galway, Feb.). (E) L.S. of spermatangial receptacle containing numerous spermatangial axes, some terminating in large sterile cells (arrow), with clouds of released spermatia (s) (Down, Jan.). (F) L.S. of cystocarp containing large clavate carposporangia (Dublin, May).

Overall size varies greatly with habitat. On open rock, plants are less than 5 cm high, tufted and densely branched; plants in pools are laxer and up to 8.5 (-15) cm long. The bright green coloration on main axes, which is often regarded as characteristic of this species (Cotton, 1912), results mainly from green algae endophytic in the outer cortex.

L. hybrida and other terete species of *Laurencia* could be confused with *Chondria dasyphylla* (q. v.), but they lack the distinctive ring of 5 periaxial cells seen in *Chondria* species.

Laurencia obtusa (Hudson) Lamouroux (1813), p. 42.

Lectotype: BM-K, ex Herb. Hudson. Undated, unlocalized.

Fucus obtusus Hudson (1778), p. 586.
Chondria obtusa C. Agardh (1822), p. 340.

Thalli attached by tangled stolonous holdfast, consisting of dense irregularly shaped clumps of erect axes 4-16 cm high; *main axes* distinct, terete, 1-1.2 mm in diameter, bearing laterals spirally, often in subopposite pairs or threes; *first-order laterals* terete, 0.7-0.9 mm in diameter, occasionally equalling the main axis in length, shorter towards apices, producing a pyramidal outline, bearing 3 further orders of branching in the same spiral to subopposite arrangement; *ultimate branchlets* cylindrical to clavate, 0.3-0.4 µm in diameter; colour brownish-orange to bright orange; texture crisp and brittle, except for lower main axes which may be quite flexible.

Axes growing from circular terminal depression, consisting in TS of thin-walled medullary cells, 45-130 µm in diameter and lacking lenticular wall thickenings, intergrading with subcortical cells, and inner and outer cortical layers; *outer cortex* composed of obconical to radially elongate cells with a slightly to markedly curved outer wall, 40-100 µm long, elongate-polygonal in surface view, 30-130 x 20-55 µm, with secondary pit connections and a single spherical body in each cell visible in fresh and recently preserved specimens, often eroded on old main axes, allowing the subcortex to be invaded by green endophytes; *plastids* discoid in young cells, becoming narrow and reticulate.

Spermatangial receptacles terminal on lateral branchlets, often in compound clusters, swollen, 1.1-1.5 mm in diameter, with a conspicuous axial filament and numerous dichotomously branched spermatangial axes bearing ovoid spermatangia measuring 8 x 6 µm. *Cystocarps* borne on last 2 orders of branching, sessile, globular, 600-1000 µm diameter, with prolonged ostiole 100-120 µm long; carposporangia 130-160 x 55-65 µm. *Tetrasporangia* borne in last 3 orders of branching, mostly in little-differentiated short clavate to cylindrical ultimate branchlets, cut off abaxially, 120-145 µm in diameter when mature.

Epiphytic, very rarely on dead shells, typically forming dense stands in extensive shallow mid-shore pools and occasionally near extreme low water on fairly sheltered to fairly exposed shores; rarely subtidal to 6 m depth at extremely sheltered sites, where plants may be loose-lying.

South and west coasts northwards to Wigtown (L. Ryan) and Ayr, rare on east coasts (Norfolk); reports from localities northwards to Orkney and on north-east coasts (Batters

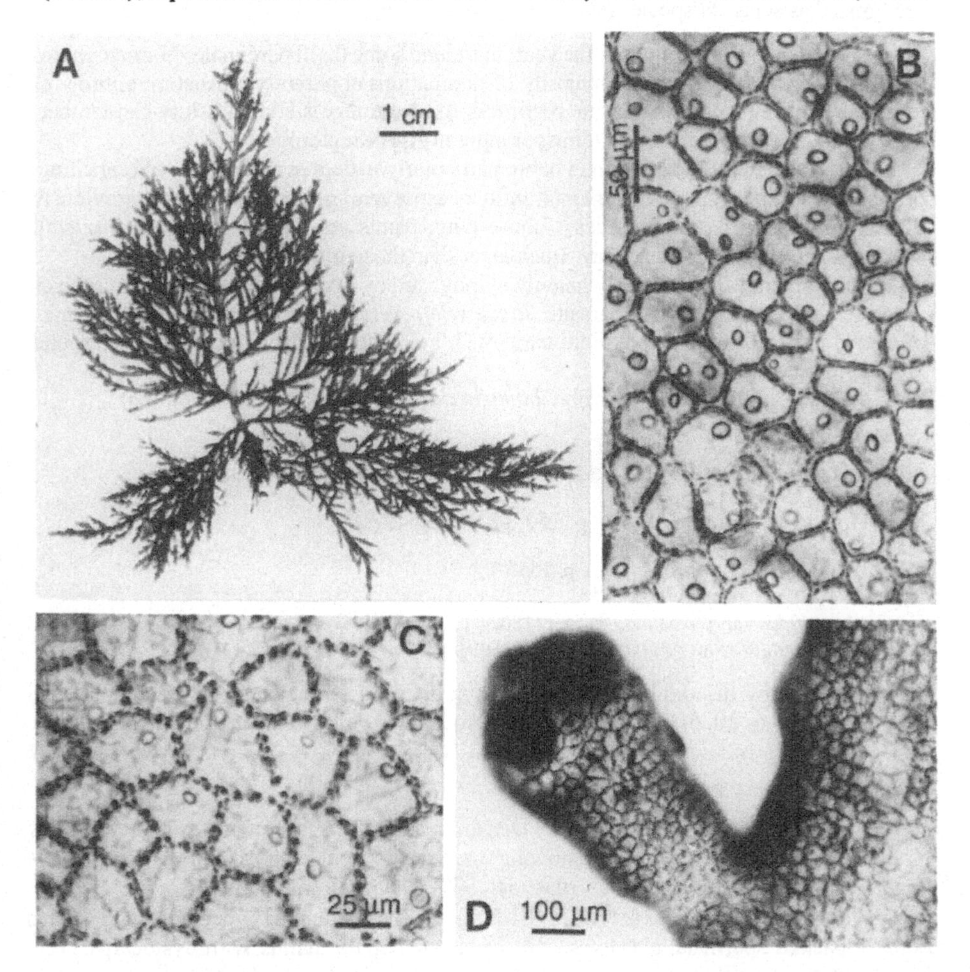

Fig. 125. *Laurencia obtusa*

(A) Habit (Donegal, Nov.). (B) Outer cortex focused on spherical hyaline bodes ("corps en cerise") (B-D Donegal, July). (C) Outer cortex focused on beaded plastids. (D) Tetrasporangia borne at apex of lateral.

1902) require confirmation; widely distributed in Ireland although at few localities; Channel Isles.

British Isles to Canaries (Ardré, 1970); Mediterranean. Wider distribution impossible to assess because several algae previously considered as varieties of *L. obtusa* are now recognized as separate species (see Saito & Womersley, 1974).

Large plants found throughout the year; in Ireland some thalli perennate. Gametophytes are usually very rare, the great majority of populations apparently consisting entirely of tetrasporophytes. Spermatangia recorded for February-May and July-September; cystocarps in July-October; and tetrasporangia in April-September.

There is considerable variation in habit, particularly in degree and pattern of branching, probably resulting from environmental influences; several growth forms were previously distinguished as separate varieties. Loose-lying plants are almost spherical in overall shape, and subtidal individuals are much laxer than those in tide-pools.

Intertidal plants are frequently heavily epiphytized by cyanobacteria and algae such as crustose Corallinaceae. The parasite *Janczewskia verrucaeformis* Solms occurs on *L. obtusa* in the Mediterranean (J. Feldmann & G. Feldmann, 1958) but has not been found on Atlantic coasts.

L. obtusa could be confused with *Chondria dasyphylla* (q. v. for distinctions).

Laurencia osmunda (S. G. Gmelin) Maggs & Hommersand, comb. nov.

Lectotype: S. G. Gmelin (1768), p. 155, pl. 16, fig. 2*.

Fucus osmundus S. G. Gmelin (1768), p. 155.
Fucus multifidus Hudson (1778), p. 581.
Fucus pinnatifidus var. β *osmunda* Turner (1802), p. 267.
Laurencia pinnatifida var. *osmunda* Kützing (1849), p. 856.

Thalli attached by discoid holdfast 5-10 mm in diameter; erect axes compressed, arising in groups of up to 20, 5-14(20) cm high and to 25 cm broad; *main axes* distinct, 2-4 mm

Fig. 126. *Laurencia osmunda*

(A) Habit (A & C-F Antrim, Feb.). (B) Discoid holdfast bearing several young erect axes (Galway, Feb.). (C) T.S. of cortex, showing two outer cortical layers of cells approximately equal in size, with conspicuous plastids. (D) Detail of male thallus with urn-like spermatangial receptacles (arrows). (E) L.S. of spermatangial receptacle containing densely-packed spermatangial axes. (F) Detail of female thallus with cystocarps. (G) Detail of tetrasporophyte with conspicuous tetrasporangia (Down, June).

* S. G. Gmelin's herbarium is destroyed.

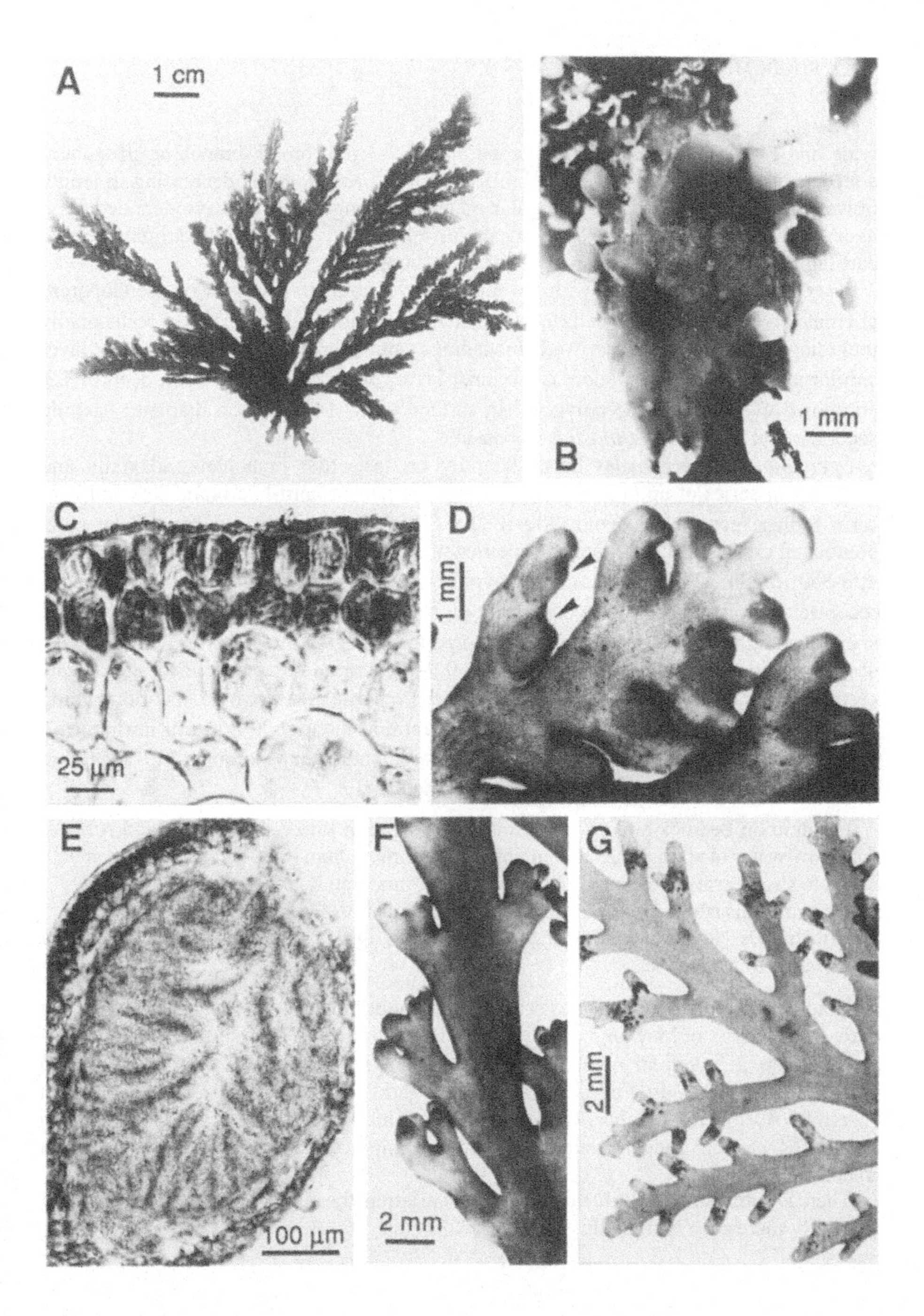
A
1 cm
B
1 mm
C
25 µm
D
1 mm
E
100 µm
F
2 mm
G
2 mm

wide and c. 1.2 mm thick, with flattened apices 2-3 mm broad, branching irregularly alternate-distichously to 6 orders of branching; *first-order laterals* decreasing in length upwards, producing a broadly triangular outline, eroding below to leave denuded main axes; colour brownish-pink to deep purple, apices red in transmitted light; texture cartilaginous, thalli decaying rapidly after collection.

Axes growing from apices with terminal groove, composed in TS of pseudoparenchymatous medulla of ovoid cells up to 120 x 70 µm, with walls 2 µm thick, occasionally including lenticular refractive wall thickenings, and 2-layered cortex, the outer layer similar in size and pigmentation to the inner layer; *outer cortex* 16-28 µm deep in TS, composed of cells elongate-polygonal in surface view, 14-35 µm in diameter, lacking secondary pit connections; *plastids* ribbon-like.

Spermatangial receptacles borne laterally on last-order branchlets, adaxially and abaxially in series of up to 3, or singly between a pair of horn-like branchlets that later form further receptacles, ovoid, 900-1080 x 600-900 µm, with a narrow apical pore, containing densely packed branched spermatangial axes; spermatangial axes cylindrical, 200-350 µm long, bearing numerous ovoid spermatangia 10 x 7 µm, a few in each receptacle terminating in 1-4 obconical non-pigmented sterile cells c. 25 µm long; released spermatia slightly ovoid, 7 x 5 µm. *Cystocarps* formed laterally on last-order branchlets, sessile, ovoid with non-protuberant ostiole, 0.9-1.2 mm in diameter, containing clavate carposporangia 140-170 x 100 µm. *Tetrasporangia* borne in last two orders of branching in branchlets that are initially compressed, becoming terete, 600-850 µm in diameter; tetrasporocytes cut off adaxially near branch apices; mature tetrasporangia 110-145 µm in diameter; released tetraspores 70-90 µm in diameter.

Epilithic on bedrock and epiphytic on larger algae in lower-shore pools and subtidal from low water of spring tides to 5 m, intolerant of more than brief emersion, at sites with slight to strong wave exposure, most abundant at moderately exposed sites.

Generally distributed in the British Isles, northwards to Orkney.

British Isles, Netherlands (Stegenga & Mol, 1983, fig. 106.3), Atlantic France, possibly southwards to Morocco.

Holdfasts perennial, new blades arising throughout the year, reaching their maximum size in January to May and decaying in late summer. Tetrasporophytes usually predominate in intertidal populations. Spermatangia present in December-June and October, cystocarps in February-September and November and tetrasporangia in December-October. A *Polysiphonia*-type isomorphic life history has been observed in culture (M. D. Guiry, unpublished). We have determined the chromosome number in Irish material as $n = 29$.

There is comparatively little morphological variation, the number of orders of branching generally increasing with thallus size.

Laurencia pinnatifida (Hudson) Lamouroux (1813), p. 42.

Lectotype: BM-SL, Herb. Petiver no. 405 (see Irvine & Dixon 1982). Essex (Harwich)*.

Fucus pinnatifidus Hudson (1762), p. 473.
Chondria pinnatifida (Hudson) C. Agardh (1822) p. 337.
Gelidium pinnatifidum (Hudson) Lyngbye (1819), p. 40.

Thalli forming extensive turfs attached by closely interwoven stolonous holdfasts; erect axes 2-8 cm high; *main axes* compressed, 1-2.5 mm wide and 0.6-0.8 mm thick, branching irregularly alternate-distichously to 4-5 orders, complanate, linear to flabellate in outline depending on length of first-order laterals, 1-8 cm broad, sometimes curving down towards substratum and giving rise to further stolons, frequently denuded below by erosion of laterals; almost black or brownish-purple to bleached yellow in colour, with apices greenish-brown in transmitted light; texture cartilaginous; fresh plants have a strong chemical smell and flavour.

Axes growing from apices with terminal groove, consisting in TS of pseudoparenchymatous medulla of thick-walled cells up to 80 μm diameter, with or without lenticular wall thickenings, merging through subcortex into 2-layered cortex; *outer cortical layer* conspicuously less pigmented and smaller than inner cortex; cells rounded to radially elongate, 8-16 μm deep, elongate-polygonal in surface view, 14-32 x 7-24 μm, lacking secondary pit connections; *plastids* few, discoid, in outer cortex.

Spermatangial receptacles borne laterally on ultimate branchlets, adaxially and abaxially in series of 1-3, or singly between a pair of horn-like branchlets that later form further receptacles, ovoid, 1000-1200 x 720-1120 μm, with a narrow apical pore, containing densely-packed branched spermatangial axes; spermatangial axes cylindrical, 130-465 μm long x 50-85 μm wide, covered with ovoid spermatangia measuring 10 x 4-6 μm, and terminating in 1-7 cylindrical to pyriform sterile cells. *Cystocarps* formed laterally on ultimate branchlets, sessile, often fused laterally to surrounding branchlets, slightly ovoid, 840-960 μm high x 660-840 μm wide; carposporangia 160-193 x 33-60 μm. *Tetrasporangia* borne in last three orders of branching, in branchlets that are initially compressed, becoming terete, 420-540 μm wide; tetrasporocytes cut off adaxially; mature tetrasporangia 75-125 μm in diameter.

Forming extensive turfs on bedrock in pools and in open situations, occasionally binding sand over rock, extending from shaded mid-shore crevices to low water of spring tides, never subtidal; common on exposed to moderately sheltered shores.

Generally distributed in the British Isles, northwards to Shetland.

British Isles to Portugal; Mediterranean. Wider distribution of *L. pinnatifida* is difficult to assess owing to taxonomic confusion. Norwegian, Swedish and Danish records appear

* Not Deal in Kent, see Dale (1732, p. 347).

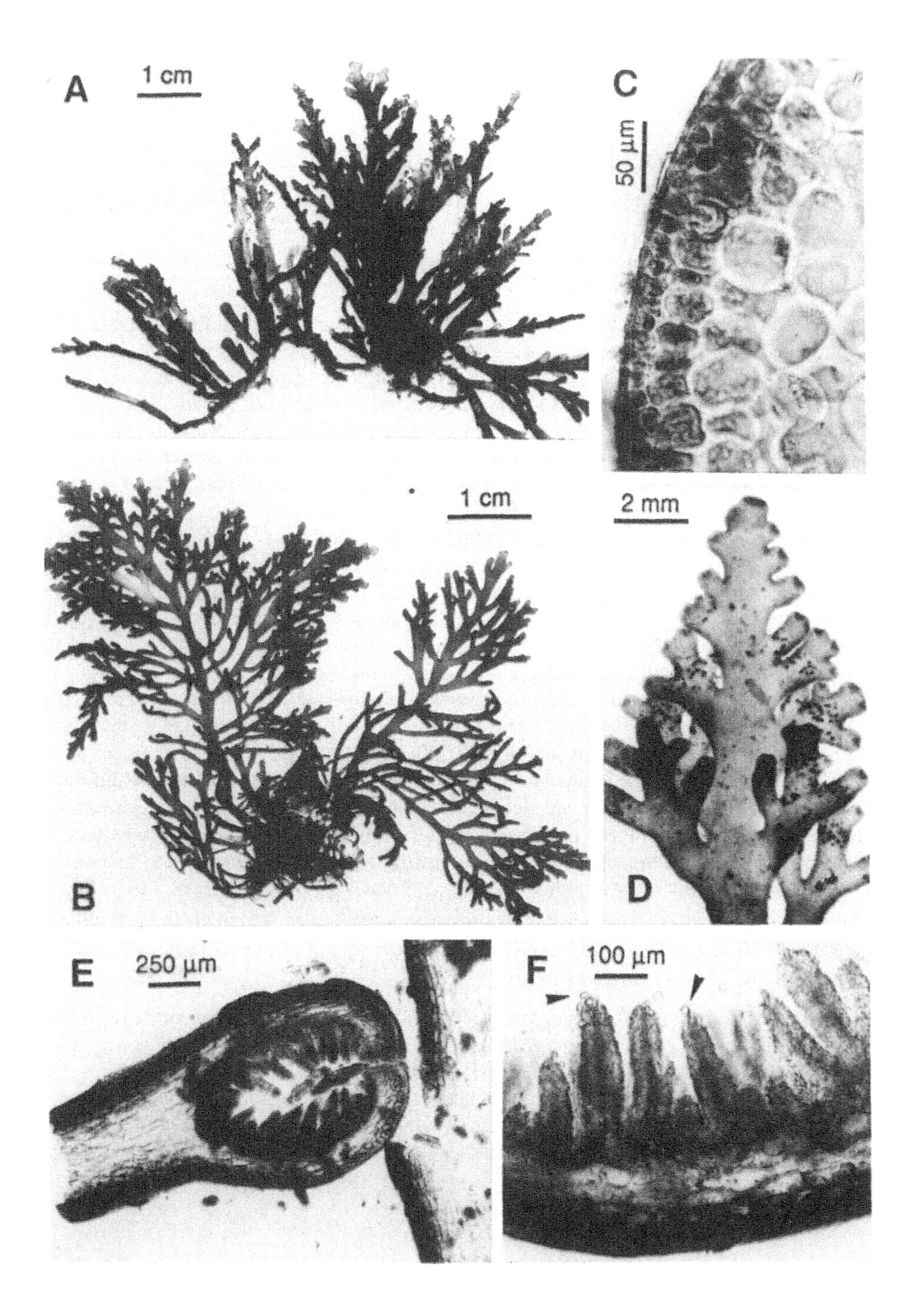
A
1 cm
C
50 µm
B
1 cm
2 mm
D
E
250 µm
F
100 µm

to be based on specimens of *L. truncata*; reported collections from the Faroes (see D. Irvine 1982) require re-examination. *L. pinnatifida* has also been reported southwards to Mauritania (Ardré 1970), but records south of Portugal require confirmation as some herbarium specimens resemble *L. osmunda.*

Turfs perennate in June-September as creeping mats of prostrate axes that continue to grow outwards, increasing the size of the turfs. Erect fronds develop from October onwards, reach their maximum size in February-May, and most are lost by the end of June. Spermatangia recorded in February-June; female plants are rare and cystocarps have been noted only in March and June; tetrasporangia present in October and December-June. Some reports of reproduction outside these periods may be based on observations of *L. osmunda* or *L. truncata*. Chromosome number in French material was reported as $n = 29$ (Magne 1964); we have confirmed this for Irish plants.

Morphology varies according to position on the shore. Main axes of lower-shore plants are usually 2-3 mm wide, whereas in mid-shore they are only 1 mm, and mat-forming thalli on sandy or muddy rock may be almost terete. Lower-shore stands are up to 8 cm high but mid-shore turfs are usually less than 3 cm thick.

Laurencia pyramidalis Bory ex Kützing (1849), p. 854.

Probable syntypes: L, LD, BM.* France (Cherbourg), *Lenormand.*

Laurencia obtusa var. *pyramidata* Bory ex J. Agardh (1852), p. 751.

Thalli attached by extensive tangled stolonous holdfast from which new axes arise, consisting of elongate, densely branched clumps of 3-10 erect axes 12-20 cm high; *main axes* distinct, terete, 900-1700 µm in diameter, bearing laterals spirally, often in groups resembling whorls separated by lengths of unbranched main axis; *first-order laterals* 850-1200 µm wide, decreasing in length upwards, bearing three further orders of branching

Fig. 127. *Laurencia pinnatifida*

(A) Habit, showing erect axes borne on prostrate stolon-like axes (Antrim, Feb.). (B) Habit of plant that may be a hybrid between *L. osmunda* and *L. pinnatifida* (Dorset, Nov.). (C) T.S. of cortex, showing cells of outermost cortical layer much smaller than underlying layer (Devon, March). (D) Detail of tetrasporophyte with conspicuous tetrasporangia (Donegal, May). (E) L.S. of urn-like spermatangial receptacle with narrow ostiole, containing densely-packed spermatangial axes (E-F Donegal, Feb.). (F) Part of spermatangial receptacle, showing spermatangial axes covered with spermatangia and terminating in sterile cells (arrows).

* No specimens have been found in L that were obviously examined by Kützing.

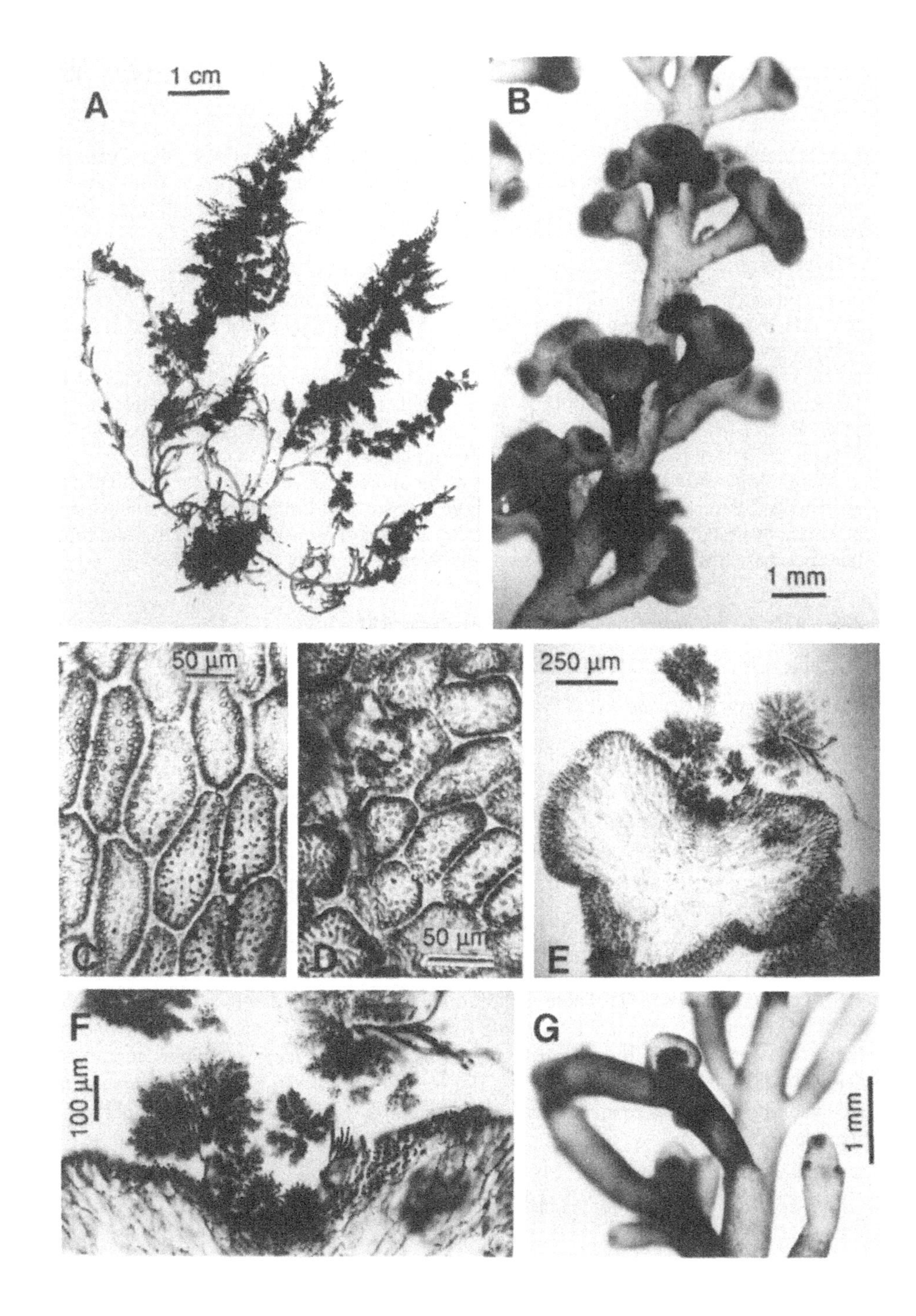
1 cm
A
B
1 mm
50 µm
50 µm
250 µm
C
D
E
F
100 µm
G
1 mm

in the same arrangement, with a conical outline; *ultimate branchlets* cylindical to clavate, 270-450 µm in diameter; colour brownish-red when fresh, becoming purple when preserved; texture soft and very flexible.

Axes growing from apices with circular terminal depression; consisting in TS of thick-walled rounded medullary cells up to 225 µm in diameter, without lenticular wall thickenings, intergrading through subcortex into cuboid to ovoid outer cortical cells 47-93 µm deep; *outer cortex* composed in surface view of elongate-polygonal cells 67-226 x 26-93 µm, with conspicuous secondary pit connections and a single spherical hyaline body in each cell, visible only in fresh and recently preserved specimens; outer walls strongly curved in young axes, slightly curved in main axes; *plastids* in young cells both discoid and ribbon-like, becoming beaded and joined in strings by narrow threads.

Spermatangial receptacles terminating all ultimate branchlets, later also developing in a ring around the first to form dense clusters, capitate, 720-1080 µm diameter, with a central axial filament conspicuous in LS, terminating in a shallow depression from which protrude sterile and spermatangial trichoblasts; spermatangial axes forming spherical clusters, 170-265 µm wide, of densely branched filaments, a few terminating in single pyriform sterile cells, branching pseudodichotomous basally, alternate to spiral distally, covered with densely packed ovoid spermatangia, 8-10 x 5-6.5 µm; released spermatia spherical, 4.5-5 µm diameter. *Cystocarps* formed on penultimate branches, slightly pyriform, with a protruding ostiole, 780-870 µm high x 750-930 µm wide, sessile but not fused to branchlet; carposporangia 40-76 x 12-18 µm. *Tetrasporangia* borne in last two orders of branching, in cylindrical branchlets up to 2.1 mm long; tetrasporocytes cut off abaxially; mature tetrasporangia 72-125 µm in diameter.

Epiphytic on small perennial algae, holdfasts occasionally spreading onto nearby hard substrata, on moderately sheltered shores, usually near extreme low water of spring tides; rarely isolated thalli are found in lower-shore pools.

Confined to extreme southern coasts of the British Isles, rare in Dorset (Batters, 1902) but common in the Isles of Scilly and Channel Isles.

England to N. France, ?S. Spain (LD 36845-7, 9 from Cadiz, previously determined as *L. obtusa* var. *gracilis* C. Agardh). Wider distribution cannot be assessed as *L. pyramidalis* was previously considered conspecific with *L. obtusa*.

In the British Isles, this species has been collected only in August-October.

Fig. 128. *Laurencia pyramidalis*

(A) Habit (A-G Channel Isles, Sep.). (B) Main axis bearing short laterals that terminate in enlarged spermatangial receptacles. (C) Surface view of cortex with beaded plastids in outermost cortical layer. (D) Outer cortex with radial plastids. (E) L.S. of spermatangial receptacle, showing conspicuous axial filament that terminates in receptacle, and globular, much-branched spermatangial axes. (F) Spermatangial axes bearing spermatangia. (G) Cystocarps.

Spermatangia have been observed in September; cystocarps in August-September, and tetrasporangia in August-October. Tetrasporophytes predominate; male and female gametophytes are relatively rare.

Little morphological variation has been observed.

L. pyramidalis can be distinguished from *Chondria dasyphylla* by the lack of a distinct ring of 5 periaxial cells in TS.

Laurencia truncata Kützing (1865), p. 19, pl. 51c.

Holotype: L 941.99.271. Croatia (Pirano), *Pius Titius*.

Thalli attached by discoid holdfast up to 10 mm diameter, lacking stolons; erect axes arising in tufts, 3-9 (-12) cm high; *main axes* distinct, compressed, 0.7-1 mm wide, branched to 5 orders, with a linear or irregular outline, to 10 cm broad, initially complanate, becoming bushy as laterals twist out of the initial branching plane; *first-order laterals* compressed, often equal in length to main axis, up to 2.5 mm wide, bearing a distichous arrangement of second-order laterals, often with a lanceolate outline; *lower-order laterals* compressed or terete, with flattened or terete apices 1-1.5 mm broad; lower laterals eroding; colour deep pink to yellowish-brown, texture brittle, plants decaying rapidly after collection.

Axes growing from apices with circular or oval depression, consisting in TS of pseudoparenchymatous medulla of oval cells, up to 100 x 82 µm, occasionally with lenticular wall thickenings, intergrading through 1-2 subcortical layers into two cortical layers; *outer cortex* 14-32 µm deep of cuboid to obconical cells, polygonal to elongate in surface view, with obvious secondary pit connections, 20-108 x 8-44 µm; *plastids* ribbon-like or discoid.

Spermatangial receptacles terminal on lateral branches that later develop branchlets on either side of the receptacle and may form further receptacles, 900-1700 µm wide, developing as shallow cups and maturing to discoid plates with a narrow rim, containing numerous densely packed cylindrical spermatangial axes; spermatangial axes branched or unbranched, 170-250 µm long x 33-65 µm wide, terminating in one or more spherical sterile cells that contain bright yellow pigment when fresh; spermatangia ovoid, 7-8 x 5-6 µm; released spermatia spherical, 4-6 µm in diameter. *Cystocarps* sessile, ovoid with protruding ostiole, 750-1300 µm wide; carposporangia 120-180 x 45-65 µm.

Fig. 129. *Laurencia truncata*

(A) Habit of male plant from the shallow subtidal (A, C & E-G Cork, Nov.). (B) Surface view of cortex, stained to show secondary pit connections between cells of outermost cortical layer (scale bar = 50 µm) (Cornwall, March). (C) T.S. of medulla with densely-stained wall thickenings. (D) Habit of plant collected intertidally (Pembroke, Mar.). (E) Cup-like spermatangial receptacles (arrows). (F) L.S. of spermatangial receptacle, showing numerous spermatangial axes covered with spermatangia and terminating in large sterile cells. (G) Cystocarps with protruding ostioles.

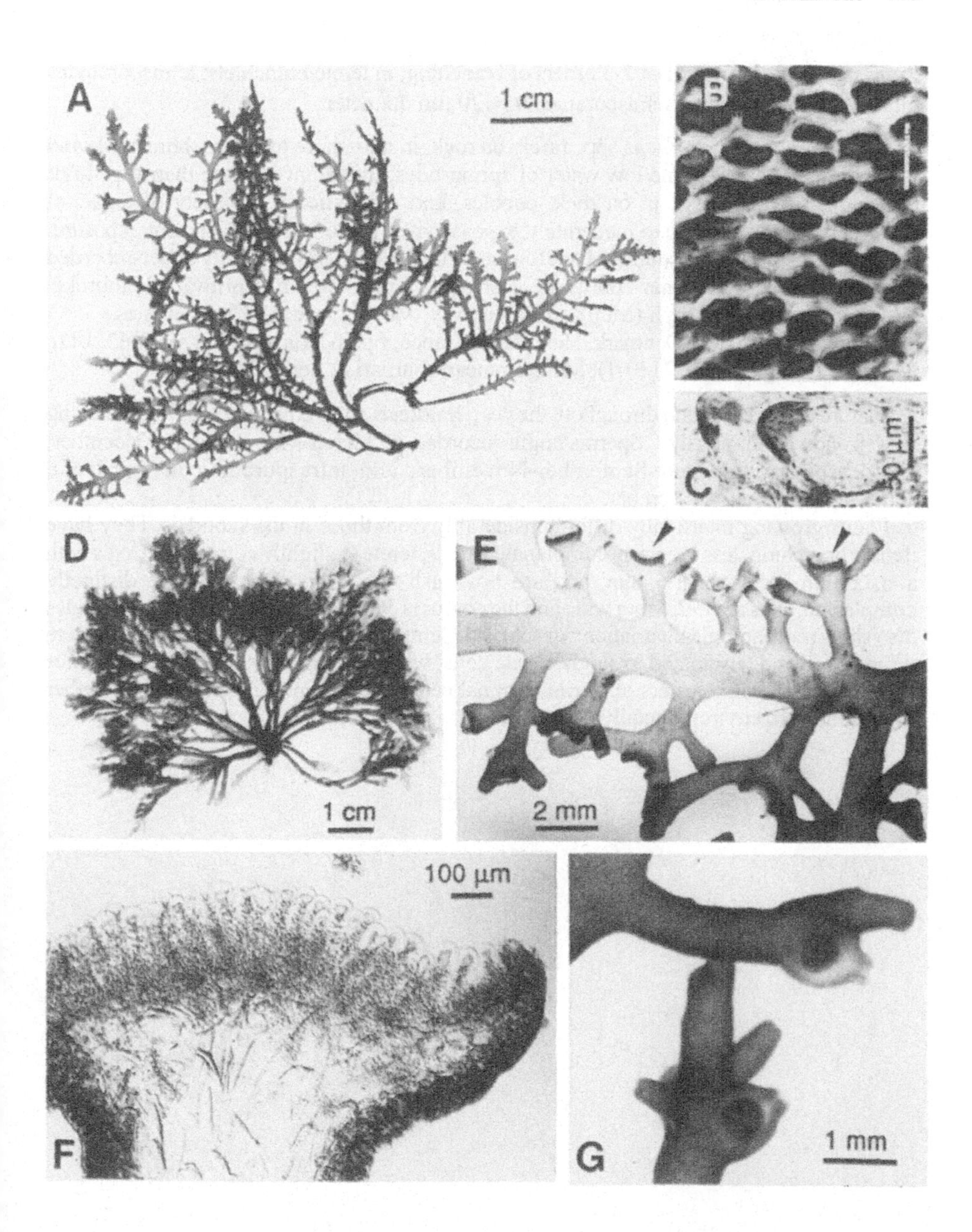
A
1 cm
B
C
D
1 cm
E
2 mm
F
100 µm
G
1 mm

Tetrasporangia borne in last 2-3 orders of branching, in terete branchlets; tetrasporocytes cut off abaxially; mature tetrasporangia 95-120 µm diameter.

Usually epiphytic on *Fucus* spp., rarely on rock, in mid-shore to lower-shore pools and on open rock near extreme low water of spring tides, intolerant of more than very brief emersion, subtidal to 12 m on rock, pebbles, and epiphytic on a variety of algae, at extremely wave-sheltered to moderately exposed sites, sometimes with current exposure.

Widely distributed on south and west coasts of the British Isles, probably under-recorded due to previous taxonomic confusion: Norfolk, Dorset, Devon, Cornwall, Pembroke, Wigtown (L. Ryan), Argyll (Mull); Cork, Galway, Mayo, Donegal; Channel Isles.

S. Norway, Sweden, Denmark, Helgoland, France, Spain (Ria Arosa, 4 vii 1963, LD), Morocco (Tangier, LD 37110-1); Mediterranean (Furnari & Serio, 1993).

Mature plants are found throughout the year, holdfasts apparently perennial and forming new fronds continuously. Spermatangia recorded in March-May and July-December; cystocarps in June and September-November; and tetrasporangia in April and June-September and November.

Plants growing intertidally differ considerably from those in the subtidal. They have denser branching, less regularly complanate fronds, terete to slightly compressed axes with a maximum width of 1.5 mm, and are brownish-red. Subtidal plants are distinctly complanate, axes are to 2.5 mm wide and the colour is clear red. Spermatangial receptacles also show morphological variation: in subtidal plants from Cork, spermatangial axes were unbranched and terminated in single sterile cells, whereas in most other collections, axes were branched with clusters of sterile terminal cells. At present it is not known whether these forms are environmentally determined.

Glossary

Abaxial — The side of a lateral away from the main axis (e.g. abaxial branchlet, Fig. 32B).
Adaxial — The side of a lateral towards the main axis.
Adventitious — Arising from an abnormal position (e.g. a branch not formed at the apex of the thallus but on a mature axis).
Anastomosing — Branched and fused into a network.
Anterior — Towards the apex.
Apex — The tip or youngest part of a thallus or branch.
Apical cell — Cell at the apex of an axis, from which the axis is derived by cell division.
Auxiliary cell — Cell that receives the diploid nucleus from the carpogonium and gives rise to the gonimoblast.
Axial cell — One cell of axial filament, sometimes distinguishable among medullary cells in transverse section of axes.
Axial filament — Central filament of axis or entire monosiphonous axis.
Axil — The angle between two structures, e.g. main and lateral axis.
Axillary — Formed in the angle between two structures (e.g. branch formed between trichoblast and main axis, Fig. 97C).

Bacilloid — Resembling bacteria, minute and elongate.
Basal cell (in context of gonimoblast development) — Lower cell (also called foot cell) formed by division of auxiliary cell into two.
Binucleate — With two nuclei.
Bisporangium — Sporangium containing two spores.
Bispore — One of two spores released from a bisporangium.
Bisporophyte — Phase of life history that forms bisporangia.
Blade — Leaf-like part of thallus.
Bladelet — Small blade, often specialized for reproduction.
Branchlet — Small, higher-order branch.

Capitate — With a distinct head.
Carpogonial branch — Filament bearing the female gametangium, the carpogonium.

Carpogonium — Female gametangium.
Carposporangium — Sporangium formed by gonimoblast tissue.
Carpospore — Spore released from carposporangium.
Cartilaginous — With a tough texture rather like cartilage.
Clavate — Club-shaped.
Coalescence — Fusion of main and lateral axes over a number of segments (e.g. in *Pterosiphonia*).
Complanate — Flattened or branched in one plane.
Connecting cell — Cell that links two structures and usually carries a nucleus, e.g. after fertilization of carpogonium.
Cordate — Heart-shaped.
Cortex — Outer cell layers of axis, usually small and well-pigmented.
Cortical band — Ring of periaxial cells, with varying amounts of cortical tissue, formed around nodes (e.g. in *Ceramium*).
Corticated — With a secondary cortex formed by development of filaments within, and sometimes outside, the outer walls of primary filaments.
Cruciate — Division of tetrasporangium in which the first and second cleavages are perpendicular to each other.
Cystocarp — The gonimoblast tissue and surrounding gametophytic pericarp tissue; structure from which carpospores are released.

Decumbent — Creeping or growing parallel to substratum.
Decussate — With successive pairs formed at right angles to each other; in tetrasporangia the planes of longitudinal division are perpendicular to each other.
Dentate — With teeth.
Determinate — Of limited growth.
Dichotomous — Bearing two equal branches at each branch point.
Diploid — With two sets of chromosomes in each nucleus; usually sporophytic.
Diploidization — Becoming diploid, usually by fusion of two haploid gametes (syngamy).
Digitate — Shaped like the fingers of a hand.
Disposal cell — Cell cut off from auxiliary cell during post-fertilization development that contains a haploid nucleus.
Distichous — Formed in two rows along axis.
Dioecious — Forming male and female gametangia on separate male and female individuals.
Discoid — Shaped like a solid disc.
Distal — Away from the base of the thallus.
Distromatic — Composed of two layers of cells.

Ecorticate — Lacking a cortex.
ELWST — Extreme low water level of spring tides.
Endophytic — Growing within the thallus of another plant.
Endopolyploidy — Becoming diploid or polyploid without syngamy, usually by failure

of nuclei to separate after DNA replication.
Epiphytic — Growing on a plant or seaweed.
Epizoic — Growing on an animal.
Erect — Growing upwards away from substratum.

Facultative — Not necessarily taking place.
Fertile — Bearing reproductive structures.
Filiform — String-like.
Foot cell — Lower cell formed by division of auxiliary cell into two during post-fertilization development.
Fusion cell — Cell derived by fusion of various cells of post-fertilization tissue, sometimes also including gametophytic tissue (e.g. Fig. 76E).

Gametophyte — Phase of life history, usually haploid, that forms gametangia.
Gland cell — An almost colourless cell with highly refractive contents usually occupying a definite position in the thallus (e.g. Fig. 6B).
Globose — Rounded, globular.
Gonimoblast — Tissue which develops on the female plant after fertilization and which ultimately produces carposporangia.
Gonimoblast initial — The first cell arising from a fertilized carpogonium or diploidized auxiliary cell, which develops into the gonimoblast.
Gonimolobe — Element of gonimoblast that develops from a single initial cell and gives rise to carposporangia; usually several formed sequentially in cystocarp.

Hair — Colourless, elongate cell or multicellular structure; often ephemeral in the Ceramiales.
Haploid — With a single set of chromosomes in each nucleus; usually gametophytic.
Haematoxylin — Stain used to show up nuclei and other cellular structures.
Holdfast — Structure attaching thallus to substratum.
Holotype — The one specimen or other element used by the author or designated by him as the nomenclatural type.
Hyaline — Colourless and transparent.

Imbricate — Overlapping in series.
Indeterminate axis — Axis of potentially unlimited growth.
Intertidal — Growing bewteen extreme high water and low water levels.
Involucre — Ring of axes that form a partial enclosure around reproductive structure (e.g. Fig. 14C).
Involute — Curved inwards.
Isomorphic — Of the same morphology, e. g. of different life-history phases.
Isotype — Any duplicate (part of a single gathering made by a collector at one time) of the holotype.

Lateral — Axis formed as a branch from the main axis or from another lateral axis.
Lateral cell (in context of gonimoblast development) — Cell in which the haploid nucleus is disposed of during post-fertilization development.
Lectotype — A specimen or other element selected from the orginal material to serve as a nomenclatural type when no holotype was designated at the time of publication or as long as it is missing.
L.S. — Longitudinal section.

Macroscopic — Clearly visible without magnification.
Maerl — Loose-lying coralline algae often accumulated into large masses.
Medulla — The internal region of a thallus, usually consisting of almost colourless tissue.
Membranous — Delicate thin sheet of cells resembling a membrane.
Midrib — Conspicuously thickened central rib extending the entire length of a blade.
Mixed phases — Thallus bearing reproductive structures normally found only on separate individuals, e.g. both gametangia and tetrasporangia.
MLWNT — Mean low water level of neap tides.
Monoecious — Forming male and female gametangia on the same thallus.
Monopodial — Development in which the primary axis is maintained as the main line of growth and other axes are produced from it.
Monosiphonous — Consisting of a single row of cells, with or without cortication but without periaxial cells (e.g. Fig. 2C) (cf. polysiphonous).
Monosporangium — Sporangium in which a single cell is formed.
Monostromatic — Composed of a single layer of cells.
Mucilaginous — Slippery and slimy.
Mucronate — Terminating in a point.
Multinucleate — Containing more than two nuclei (e.g. Fig. 41C).

Neotype — A specimen or illustration selected to serve as nomenclatural type as long as all of the material on which the name of the taxon was based is missing.
Node — Joint in a segmented axis; end-walls between contiguous cells of a filament.
Nomenclatural type — That element to which the name of a taxon is permanently attached, whether as a correct name or as a synonym.

Ob (prefix) — Inverted, e.g. obconical.
Octosporangium — Sporangium that contains eight spores at maturity, usually derived from a binucleate cell.
Ostiole — Small opening in a reproductive structure through which spore release occurs.

Parasporangium — Structure of irregular shape usually containing more than four spores, not homologous with tetrasporangia (e.g. Fig. 35D).
Paraspore — Spore released by a parasporangium.
Paratype — A specimen or illustration cited in the protologue that is neither the holotype nor an isotype, nor one of the syntypes if two or more specimens were simultaneously

designated as types.
Pedicel — Stalk of a reproductive structure.
Pedicellate — Having a pedicel.
Percurrent — Extending through the entire length or axis of a plant.
Periaxial (= pericentral) cell — One of a number, often a ring, of cells cut off from and surrounding an axial cell (e.g. Fig. 97E).
Pericarp — That part of the cystocarp produced by the female gametangial thallus and forming a covering to the developing gonimoblast.
Phenotype — Morphology resulting from combined effects of genotype and environment.
Pit connection — Small pore between adjacent cells, occluded by a plug.
Plastid — Cell organelle containing pigment with photosynthetic functions.
Polymorphic — Having more than one form or phenotype.
Polyploid — Having three or more sets of chromosomes in each nucleus.
***Polysiphonia*-type life history** — With two independent, isomorphic, life-history phases, the haploid gametophyte and the diploid tetrasporophyte, and a further diploid phase, the carposporophyte, that grows on the female plant and is usually surrounded by protective tissue formed by the female.
Polysiphonous — With each axial cell surrounded, usually over its entire length, with a particular number of periaxial cells (e.g. Fig. 104E-G).
Polysporangium — Sporangium containing more than 8 spores, usually developing from a multinucleate cell.
Polystromatic — Composed of several layers of cells.
Posteriolateral — In a lateral and posterior position.
Posterior — Away from the apex.
Primary gonimoblast cell — First-formed cell of gonimoblast.
Procarp — Carpogonial branch and auxiliary cell in close proximity in branching system.
Propagule — Structure that becomes detached from parent plant and gives rise to a new individual.
Prostrate — Parallel to substratum.
Proximal — Towards the base of the thallus.
Pseudolateral — Branched pigmented monosiphonous filaments in the Dasyaceae (e.g. Fig. 84F).
Pseudodichotomous — Apparently forming two equal branches at branch points, but one of the branches being derived from a lateral branch.
Pyriform — Pear-shaped.

Quadrinucleate — Containing four nuclei.
Quadrichotomous — Forming four branches at each branch point.

Repent — Creeping.
Rhizoid — Cell or filament potentially involved in attachment to the substratum.
Rhizoidal filament — Filament resembling a rhizoid but not necessarily involved in attachment, sometimes formed internally within thallus.

Secondary pit connection — Pore formed between cells of different filaments, established by formation of a connector (= conjunctor) cell that fuses with one of the cells.
Secund — Arranged on one side only (e.g. branchlets, Fig. 2B).
Seirosporangia — Monosporangia produced terminally in branched or unbranched linear series (Fig. 40H).
Segment — Length of thallus consisting of one axial cell and its surrounding periaxial cells and cortex if present.
Septum — Division.
Sessile — Without a stalk (e.g. tetrasporangia, Fig. 34E).
Sinusoidal — Curved backwards and forwards like a sine wave (e.g. Fig. 5C).
Sorus — Aggregation of reproductive structures (e.g. tetrasporangia, Fig. 63G).
Spermatangial branch, axis or head — Aggregation of spermatangial mother cells into a compact structure bearing numerous spermatangia (e.g. Fig. 49D).
Spermatangial mother cell — Cell that bears spermatangia.
Spermatangium — Gametangium that releases a single male gamete = spermatium (e.g. Fig. 30C).
Spermatium — Non-pigmented, non-flagellated male gamete.
Spine — Pointed, more-or-less colourless unicellular or multicellular structure (e.g. Fig. 14B).
Sporangium — Cell that releases one or more spores.
Sporophyte — Plant, usually diploid, on which sporangia are formed.
Sterile group — Cells formed during post-fertilization development that do not contribute to gonimoblast development.
Stichidium — Specialized pod-like branchlet bearing series of tetrasporangia (e.g. Fig. 85D).
Striae — Rays or rows of differentiated cells.
Subtending — Below.
Subtidal — Growing below extreme low water of spring tides.
Supporting cell — Cell bearing one or more carpogonial branches.
Sympodial — Development in which the primary axis is continually being replaced by lateral axes which become temporarily dominant, but soon are replaced by their own laterals.
Syntype — Any one or two or more specimens cited by the author when no holotype was designated; or any one of two specimens simultaneously designated as types.

Terete — Circular in transverse section.
Tetrahedral — Division of tetrasporangium in which all radial walls are at 120° to each other.
Tetrasporangium — Sporangium that releases four spores, usually after meiosis.
Tetraspore — Spore released by tetrasporangium.
Tetrasporocyte — Cell during development into a tetrasporangium, prior to cleavage into spores.
Tetrasporophyte — Phase of life history, usually diploid, that bears tetrasporangia

(sometimes mixed with octosporangia).
Tetrastichous — In four rows along axis.
Thallus — Individual seaweed.
Tribe — Phylogenetic grouping of genera at a level below that of subfamily.
Trichoblasts — Simple or branched filaments, pigmented or colourless, formed in particular positions near apices, often on every segment of axis, usually shed rapidly (deciduous).
Trichogyne — Elongate extension of female gametangium (carpogonium) on which spermatia lodge.
Trichotomous — Bearing three branches at each branch point.
TS — Transverse section.
Tristichous — In three rows along axis.
Tristromatic — Composed of three layers of cells.
Turbinate — Inversely conical, with a contraction towards the point.

Uniaxial — Containing only a single axial filament.
Uninucleate — Containing only a single nucleus (e.g. Fig. 31D).
Urceolate — Urn-shaped.

Veins — Parallel or radiating thickenings or differentiated cell rows in blade or flattened axis.
Vestigial — Structure only partly-formed and apparently non-functional.

Whorl — Ring of branches or structures inserted at one level on axis.
Whorl-branch — One branch of a whorl or pair of branches of limited growth.

Zonate — Division of tetrasporangium in which all cleavages are parallel to each other and perpendicular to the long axis of the sporangium.

References

Abbott & Hollenberg, J. G. 1976. *Marine algae of California*. Stanford.
Adams, N. M. 1991. The New Zealand species of *Polysiphonia* Greville (Rhodophyta). *New Zealand J. Bot.* **29**: 411-427.
Agardh, C.A. 1817. *Synopsis algarum scandinaviae.....* Lund.
—— 1822. *Species algarum....* **1** (2). Lund.
—— 1824. *Systema algarum*. Lund.
—— 1827. Aufzählung einiger in den österreichischen Ländern gefundenen neuen Gattungen und Arten von Algen, nebst ihrer Diagnostik und beigefügten Bemerkungen. *Flora, Regensburg* **10**: 625-646.
—— 1828. *Species algarum....* **2** (1). Greifswald.
Agardh, J. G. 1841. In historiam algarum symbolae. *Linnaea* **15**: 1-50.
—— 1842. *Algae maris mediterranei et adriatici .....* Paris.
—— 1844. *In systemata algarum hodierna adversaria*. Lund.
—— 1851. *Species genera et ordines algarum ...* **2** (1). Lund.
—— 1852. *Species genera et ordines algarum ...* **2** (2). Lund.
—— 1863. *Species genera et ordines algarum ...* **2** (3). Lund.
—— 1876. *Species genera et ordines algarum ...* **3** (1). Leipzig.
—— 1898. *Species genera et ordines algarum ...* **3** (3). Lund.
Aponte, N. E. & Ballantine, D. L. 1990. The life history in culture of *Callithamnion boergesenii* sp. nov. (Ceramiaceae, Rhodophyta) from the Caribbean. *Phycologia* **29**: 191-199.
Ardissone, F. 1871. Rivista dei Ceramii della flora Italiana. *Nuovo G. Bot. Ital.* **3**: 32-50.
—— 1883. *Phycologia mediterranea, 1, Floridee*. Varese.
Ardré, F. 1967. Remarques sur la structure des *Pterosiphonia* (Rhodomélacées, Céramiales) et leurs rapports systématiques avec les *Polysiphonia*. *Rev. algol.* **1**: 37-77.
—— 1970. Contribution à l'étude des algues marines du Portugal. 1. La flore. *Port. Acta Biol., ser. B* **10**: 1-423.
—— 1973. Observations sur deux Délesseriacées (Rhodophycées, Céramiales) des côtes de Bretagne: le *Nitophyllum heterocarpum* (Chauvin) Ernst et Feldmann et le

Radicilingua thysanorhizans (Holmes) Papenfuss. *Bull. Soc. phycol. Fr.* **18**: 33-46.

Areschoug, J. E. 1847. Phycearum, quae in maribus Scandinaviae crescunt, enumeratio. Sectio prior, Fucaceas continens. *Nova Acta Reg. Soc. Scient. upsal.*, ser. 2 **13**: 223-382.

—— 1850. Phyceae scandinavicae marinae .... [also issued as Phycearum, quae in maribus Scandinaviae crescunt, enumeratio. Sectio posterior, Ulvaceas continens]. *Ups. Soc. sci. Nova Acta*, ser. 2 **14**: 385-454.

Athanasiadis, A. 1985a. The taxonomic recognition of *Pterothamnion crispum* (Rhodophyta, Ceramiales), with a survey of the carposporophyte position in genera of the Antithamnieae. *Br. phycol. J.* **20**: 381-389.

—— 1985b. North Aegean marine algae. I. New records and observations from the Sithonia Peninsula, Greece. *Botanica marina* **28**: 453-468.

—— 1986. A comparative study of *Antithamnion tenuissimum* and three varieties of *A. cruciatum*, including var. *scandinavicum* var. nov. (Rhodophyceae). *Nord. J. Bot.* **6**: 703-709.

—— 1987. *A survey of the seaweeds of the Aegean Sea with taxonomic studies on species of the tribe Antithamnieae (Rhodophyta).* Göteborg.

—— 1988. The status and typification of *Antithamnion cruciatum* var. *pumilum* and its relationship to *A. cruciatum* var. *scandinavicum* (Rhodophyta). *Br. phycol. J.* **23**: 395-398.

—— 1990. Evolutionary biogeography of the North Atlantic antithamnioid algae. In: Garbary, D.J. & South, G.R. (eds) *Evolutionary biogeography of the marine algae of the North Atlantic*, pp. 219-240. Berlin.

—— **& Rueness, J.** 1992. Biosystematic studies of the genus *Scagelia* (Rhodophyta, Ceramiales) from Scandinavia: genetic variation, life histories and chromosome numbers. *Phycologia* **31**: 1-15.

Austin, A. P. 1956. Chromosome counts in the Rhodophyceae. *Nature, Lond.* **178**: 370-371.

—— 1959. Iron-alum aceto-carmine staining for chromosomes and other anatomical features of Rhodophyceae. *Stain Technol.* **34**: 69-75.

Bailey, J. W. 1848. Continuation of the list of localities of algae in the United States. *Amer. J. Sci. Arts*, ser. 2 **6**: 37-42.

Baldock, R. N. 1976. The Griffithsieae group of the Ceramiaceae (Rhodophyta) and its southern Australian representatives. *Aust. J. Bot.* **24**: 509-593.

——**& Womersley, H. B. S.** 1968. The genus *Bornetia* (Rhodophyta, Ceramiaceae) and its southern Australian representatives, with a description of *Involucrana* gen. nov. *Aust. J. Bot.* **16**: 197-216.

Batten, L. 1923. The genus *Polysiphonia* Grev., a critical revision of the British species, based upon anatomy. *J. Linn. Soc. Bot.* **46**: 271-311.

Batters, E. A. L. 1892. *Gonimophyllum buffhami*: a new marine algae. *J. Bot., Lond.* **30**: 65-67.

—— 1896. Some new British marine algae. *J. Bot., Lond.* **34**: 6-11.

—— 1902. A catalogue of the British marine algae. *J. Bot., Lond.* **40** (suppl.): 1-107.

Bird, C. J. & Johnson, C. R. 1984. *Seirospora seirosperma* (Harvey) Dixon (Rhodophyta, Ceramiaceae) - a first record for Canada. *Proc. Nova Scotian Inst. Sci.* **34**: 173-175.

—— **& McLachlan, J.** 1992. *Seaweed flora of the Maritimes. 1. Rhodophyta - the red algae*. Bristol.

Blackler, H. 1974. Flora. In: Laverack, M. S. & Blackler, H. (eds) *Fauna and flora of St. Andrews Bay*, pp. 167-295. Edinburgh.

Boddeke, R. 1958. The genus *Callithamnion* Lyngb. in the Netherlands. A taxonomic and oecological study. *Acta bot. neerl.* **7**: 589-604.

Bonnemaison, T. 1828. Essai sur les Hydrophytes loculées (ou articulées) de la famille des Epidermées et des Céramiées. *Mem. Mus. Hist. nat. Paris* **16**: 49-158.

Boo, S. M. & Lee, I. K. 1983. A life history and hybridization of *Antithamnion sparsum* Tokida (Rhodophyta, Ceramiaceae) in culture. *Korean J. Bot.* **26**: 141-150.

—— **&** —— 1985. A taxonomic reappraisal of *Ceramium fastigiatum* Harvey (Rhodophyta, Ceramiaceae). *Korean J. Bot.* **28**: 217-224.

Børgesen, F. 1916. The marine algae of the Danish West Indies, 2. Rhodophyceae. *Dansk Bot. Ark.* **3**(1b): 81-144.

—— 1930. Marine algae from the Canary Islands especially from Teneriffe and Gran Canaria III. Rhodophyceae Part II Cryptonemiales, Gigartinales and Rhodymeniales. *Biol. Meddr* **8**: 1-97.

Bornet, E. 1892. Les algues de P.K.A. Schousboe, récoltées au Maroc et dans la Méditerranée de 1815 à 1829 et déterminées par M. Edouard Bornet. *Mém. Soc. nat. Sci. nat. Math. Cherbourg*. **28**: 165-376.

Bory [de Saint-Vincent], J. B. G. M. 1822. Dictionnaire classique d'histoire naturelle ...., **2**. Paris.

—— 1828. Cryptogamie. In: Duperrey, L.I. (ed.) *Voyage autour du monde ... "La Coquille"*, pp. 1-96. Paris.

Boudouresque, C.-F. & Coppejans, E. 1982. Végétation marine de l'Ile de Port-Cros (Parc National). XXIII. Sur deux espèces de *Griffithsia* (Ceramiaceae, Rhodophyta). *Bull. Soc. Roy. Bot. Belg.* 115: 43-52.

Broadwater, S. T., Scott, J. L. & West, J. A. 1991. Spermatial appendages of *Spyridia filamentosa* (Ceramiaceae, Rhodophyta). *Phycologia* **30**: 189-195.

Buffham, T. H. 1884. Notes on the Florideae and on some newly-found antheridia. *J. Queckett microsc. Club*, ser. 2 **1**: 337-344.

—— 1888. On the reproductive organs, especially the antheridia, of some of the Florideae. *J. Queckett microsc. Club*, ser. 2 **3**: 257-266.

—— 1891. On the reproductive organs, especially the antheridia, of some of the Florideae. *J. Queckett microsc. Club*, ser. 2 **4**: 246-253.

—— 1896. Notes on some Florideae. *J. Queckett microsc. Club*, ser. 2 **6**: 183-190.

Cabioch, J. & Boudouresque, C.-F. 1992. *Guide des algues des mers d'Europe*. Lausanne.

Candolle, A. P. de 1806. *Synopsis plantarum in flora gallica descriptarum*. Paris.
—— 1815. *Flore française* .... **2**, ed. 3 [reissue]. Paris.
Castagne, J. L. M. 1845. *Catalogue des plantes que croissent naturellement aux environs de Marseille*. Aix.
Christensen, T. 1967. Two new families and some new names and combinations in the algae. *Blumea* **15**: 91-94.
Clemente [y Rubio], S. de Rojas. 1807. *Ensayo sobre la variedades de la vid commun que vegetan en Andalucia*. Madrid.
Clokie, J. J. P. & Boney, A. D. 1979. Check-list of the marine algae of the Firth of Clyde. *Scott. Fld Stud.* 1979: 3-13.
Cole, K. M. 1990. Chromosomes. In: Cole, K. M. & Sheath, R. G. (eds) *Biology of the red algae*, pp. 73-102. Cambridge.
Cole K. M. & Sheath, R. G. (eds) 1990. *Biology of the red algae*. Cambridge.
Collins, F. S. & Hervey, A. B. 1919. The algae of Bermuda. *Proc. Amer. Acad. Arts Sci.* **53**: 1-195.
Coomans, R. J. & Hommersand, M. H. 1990. Vegetative growth and organization. In: Cole, K. M. & Sheath, R. G. (eds) *Biology of the red algae*, pp. 275-304. Cambridge.
Cormaci, M. & Motta, G. 1987. Osservazioni sulla morfologia di *Ceramium rubrum* (Hudson) C. Agardh (Rhodophyta, Ceramiales) in coltura e considerazioni sulla sua tassonomia. *Boll. Acc. Gioenia Sci. nat.* **20**: 239-251.
Cotton, A. D. 1912. Marine algae. In: Praeger, R. L. (ed.) A biological survey of Clare island in the county of Mayo, Ireland and of the adjoining district. *Proc. R. Ir. Acad.* **31**: 1-178.
Cramer, C. 1863. Physiologisch-systematische Untersuchungen über die Ceramiaceen. *Neue Denkschr. Allg. Schweiz. Ges. Naturwiss.* **20**: 1-131.
Cremades, J. & Pérez-Cirera, J. L. 1990. Nuevas combinaciones de algas bentónicas marinas, como resultado del estudia del herbario de Simón de Rojas Clemente y Rubio (1777-1827). *Anal. Jardín Bot. Madrid.* **47**: 489-492.
—— 1993. Contribucion al conocimiento de la obra ficologica de Simon de Rojas Clemente (1777-1827) y tipificacion de los nuevos nombres de su *Ensayo*. *Anal. Jard. Bot. Madrid* **51(1)** (in press).
Crouan, P. L. & Crouan, H. M. 1848. Sur l'organisation, la fructification et la classification du *Fucus wigghii* de Turner et de Smith, et de l'*Atractophora hypnoides*. *Ann. Sci. nat. Bot.*, sér. 3 **12**: 361-376.
——**&**—— 1851. Études microscopiques sur quelques algues nouvelles ou peu connues constituant un genre nouveau. *Ann. Sci. nat. Bot.*, sér. 3, **15**: 365-366.
——**&**—— 1867. *Florule du Finistère*. Paris & Brest.
Cullinane, S. P. 1970. The occurrence of the alga *Lophosiphonia* on the Cork coast. *Proc. R. Ir. Acad.*, ser. B **70**: 1-4.

Dale, S. 1732. *In* Taylor, S. & Dale, S. (eds) *The history and antiquities of Harwich and Dovercourt, in the county of Essex* ... Ed. 2. London.

De Toni, G. B. 1903. *Sylloge algarum omnium hucusque cognitarum, 4 Florideae, 3 Rhodomelaceae, Ceramiaceae*. Padua.

——**& Levi, D.** 1888. *Civico Museo e raccolta correr in Venezia. Collezioni de storia naturale. 1. Collezione botaniche. L'algarium Zanardini*. Venice.

De Valéra, M. 1939. Some new or critical algae from Galway Bay, Ireland. *K. fysiogr. Sällsk. Lund Förh.* **9**: 91-104.

Dillwyn, L. W. 1802-1809. *British Confervae*. London.

Dixon, P. S. 1958. *Ceramium codii* (Richards) Mazoyer: an addition to the British marine algal flora. *Ann. Mag. Nat. Hist.*, ser. 13, **1**: 14-16.

—— 1960a. Studies on marine algae of the British Isles: the genus *Ceramium*. *J. mar. biol. Ass. U.K.* **39**: 331-374.

—— 1960b. Studies on marine algae of the British Isles: *Ceramium shuttleworthianum* (Kütz.) Silva. *J. mar. biol. Ass. U.K.* **39**: 375-390.

——1960c. Taxonomic and nomenclatural notes on the Florideae, II. *Bot. Notiser* **113**: 295-319.

Dixon, P. S. 1962a. Taxonomic and nomenclatural notes on the Florideae, III. *Bot. Notiser* 115: 245-260.

—— 1962b. The genus *Ptilothamnion* in Europe and North Africa. *Br. phycol. Bull.* **2**: 154-161.

—— 1963. *Sphondylothamnion multifidum* (Huds.) Näg. in western Europe. *Br. phycol. Bull.* **2**: 219-223.

—— 1964. Taxonomic and nomenclatural notes on the Florideae, IV. *Bot. Notiser* **117**: 56-78.

—— 1965. Perennation, vegetative propagation and algal life histories, with special reference to *Asparagopsis* and other Rhodophyta. *Bot. Gothoburg.* **3**: 67-74.

—— 1971. Studies of the genus *Seirospora*. *Botaniste* **54**: 35-48.

—— 1973. *Biology of the Rhodophyta*. Edinburgh.

—— 1983. The algae of Lightfoot's *Flora Scotica*. *Bull. Br. Mus. nat. Hist. (Bot.)* **11**: 1-15.

——**& Irvine, L. M.** 1970. Miscellaneous notes on algal taxonomy and nomenclature, III. *Bot. Notiser* **123**: 474-487.

——**& Parkes, H. M.** 1968. Miscellaneous notes on algal taxonomy and nomenclature, II. *Bot. Notiser* **121**: 80-88.

—— **& Price, J. H.** 1981. The genus *Callithamnion* (Rhodophyta, Ceramiaceae) in the British Isles. *Bull. Br. Mus. nat. Hist. (Bot.)* **9**: 99-141.

Drew, K. M. 1939. An investigation of *Plumaria elegans* (Bonnem.) Schmitz with special reference to triploid plants bearing parasporangia. *Ann. Bot.*, N.S. **3**: 347-367.

—— 1943. Contributions to the cytology of *Spermothamnion Turneri* (Mert.) Aresch. II. The haploid and triploid generations. *Ann. Bot.*, N.S. **7**: 23-30.

—— 1955. Sequence of sexual and asexual phases in *Antithamnion spirographidis* Schiffner. *Nature, Lond.* **175**: 813-814.

Duby, J. E. 1830. *Botanicon Gallicum* .... **2.**, ed. 2. Paris.

Ducluzeau, J. A. P. 1806. ['1805'] *Essaie sur l'histoire naturelle des conferves des environs de Montpellier.* Montpellier.

Dumortier, B. C. J. 1822. *Commentationes botanicae. Observations botaniques, dédiées à la Société d'Horticulture de Tournay.* Tournay.

Edwards, P. 1969. The life history of *Callithamnion byssoides* in culture. *J. Phycol.* **5**: 266-268.

—— 1970. Field and cultural observations on growth and reproduction of *Polysiphonia denudata* from Texas. *Br. phycol. J.* **5**: 145-153.

—— 1973. Life history studies of selected British *Ceramium* species. *J. Phycol.* **9**: 181-184.

—— 1979. A cultural assessment of the distribution of *Callithamnion hookeri* (Dillw.) S. F. Gray (Rhodophyta, Ceramiales) in nature. *Phycologia*, **18**: 251-263.

Ellis, J. 1768. Extract of a letter from John Ellis, Esquire, F.R.S. to Dr. Linnaeus, of Upsal, F.R.S. on the animal nature of the genus of zoophytes, called *Corallina. Phil. Trans. R. Soc.* **57**: 404-427.

Ernst, J. & Feldmann, J. 1957. Une nouvelle Delessériacée des côtes de Bretagne: *Drachiella spectabilis*, nov. gen. nov. sp. *Rev. gen. Bot.* **64**: 446-459.

Falkenberg, P. 1901. *Die Rhodomelaceen des Golfes von Neapel und der angrenzenden Meeresabschnitte.* Berlin.

Feldmann, G. 1964. Sur une nouvelle espèce iridescente de *Chondria* (Rhodophyceae, Rhodomelaceae). *Rev. gen. Bot.* **71**: 45-55.

—— 1970. Sur l'ultrastructure des corps irisants des *Chondria* (Rhodophycées). *C. r. Acad. Sci. Paris, sér. D* **271**: 945-946.

Feldmann, J. 1937. Recherches sur la végétation marine de la Méditerranée. La Côte des Albères. *Revue algol.* **10**: 1-139.

—— 1954. Inventaire de la flora marine de Roscoff. *Travaux de la Station Biologique de Roscoff*, suppl. **6**: 1-152.

—— **& Feldmann, G.** 1951. Un nouveau genre de Rhodophycée parasite d'une Delesseriacée. *C. r. Acad. Sci. Paris, sér. D* **223**: 1137-1139.

—— & —— 1958. Recherches sur quelques Floridées parasites. *Rev. gén. Bot.* **65**: 49-128.

—— & —— 1961. Une nouvelle espèce de Delesseriacée adelphoparasite: *Gonimocolax roscoffensis* sp. nov. *Bull. Soc. Bot. Fr.* **108**: 18-24.

Feldmann-Mazoyer, G. 1941 ['1940']. *Recherches sur les Céramiacées de la Méditerranée occidentale.* Algiers.

—— **& Meslin, R.** 1939. Note sur le *Neomonospora furcellata* (J. Ag.) comb. nov. et sa naturalisation dans la Manche. *Rev. gén. Bot.* **51**: 193-203.

Furnari, G. & Serio, D. 1993. The distinction of *Laurencia truncata* (Ceramiales, Rhodophyta) in the Mediterranean Sea from *Laurencia pinnatifida. Phycologia* **23** (in press).

Gallardo, T., Garreta, A. G., Ribera, M. A., Alvarez, M. & Conde, F. 1985. *A preliminary checklist of Iberian benthic marine algae*. Madrid.

Garbary, D. J., Grund, D. & McLachlan, J. 1978. The taxonomic status of *Ceramium rubrum* (Huds.) C. Ag. (Ceramiales, Rhodophyceae) based on culture experiments. *Phycologia*, **17**: 85-94.

Gardner, N. L. 1927. New Rhodophyceae from the Pacific coast of North America, IV. *Univ. Calif. Publs Bot.* **13**: 373-402.

Gatty, M. 1872. *British sea-weeds* **1** & **2**. London.

Gmelin, S. G. 1768. *Historia fucorum*. Petropolis.

Gobi, C. 1878. Die Algenflora des Weissen Meeres .... . *Mém. Acad. Imp. Sci. St-Pétersbourg*, ser. 7 **26**: 1-92.

Goff, L. J. 1982. The biology of parasitic red algae. In: Round, F. & Chapman, D. (eds) *Progress in phycological research* **1**, pp. 289-369.

Gonzales, M. A. & Goff, L. J. 1989. The red algal epiphytes *Microcladia coulteri* and *M. californica* (Rhodophyceae, Ceramiaceae) 1. Taxonomy, life history and phenology. *J. Phycol.* **25**: 446-454.

Goodenough, S. & Woodward, T. J. 1797. Observations on the British Fuci, with particular descriptions of each species. *Trans. Linn. Soc. Lond.* **3**: 84-235.

Gordon, E. M. 1972. Comparative morphology and taxonomy of the Wrangelieae, Sphondylothamnieae, and Spermothamnieae (Ceramiaceae, Rhodophyta). *Aust. J. Bot.*, suppl. ser., **4**: 1-180.

—— **& Womersley, H. B. S.** 1966. The morphology and reproduction of *Sphondylothamnion multifidum* (Hudson) Naegeli (Ceramiaceae). *Br. phycol. Bull.* **3**: 23-30.

Gordon-Mills, E. 1977. Two new species of marine algae from Stewart Island, New Zealand, *Mediothamnion norrisii* and *Ptilothamnion rupicolum* (Ceramiaceae, Rhodophyta). *Phycologia* **16**: 79-85.

—— 1987. Morphology and taxonomy of *Chondria tenuissima* and *Chondria dasyphylla* (Rhodomelaceae, Rhodophyta). *Br. phycol. J.* **22**: 237-255.

—— **& Wollaston, E. M.** 1990. *Compsothamnionella huismanii* sp. nov. (Ceramiales, Rhodophyta) from Southern Australia. *Bot. mar.* **33**: 9-17.

—— **& Womersley, H. B. S.** 1987. The genus *Chondria* C. Agardh (Rhodomelaceae, Rhodophyta) in southern Australia. *Aust. J. Bot.* **35**: 477-565.

Granja, A., Cremades, J. & Barbara, I. 1992. Catálogo de las algas bentónicas marinas de la Ría de Ferrol (Galicia, N.O. de la Península Ibérica) y consideraciones biogeográficas sobre su flora. *Nova Acta Cient. Compostelana (Bioloxía)* **3**: 3-11.

Grateloup, J. P. 1807. Dissertation II. *Hist. Soc. Med. Montpelier* **1807**: 34.

Gray, S. F. 1821. *A natural arrangement of British plants* .... **1**. London.

Greville, R. K. 1823-4. *Scottish cryptogamic flora* .... **2**. Edinburgh & London.

—— 1824. *Flora edinensis*.... . Edinburgh & London.

—— 1827-8. *Scottish cryptogamic flora* .... **6**. Edinburgh.

—— 1830. *Algae britannicae*.... . Edinburgh & London.

Grubb, V. M. 1926. The male organs of the Florideae. *J. Linn. Soc. Bot.* **47**: 177-255.

Guiry, M. D. 1978. *A concensus and bibliography of Irish seaweeds. Cramer, Vaduz.*
—— 1990. Sporangia and spores. In: Cole, K. M. & Sheath, R. G. (eds) Biology of the red algae, pp. 347-376. Cambridge.
—— **& Maggs, C. A.** 1991. *Antithamnion densum* (Suhr) Howe from Clare Island, Ireland: a marine red alga new to the British Isles. *Cryptogamie algol.* **12**: 189-194.
Gunnerus, J. E. 1772. *Flora Norvegica* .... **2**. Trondhjem.

Halos, M. -T. 1965a. Sur l'*Aglaothamnion decompositum* (Grateloup ex J. Ag.) comb. nov. et sa position systématique. *Bull. Soc. phycol. Fr.* **10**: 18-19.
—— 1965b. Sur trois Callithamniées des environs de Roscoff. *Cah. Biol. mar.* **6**: 117-134.
Hansen, G. I. & Scagel, R. F. 1981. A morphological stutdy of *Antithamnion boreale* (Gobi) Kjellman and its relationship to the genus *Scagelia* Wollaston (Ceramiales, Rhodophyta). *Bull. Torrey Bot. Club* **108**: 205-212.
Harris, R. E. 1962. Contribution to the genus *Callithamnion* Lyngbye emend. Naegeli. *Bot. Notiser* **115**: 18-28.
—— 1966. Contribution to the genus *Callithamnion* Lyngbye emend. Naegeli: taxonomy of the species indigenous to the British Isles. *Advg Front. Pl. Sci.* **14**: 109-131.
Harvey, W. H. 1834. Algological illustrations. No. 1. Remarks on some British algae, and descriptions of new species recently added to our flora. *J. Bot., Hooker* **1**: 296-305.
Harvey, W. H. 1841. *A Manual of the British Algae.* London.
—— 1844. Description of a new British species of *Callithamnion* (*C. pollexfenii*). *Ann. Nat. Hist.* **14**: 109-131.
—— 1846. *Phycologia Britannica,* pls 1-72. London.
—— 1847. *Phycologia Britannica,* pls 73-144. London.
—— 1848. *Phycologia Britannica,* pls 145-216. London.
—— 1849a. *Phycologia Britannica,* pls 217-252. London.
—— 1849b. *A Manual of the British Algae,* ed. 2. London.
—— 1850. *Phycologia Britannica,* pls 253-306. London.
—— 1851. *Phycologia Britannica,* pls 307-360. London.
—— 1853. Nereis boreali-americana. Part II. Rhodospermeae. *Smithsonian Contr. Knowl.* **5**(5): 1-258.
—— 1855. Some account of the marine botany of the colony of Western Australia. *Trans. R. Ir. Acad.* **22**: 525-566.
—— 1862. Notice of a collection of algae made on the northwest coast of North America, chiefly at Vancouver's Island, by David Lyall, Esq., M. D., R. N., in the years 1859-61. *J. Linn. Soc. Bot.* **6**: 157-177.
—— **& Hooker, J. D.** 1845. *Botany of the Antarctic voyage of H. M. discovery ships Erebus and Terror in the years 1839-1843* ..... 1. Flora Antarctica. Part 1. Algae, pp. 175-193.
Hassinger-Huizinga, H. 1952. Generationswechsel und Geschlechtsbestimmung bei *Callithamnion corymbosum* (Sm.) Lyngb. *Arch. Protistenk.* **98**: 91-125.

Hauck, F. 1885. *Die Meeresalgen Deutschlands und Oesterreichs*. Leipzig.
Hiscock, S. 1986. *A field key to the British red seaweeds*. Taunton.
—— **& Maggs, C. A.** 1984. Notes on the distribution and ecology of some new and interesting seaweeds from south-west Britain. *Br. phycol. J.* **19**: 73-87.
Holmes, E. M. 1873. New British algae. *Grevillea* **2**: 1-3.
—— **& Batters, E. A. L.** 1891. A revised list of the British marine algae. *Ann. Bot., Lond.* **5**: 63-107.
Holmgren, P. K., Keuken, W. & Schofield, E. K. 1981. Index herbariorum. 1. Herbaria of the world, ed. 7. *Regnum veg.* **106**: 1-452.
Hommersand, M. H. 1963. The morphology and classification of some Ceramiaceae and Rhodomelaceae. *Univ. Calif. Publs Bot.* **35**: 165-366.
—— **& Fredericq, S.** 1990. Sexual reproduction and cystocarp development. In: Cole, K. M. & Sheath, R. G. (eds) *Biology of the red algae*, pp. 305-346. Cambridge.
Hooker, J. D. & Harvey, W. H. 1845a. Algae antarcticae, being characters and descriptions of the hitherto unpublished species of algae, discovered in Lord Auckland's group ... . *J. Bot., Lond.* **4**: 249-276, 293-298.
—— **&** —— 1845b. Algae Novae Zelandiae *J. Bot., Lond.* **4**: 521-551.
Hooker, W. J. 1833. Cryptogamia Algae ... Div. 1. Inarticulatae. In, Hooker, W. J., *The English flora of Sir James Edward Smith. Class XXIV. Cryptogamia.* London.
Howe, M. A. 1914. The marine algae of Peru. *Mem. Torrey bot. Club* **15**: 1-185.
Hudson, W. 1762. *Flora Anglica.* London.
—— 1778. *Flora Anglica*, ed. 2. London.
Huisman, J. M. & Kraft, G. T. 1982. *Deucalion* gen. nov. and *Anisoschizus* gen. nov. (Ceramiaceae, Ceramiales), two new propagule-forming red algae from southern Australia. *J. Phycol.* **18**: 177-192.
—— **&** —— 1992. Disposal of auxiliary cell haploid nuclei during post-fertilization development in *Guiryella repens* gen. et sp. nov. (Ceramiaceae, Rhodophyta). *Phycologia* **31**: 127-137.
Huvé, P. & Riouall, R. 1970. Présence dans l'étang de Berre (Bouches-du-Rhône) d'une algue atlantique intéressante *Radicilingua thysanorhizans* (Homes) Papenfuss (Rhodophycée, Ceramiale, Delesseriacée). *Bull. Mus. Hist. nat. Marseille* **30**: 135-144.

Irvine, D. E. G. 1982. Seaweeds of the Faroes. 1. The flora. *Bull. Br. Mus. nat. Hist. (Bot.)* **10**: 109-131.
Irvine, L. M. & Dixon, P. S. 1982. The typification of Hudson's algae: a taxonomic and nomenclatural reappraisal. *Bull. Br. Mus. nat. Hist. (Bot.)* **10**: 91-105.
Itono, H. 1977. *Studies on the ceramiaceous algae (Rhodophyta) from southern parts of Japan*. Vaduz.

Jacobsen, T., Rueness, J. & Athanasiadis, A. 1991. *Antithamnionella floccosa* (Rhodophyta) in culture: distribution, life history and chromosome number. *Bot. mar.* **34**: 491-499.

Kain, J. M. 1982. The reproductive phenology of nine species of Rhodophyta in the subtidal region of the Isle of Man. *Br. phycol. J.* **17**: 321-331.

—— 1984. Seasonal growth of two subtidal species of Rhodophyta off the Isle of Man. *J. exp. Mar. Biol. Ecol.* **82**: 207-220.

—— 1987. Photoperiod and temperature as triggers in the seasonality of *Delesseria sanguinea*. *Helgol. Meeresunters.* **41**: 355-370.

—— 1991. The dithering males of *Delesseria*. *Br. phycol. J.* **26**: 90.

Kapraun, D. F. 1977a. Studies on the growth and reproduction of *Antithamnion cruciatum* (Rhodophyta, Ceramiales) in North Carolina. *Norw. J. Bot.* **24**: 269-274.

—— 1977b. Asexual propagules in the life history of *Polysiphonia ferulacea* (Rhodophyta, Ceramiales). *Phycologia* **16**: 417-426.

—— 1977c. The genus *Polysiphonia* in North Carolina. *Bot. mar.* **20**: 143-153.

—— 1978a. A cytological study of varietal forms in *Polysiphonia harveyi* and *P. ferulacea* (Rhodophyta, Ceramiales). *Phycologia* **17**: 152-156.

—— 1978b. Field and culture studies on selected North Carolina *Polysiphonia* species. *Bot. mar.* **21**: 143-153.

—— 1979. Comparative studies of *Polysiphonia urceolata* from three North Atlantic sites. *Norw. J. Bot.* **26**: 269-276.

—— 1980. *An illustrated guide to the benthic marine algae of coastal North Carolina. 1. Rhodophyta.* Chapel Hill.

—— **& Norris, J. N.** 1982. The red alga *Polysiphonia* Greville (Rhodomclaceae) from Carrie Bow Cay and vicinity, Belize. In: Ruetler, K. & MacIntyre, I. G. (eds). *Atlantic barrier reef ecosystem at Carrie Bow Cay, Belize. 1. Structure and communities*, pp. 225-238. Washington.

Kapraun, D. F. & Rueness, J. 1983. The genus *Polysiphonia* (Ceramiales, Rhodomelaceae) in Scandinavia. *G. Bot. ital.* **117**: 1-30.

Kim, H. S. & Lee, I. K. 1989. Morphology and asexual reproduction of *Monosporus indicus* Børgesen (Rhodophyta, Ceramiaceae) in Korea. *Korean J. Phycol.* **4**: 11-17.

King, R. J. & Puttock, C. 1989. The morphology and taxonomy of *Bostrychia* Montagne and *Stictosiphonia* J. D. Hooker et Harvey (Rhodomelaceae/ Rhodophyta). *Aust. Syst. Bot.* **2**: 1-73.

Kjellman, F. R. 1883. Norra Ishafvets Algflora. *Vega-Exped. Vetensk. Arbeten* **3**: 1-431.

Kleen, E. 1874. Om Nordlandens högre hafsalger. *Ofvers. K. Svensk Vetenskapsakad. Forh.* **31**: 1-46.

Koch, C. 1986. Attempted hybridization between *Polysiphonia fibrillosa* and *P. violacea* (Bangiophyceae) from Denmark; with culture studies primarily on *P. fibrillosa*. *Nord. J. Bot.* **6**: 123-128.

Kornmann, P. & Sahling, P.-H. 1978. Meeresalgen von Helgoland. Benthische Grün-, Braun- und Rotalgen. *Helgol. wiss. Meeresunters.*, **29**: 1-289.

Kuntze, O. 1891. *Revisio generum plantarum* .... **1**. Leipzig.

Kützing, F.T. 1842. Ueber *Ceramium* Ag. *Linnaea* **15**: 727-746.

—— 1843. *Phycologia generalis (oder Anatomie, Physiologie und Systemkunde der Tange).* Leipzig.

—— 1847. Diagnosen und Bemerkungen zu neuen oder kritischen Algen (Fortsetzung). *Bot. Zeit.* **5**: 33-38.

—— 1849. *Species algarum.* Leipzig.

—— 1861. *Tabulae phycologicae ....,* **11**. Nordhausen.

—— 1862. *Tabulae phycologicae ....,* **12**. Nordhausen.

—— 1863. *Tabulae phycologicae ....,* **13**. Nordhausen.

—— 1864. *Tabulae phycologicae ....,* **14**. Nordhausen.

—— 1865. *Tabulae phycologicae ....,* **15**. Nordhausen.

—— 1869. *Tabulae phycologicae ....,* **16**. Nordhausen.

Kylin, H. 1914. Studien über die Entwicklungsgeschichte von *Rhodomela virgata* Kjellm. *Svensk. Bot. Tidskr:.* **8**: 33-69.

—— 1916. Die Entwicklungsgeschichte von *Griffithsia corallina* (Lightf.) Ag. *Zeitschr. Bot.* **8**: 97-124.

—— 1923. Studien über die Entwicklungsgeschichte der Florideen. *K. Svensk Vetensk.-Akad. Handl.* **63** (11): 1-139.

—— 1924. Studien über die Delesseriaceen. *Acta Univ. lund.*, Ny Följd, Avd. 2 **20** (6): 1-111.

—— 1925. The marine red algae in the vicinity of the Biological Station at Friday Harbor, Wash. *Acta Univ. lund.*, Ny Följd, Avd. 2 **21** (9): 1-87.

—— 1928. Entwicklungsgeschichtliche Florideenstudien. *Acta Univ. lund.*, Ny Följd, Avd. 2 **24** (4): 1-127.

—— 1930. Über die Entwicklungsgeschichtliche der Florideen. *Acta Univ. lund.*, Ny Följd, Avd. 2 **26** (6): 1-104.

—— 1934. Bermerkungen über einiger Nitophyllaceen. *K. Fysiogr. Sällsk. Lund Förhandl.* **4**: 1-8.

Kylin, H. 1956. *Die Gattungen der Rhodophyceen.* Lund.

Lamouroux, J. V. 1813. Essai sur les genres de la famille des thalassiophytes non articulées. *Annls Mus. Hist. nat. Paris* **20**: 21-47, 115-139, 267-293.

Lauret, M. 1967. Morphologie, phénologie, répartition des *Polysiphonia* marins du littoral languedocien. I. Section *Oligosiphonia. Naturalia monspeliensa, Bot.* **18**: 347-373 + 14 pls.

—— 1970. Morphologie, phénologie, répartition des *Polysiphonia* marins du littoral languedocien. II. Section *Polysiphonia. Naturalia monspeliensa, Bot.* **21**: 121-163 + 14 pls.

Le Jolis, A. 1863. Liste des algues marines de Cherbourg. *Mém. Soc. nat. sci. Cherbourg.* **10**: 6-168.

Lewis, I. F. 1909. The life history of *Griffithsia bornetiana. Ann. Bot.* **23**: 639-690.

L'Hardy-Halos, M.-T. 1967. Le croissance de *Neomonospora pedicellata* (Smith) G. Feldmann et Meslin (Floridées - Ceramiaceae) et son intérêt morphologique. *Bull. Soc. Bot. Fr.* **114**: 281-285.

—— 1968a. Observations sur la morphologie du *Neomonospora furcellata* (J. Ag.) G. Feldmann et Meslin (Rhodophyceae - Ceramiaceae) et sur sa position taxinomique. *Bull. Soc. bot. Fr.* **115**: 523-528.

—— 1968b. Les Ceramiaceae (Rhodophyceae, Florideae) des côtes de Bretagne. 1. Le genre *Antithamnion* Nägeli. *Revue algol.* **9**: 152-183.

—— 1970. Recherches sur les Céramiacées (Rhodophycées - Céramiales) et la morphogénèse. *Rev. Gen. Bot.* **77**: 211-287.

—— 1971. Recherches sur les Céramiacées (Rhodophycées - Céramiales) et la morphogénèse. II. Les modalités de la croissance et les remaniements cellulaires. *Rev. Gen. Bot.* **78**: 201-256.

—— 1973. Observations sur deux Délessériacées (Rhodophycées, Céramiales) des côtes de Bretagne: le *Nitophyllum heterocarpum* (Chauvin) Ernst et Feldmann et le *Radicilingua thysanorhizans* (Holmes) Papenfuss. *Bull. Soc. phycol. Fr.* **18**: 33-46.

—— 1985. Les Céramiacées (Rhodophycées, Floridées) des côtes de Bretagne. II. Particularités biologiques de l'*Antithamnion sarniense* (Lyle) G. Feldmann et de l'*Antithamnion spirographidis* Schiffner. *Rev. Cytol. Biol. vég. Bot.* **8**: 89-116.

—— 1986. Observations on two species of *Antithamnionella* from the coast of Brittany. *Botanica mar.* **24**: 37-42.

—— **& Rueness, J.** 1990. Comparative morphology and crossability of related species of *Aglaothamnion* (Rhodophyta). *Phycologia* **29**: 351-366.

—— **& Maggs, C. A.** 1991. A novel life history in *Aglaothamnion diaphanum* sp. nov. (Ceramiaceae, Rhodophyta) from Brittany and the British Isles. *Phycologia* **30**: 467-479.

Lightfoot, J. 1777. *Flora scotica....*, **2**. London.

Lindstrom, S. C. & Gabrielson, P. W. 1989. Taxonomic and distributional notes on northeast Pacific Antithamnionieae (Ceramiales, Rhodophyta). *Jap. J. Phycol. (Sôrui)* **37**: 221-235.

Lining, T. & Garbary, D.J. 1992. The *Ascophyllum/Polysiphonia/Mycosphaerella* symbiosis III. Experimental studies on the interactions between *P. lanosa* and *A. nodosum*. *Bot. marina* **35**: 341-349.

Linnaeus, C. 1753. *Species plantarum* **2**. Stockholm.

—— 1759. *Systema naturae per regna tria naturae* **2**. *Regnum vegetabile*, ed. 10. Stockholm.

—— 1767. *Systema naturae per regna tria naturae* **2**. Ed. 12. Stockholm.

Lucas, J. A. W. 1953. *Ceramium diaphanum* (Lightf.) Roth, its varieties and forms as found in the Netherlands. *Acta Bot. Neerl.* 2: 316-326.

Lüning, K. 1990. *Seaweeds. Their environment, biogeography and ecophysiology*. New York.

Lyle, L. 1922. *Antithamnionella*, a new genus of algae. *J. Bot., Lond.* **60**: 346-350.

Lyngbye, H. C. 1819. *Tentamen hydrophytologiae danicae* .... Copenhagen.

Mackay, H. T. 1836. *Flora hibernica*. Dublin.

Maggs, C.A. & Hommersand, M.H. 1990. *Polysiphonia harveyi*: a recent introduction to the British Isles? *Br. phycol. J.* **25**: 92.

—— **& L'Hardy-Halos, M.-T.** (1993) Nuclear staining in algal herbarium material: a reappraisal of the holotype of *Callithamnion decompositum* J. Agardh (Rhodophyta). *Taxon*, **42**: 521-530.

——, **Guiry, M. D. & Rueness, J.** 1991. *Aglaothamnion priceanum* sp. nov. (Ceramiaceae, Rhodophyta) from the North-eastern Atlantic: morphology and life history of parasporangial plants. *Br. phycol. J.* **26**: 343-352.

Magne, F. 1964. Recherches karyologiques chez les Floridées (Rhodophycées). *Cah. Biol. mar.* **5**: 461-671.

—— 1957. Sur le "*Myriogramme minuta*" Kylin. *Rev. algol.*, N.S. **3**: 16-25.

—— 1980. *Laurencia platycephala* Kützing (Rhodophycée), espèce méconnue des côtes de la Manche. *Cah. Biol. mar.* **21**: 227-237.

—— 1986a. Anomalies du développement chez *Antithamnionella sarniensis* (Rhodophyceae, Ceramiaceae) I: formation et début du développement des tetraspores. *Cryptogamie Algol.* **7**: 135-147.

—— 1986b. Anomalies du développement chez *Antithamnionella sarniensis* (Rhodophyceae, Ceramiaceae) II: nature des individus issus des tetraspores. *Cryptogamie Algol.* **7**: 215-229.

—— 1991. Sur la répartition géographique de l'*Antithamnionella sarniensis* (Rhodophyceae, Ceramiales). *Cryptogamie Algol.* **12**: 121-124.

Masuda, M. 1973. The life history of *Pterosiphonia pennata* (Roth) Falkenberg (Rhodophyceae, Ceramiales) in culture. *J. Jap. Bot.* **48**: 122-127.

—— 1982. A systematic study of the tribe Rhodomeleae (Rhodomelaceae, Rhodophyta). *J. Fac. Sci. Hokkaido Univ., Ser. 5 (Bot.)* **12**: 209-400.

—— **& Sasaki, M.** 1990. Taxonomic notes on Japanese *Ptilota* (Ceramiales, Rhodophyta). *Jap. J. Phycol. (Sôrui)* **38**: 345-354.

Mazoyer, G. 1938. Les Céramiées de l'Afrique du Nord. *Bull. Soc. Hist. nat. Afrique Nord* **29**: 317-331.

McAllister, H. A., Norton, T. A. & Conway, E. 1967. A preliminary list of sublittoral marine algae from the west of Scotland. *Br. phycol. Bull.* **3**: 175-184.

Mendoza, M. L. 1970. The Delesseriaceae (Rhodophyta) from Puerto Deseado, Santa Cruz Province, Argentina. II. Systematic study of the genus *Gonimophyllum* Batters. *Physis*, sect. A **29**: 372-378.

Millar, A. J. K. 1986. *Baldockia verticillata* (Griffithsieae, Ceramiales), a new red algal genus and species from eastern Australia. *Phycologia* **25**: 87-97.

—— 1990. Marine algae of Coff's Harbour ..... *Aust. Syst. Bot.* **3**: 293-593.

Miranda, F. 1932. Algues marines des côtes de la Manche. *Rev. Algol.* **6**:281-292.

Miranda, F. 1936. Nuevas localidades de algas de las costas septentrionales y occidentales de España. *Bol. Soc. España Hist. Nat.* **36**: 367-381.

Moe, R. L. & Silva, P. C. 1980. Morphological and taxonomic studies on Antarctic Ceramiaceae (Rhodophyceae). II. *Pterothamnion antarcticum* (Kylin) comb. nov. (*Antithamnion antarcticum* Kylin). *Br. phycol. J.* **15**: 1-17.

Montagne, C. 1837. Centurie de plantes cellulaires exotiques nouvelles. *Ann. Sci. nat. Bot.* **8**: 345-370.

—— 1841. Plantae cellulares. In: Barker-Webb, P. & Berthelot, S. *Histoire naturelle des Iles Canaries*, **3**, pp. 161-208. Paris.

—— 1842. Botanique. Plantes cellulaires. In: de la Sagra, R. *Histoire physique, politique et naturelle de l'Ile de Cuba.* Paris.

Moris, J. [G. G.] & De Notaris, J. [G.] 1839. Florula Capreriae..... *Mem. Accad. Torino*, ser. 2 **2**: 59-130.

Morison, R. 1680-1699. *Plantarum historiae universalis oxoniensis* **2** & **3**. Oxford.

Müller, O. F. 1782. Flora Danicae Iconum, 5, tab. 828, Fig. 1. In: Oeder, B. C. (ed.) *Icones Plantarum sponte nascentium in regnis daniae et Norvegiae*. Hafniae.

Nägeli, C. 1847. Die neueren Algensysteme. Zurich.

—— 1855. Wachstumsgeschichte von *Pterothamnion plumula* und *floccosum*. In: Nägeli, C. & Cramer, C. (eds) *Pflanzenphysiologische Untersuchungen*, pp. 54-68. Zürich.

—— 1862. Beiträge zur Morphologie und Systematique der Ceramiaceae. *Sber. bayer. Akad. Wiss. Jb.* **1**: 297-415.

Newton, L. 1931. *A handbook of the British seaweeds.* London.

Norris, R. E. 1985. Studies on *Pleonosporium* and *Mesothamnion* (*Ceramiaceae*, Rhodophyta) with a description of a new species from Natal. *Br. phycol. J.* **20**: 59-68.

—— **& Aken, M. E.** 1985. Marine benthic algae new to South Africa. *S. Afr. J. Bot.* **51**: 55-65.

Norton, T. A. 1975. Growth-form and environment in cave-dwelling plants of *Plumaria elegans*. *Br. phycol. J.* **10**: 225-233.

—— 1985. *Provisional atlas of the marine algae of Britain & Ireland.* Huntingdon.

—— **& Parkes, H. M.** 1972. The distribution and reproduction of *Pterosiphonia complanata*. *Br. phycol. J.* 7: 13-19.

Okamura, K. 1907-1942. *Icones of Japanese algae.* Tokyo.

Oltmanns, F. 1904. *Morphologie und Biologie der algen* **1**. Jena.

Oltmanns, F. 1922. *Morphologie und Biologie der algen* **1**. Ed. 2. Jena.

Öztig, F. 1959. Etude comparée de la structure morphologique et anatomique de *Boergeseniella fruticulosa* (Wulf.) Kylin de la Mediterranée et de l'Océan Atlantique. *Vie Milieu* **10**: 280-295.

Papenfuss, G. F. 1939. The development of the reproductive organs in *Acrosorium acrospermum*. *Bot. Notiser* **1939**: 11-20.

—— 1956. On the nomenclature of some Delesseriaceae. *Taxon* **5**: 158-162.

Parke, M. & Dixon, P. S. 1976. Check-list of British marine algae - third revision. *J. Mar. Biol. Ass. UK* **56**: 527-594.

Parsons, M. J. 1975. Morphology and taxonomy of the Dasyaceae and the Lophothaliae (Rhodomelaceae) of the Rhodophyta. *Aust. J. Bot.* **23**: 549-713.

—— 1980. The morphology and taxonomy of *Brongniartella* Bory sensu Kylin (Rhodomelaceae, Rhodophyta). *Phycologia* **19**: 273-295.

Pearson, G. A. & Evans, L. V. 1990. Settlement and survival of *Polysiphonia lanosa* (Ceramiales) spores on *Ascophyllum nodosum* and *Fucus vesiculosus* (Fucales). *J. Phycol.* **26**: 597-603.

Plattner, S. B. & Nichols, H. W. 1977. Asexual development in *Seirospora seirosperma*. *Phytomorphology* **27**: 371-377.

Price, J. H. 1978. Ecological determination of adult form in *Callithamnion*: its taxonomic implications. In: Irvine, D. E. G. & Price, J. H. (eds) *Modern approaches to the taxonomy of red and brown algae*, pp. 263-300. London & New York. .

—— **& Tittley, I.** 1978. Marine algae (excluding diatoms). In: Jermy, A. C. & Crabbe, A. C. (eds) *The island of Mull: a survey of its flora and environment*, pp. 19.1-19.36. London.

——**, Hepton, C. E. L., & Honey, S. I.** 1981. The inshore benthic biota of the Lizard peninsula, south west Cornwall. II. The marine algae: Rhodophyta; discussion. *Cornish Stud.* **8**: 5-36.

——**, John, D. M. & Lawson, G. W.** 1986. Seaweeds of the western coast of tropical Africa and adjacent islands: a critical assessment, III. Rhodophyta (Florideae). 1. Genera A-F. *Bull. Br. Mus. nat. Hist. (Bot.)* **15**: 1-122.

——**, John, D. M. & Lawson, G. W.** 1988. Seaweeds of the western coast of tropical Africa and adjacent islands: a critical assessment, IV. Rhodophyta (Florideae). 2. Genera G. *Bull. Br. Mus. nat. Hist. (Bot.)* **18**: 195-273.

Pringsheim, N. 1861. Beiträge zur Morphologie der Meeresalgen. *Abh. Acad. Wissensch. Berlin:* 323-358.

Prud'homme van Reine, W. F. & Sluiman, H.J. 1980. Red algae found on European salt-marshes. 1. *Bostrychia scorpioides* (Rhodomelaceae). *Aquatic Bot.* **9**: 323-342.

Rao, B. G. S., Mantha, S. & Rao, M. U. 1978. Chromosome behaviour at meiosis and its bearing on the cytotaxonomy of *Ceramium* species. *Bot. mar.* **21**: 123-129.

Rawlence, D. J. & Taylor, A. R. A. 1970. The rhizoids of *Polysiphonia lanosa*. *Can. J. Bot.* **48**: 607-611.

Ray, J. 1724. *Synopsis methodica stirpium britannicarum*, ed. 3. London.

Reinke, J. 1891. Die braunen und rothen Algen von Helgoland. *Ber. dt. bot. Ges.* **9** (8): 271-273.

Reinsche, P. F. 1875. *Contributiones al algologiam et fungologiam*. Leipzig.

Richards, H. H. 1901. *Ceramothamnion codii*, a new rhodophyceous alga. *Bull. Torrey Bot. Club* **28**: 257-273.

Ricker, R. W. 1987. *Taxonomy and biogeography of Macquarie Island seaweeds*. London.
Rosenberg, T. 1933. *Studien über Rhodomelaceen und Dasyaceen*. Lund.
Rosenvinge, L. K. 1893. Grønlands Havalger. *Medd. Grønland* **3**: 765-981.
—— 1923-1924. The marine algae of Denmark. Contributions to their natural history. Part III. Rhodophyceae III (Ceramiales). *K. danske Vidensk. Selsk. Skr.* **7**: 287-486.
Roth, A. W. 1797. *Catalecta botanica*, **1**. Lipsiae.
—— 1798. Novae plantarum species. *Archiv Bot. (Römer)* **1**(3): 37-52.
—— 1800a. *Catalecta botanica*, **2**. Lipsiae.
—— 1800b. *Tentamen florae germanicae*, **3**. Leipzig.
—— 1806. *Catalecta botanica*, **3**. Lipsiae.
Rueness, J. 1968. Paraspores from *Plumaria elegans* (Bonnem.) Schmitz in culture. *Nytt Mag. Bot.* **15**: 220-224.
—— 1971. The life history of *Spermothamnion repens* (Dillw.) Rosenv. in culture. *Norw. J. Bot.* **18**: 93-95.
—— 1973. Culture and field observations on growth and reproduction of *Ceramium strictum* Harv. from the Oslofjord, Norway. *Norw. J. Bot.* **20**: 61-65.
—— 1978. Hybridization in red algae. In: Irvine, D. E. G. & Price, J. H. (eds) *Modern approaches to the taxonomy of red and brown algae*, pp. 247-262. London & New York.
—— 1992. *Ceramium cimbricum* (Rhodophyceae, Ceramiales) from Scandinavia: structure, reproduction and systematics. *Nord. J. Bot.* **12**: 135-140.
—— **& Rueness, M.** 1975. Genetic control of morphogenesis in two varieties of *Antithamnion plumula* (Rhodophyceae, Ceramiales). *Phycologia* **14**: 81-85.
—— **&** —— 1978. A parasporangium-bearing strain of *Callithamnion hookeri* (Rhodophyceae, Ceramiales) in culture. *Norw. J. Bot.* **25**: 201-205.
—— **&** —— 1980. Culture and field observations on *Callithamnion bipinnatum* and *C. byssoides* (Rhodophyta, Ceramiales from Norway. *Sarsia* **65**: 29-34.
—— **&** —— 1982. Hybridization and morphogenesis in *Callithamnion hookeri* (Dillw.) S. F. Gray (Rhodophyceae, Ceramiales) from disjunct north-eastern Atlantic populations. *Phycologia* **21**: 137-144.
—— **&** —— 1985. Regular and irregular sequences in the life history of *Callithamnion tetragonum* (Rhodophyta, Ceramiales). *Br. phycol. J.* **20**: 329-333.
—— **& L'Hardy-Halos, M.-T.** 1991. *Aglaothamnion westbrookiae* sp. nov., a species previously confused under the name *Callithamnion byssoides*. *J. Phycol.* **27**: 649-652.
Ruprecht, F. J. 1851. Tange des Ochotskischen Meeres. In: Middendorff, A. T. (ed.) *Reise in den äussersten Norden und Osten Siberiens während der Jahre 1843 und 1844* ..., pp. 191-435. St Petersberg.

Saito, Y. 1967. Studies on Japanese species of *Laurencia*, with special reference to their comparative morphology. *Mem. Fac. Fish. Hokkaido Univ.* **15**: 1-81.

—— 1969. On morphological distinctions of some species of Pacific North American *Laurencia*. *Phycologia* **8**: 85-90.

—— 1982. Morphology and infrageneric position of three British species of *Laurencia* (Ceramiales, Rhodophyta). *Phycologia* **21**: 299-306.

—— **& Womersley, H. B. S.** 1974. The southern Australian species of *Laurencia* (Ceramiales, Rhodophyta). *Aust. J. Bot.* **22**: 815-874.

Scagel, R. F. 1953. A morphological study of some dorsiventral Rhodomelaceae. *Univ. Calif. Publs Bot.* **27**: 1-108.

——, Gabrielson, P. W., Garbary, D. J., Golden, L., Hawkes, M. W., Lindstrom, S. C., Oliveira, J. C. & Widdowson, T. B. 1989. *A synopsis of the benthic marine algae of British Columbia, southeast Alaska, Washington and Oregon.* Vancouver.

Schiffner, V. 1916. Studien über Algen des Adriatischen Meeres. *Wiss. Meeresunters*, N.F. **11**: 129-198.

Schiller, J. 1911. Botanische Beobachtungen, IV. *Jber. Ver. naturw. Erforsch. Adria* **6**: 89-91.

Schlech, K. E. & Abbott, I. A. 1989. Species of Dasyaceae (Rhodophyta) from Hawaii. *Pacific Sci.* **43**: 332-351.

Schmitz, F. 1889. Systematische Ubersicht der bisher bekannten Gattungen der Florideen. *Flora, Jena* **72**: 435-456.

—— 1893a. Die Gattung *Lophothalia* J. Ag. *Ber. dtsch. bot. Ges.* **11**: 212-232.

—— 1893b. Die Gattung *Microthamnion*. *Ber. dtsch. bot. Ges.* **11**: 273-286.

—— 1896. Kleinere Beiträge zur Kenntnis der Florideen. 6. *Nuova Notarisia* **7**.

—— **& Falkenberg, P.** 1897. Rhodomelaceae. In Engler, A. & Prantl, K. (eds) *Die natürlichen Pflanzenfamilien* **1** (2), pp. 421-480. Leipzig.

Schneider, C. W. & Searles, R. B. 1991. *Seaweeds of the southeastern United States: Cape Hatteras to Cape Canaveral.* Durham.

Setchell, W. A. & Gardner, N. L. 1903. Algae of Northwestern America. *Univ. Calif. Publs Bot.* **1**: 165-418.

Silva, P. C. 1952. A review of nomenclatural conservation in the algae from the point of view of the type method. *Univ. Calif. Publs Bot.* **25**: 241-324.

—— 1959. Remarks on algal nomenclature. II. *Taxon* **8**: 60-64.

——, Maggs, C. A. & Irvine, L. M. 1993. Taxonomic and nomenclatural notes on *Plumaria* Schmitz, nom. cons., and Ptilota C. Agardh, nom. cons., (Rhodophyceae), with a proposal to change the type of *Ptilota* to a conserved type. *Taxon* **42** (in press).

Smith, J. E. in Smith, J. E. & Sowerby, J. 1790-1814. *English Botany* .... London.

South, G. R. 1984. A checklist of marine algae of eastern Canada, second revision. *Can. J. Bot.* **62**: 680-704.

—— **& Tittley, I.** 1986. *A Checklist and Distributional Index of the Benthic Marine Algae of the North Atlantic Ocean.* St Andrews and London.

Sprengel, C. 1827. *C. Linnaei Systema Vegetabilium curante curtio Sprengel*, **4** (1), ed. 16. Gottingae.

Stackhouse, J. 1797. Description of *Ulva punctata*. *Trans. Linn. Soc. Lond.* **3**: 236-237.

Stackhouse, J. 1801. *Nereis britannica* .... Bath.
—— 1809. Tentamen marino-cryptogamicum.... *Mém. Soc. Natural. Moscou* **2**: 50-97, pls 5-6.
Stegenga, H. 1985a. A new species of *Bornetia* (Rhodophyta, Ceramiaceae) from southern Africa. *Br. phycol. J.* **20**: 163-168.
—— 1985b. A note on *Anotrichium tenue* (C. Ag.) Naeg. (Ceramiaceae, Rhodophyta) from southern Africa. *Acta Bot. Neerl.* **34**: 145-155.
—— 1986. *The Ceramiaceae (excl. Ceramium) (Rhodophyta) of the south west Cape Province, South Africa.* Berlin & Stuttgart.
—— **& Mol, I.** 1983. *Flora van de Nederlandse Zeewieren.* Amsterdam.
Strömfelt, H. F. G. 1888. Algae novae quas ad litora Scandinaviae indagavit. *Notarisia Anno III* **9**: 381-384.
Suhr, J.N. von. 1840. Beiträge zur Algenkunde. *Flora, Jena* **18**: 273-282.
Sundene, O. 1959. Form variation in *Antithamnion plumula.* Experiments on Plymouth and Oslofjord strains in culture. *Nytt. Mag. Bot.* **7**: 181-187.
—— 1962. Reproduction and morphology in strains of *Antithamnion boreale* originating from Spitsbergen and Scandinavia. *Norske Vidensk.-Akad. Oslo 1. Mat.-Naturv. Klasse*, N. S. **5**: 1-19.
—— 1964. The conspecificity of *Antithamnion sarniense* and *A. spirographidis* in view of culture experiments. *Nytt Mag. Bot.* **12**: 35-42.
—— 1975. Experimental studies on form variation in *Antithamnion plumula* (Rhodophyceae). *Norw. J. Bot.* **22**: 35-42.
Suneson, S. 1938. Über die Entwicklungsgeschichte von *Plumaria elegans. K. Fysiogr. Sällsk. Lund Förhandl.* **8** (9): 1-8.
—— 1940. Studies on the structure and the reproduction of *Pterosiphonia parasitica. Svensk Bot. Tidskr.* **34**: 315-333.

Tandy, G. 1931. Notes on phycological nomenclature - 1. *J. Bot. Lond.* **69**: 225-227.
Taylor, W. R. 1957. *Marine algae of the northeastern coast of North America*, ed. 2. Ann Arbor.
—— 1960. *Marine algae of the eastern tropical and subtropical coasts of the Americas.* Ann Arbor.
Thuret, G. 1855. Note sur un nouveau genre d'algues de la famille de Floridées. *Mém. Soc. sci. nat. Cherbourg* **3**: 455-460.
—— **& Bornet, E.** 1878. *Etudes phycologiques.* Paris.
Tokida, J. 1932. On two new species of *Antithamnion* from Japan. *Trans. Sapporo Nat. Hist. Soc.* **12**: 105-113.
Trévisan [di San Leon], V. B. A. 1845. *Nomenclator Algarum ... 1.* Padua.
Turner, D. 1802. *A synopsis of the British Fuci.* Yarmouth.
—— 1807-1808. *Fuci; or coloured figures and descriptions of the plants referred by botanists to the genus Fucus* **1.** London.

Wagner, F. S. 1954. Contributions to the morphology of the Delesseriaceae. *Univ. Calif. Publs Bot.* **27**: 279-346.
Ward, B. A. 1992. Molecular approaches to taxonomy in *Ceramium* (Rhodophyta, Ceramiales). *Br. phycol. J.* **27**: 101.
West, J. A. & Norris, R. E. 1966. Unusual phenomena in the life histories of florideae in culture. *J. Phycol.* **2**: 54-57.
—— **& Calumpong, H. P.** 1989. On the reproductive biology of *Spyridia filamentosa* (Wulfen) Harvey (Rhodophyta) in culture. *Botanica mar.* **32**: 379-387.
Westbrook, M. A. 1927. *Callithamnion scopulorum* C. Ag. *J. Bot. Lond.* **65**: 129-138.
—— 1930a. *Callithamnion tetricum* (Dillw.) Ag. *J. Bot. Lond.* **68**: 193-203.
—— 1930b. Notes on the distribution of certain marine red algae. *J. Bot. Lond.* **68**: 257-264.
—— 1930c. *Compsothamnion thuyoides* (Smith) Schmitz. *J. Bot. Lond.* **68**: 353-364.
—— 1934. *Antithamnion spirographidis* Schiffner. *J. Bot. Lond.* **72**: 65-68.
Whittick, A. 1977. The reproductive ecology of *Plumaria elegans* (Bonnem.) Schmitz (Ceramiaceae: Rhodophyta) at its northern limits in the western Atlantic. *J. exp. mar. Biol. Ecol.* **29**: 223-230.
—— 1980. *Antithamnionella floccosa* (O. F. Müll.) nov. comb.: a taxonomic re-appraisal of *Antithamnion floccosum* (O. F. Müll.) Kleen (Rhodophyta, Ceramiaceae). *Phycologia* **19**: 74-79.
—— 1981. Culture and field studies on *Callithamnion hookeri* (Dillw.) S. F. Gray (Rhodophyta: Ceramiaceae) from Newfoundland. *Br. phycol. J.* **16**: 289-295.
—— 1983. Spatial and temporal distributions of dominant epiphytes on the stipes of *Laminaria hyperborea* (Gunn.) Fosl. (Phaeophyta: Laminariales) in S. E. Scotland. *J. exp. mar. Biol. Ecol.* **73**: 1-10.
—— 1984. The Newfoundland Ceramiaceae, why are there so many tetrasporophytes? *Br. phycol. J.* **19**: 201.
—— **& Hooper, R. G.** 1977. The reproduction and phenology of *Antithamnion cruciatum* (Rhodophyta: Ceramiaceae) in insular Newfoundland. *Can. J. Bot.* **55**: 520-524.
—— **& West, J. A.** 1979. The life history of a monoecious species of *Callithamnion* (Rhodophyta: Ceramiaceae) in culture. *Phycologia* **18**: 30-37.
Withering, W. 1796. *An arrangement of British plants* **4**, ed. 3. Birmingham & London.
Wollaston, E.M. 1968. Morphology and taxonomy of southern Australian genera of Crouanieae Schmitz (Ceramiaceae, Rhodophyta). *Aust. J. Bot.* **16**: 217-417.
—— 1972a ['1971']. *Antithamnion* and related genera occurring on the Pacific coast of North America. *Syesis* **4**: 73-92.
—— 1972b. Generic features of *Antithamnion* (Ceramiaceae, Rhodophyta) in the Pacific Region. *Proc. intl Seaweed Symp.* **7**: 142-5.
—— 1984. Species of Ceramiaceae (Rhodophyta) recorded from the International Indian Ocean Expedition, 1962. *Phycologia* **23**: 281-299.
Womersley, H. B. S. 1978. Southern Australian species of *Ceramium* Roth (Rhodophyta).

Aust. J. Mar. Freshwater Res. **29**: 205-257.

Womersley, H.B.S. 1979. Southern Australian species of *Polysiphonia* Greville (Rhodophyta). *Aust. J. Bot.* **27**: 459-528.

—— **& Cartledge, S. A.** 1975. The southern Australian species of *Spyridia* (Ceramiaceae: Rhodophyta). *Trans. R. Soc. Aust.* **99**: 221-233.

—— **& Shepley, E. A.** 1982. Southern Australian species of *Hypoglossum* (Delesseriaceae, Rhodophyta). *Aust. J. Bot.* **30**: 321-346.

Woodward, T. J. 1791. The history and description of a new species of *Fucus*. *Trans. Linn. Soc. Lond.* **1**: 131-134, pl. 12.

—— 1794. Descriptions of two new British Fuci. *Trans. Linn. Soc. Lond.* **2**: 29-31.

Wulfen, F. X. 1789. Plantae rariores carinthiacae. In: Jacquin, N. J. (ed.) *Collectanea ad botanicam, chemica et historiam naturalem*, **3**. Vindobonae.

—— 1803. Cryptogama aquatica. *Archiv. Bot. (Römer)* **3**: 1-64.

Wynne, M. J. 1983. The current status of genera in the Delesseriaceae. *Bot. mar.* **26**: 437-450.

—— 1984a. The correct name for the type of *Hypoglossum* Kützing (Delesseriaceae, Rhodophyta). *Taxon* **33**: 85-87.

—— 1984b. The occurrence of *Apoglossum* and *Delesseria* (Ceramiales, Rhodophyta) in South Africa. *S. Afr. J. Bot.* **3**: 136-45.

—— 1985. Concerning the names *Scagelia corallina* and *Heterosiphonia wurdemanii* (Ceramiales, Rhodophyta). *Cryptogamie algol.* **6**: 81-90.

—— 1986. Report on a collection of benthic marine algae from the Namibian coast (southwestern Africa). *Nova Hedwigia* **43**: 311-355.

—— 1988. A reassessment of the *Hypoglossum* group (Delesseriaceae, Rhodophyta), with a critique of its genera. *Helgoländer. Meeresunters.* **42**: 511-534.

—— 1989. Towards the resolution of taxonomic and nomenclatural problems concerning the typification of *Acrosorium uncinatum* (Delesseriaceae: Rhodophyta). *Br. phycol. J.* **24**: 245-252.

—— 1991. A change in the name of the type of *Chondria* C. Agardh (Rhodomelaceae, Rhodophyta). *Taxon* **40**: 316-318.

—— **& Scott, F. J.** 1989. *Phitycolax*, a new genus of adelphoparasitic red algae from Ile Amsterdam, southern Indian Ocean. *Cryptogamie Algol.* **10**: 23-32.

Yabu, H., Notoya, M. & Fukui, K. 1981. Nuclear divisions in *Ceramium fastigiatum* Harvey and *Sorella repens* (Okamura) Hollenberg (Ceramiales, Rhodophyta). *Jap. J. Phycol.* **32**: 221-224.

Zanardini, G. A. M. 1841. Synopsis algarum in mari Adriatico huiusque collectarum.... *Mem. Realle Acad. Sci. Torino* ser. 2 **4**: 1-105.

—— 1847. Notizie intorno alle cellulari marine delle lagune e delitorali de Venezia. *Atti Ist. Veneto Sci. Lett. Cl. Sci. Mat. Nat.* **6**: 185-262.

—— 1866. Scelta di ficee nuove o piú rare dei Mari Mediterraneo ed Adriatic. *Mem. Ist. eneto Sci. Lett. Arti.* **13**: 143-176.

Zinova, A. D. 1981. De positione systematica nitophylli (myriogrammes) yezoensis (Yamada et Tokida) Mikami (Delesseriaceae). *Novit. System. Plant. non Vascul.* **18**: 10-15.

Zoega, J. 1775. *Flora islandica.* Copenhagen & Leipzig.

Index

Taxa of the rank of tribe and Delesseriacean group subfamily and family are shown in capitals; genera and species covered in this volume are shown in Roman type; and names of species and genera that are believed to be synonyms or not to occur within our geographical area are given in italics.

SHETLAND Is.
ORKNEY Is.
CAITH-NESS
SUTHER-LAND
OUTER HEBRIDES
ROSS & CROMARTY
SKYE
NAIRN
MORAY
BANFF
ABER-DEEN
INVERNESS
KINCARDINE
INNER HEBRIDES
ANGUS
PERTH
ARGYLL
FIFE
E.LOTHIAN
RENFREW
W.
MID-LOTHIAN
BER-WICK
AYR
DUMFRIES
NORTH-UMBERLAND
KIRK-CUDBRIGHT
WIG-TOWN
DONEGAL
LONDON-DERRY
CUMBER-
DURHAM

SLIGO
LEITRIM
MAYO
DOWN
I. of MAN
YORKSHIRE
LOUTH
MEATH
GALWAY
DUBLIN
LANCA-SHIRE
FLINT
LINCOLN
CHESHIRE
CAERNARVON
DEN-BIGH
CLARE
WICK-LOW
NORFOLK
MERIONETH
LIMERICK
WEX-FORD
SUFFOLK
CARDIGAN
WATER-FORD
CORK
KERRY
CAR-MARTHEN
ESSEX
PEMBROKE
MON-MOUTH
GLOUCESTER
GLAMORGAN
HAMP-SHIRE
KENT
SOMERSET
SUSSEX
DEVON
DORSET
I. of WIGHT
CORNWALL
SCILLY Is.
CHANNEL Is

CPSIA information can be obtained at www.ICGtesting.com
Printed in the USA
BVOW06s1122011115

424877BV00005B/14/P